EMC 2008
14th European Microscopy Congress
1–5 September 2008, Aachen, Germany

Anke Aretz · Benita Hermanns-Sachweh
Joachim Mayer
Editors

EMC 2008

14th European Microscopy Congress
1–5 September 2008, Aachen, Germany

Volume 3:
Life Science

Springer

Anke Aretz
RWTH Aachen
Central Facility for Electron Microscopy
Ahornstr. 55
52074 Aachen
Germany
aretz@gfe.rwth-aachen.de

Benita Hermanns-Sachweh
RWTH Aachen
University Hospital
Department of Nuclear Medicine
Pauwelsstraße 30
52074 Aachen
Germany
bhermanns@ukaachen.de

Prof. Dr. Joachim Mayer
RWTH Aachen
Central Facility for Electron Microscopy
Ahornstr. 55
52074 Aachen
Germany
mayer@gfe.rwth-aachen.de
and
Ernst Ruska-Centre for Microscopy and
Spectroscopy with Electrons
Institute of Solid State Research
Research Centre Jülich
52425 Jülich
Germany

ISBN 978-3-642-09897-0 e-ISBN 978-3-540-85228-5

DOI 10.1007/978-3-540-85228-5

Softcover reprint of the hardcover 1st edition 2008

Cover design: eStudioCalamar S.L., F. Steinen-Broo, Girona, Spain

Printed on acid-free paper

9 8 7 6 5 4 3 2 1

springer.com

Preface Volume 3: Life Science

This volume contains all life science related contributions presented at the European Microscopy Congress 2008 held in Aachen on 1 – 5 September 2008. Fourteenth in the series of quadrennial conferences, the meeting focused on latest developments in the field of advanced electron microscopy techniques together with the application to materials science and life science. The international character of this series of congresses was once again emphasised – over 1000 contributions from 44 countries representing all areas of Europe and the globe have been submitted for presentation to the programme committee.

The contributions to EMC 2008 dedicated to life science covered diversified topics like analysis of macromolecules, cell structure and dynamics, sample preparation and identification of molecular targets, hybrid methods as well as microscopy advances in life science.

The present hardcopy of volume 3 life science contains all abstracts of the plenary lectures related to these topics, the invited presentations and all contributed papers. All abstracts are included in their original form including black/white figures and the e-mail-address of each main author. We encourage further exchange with the authors after the meeting.

The European Microscopy Congress 2008 was conjointly organised by the European Microscopy Society (EMS), the German Society for Electron Microscopy (DGE), and local microscopists from RWTH Aachen University and the Research Centre Jülich. Hosted by the Eurogress Conference Centre, the congress organisation was supported by the thorough work carried out by the staff of the Central Facility for Electron Microscopy as well as the staff of Aachen Congress who deserve very special thanks.

Anke Aretz, Benita Hermanns-Sachweh, and Joachim Mayer

Editors, Volume 3 of the EMC 2008 Proceedings

Content

L Life Science

Cryoelectron microscopy: from molecules to systems 1
W. Baumeister

Progress in High-resolution Scanning Probe Microscopy 3
A. Engel, P. Bosshart, P. Frederix, and S.A. Müller

L1 Analysis of macromolecules and their supramolecular assemblies

L1.1 Cryo-electron microscopy of macromolecular complexes and supramolecular assemblies/Fitting atomic structures into TEM density maps

Modular organization of RNase P 5
B. Böttcher, C. Batisse, and K. Galani

Molecular anatomy of the human pathogen *Leptospira interrogans* 7
M. Beck, J. Malmström, V. Lange, A. Schmidt, E. Deutsch, and R. Aebersold

Unveiling conformational changes of biological molecules using multiscale modeling and multiresolution experiments 9
F. Tama

Nicolas Boisset: In memoriam 11
J.L. Carrascosa on behalf of his friends and colleagues

Cryo TEM-based 3D reconstruction of the recombinant expressed human zinc peptidase Meprin β 13
P. Arnold, A. Moeller, F. Depoix, C. Becker-Pauly, W. Stöcker, and U. Meissner

Origin melting by SV40 Large T antigen-new insights from 3D-EM-MLF3D classification method 15
J.M. Carazo, I. Cuesta, R. Nuñez-Ramirez, and S.H. Scheres

Sequence analysis and modelling of the two large subunits of Phosphorylase Kinase 17
C. Carrière, J.-P. Mornon, S. Jonic, E. Larquet, C. Venien-Bryan, N. Boisset, and I. Callebaut

Cellular Uptake of Polymer Nanoparticles Imaged by Electron Microscopy 19
M. Dass, A. Musyanovych, K. Landfester, and P. Walther

9 Å cryo-EM structure and molecular model of a gastropod hemocyanin didecamer (KLH1) reveals the architecture of the asymmetric collar 21
C. Gatsogiannis, K. Büchler, F. Depoix, and J. Markl

Microstructure of model systems for sauces based in polysacharides observed by Cryo-SEM 23
I. Hernando, L.M. Guardeño, E. Llorca, A. Quiles, and I. Pérez-Munuera

Three-dimensional architecture of outer- and inner-dynein arms in flagella revealed by cryo-electron tomography and single particle analysis 25
T. Ishikawa, K.H. Bui, T. Movassagh, H. Sakakibara, and K. Oiwa

Structural and functional studies of rabbit skeletal muscle Phosphorylase Kinase 27
S. Jonic, C. Carrière, E. Larquet, V. Skamnaki, L. Johnson, C. Venien-Bryan, J.-P. Mornon, I. Callebaut, and N. Boisset

Single-particle analysis of the Cdc6/Orc1, archaeal DNA replication initiator protein 29
N. Kajimura, H. Suzuki, K. Mayanagi, Y. Ishino, and K. Morikawa

A high-throughput acquisition toolbox in Matlab for cryo-electron microscopy studies 31
A. Korinek, F. Beck, S. Nickell, W. Baumeister, and J.M. Plitzko

Architecture of bacterial glutamate synthase complexe using hybrid approaches 33
M. Cottevieille, E. Larquet, S. Jonic, I. Callebaut, D. Svergun, M.A. Vanoni, and N. Boisset

3D Structure of DNA Repair Macromolecular Complexes Participating in Non-Homologous End-Joining (NHEJ) 35
O. Llorca, A. Rivera-Calzada, L. Spagnolo, M.A. Recuero-Checa, and L.H. Pearl

Quaternary structure of recombinant human meprin β, a zinc peptidase of the astacin family 37
U. Meissner, P. Arnold, A. Moeller, F. Depoix, C. Becker-Pauly, and W. Stöcker

Allosterism of *Nautilus pompilius* hemocyanin as deduced from 8 Å cryo-EM structures obtained under oxy and deoxy conditions 39
A. Moeller, C. Gatsogiannis, F. Depoix, U. Meissner, and J. Markl

High resolution structure of a 6 MDa protease by xray-crystallography and cryo-EM 41
B. Rockel, C.K. Chuang, G. Seyit, P.J. Walian, J. Peters, W. Baumeister, and B.K. Jap

Cryo-TEM of liquid crystalline particles: Application of materials science techniques to study liquid crystalline particle structures. 43
L. Sagalowicz

Structural basis for the concerted integration of HIV-1 DNA in the human genome, role of the human cofactor LEDGF/p75. 45
P. Schultz, F. Michel, R. Benarous, M. Gottikh, D. Moras, and M. Ruff

Titan Krios: Automated 3D imaging at ambient and cryogenic conditions 47
M. Storms, R. Wagner, U. Luecken, W. Hax, and M. Harris

Self assembly and phase behaviour of new sugar based gemini amphiphiles 49
M.C.A. Stuart, M. Scarzello, J.E. Klijn, E.J. Boekema, and J.B.F.N Engberts

Peptide-nanotube formation by lysine based lipoamino acids 51
L. Ziserman, A. Mor, and D. Danino

L1.2 New 2D and 3D image processing and analysis developments

Advances in biological microanalysis using event streamed spectrum imaging and programmed beam acquisition 53
P. Ingram, S. Davilla, and A. LeFurgey

3D-correlation-averaging for membrane-protein-crystals 55
P.J.B. Koeck, P. Purhonen, R. Alvang, B. Grundberg, and H. Hebert

8 Å cryo-EM structure of the giant hemoglobin from the planorbid snail *Biomphalaria glabrata* 57
J. Markl, F. Depoix, A. Moeller, and W. Gebauer

Structure of Pex5p and Pex5-20 complexes in the yeast *Hansenula polymorpha*. Pex20p causes a conformational change upon binding to Pex5p tetramers involved in peroxisomal protein transport. 59
K. Moscicka, S.H. Klompmaker, D. Wang, I.J. van der Klei, and E.J. Boekema

Structure of OprM-MexA interacting complex revealed by cryo electron tomography 61
S. Trépout, JC. Taveau, H. Benabdelhak, A. Ducruix, A.S. Frangakis, and O. Lambert

L2 Cell Structure and Dynamics

L2.1 Tomography of cells and organelles

Pleiomorphic viruses revealed by cryo tomography: the structure of coronaviruses 63
M. Bárcena, W. Baterlink, G.T. Oostergetel, A. Verklijk, P.J.M. Rottier, A.J. Koster, and B.J. Bosch

Combined structural and chemical analysis of unique anammox bacteria that contain a prokaryotic organelle 65
L. van Niftrik, W.J.C. Geerts, E.G. van Donselaar, B.M. Humbel, A.J. Verkleij, M. Strous, and M.S.M. Jetten

Structural and functional considerations on the 3-D organization of the fenestrated cytoplasm of hepatic endothelial cells 67
F. Braet, W. Geerts, J. Riches, S. Satchell, E. Wisse, and P. Frederik

Molecular architecture of the presynaptic compartment studied by cryo-electron tomography 69
R. Fernández-Busnadiego, B. Zuber, U. Maurer, G.W.K. Kong, M. Cyrklaff, J. Dubochet, W. Baumeister, and V. Lučić

Synapses in high pressure frozen *Caenorhabditis elegans* 71
J. Hegermann, and S. Eimer

High resolution EM-tomography of a whole vitrified cell 73
H. Heijnen, H. van Nispen tot Pannerden, and F. de Haas

Evaluation of beam damage in catalase crystals observed in vitrified sections 75
H.-M. Han, B. Gibson, and J. Dubochet

Urothelial fusiform vesicles are formed in post-Golgi compartment 77
S. Hudoklin, R. Romih, J. Neumüller, M. Pavelka, and K. Jezernik

Localisation of GFP and GFP tagged PKD2 in cultivated pancreatic cancer cells using high-pressure freezing 79
H. Krisp, G. von Wichert, T. Seufferlein, and P. Walther

Cytomegalovirus membrane envelopment revisited – a STEM-tomography study based on high-pressure freezing and freeze substitution. 81
S. Landwehr, J. von Einem, T. Mertens, and P. Walther

Electron tomography & template matching of biological membranes 83
M.N. Lebbink, W.J. Geerts, E. van Donselaar, B.M. Humbel, J.A. Post, L.O. Hertzberger, A.J. Koster, and A.J. Verkleij

Correlated Microscopy: From Dynamics to Structure 85
S. Lepper, M. Kudryashev, A. Sartori, S. Bohn, M. Cyrklaff, and F. Frischknecht

3DEM-analyses of Golgi apparatus subunits during physiologic and pathologic reorganizations 87
J. Neumüller, M. Vetterlein, C. Meißlitzer-Ruppitsch, C. Weiss, A. Ellinger, and M. Pavelka

Visualization of the 80s ribosome *in situ* using cryo electron tomography of vitreous sections 89
J. Pierson, J.J. Fernández, E. Bos, H. Gnäegi, J.L. Carrascosa, M. Vos, and P.J. Peters

Three-dimensional analysis of the intermediate filament network using SEM-tomography 91
M. Sailer, S. Lück, V. Schmidt, M. Beil, G. Adler, and P. Walther

The apical vesicles of the tuft cells in the main excretory duct of the rat submandibular gland by EFTEM-TEM tomography 93
A. Sato, J. Kodama, Y. Inoue, Y. Sawa, and T. Oikawa

(S)TEM Dual Axis Tomography **95**
A. Voigt, A. Yakushevska, F. de Haas, W. Voorhout, and R. Schoenmakers

STEM tomography of high-pressure frozen cell monolayers **97**
P. Walther, S. Landwehr, H. Krisp, T. Seufferlein, T. Mertens, and G. Adler

L2.2 Functional organization of micro-organisms, cells and tissues

Multimodality of pericellular proteolysis in cancer cell invasion **99**
P. Friedl

Photonic crystal type nanoarchitectures in butterfly wing scales **101**
Z. Vértesy, K. Kertész, Zs. Bálint, G. Márk, L. Tapasztó, J.P. Vigneron, and L.P. Biró

Iridovirus-like viruses of *Lacerta monticola* from Serra da Estrela, Portugal **103**
A.P. Alves de Matos, M.F. Caeiro, G. Costa, T. Granja, T. Papp, and R. Marschang

Structural and ultrastructural investigations on the *in vitro* effects of plant extracts in chicken and rabbit whole blood cultures **105**
L. Barbu-Tudoran, M. Spanu, C. Craciun, and C.V. Mihali

Fine structural characterization of cloned human endogenous retrovirus HERV-K113 .... **107**
K. Boller, K. Schönfeld, and R.R. Tönjes

The procaine hydrochlorate effect onto the corpuscular anthocyans from the vacuolar sap of different plant cells **109**
D. Cachita, A. Ardelean, C. Craciun, V. Turcus, and L. Barbu-Tudoran

Comparative anatomical and ultrastructural investigations on normal and vitroplantlets leaves from *Pistia stratiotes* L. **111**
D. Cachiță-Cosma, D. Beles, C. Craciun, and L. Barbu-Tudoran

Autophagy of mitochondria in brown adipocytes of chemically thyroidectomised rats **113**
M. Čakić-Milošević, A. Korać, and M. Ukropina

Ultrastructural studies of immune organs in antigen primed chickens treated with vegetal extractions **115**
C. Craciun, M. Spanu, and A. Ardelean

Structural features of HIV-1 immunologic activation **117**
A. Foglia

Functions of actin-binding proteins in the cell nucleus **119**
J. Hofmannová, M. Maninová, V. Filimonenko, and P. Hozák

Ultrastructural mechanisms of entamoebae movement system **121**
M. Hovnanyan, K. Hovnanyan, and R. Asatryan

White adipocytes transdifferentiation into brown adipocytes induced by triiodothyronine 123
A. Korac, M. Cakic-Milosevic, M. Ukropina, M. Grubic, K. Micunovic, V. Petrovic, B. Buzadzic, A. Jankovic, A. Vasilijevic, and B. Korac

Ultrastructural studies of the tegument of cestodes (Platyhelminthes): phylogenetic implications 125
C. Levron

Crystallization stages of the $CaCO_3$ deposits in the earthworm's calciferous gland 127
J. Méndez, J. B. Rodríguez-González, R. Alvarez-Otero, M.J.I. Briones, and L. Gago-Duport

Scanning electron microscopic analysis of heavy metal resistant microorganisms 129
S. Nietzsche, A. Schmidt, E. Kothe, M. Westermann, and W. Richter

Characterization of mouse embryoid bodies by Scanning Electron Microscopy (SEM) and Confocal Laser Scanning Microscopy (CLSM) 131
C. Nogués, R. Arroyo, and N. Gaztelumendi

Stage-dependent localization of mitochondrial DNA during the cell cycle of Euglena gracilis Z by immunogold electron microscopy 133
T. Osafune, N. Kiyohara, I. Watanabe, and T. Ehara

The virus-inducted structures in RNA virus infected macrophages 135
N.G. Plekhova, L.M. Somova, and E.V. Pustovalov

Surface layers of ore-leaching Bacteria and Archaea 137
R. Rachel, A. Klingl, C. Neuner, W. Depmeier, G. Schmalz, M. Thomm, and H. Huber

Structure of *Bacillus anthracis* spores analyzed by CEMOVIS 139
M. Sachse, G. Pehau-Arnaudet, E. Couture-Tosi, J.-L. Ranck, P. Sylvestre, and M. Mock

Insights into the role of pUL71 during HCMV morphogenesis 141
M. Schauflinger, S. Landwehr, P. Walther, Th. Mertens, and J. von Einem

Fine structure of the genital chamber of female *Chrysomya megacephala* (F.) (Diptera: Calliphoridae) 143
S. Siriwattanarungsee, T. Chaiwong, K. Sukontason, U. Chaisri, and K.L. Sukontason

Ultrastructure of the testes and spermatozoa of the blow fly *Chrysomya megacephala* (F.) (Diptera:Calliphoridae) 145
T. Chaiwong, K. Sukontason, U. Chaisri, H. Kurahashi, J.K. Olson, U. Chaithong, and K.L. Sukontason

Biosynthesis of cellulose in red algae 147
I. Tsekos

Light and electron microscopy aspects of the glandular sessile hairs from the vitroplantlet leave of *Drosera rotundifolia* 149
V. Turcuş, D. Cachiţă, C. Crăciun, A. Ardelean, L. Barbu-Tudoran, and C. Mihali

Ultrastructure of selected *Euglena* species in relation to their taxonomy 151
M. Weidinger, and E. Kusel-Fetzmann

L2.3 Tracking molecular events in vivo/Tracking the dynamics of cells and organelles

Confocal Microscopy Reveals Molecular Interactions Between DNA-binding Drugs and Chromatin in Live Cells 153
J.W. Dobrucki, and K. Wojcik

Fluorescence microscopy methods for measuring the mobility and stability of molecules in 3-D samples 155
K. Braeckmans, D. Vercauteren, H. Deschout, D. Mazza, A. Diaspro, S.C. De Smedt, and J. Demeester

Phasor-FLIM analysis of FRET for homotypic and heterotypic non-covalent interactions: stimulated membrane receptors in live cells 157
V.R. Caiolfa, E. Gratton, and M. Zamai

***In vitro* immunofluorescence and ultrastructural analysis of the expression of noncollagenous matrix proteins in human pulpal cells induced by mineralizing factors** 159
V.E. Arana-Chavez, E.F. Martinez, and T.A.G. Donato

Differentiation-dependent Golgi fragmentation in the bladder urothelial cells in vitro 161
K. Jezernik, M.E. Kreft, and A.A. Mironov

Monitoring Mitochordria Dynamics in Living Cell System by Confocal Microscopy 163
C.-Y. Lin, and L.-T. Li

The freeze-fracture replica immunolabeling technique for the study of lipid droplet biogenesis and function 165
H. Robenek, I. Buers, O. Hofnagel, M.J. Robenek, D. Troyer, and N.J. Severs

L2.4 Imaging cells and tissues in vivo

Dynamics of epigenome duplication, accessibility and translation 167
M.C. Cardoso

Controlled light exposure microscopy reveals telomeric microterritories throughout the cell cycle 169
W.H. De Vos, R.A. Hoebe, G. Joss, E. Manders, and P. Van Oostveldt

Calibration in fluorescence correlation spectroscopy for measurements of stem cell differentiation kinetic 171
D. Dumas, B. Riquelme, H. Castellini, L. Basciano, N. de Isla, and JF Stoltz

Spectral methods for cells and particles identifications 173
N. Koltovoy

3D realistic visualization of supramolecular assemblies 175
D. Larivière, and E. Fourmentin

Heterosis in Arabidopsis thaliana: Structural aspects in mature and germinating seeds from hybrid and parental lines 177
E. Maximova, A. Blacha, Th. Altmann, and M. Udvardi

Colour visualization of red blood cells in native smears by the new method reflected light microscopy 179
A. Paiziev, and V. Krakhmalev

***In vitro* culture of *Trigonella foenum-graecum* plantules and their anatomic characterization** 181
C. Pribac, and A. Ardelean

Live cell imaging by SEM-hosted X-ray microscope in "water window" energy range 183
A. Sasov

Microscopic anatomy and mineral composition of cuticle in amphibious isopods *Ligia italica* and *Titanethes albus* (Crustacea:Isopoda) 185
J. Štrus, N. Žnidaršič, S. Hild, and A. Ziegler

In vivo imaging and quantification of the continuous keratin filament network turnover 187
R. Windoffer, A. Kölsch, S. Wöll, T. Würflinger, T. Aach, and R.E. Leube

L3 Microscopy advances in the life sciences

L3.1 High-resolution fluorescence and confocal microscopy/Nonlinear Microscopy

Necrotic cell death, a controlled way of cellular explosion 189
T. Vanden Berghe, N. Vanlangenakker, E. Parthoens, W. Deckers, M. Devos, N. Festjens, and P. Vandenabeele

Dissecting mitochondrial protein distributions using sub-diffraction resolution fluorescence microscopy 191
S. Jakobs

High Thoughput, High Content Tissue Cyotometry 193
P.T.C. So

Focus on the vascular wall: Imaging of large arteries using two-photon microscopy 195
M.A.M.J. van Zandvoort, R.T.A. Megens, D.W. Slaaf, and M.G.A. oudeEgbrink

High resolution confocal Ca^{2+} imaging of the pulmonary neuroepithelial body microenvironment in lung slices 197
I. Brouns, I. De Proost, I. Pintelon, R. Lembrechts, J.-P. Timmermans, and D. Adriaensen

Two photon excitation microscopy and SHG imaging as a tool for visualisation of type I and type II collagen, and their use in tissue engineering 199
Z. Burdíková, E. Filová, M. Rampichová, P. Bianchini, M. Čapek, E. Amler, and L. Kubínová

Testing calibration standards for confocal and two-photon microscopy 201
O. Chernyavskiy, and L. Kubínová

Distribution of the voltage-gated delayed-rectifier K^+ subunits, $K_v1.1$ and $K_v1.2$, in the adult murine enteric nervous system 203
A. Costagliola, L. Van Nassauw, and J.-P. Timmermans

Observation of Ventricular Myocyte Morphology in Long Term Culture using High Resolution Confocal Imaging 205
K. Hammer, S. Ruppenthal, M. Oberhofer, A. Vecerdea, L. Kaestner, and P. Lipp

A magnetic field enhances the inmunofluorescence signal in a lymphocytic cell 207
N.V. Joshi, Reyes, L. Barboza, and L. Berrueta

Novel 515 nm DPSS-laser brings excitement in life cell imaging 209
L. Kaestner, P. Lipp, D. von Gegerfelt, H. Karlsson, E. Illy, and J. Hellström

Confocal laser scanning microscopy study on human fissure caries stained with alizarin red 211
S.M. Masudi, and N.I.M. Salleh

Non-invasive skin tissue characterization using non-linear spectral imaging microscopy 213
J.A. Palero, H.S. de Bruijn, A. van der Ploeg van den Heuvel, H.J.C.M Sterenborg, and H.C. Gerritsen

L3.2 Biological nanomachines and nanomanipulation

Force Spectroscopy reveals mechanical structure of capsids 215
G.J.L. Wuite, W.H. Roos, C. Uetrecht, N. Watts, P. Wingfield, A.C. Steven, A.J.R. Heck, J.L. Carrascosa, and C.F. Schmidt

Manipulating and imaging molecular motors with optical traps, single-molecule fluorescence and atomic force microscopy 217
C.F. Schmidt, L.C. Kapitein, B.H. Kwok, J.S. Weinger, T.M. Kapoor, E.J.G. Peterman, I.A.T. Schaap, C. Carrasco, Pedro J. de Pablo, and Stefan Lakämper

Mechanical stability of ethanol-induced A-DNA 219
J.R. Arias-Gonzalez, S. Hormeño, B. Ibarra, J.M. Valpuesta, and J.L. Carrascosa

Manipulation of the mechanical properties of a virus by protein engineering 221
P.J. de Pablo, C. Carasco, M. Castellanos, and M.G. Mateu

A single-molecule mechanical assay to study DNA replication coupled to DNA unwinding 223
B. Ibarra, J.A. Morin, R. Arias, S. Hormeño, M. Salas, J.M. Valpuesta, and J.L. Carrascosa

Generation of multiple DNA lesions at subnuclear resolution by multi-photon irradiation 225
D. Träutlein, M. Deibler, E. Ferrando-May, and A. Leitenstorfer

L3.3 Microscopy in human health and disease

Light and electron microscopy of phagosomes 227
G. Griffiths, D. Liebl, C. Bleck, A. Merz, P. Walther, J. Dubochet, and B. Zuber

The Use of Electron Microscopy in the Diagnosis of Human Microsporidial Infection – the Manchester, UK, Experience 229
A. Curry

Overview of the impact of biological microanalysis in health and disease 231
A. LeFurgey, and P. Ingram

Immunolocalization of FOXP3 in HCV-infected liver biopsies. Preliminary observations. 233
F. D'Amico, A. Amoroso, M. Consolo, E. Skarmoutsou, B. Mauceri, G. Malaponte, F. Stivala, C.M. Bruno, S. Neri, and M.C. Mazzarino

Ultrastructural and Intracellular Elemental Composition Analysis of Human Hematopoietic Cells During Cold Storage in Preservation Solutions 235
F. Arrebola, F.J. Cañizares, M.A. Cubero, M.M. Serrano, M.A. Robles, and E. Fernández-Segura

Reconstitution and morphological characterisation of an original Human Endogenous Retrovirus 237
N. Bannert, N. Beimforde, L. Möller, G. Holland, M. Laue, and R. Kurth

Cell death in human articular chondrocyte: an ultrastructural study in micromass model 239
M. Battistelli, A. D'Emilio, E. Olivotto, S. Pagani, R. Borzì, A. Facchini, and E. Falcieri

Pathomorphological diagnostic of paraffin embedded versus epon embedded cardiac tissues with transmission electron microscope analysis 241
J. Bornemann, B. Hermanns-Sachweh, and N. Gaßler

Interaction of carbon nanotubes with macrophages: STEM and EELS approach 243
C. Bussy, N. Brun, J. Cambedouzou, S. Lanone, A. Grodet, V. Heresanu, J. Boczkowski, P. Launois, and C. Mory

The ultastructural effect of nitrogen mustard gas on rat brain and therapeutic value of proanthocyanidin 245
B. Can, A. Tekiner, S.A. Köse, O. Yücel, and M.A. Bayar

Assembly and maturation of pestiviruses 247
S. Deike, M. König, and H.-J. Thiel

Analysis of radiation-induced bystander effects using high content screening 249
B. Dieriks, W. De Vos, and P. Van Oostveldt

Electron microscopic investigations on normal and dexamethasone applied rat placentas 251
H. Er, E.T. Korgun, Ç. Çelik-Özenci, M. Saçan, and M. Asar

Changes in dental enamel birefringence after CO_2 laser irradiation through fluoride gel –a pilot study 253
M. Esteves-Oliveira, C.P. Eduardo, F. Lampert, and C. Apel

SEM and light microscopic examination and morphometric analysis of temporal changes in intimal and medial connective tissue component of human carotid arteries 255
O. Evirgen, T. Zeyrek, N. Apaydin, H. Ilkay, and O. Semiz

Ultrastructure of human amniotic membrane covering villous chorion and smooth chorion in preeclampsia 257
O. Evirgen, H. İlkay, S. Akkuş, O. Semiz, O.A. Heper, R. Aytaç

Changes in Intracellular Sodium, Chlorine, and Potassium Content in Hematopoietic Cells after Hypotermic Storage 259
E. Fernández-Segura, F. Arrebola, M.A. Cubero, F.J. Cañizares, M.A. Robles, and P. Navarrete

Requirements for a bright future of electron microscopy in the rapid laboratory diagnosis of infectious diseases 261
H.R. Gelderblom

Changed thickness of zona pellucida as a result of stimulated ovulation 263
B. Jodłowska-Jędrych, M. Jędrych, W. Matysiak, and K. Czerny

Mozaic microscopy of pancreatic islets 265
D. Kalicharan, K.A. Sjollema, F. Dijk, J.J.L. van der Want, and B.N.G. Giepmans

Microscope history Database 267
N. Koltovoy

Interaction of H-Ras Transformed Cell Line with Folic Acid Modified Magnetic Nanoparticles and Detection by Transmission Electron Microscope ... 269
M. Kutlu, N. Saltan, R. Say, and A. İşcan

Sedimentation of suspensions for diagnostic thin section electron microscopy 271
M. Laue, and N. Bannert

Evaluation of rofecoxib administration influence on ultrastructural image of the kidney 273
W. Matysiak, B. Jodłowska-Jędrych, E. Kifer-Wysocka, J. Romanowska-Sarlej, and K. Czerny

Microstructural changes produced by Pulsed Electric Fields in liquid whole egg studied by Transmission Electron Microscopy (TEM) 275
I. Pérez-Munuera, R. Marco-Molés, I. Hernando, A. Quiles, and A. Puig

Microstructural changes in dough treated by Glucose oxidase (GOX) and Transglutaminase (TG), studied by Scanning Electron Microscopy (SEM) 277
I. Pérez-Munuera, A. Quiles, A. Puig, and I. Hernando

Cytoplasmic K/Na balance in the cardiomyocyte of young and old rats 279
M.A. Pogorelova, and A.G. Pogorelov

Hyperspectral imaging – a novel concept for marker free chromosome characterization 281
K. Rebner, T. Merz, and R.W. Kessler

Poxviruses: Morphogenesis of an Orthopoxvirus (CPXV) and an Avipoxvirus (FPXV) in host tissue 283
S. Richter, and P. Wernsdorf

Ultrastructural alterations of methotrexate in mouse kidney 285
A.K. Sargın, E.O. Oğuz, B. Can, and Y. Saran

Ultrastructural gadolinium evidence in skin of a patient with Nephrogenic Fibrosing Dermopathy 287
J.A. Schroeder, C. Weingart, T. Vogt, I. Hausser, F. Hofstaedter, and B.K. Krämer

Nephrotoxicity induced by inorganic Hg(II) and Pb(II): a microscopic and biochemical in vitro study 289
A. Stacchiotti, F. Morandini, F. Bettoni, L.F. Rodella, P. Grigolato, A. Lavazza, and M.F. Aleo

Tissue distribution of peroxisomes in zebrafish 291
M. Stevens, O. Krysko, M. Baes, and M. Espeel

An analysis of the ultrastructure of matured HIV 293
I. Takahashi, M. Takama, and A.M. Ladhoff

***Trichinella spiralis* and *Trichinella britovi* cuticle and hypodermal glands Ultrastructure** 295
Tosti M.

Ultrastructural analysis of lysosomal storage diseases: effects of therapy 297
M.A. van den Bergh Weerman, J. Cox-Brinkman, A.C. Vedder, F.A. Wijburg, C.E.M. Hollak, and S. Florquin

L4 Sample preparation and identification of molecular targets

L4.1 Ultrastructural localization of molecules and elements

Immunolabeling for Electronmicroscopy 299
G. Posthuma

Molecular Organisation of Cadherins in Intercellular Adhesion Junctions by Cryo-Electron Tomography of Vitreous Sections 301
A. Al-Amoudi, and A. Frangakis

Ultrastructural observations of indium in the lactating mammary gland cells 303
A. Ayadi, S. Maghraoui, A. El Hili, and L. Tekaya

An active mechanism flanks and modulates the export of the small ribosomal subunits 305
B. Cisterna, M. Malatesta, J. Dieker, E. Prosperi, S. Muller, and M. Biggiogera

New methods for micro-domain detection in bacterial sacculi 307
M.A. de Pedro, and H. Schwarz

Subcellular localization of Myogenic Regulatory Factors along skeletal muscle development 309
E. Falcieri, S. Burattini, L. Biagiotti, P. Ferri, A. D'Emilio, M. Guescini, E. Barbieri, and V. Stocchi

Immunocytochemical Strategies: LR Resins or Lowicryl, Gold or Peroxidase – Which is Better? 311
J.A. Hobot, and C.J. von Ruhland

Elemental Analysis in Electron Microscopy for Medical Diagnostics 313
L. Jonas, and S. Schäd

Synaptic localization of KCa1.1 potassium channels in central neurons revealed by postembedding immunogold and SDS-digested freeze-fracture replica labelling 315
W.A. Kaufmann, R. Shigemoto, H.-G. Knaus, and F. Ferraguti

Ultrastructural studies of rod photoreceptor nuclei from SCA7 mouse 317
C. Kizilyaprak, D. Spehner, D. Devys, L. Tora, and P. Schultz

Application of electron energy loss spectroscopy for heavy metal localization in unicellular algae 319
U. Lütz-Meindl, M. Milla, and B. Stöger

Localization of "flagella" proteins in the Archaeon *Ignicoccus hospitalis* 321
C. Meyer, D. Müller, S. Gürster, H. Huber, R. Wirth, and R. Rachel

Detection of gold nanoparticles by Autrata YAG detector in FE SEM working in cryo mode 323
J. Nebesarova, M. Vancova, S. Hucek, and J. Vanecek

Correlative cryo-fluorescence and electron microscopy 325
A. Rigort, A. Leis, M. Gruska, A. Sartori, W. Baumeister, and J.M. Plitzko

The effect of hypometabolizing molecules on transcription as shown by a new two-step pulse-chase method 327
A. Spedito, M. Malatesta, and M. Biggiogera

Cryofixation and Freeze-Substitution Combined with Tokuyasu Cryo-section Labelling 329
Y.-D. Stierhof, and H. Schwarz

The *Drosophila* active zone architecture: combining confocal, STED and transmission electron microscopy 331
C. Wichmann, W. Fouquet, S. Mertel, D. Owald, M. Dyba, S. Eimer, and S.J. Sigrist

Distribution and function of amorphous $CaCO_3$ and calcite within the tergite cuticle of *terrestrial isopods* (Crustacea) 333
A. Ziegler, and S. Hild

L 4.2 Recombinant in situ labeling for light and electron microscopy/ Advanced methods of sample preparation for the life sciences

A symbiosis: tracking cell signaling with expression probes, quantum dots and a programmable array microscope (PAM) 335
M.G. Botelho, W. Caarls, S. Kantelhardt, A. Cambi, G. Hagen, D.J. Arndt-Jovin, and T.M. Jovin

Golgi twins in mitosis revealed by genetically encoded tags for live cell imaging and correlated electron microscopy 337
B.N.G. Giepmans, G.M. Gaietta, T.J. Deerinck, S.R. Adams, R.Y. Tsien, and M.H. Ellisman

Understanding the 3D architecture of organelle bound protein complexes using cryo-electron tomography of frozen hydrated sections and immunogold; our next great challenge 339
P.J. Peters, J. Pierson, D. Houben, M. Sani, S. Godsave, P. Kujala, M. Vos, E. Bos, H. Jansen, M. van Zon, and N. Van der Wel

Cryo-electron microscopy of vitreous sections 341
B. Zuber, M. Chami, C. Houssin, J. Dubochet, G. Griffiths, and M. Daffé

The visualization of viruses in the low-voltage electron microscope 343
H. Bielniková, J. Nebesářová, M. Vancová

Exploring HPF/FS methods for sensitive immunogold labeling on resin sections 345
P. Hozák, A.A. Philimonenko, J. Janáček, and P. Janda

Low contrast of the ER-membranes in the high pressure frozen, freeze substituted specimens 347
M. Lindman, H. Vihinen, and E. Jokitalo

HPF of cultured cell monolayers: towards a standard method for high pressure freezing, freeze substitution and electron tomography acquisition 349
C. López-Iglesias, J. Fontana, S. Ruíz, and C. Risco

Life-like physical fixation of large samples for correlative microscopy 351
M.S. Lucas, M. Wölfel, F. Lucas, and R. Wepf

Integrating automated single, double and triple *in situ* molecular detection with imaging for the analysis of inter- and intracellular events in the seminiferous tubule 353
K. Manova, Y. Romin, A. Barlas, S. Gonzalez, M. Turkekul, T. Tong, S. Suh, V. Gueorguiev, and Z. Lazar

Freeze-substitution in Epon: An attempt to combine immunolabeling and improvement of structural preservation 355
E.M. Schraner, S. Leisinger, M. Müller, and P. Wild

The Shape of Caveolae is Omega-like after Chemical Fixation and Cup-like after Cryofixation 357
M. Westermann, F. Steiniger, S. Nietzsche, G. Hause, C. Lemke, and W. Richter

L5 Hybrid methods and approaches in microscopy

L5.1 Correlative light, electron and scanning probe microscopy

Integrated laser and transmission electron microscopy 359
A.V. Agronskaia, J.A. Valentijn, L.F. van Driel, B.M. Humbel, A.J. Verkleij, A.J. Koster, and H.C. Gerritsen

Correlative 3D microscopy: CLSM and FIB/SEM tomography used to study cellular entry of vaccinia virus 361
M.S. Lucas, P. Gasser, M. Günthert, J. Mercer, A. Helenius, and R. Wepf

Strategies for the morphological analysis of hydrated and life-like preserved biomedical material in the SEM 363
H. Hohenberg, R. Reimer, M. Warmer, and B. Holstermann

Denaturation of metaphase chromatin plates observed by transmission electron microscopy and atomic force microscopy in aqueous solution 365
I. Gállego, X. Sisquella, X. Fernàndez-Busquets, and J.R. Daban

Discovery of a Nuclear Structure in UVC-induced Apoptotic Cells by Integrated Laser Electron Microscopy 367
B.M. Humbel, M.A. Karreman, A.V. Agronskaia, A.J. Verkleij, H.C. Gerritsen, and A.F.M. Cremers

Analysis of biomineral formation in three-dimensional micro-mass stem cell cultures ... 369
L. Lammers, U. Meyer, J. Handschel, C. Naujoks, and H.P. Wiesmann

Comparative structural studies of highly-filled polymeth-acrylates with light microscopy, SEM, TEM and AFM ... 371
C. Lurtz, L. Schmitt, D. Behrend, K. Sternberg, and K.-P. Schmitz

Cell surface characteristics as reporters for cellular energy state ... 373
H. Pluk, H. Croes, B. Lich, B. Wieringa, and J. Fransen

Analysis of multimodal 3D microscopy measurements ... 375
M. Ritter, A. Kranzmann, and M. Hemmleb

Permanent plastid – nuclear complexes (PNCs) in plant cells ... 377
T. Selga, and M. Selga

L5.2 Cellular imaging and nano-analytics

Cell imaging by dynamic secondary ion mass spectrometry (SIMS): basic principles and biological applications ... 379
A. Croisy, J.-L. Guerquin-Kern, and T.-D. Wu

Water imaging in cells cryosections by electron energy loss spectroscopy ... 381
J. Michel, C. Terryn, and G. Balossier

Study of the cellular uptake of Pt nanoparticles in human colon carcinoma cells ... 383
H. Blank, R. Schneider, P. Brenner, D. Gerthsen, M. Esselen, D. Marko, M. Crone, M. Türk, T. Muller, and S. Bräse

Ion spectrometry and electron microscopy correlative imaging for accurate comprehensive trace element localization at subcellular level, in normal and pathological keratinocytes ... 385
J. Degrouard, E. Larras-Regard, and D. Jaillard

Calcium Spark Detection and Analysis in Time Series of Two-Dimensional Confocal Images ... 387
K. Hammer, A. Flockerzi, A. Zeug, P. Lipp, and L. Kaestner

Visualisation of the attachment, possible uptake and distribution of technical nanoparticles in cells with electron microscopy methods ... 389
A. Springer, W. Pompe, S. Bastian, M. Iwe, H. Ikonomidou, W. Busch, D. Kühnel, A. Pothoff, V. Richter, and M. Gelinsky

Author Index ... 391

Subject Index ... 401

Cryoelectron microscopy: from molecules to systems

W. Baumeister

Max-Planck-Institute of Biochemistry, Am Klopferspitz 18, D-82152 Martinsried, Germany

baumeist@biochem.mpg.de
Keywords: electron crystallography, electron tomography, single-particle analysis

Cryo electron microscopy is a generic term referring to different imaging modalities as applied to samples embedded in vitreous ice. Thereby it combines a close-to-life preservation of biological materials with the potential of high-resolution three-dimensional imaging [1].

Electron crystallography *sensu strictu* refers to the study of two-dimensional cystals, natural or synthetic, and can attain atomic resolution. It is particularly powerful in applications to membrane proteins. Although there is now a number of very impressive applications, its use has not been pervasive, mainly because the acquisition of high resolution data sets remains very cumbersome.

Cryo EM of single particles plays an increasingly important role in studying the structure of complexes – the larger the better. Although attaining high (better than 5Å) resolution is possible in principle, this remains an elusive goal in most applications. Nevertheless, medium resolution structures, as can be obtained routinely, provide an excellent platform for hybrid approaches and often allow to obtain rather accurate pseudoatomic resolution maps [1,2,3].

Electron Tomography is uniquely suited to obtain three-dimensional (3-D) images of large pleiomorphic structures, such as supramolecular assemblies, organelles or even whole cells. While the principles of ET have been known for decades, its use has gathered momentum only in recent years. Technological advances (namely computer controlled transmission electron microscopes and large area CCD cameras) have made it possible to develop automated data acquisition procedures. This, in turn, allowed to reduce the total electron dose to levels low enough for studying radiation sensitive biological materials embedded in vitreous ice. As a result, we are now poised to combine the power of high-resolution 3-D imaging with the best possible preservation of the specimen [4].

In the past, ET has mainly been used to examine thin sections of plastic-embedded materials. This approach has provided valuable insights into cellular ultrastructure, but it falls short of revealing the macromolecular organization inside cells. Chemical fixation, staining with contrast enhancing heavy atom compounds, dehydration and plastic embedding affect the specimen significantly and make the interpretation of such tomograms at the molecular level very problematic, if not impossible. The use of high-pressure freezing instead of chemical fixation improves specimen preservation significantly but it does not eliminate the problems arising from the intricate interactions between heavy atom stains and molecular structures. Obviously, cryo-sectioning of

A. Aretz, B. Hermanns-Sachweh, J. Mayer (Eds.): EMC 2008, Vol. 3: Life Science, pp. 1–2,
DOI: 10.1007/978-3-540-85228-5_1,

vitrified material is the method of choice for large (>1 μm) objects (cells, tissues), but it remains a challenging task despite recent progress.

ET of frozen-hydrated whole prokaryotic cells or thin eukaryotic cells grown directly on EM grids provides 3-D images of macromolecular structures unperturbed and in their functional environment [5,6,7,8,9]. Currently resolution is limited to 4-5 nm but with instrumental advances (dual-axis tilting, better detectors) we are now entering the realm of molecular resolution (2-4 nm).

High-resolution tomograms of organelles or cells contain vast amounts of information; essentially they are 3-D images of the cell's entire proteome and they should ultimately enable us to map the spatial relationships of macromolecules in a cellular context, the 'interactome'. However, it is no trivial task to retrieve this information because of the poor signal-to-noise ratio of such tomograms and the crowded nature of the cytoplasm and many organelles. Denoising procedures can help to combat noise and to facilitate visualization, but advanced pattern recognition methods are needed for detecting and identifying with high fidelity specific macromolecules based on their structural signature (size and shape, for example). Provided that high- or medium-resolution structures of the molecules of interest are available, they can be used as templates for a systematic interrogation of the tomograms) [10,11].

1. A. Sali, R. Glaeser, T. Earnest and W. Baumeister, Nature **422** (2003), 216.
2. C.V. Robinson, A. Sali and W. Baumeister, Nature **450** (2007), 973.
3. B. Rockel, J. Peters, S.A. Müller, G. Seyit, P. Ringler, R. Hegerl, R.M. Glaeser and W. Baumeister, P. Natl. Acad. Sci. USA. **102** (2005), 10135.
4. V. Lucic, F. Förster and W. Baumeister, Annu. Rev. Biochem. **74** (2005), 833.
5. O. Medalia, I. Weber, A.S. Frangakis, D. Nicastro, G. Gerisch and W. Baumeister, Science **298** (2002), 1209.
6. J. Kürner, A.S. Frangakis and W. Baumeister, Science **307** (2005), 436.
7. V. Lucic, T. Yang, G. Schweikert, F. Förster and W. Baumeister, Structure **13** (2005), 423.
8. M. Beck, V. Lucic, F. Förster, W. Baumeister and O. Medalia, Nature. **449** (2007), 611.
9. O. Medalia, M. Beck, M. Ecke, I. Weber, R. Neujahr, W. Baumeister and G. Gerisch, Curr. Biol. **17** (2007), 79.
10. J. Ortiz, F. Förster, J. Kürner, A. Linaroudis and W. Baumeister, J. Struct. Biol. **156** (2006), 334.
11. S. Nickell, C. Kofler, A. Leis and W. Baumeister, Nat. Rev. Mol. Cell Biol. **7** (2006), 225.

Progress in High-resolution Scanning Probe Microscopy

A. Engel[1], P. Bosshart [1], P. Frederix [1] and S.A. Müller[1]

1. Maurice E. Müller Insitute, Biozentrum, Univeristy of Basel, Klingelbergstr. 70, 4056 Basel, Switzerland

andreas.engel@unibas.ch
Keywords: AFM, STEM, native membrane

In scanning probe microscopy, a probe is raster-scanned across an object. The first high-resolution scanning probe microscope has been developed by Crewe in the late sixties, and his group built a scanning transmission electron microscope (STEM), which attained atomic scale resolution [1]. This was achieved by using a cold field-emitter based on an etched single crystal tungsten wire that produced electrons emerging from a 100 Å spot. A single lens demagnifying this spot 50x and double deflection scan coils produced exciting images of biological macromolecules that fostered the development of multiple STEMs around the world. This type of instrument allows the mass of single macromolecules to be determined and images of unprecedented clarity to be acquired [2] – possibilities of great potential for biological studies. However, only few STEMs are currently operated for biological applications.

Etched tungsten wires were also key to the scanning tunnelling microscope (STM) built by Binnig and Rohrer in the eighties. The tip is simply brought so close to a conducting surface that electron tunnelling between tip and sample occurs. The steep dependence of tunnelling current vs tip-sample distance allowed surfaces to be contoured at atomic scale resolution. The STM was yet another novel instrument that was applied in biology [3], and that spurred the hope of biologists and physicists to achieve such goals as direct reading of DNA sequences, just like the STEM did a decade before. But the breakthrough came with the invention of the atomic force microscopy (AFM), again by Binnig [4]. In the AFM the deflection of a highly flexible cantilever to which the tip is attached is monitored while the tip is raster-scanned over the sample. By keeping this deflection constant, forces acting on the sample can be reduced to a level that prevents significant deformations of soft biological material. Importantly, this operation can be executed while the sample is submerged in a physiological salt solution and kept a room temperature [5]. Biomacromolecules can thus be observed at work [6].

Progress in imaging biological samples with the AFM was not only a matter of progress in the instrumentation, but required advances in sample preparation [7] and imaging conditions [8]. While such advances were made on various regular membranes (e.g. reconstituted 2D crystals of bacterial porin OmpF purple membranes), more recent experiments provided new insights into the organization of different native biological membranes [9, 10]. Microfabrication has provided cantilevers with reproducible specifications and pyramidal tips that allow a lateral resolution of better that 10 Å to be achieved routinely, while the vertical resolution is close to 1 Å. Microfabrication has created scanning probes that are not only suitable for high resolution imaging, but also

A. Aretz, B. Hermanns-Sachweh, J. Mayer (Eds.): EMC 2008, Vol. 3: Life Science, pp. 3–4,
DOI: 10.1007/978-3-540-85228-5_2,

capable of measuring small currents, such as those provided by single membrane channels [11].

The possibility to manipulate single molecules using the AFM tip emerged in the nineties, when Gaub and co-workers measured the forces required to disrupt the biotin-avidin interaction [12] or the unfolding of titin molecules [13]. In these experiments the tip is pressed down on the sample to allow the tip to pick up the end of a single molecule that is chemically tethered to the support at its other end. A few years later, a single bacteriorhodopsin molecules was pulled out of the purple membrane and the related vacancy imaged [14]. The technology of single molecule force spectroscopy (SMFS) was developed to a truly useful tool for studying protein unfolding, and measuring molecular interactions [15].

New avenues emerge for assessing not only the surface topography at high spatial resolution with the AFM but also the local electronic properties of single membrane proteins. Protein unfolding experiments can now be analyzed quantitatively and understood to the single amino acid level and provide indirect information on the 3D atomic structure of the addressed proteins and their interaction with ligands, adjacent proteins, and lipids. Recent advances in high resolution imaging of native membranes suggest exciting applications of AFM technology for the study of biological systems. This development fosters great expectations for future progress in assessing membrane proteins in their native environment by AFM, identifying specific proteins by their surface structure, and assessing their nanomechanical as well as electronic properties in situ.

1. Wall, J., Langmore, J., Isaacson, M., and Crewe, A.V. (1974) Proc Natl Acad Sci U S A *71*, 1-5.
2. Müller, S.A., and Engel, A. (2006) CHIMIA *60*, 749-753.
3. Baro, A.M. et al. (1985) Nature *315*, 253-254.
4. Binnig, G., Quate, C.F., and Gerber, C. (1986) Phys Rev Lett *56*, 930-933.
5. Drake, B., et al. (1989) Science *243*, 1586-1589.
6. Engel, A., and Müller, D.J. (2000). Nature Structural Biology *7*, 715-718.
7. Müller, D.J., Engel, A., and Amrein, M. (1997). Bioelectrochemistry and Bioenergetics *12*, 867-877.
8. Müller, D.J., et al. (1999) Biophys J *76*, 1101-1111.
9. Fotiadis, D., et al. (2003) Nature *421*, 127-128.
10. Scheuring, S., and Sturgis, J.N. (2005) Science *309*, 484-487.
11. Frederix, P., et al. (2005) Nanotechnology *16*, 997-1005.
12. Moy, V.T., Florin, E.L., and Gaub, H.E. (1994) Science *266*, 257-259.
13. Rief, M., Pascual, J., Saraste, M., and Gaub, H.E. (1999) J Mol Biol *286*, 553-561.
14. Oesterhelt, F., et al. (2000) Science *288*, 143-146.
15. Kedrov, A., Janovjak, H., Sapra, K.T., and Muller, D.J. (2007) Annu Rev Biophys Biomol Struct *36*, 233-260.

Modular organization of RNase P

Bettina Böttcher[1], Claire Batisse[1] and Kyriaki Galani[1]

1. Structural and Computational Biology, EMBL, Meyerhofstrasse 1, 69117 Heidelberg, Germany

boettcher@embl-heidelberg.de
Keywords: RNase P, single particle analysis

Ribonuclease P (RNase P) is a ribozyme that removes the 5' precursor sequence from premature t-RNAs. In bacteria this complex consists of one RNA subunit and a single protein subunit, whereas in eukaryotes one RNA subunit and nine protein subunits contribute. We have determined the structure of eukaryotic RNase P at ca. 15 Å resolution by reconstruction of cryo-negative stain images (frozen sandwich grids). The structure of RNase P is highly modular and the protein add-ons are located around the catalytic RNA core. The composition of the modules can be assigned by placing known structures of some archaeal RNase P proteins in the EM structure and by localizing other subunits by antibody labeling experiments. The assignment reveals the gain of different functionalities during evolution. For example eukaryotic RNase P has a channel that exits from the catalytic site of the substrate binding pocket and probably coordinates the RNA leader sequence that is cleaved off during catalysis. According to subunit composition, this channel is shorter in archaea and is completely missing in bacteria. Another module, which is formed by additional protein components that are unique to eukaryotic RNase P is a lid module, which contributes to the formation of a closed binding pocket. It is conceivable that this much more specific binding pocket is required for increasing the specificity of substrate recognition.

RNase P occurs in relatively low copy numbers in yeast cells (ca. 1000 per cell), which poses a special challenge in purifying enough material for image analysis. Typical yields were around 2-5 μg of protein per 100 g of cells. Despite these limitations, it was possible to collect ca. 45000 particle images from seven grids at a Philips CM 200 FEG electron microscope using a 2kx2k camera. It turned out that the limiting step in data acquisition was the contamination rate in the microscope, which restricted the time window to 10 hours per grid.

A. Aretz, B. Hermanns-Sachweh, J. Mayer (Eds.): EMC 2008, Vol. 3: Life Science, pp. 5–6,
DOI: 10.1007/978-3-540-85228-5_3,

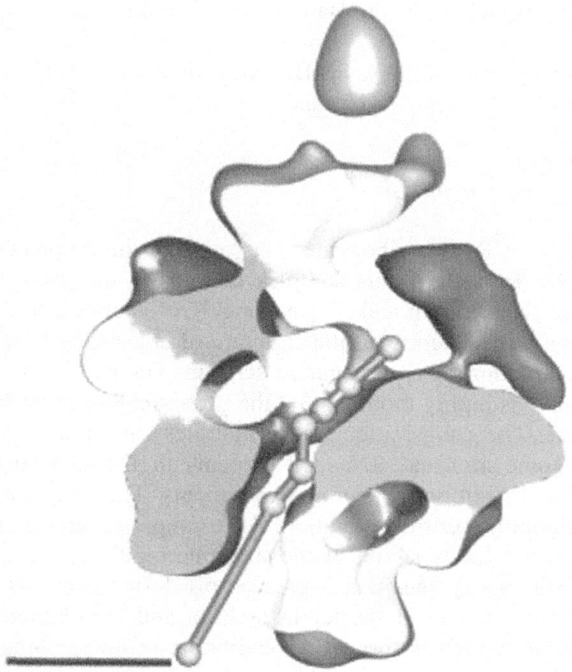

Figure 1. Image reconstruction of RNase P cut in the centre. The cut reveals the central channel. The scale bar equals 10 nm.

Molecular anatomy of the human pathogen *Leptospira interrogans*

Martin Beck[1], Johan Malmström[1], Vinzenz Lange[1], Alexander Schmidt[1], Erich Deutsch[2] and Ruedi Aebersold[1,2]

1. Institute of Molecular Systems Biology, The Swiss Federal Institute of Technology, Wolfgang Pauli-Str. 16, CH-8093 Zurich, Switzerland.
2. Institute for Systems Biology, 1441 North 34th St, Seattle, WA 98103-8904, USA.

beck@imsb.biol.ethz.ch
Keywords: cryo electron tomography, mass spectrometry, Leptospirosis

Systems biology conceptualizes biological systems as dynamic networks of interacting molecules, whereby functionally important properties are thought to emerge from the structure of such networks. Due to the ubiquitous role of complexes of interacting proteins in biological systems, their subunit composition and temporal and spatial arrangement within the cell are of particular interest. While cellular proteomics provides an average picture of the protein expression for all the cells used in a particular study, visual proteomics resembles a bridge to the observation of individual macromolecules within the context of single cells [1]. Although, the structural signature of large protein complexes can in principle be recognized within cryo electron tomograms of intact cells, this concept has so far only been applied unambiguously for ribosomes. A major difficulty is the proteome wide determination of the cellular protein concentration and its variability from cell to cell. In this study, we have tackled this problem for the human pathogen *Leptospira interrogans* by a combined strategy of cryo electron tomography and quantitative mass spectrometry (Figure 1).

We used cryo-electron tomography and template matching to observe several protein complexes involved in bacterial stress response in the cytoplasm of intact cells. Target-driven mass spectrometry, in particular inclusion list based LTQ-FT experiments and multi reaction monitoring served for relative and absolute quantification of the same protein complexes and further proteins involved in the same biological processes.

To localize protein complexes with high confidence within the cytoplasm, we employed statistical concepts used for peptide matching in proteomics [2] to template matching within tomograms of intact *L. interrogans* cells. We investigated stress response in a heat-shocked (fever), antibiotics-treatment and starved condition by targeted and visual proteomics. Our data represent snapshots of the bacterial stress response system. We have observed some of the biological processes involved at single molecule level within tomograms of intact cells. This hybrid approach of mass spectrometry and electron microscopy resembles a generic experimental design that can be applied to a wide range of biological systems in the future.

1. S. Nickell, C. Kofler, A. P. Leis et al., *Nature reviews* **7** (3), 225 (2006).
2. A. Keller, A. I. Nesvizhskii, E. Kolker et al., *Analytical chemistry* **74** (20), 5383 (2002).

A. Aretz, B. Hermanns-Sachweh, J. Mayer (Eds.): EMC 2008, Vol. 3: Life Science, pp. 7–8,
DOI: 10.1007/978-3-540-85228-5_4,

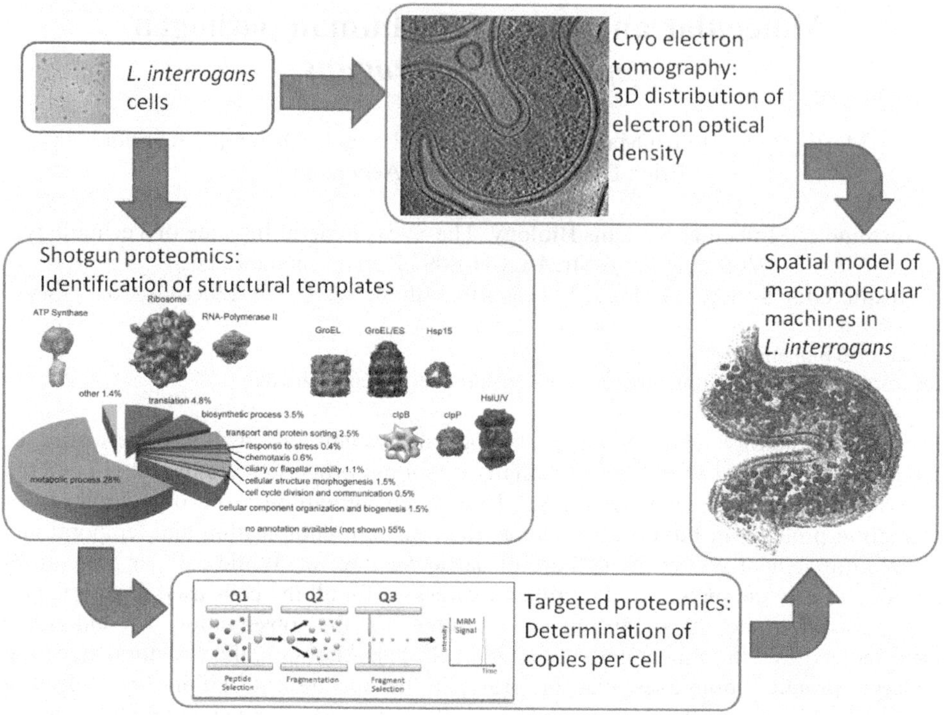

Figure 1. An expanded experimental workflow for visual proteomics. Differently stimulated cells have been subjected to shotgun MS and cryoET. A template library was built that included the protein complexes identified in the proteome if structures of satisfying homology where available in the structural database. The templates were cross correlated with the electron optical density in the tomograms by template matching as described earlier [2] and their cellular concentration was determined by targeted MS, namely MRM. Both types of information where evaluated by a novel scoring function that served for the assignment of the template structures into the spatial context of the cell.

Unveiling conformational changes of biological molecules using multiscale modeling and multiresolution experiments

Florence Tama

Department of Biochemistry & Molecular Biophysics, University of Arizona, Tucson, 85721, AZ, USA

ftama@u.arizona.edu
Keywords: flexible fitting, molecular mechanics

Characterization of the structure and functional states of large biological complexes is of great interest as those complexes perform many vital cellular functions, such as replication, transcription, protein synthesis or regulation of cellular transport, and dysfunctions of those results in severe diseases.

In order to understand diseases and find suitable treatments, the functional roles and mechanisms of these biological molecules need to be elucidated. Characterization of the structure and dynamics of these complexes is a crucial step in this process. However, for those macromolecules, structure determination is an extremely difficult task. Today, the number of high-resolution solved structures remains small compared to the number of possible complexes indicated from genomics studies.

Multipronged approaches have recently gained interest for tackling structural problems related to large biological complexes. Information on the structure and dynamics are often obtained by alternative low-resolution experimental techniques, such as cryo Electron Microscopy (cryo-EM), Small Angle X-ray Scattering (SAXS) and Fluorescence Resonance Energy Transfer (FRET). Each of these experimental approaches offers different advantages and meet with different pitfalls, artifacts and limitations. For example, some techniques reveal the overall shape of the structure, and others extract the distances between specific sites. Therefore more accurate descriptions could be obtained if all pieces of experimental data were taken together to annotate conformational states. Especially, interpretation of these experimental data would advance further by constructing structural models that are consistent with all the data. However, such operation can be done only within mathematical and computational frameworks.

To construct a structural model by deforming or rearranging a known atomic structure to fit the data, it is necessary to consider the physical properties of the molecule, such as chain connectivity and flexibility. Therefore to deform the original structure to construct a model with reasonable physical properties, we need to employ theoretical models that can analyze and predict the dynamics of the system. To access dynamics of biological molecules, which is rather complex since ranging from bonds vibration to large-scale domains motions, several approaches can be used such as normal mode analysis (NMA) or molecular dynamics (MD) simulations. Each method is suited to describe different level of conformational dynamics.

We will present our current developments of multi-resolution/multi-scale computational tools to interpret conformational changes of biological molecules based on cryo-

A. Aretz, B. Hermanns-Sachweh, J. Mayer (Eds.): EMC 2008, Vol. 3: Life Science, pp. 9–10,
DOI: 10.1007/978-3-540-85228-5_5,

EM, SAXS or distance constraints. We have successfully implemented new algorithms to reconcile low-resolution data with high-resolution structure. Using simulated data, we will show that our approaches are successful to predict structures in the range of 2~3 Å resolution (see Figure 1).

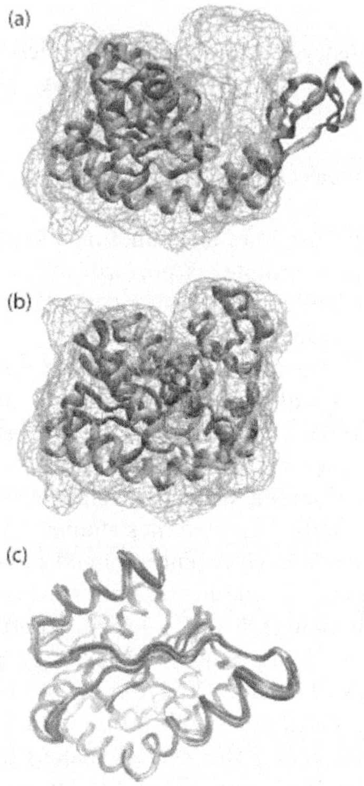

Figure 1. Initial superposition of the X-ray structure of the Adenylate Kinase onto its low-resolution structure before (a) and after (b) flexible fitting using molecular dynamics simulation. A superposition of structures displaying a high correlation coefficient (c) taken at regular intervals during the MD simulations reveals similar conformations.

Nicolas Boisset: In memoriam

José L. Carrascosa on behalf of his friends and colleagues

Centro Nacional de Biotecnologia. CSIC. Campus Cantoblanco. 28049 Madrid, Spain.

jlcarras@cnb.csic.es
Keywords: Nicolas Boisset, Cryoelectron microscopy, three-dimensional reconstruction

The disappearance of a colleague is always sad for any community. But when that person is a young, talented scientist in the best of his scientific career, the lost has more profound implications. Nicolas Boisset passed away in Paris on January 2008 when he was involved in many ongoing collaborations across Europe and USA, leading a very active CNRS group in the University of Paris 6 and Paris 7, and in a moment of excellent scientific productivity.

Achieving this seniority was the consequence of a long, well planned career that involved full dedication from Nicolas. His work was always related to the analysis of biological structure using improved electron microscopy and image processing methods. This was the main topic of his PhD thesis, obtained in the group of J. Lamy at Tours (1). Then he moved to USA to join J. Frank. From this period on, his involvement in the deep understanding of processing methods, as well as his fine abilities as a microscopist, led to a boost in his scientific productivity and in the relevance of his work, leading to publications that settled his work as a reference in the area of cryo microscopy and single particle reconstruction (2-3). During this period, a long standing relationship with Frank and several of the outstanding postdocs that were in that laboratory during those days started, a relation that was maintained over the years fed by the generous and positive character of Nicolas. His integrative attitude is evident from his publications during that period on, where co-authoring of different components from the different laboratories he was working was a constant along his career.

After moving back to Tours, he became a Research Associate of the CNRS and settled the roots of his independent career, that formally started when he moved to Paris in 2000 to set up the Structural Biology CNRS Group at the "Institut de Minéralogie et de Physique des Milieux Condensés" at the Universités Paris 6 and 7. His new position allowed him to emphasize a wider range of activities, including teaching at different levels (University Masters, International Courses and Technical workshops), the organization of French microscopists working in advanced methods for three-dimensional microscopy analyses, and the establishment of several collaborations with French, European and American laboratories (4-6). He was a permanent reference in European projects involving advanced electron microscopy and EMBO courses on these topics.

His generous attitude to his many collaborators was also reflected in his laboratory at Paris, where he managed to set up a group of young and enthusiastic scientist around a number of activities revealing his profound involvement in the development of several of the most interesting aspects in 3D-EM. Among his current projects, the analysis of structure and function relationship in multi-enzymatic complexes, and the study of self-

A. Aretz, B. Hermanns-Sachweh, J. Mayer (Eds.): EMC 2008, Vol. 3: Life Science, pp. 11–12,
DOI: 10.1007/978-3-540-85228-5_6,

assembly and maturation of viruses, were correlated with methodological projects on the use of quantum dots and new composite nanoparticles for cryo-electron microscopy, and CTF estimation and correction. Furthermore, he had the vision and strength to also address more technically-oriented ventures, as the development of a computer controlled freezing system for sample preparation, and the implementation of a prototype for automated digitization of micrographs. His disappearance is a great loss for the European microscopy community and an invaluable absence for his many friends.

1. O. Lambert, N. Boisset, JC. Taveau, G. Preaux and JN. Lamy. J. Mol. Biol. 248 : 431-448 (1995).
2. N. Boisset, P. Penczek, F. Pochon, J. Frank and J. Lamy. J. Mol. Biol. 232: 522-529 (1993).
3. N. Boisset, P. Penczek,, JC. Taveau, J. Lamy, J. Frank. J. Struct. Biol. 115: 16-29 (1995).
4. S. Jonic, CO. Sorzano, M. Cottevielle, E. Larquet and N. Boisset. J. Struct. Biol. 156-67 (2006)
5. J. Busselez, M. Cottevieille, P. Cuniasse, F. Gubellini, N. Boisset and D. Levy. Structure15 : 1674-1683 (2007).
6. M. Cottevieille, E. Larquet, S. Jonic, MV. Petoukhov, G. Caprini, S. Paravisi, DL. Svergun, MA. Varoni, and Boisset. J. Biol. Chem. 283: 8237-8249 (2008).

Cryo TEM-based 3D reconstruction of the recombinant expressed human zinc peptidase Meprin β

Philipp Arnold[1], Arne Moeller[1], Frank Depoix[1], Christoph Becker-Pauly[2], Walter Stöcker[2] & Ulrich Meissner[1]

1. Department of Molecular Physiology, Institute of Zoology, Johannes von Muellerweg 6, 55128 Mainz, Germany
2. Department of Cell and Matrix Biology, Institute of Zoology, Johannes von Muellerweg 6, 55128 Mainz, Germany

philarno@students.uni-mainz.de; meissner@uni-mainz.de
Keywords: meprin, metalloprotease, cryo-TEM, 3D-EM

Meprins are astacin-type zinc endopeptidases, which have been observed so far exclusively in vertebrates. Based on the structure of their catalytic domains these enzymes are distantly related to matrix metalloproteinases (MMPs) [1]. Typically, meprins are expressed in brush border membranes of intestine and kidney tubules, intestinal leukocytes, and certain cancer cells. This suggests a role in epithelial differentiation and cell migration [2]. For human meprin two subforms are described: Meprin α and meprin β [3].

Although Meprin α and β have an amino acid sequence identity of 44%, they show marked differences in activation, substrate specificity; most drastical deviations are seen in their quaternary structures. From negatively stained electron micrographs it is known that meprin α forms ring- and chain shaped oligomers, up to mega Dalton size. This characterizes meprin α as the largest known secreted protease [4].

Since meprin β forms homodimers, it was used in the experiments. Homogeneous preparations of soluble meprin β were purified from the supernatant of HiFive insect cells that had been infected with Baculoviruses carrying a mutated meprin β cDNA resulting in a secreted protein lacking a membrane anchor.

Preliminary data of meprin β quaternary structure now are available: Therefore specimens were plunge-frozen in liquid ethane and micrographs of pure recombinantly expressed Meprin β were obtained using a Tecnai-F20 electron microscope at a magnification of 62000 and an acceleration voltage of 200kV.

From the acquired data of about 4500 single asymmetric particles a preliminary 3D reconstruction was performed with a resolution of 12Å (½-bit). A first rigid body fitting of the different known domains from X-ray data showed good correlation of size and orientation (Figure 1.).

1. Stöcker and Bode (1995): Structural features of a superfamily of zinc-endopeptidases: the metzincins. Curr Opin Struct Biol. 5, 383-390
2. Herzog, C.et al. (2005): Generation of biologically active interleukin-1beta by meprin B. Cytokine. 31, 394-403

A. Aretz, B. Hermanns-Sachweh, J. Mayer (Eds.): EMC 2008, Vol. 3: Life Science, pp. 13–14,
DOI: 10.1007/978-3-540-85228-5_7,

3. Becker-Pauly, et al. (2007): The alpha and beta subunits of the metalloprotease meprin are expressed in separate layers of human epidermis, revealing different functions in keratinocyte proliferation and differentiation. J Invest Dermatol. 127, 1115-1125
4. Becker-Pauly, et al. (2007): The alpha and beta subunits of the metalloprotease meprin are expressed in separate layers of human epidermis, revealing different functions in keratinocyte proliferation and differentiation. J Invest Dermatol. 127, 1115-1125

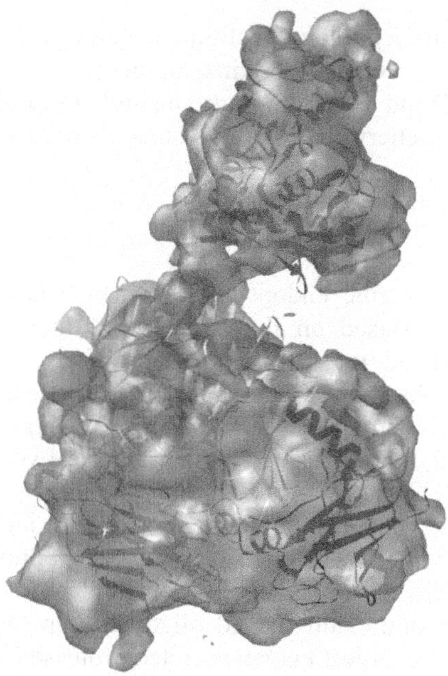

Figure 1. 12Å (1/2bit) density map with rigid body fitted domains.

Origin melting by SV40 Large T antigen-new insights from 3D-EM-MLF3D classification method

J.M. Carazo[1], I. Cuesta[1*], R. Nuñez-Ramirez[2*] and S. H. Scheres

1. Biocomputing Unit, CNB-CSIC, C/Darwin 3, 28049 Madrid, Spain
2. Centro Investigaciones Biológicas, CIB-CSIC, Madrid, Spain

carazo@cnb.cisc.es

Keywords: macromolecular complexes, cryo-electron microscopy, image classification.

A great number of important cellular processes are performed by large macromolecular complexes, known as "molecular nano-machines". Due to their complex biological functions, these complexes present a large degree of molecular flexibility, what results in very heterogeneous data sets when their structure is analyzed by three-dimensional electronic microscopy (3D-EM) of single particles.

This is the case of Large T antigen (LTAg), an ATP-driven helicase that specifically recognizes the core of the viral origin of replication (*Ori*). In the early steps of replication LTAg assembles on the *Ori* as a double hexamer, inducing the melting of the DNA, and subsequently acting as a helicase [1]. The cryo-3D-EM structure of the full-length LTAg double hexamer assembled at the *Ori* shows different degrees of bending along the DNA axis [2]. The two hexamers form an interface with their N-terminal domains, where specific hexamer-hexamer interactions are required for the cooperative full assembly of LTAg double hexamer upon binding to the *Ori* [3]. Nevertheless not much is known about this N-terminal interface structure or its implication in origin melting because of the large degree of heterogeneity that presents.

To deal with these structural heterogeneous data we have used the MLF3D image classification method developed in our group [4]. Firstly we have characterized the structure of Wt LTAg at the Ori in the presence of ADP. The LTAg-Ori structures can be classified in two main groups. One presents a straight conformation while the other shows a marked bend along the DNA axis, where the N-terminal domains are displaced one from each other and not longer parallel (Figure 1). Moreover, to override the bending heterogeneity, we classified sub-images of the C-terminal hexamer of the Wt LTAg double hexamers using MLF3D algorithm. These results revealed that, in presence of ADP, the DNA interact as a double stranded molecule in the internal cavity of the C-terminal hexamer (Figure 2).

We also studied the structure of an LTAg mutant that lacks the J-domain. This domain is located at the N-terminal end of LTAg and it is essential for initiation of replication in vivo [5]. The classification shows only one main conformation, reflecting a reduced flexibility, which is straight and with the N-terminal domains parallel and narrower in comparison with the straight form of the Wt LTAg (Figure 1).

Altogether, this results, obtained by MLF3D classification, indicate the possibly role of J-domain inducting of a bent conformation as the first step in initiation of replication.

A. Aretz, B. Hermanns-Sachweh, J. Mayer (Eds.): EMC 2008, Vol. 3: Life Science, pp. 15–16,
DOI: 10.1007/978-3-540-85228-5_8, © Springer-Verlag Berlin Heidelberg 2008

1. J. A. Borowiec and J. Hurwitz, Embo J. **7** (1988): p. 3149-58.
2. M. G. Gomez-Lorenzo, et al., Embo J. **22** (2003): p. 6205-13.
3. M. Valle, et al., J Mol Biol. **357** (2006): p. 1295-305.
4. S. H. Scheres, et al., Nat Methods. **4** (2007): p. 27-9.
5. J. M. Pipas, K. W. Peden, and D. Nathans, Mol Cell Biol. **3** (1983): p. 203-13.

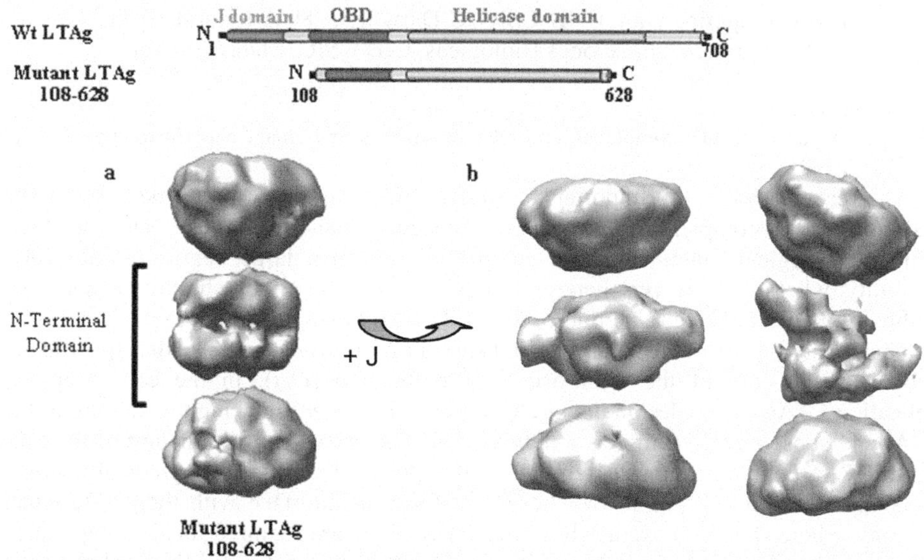

Figure 1. 3D-EM reconstruction of the major conformation of LTAg mutant lacking J-domain (panel a) and of the two main conformations adopted by LTAg-ADP complex (panel b)

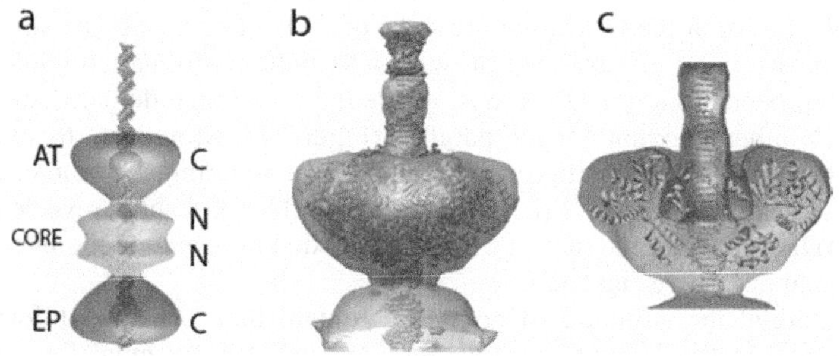

Figure 2. 3D-EM reconstruction of the C-terminal domains interacting with DNA. (a) Schema of LTAg-*Ori* complexes. (b) Map of the hexamer at the AT region of the *Ori* showing an extra row density compatible with dsDNA (c) that penetrates into the cavity of the C-terminal hexamer with the same width suggesting that also is dsDNA.

Sequence analysis and modelling of the two large subunits of Phosphorylase Kinase

C. Carrière[1], J.-P. Mornon[1], S. Jonic[1], E. Larquet[1], C. Venien-Bryan[2], N. Boisset[1] and I. Callebaut[1]

1. Département de Biologie Structurale, Institut de Minéralogie et de Physique des Milieux Condensés, UMR 7590, CNRS, Universités Paris 6 et Paris 7, Paris, France.
2. Laboratory of Molecular Biophysics, Department of Biochemistry, University of Oxford, South Parks Road, Oxford OX1 3QU, United Kingdom.

cathelene.carriere@impmc.jussieu.fr
Keywords: Phosphorylase Kinase, remote relationship, homology modelling, hydrophobic cluster analysis HCA, profile comparisons, structure, function, calcium

Phosphorylase kinase (PhK, 1.3 MDa) is a large hexadecameric complex catalyzing the phosphorylation of glycogen phosphorylase (GP). It consists in four subunits (α, β, γ and δ) present each in four copies (Figure 1). The overall 3D structure of the PhK, obtained at 0.99 nm by cryo-electron microscopy, is presented in another poster at this meeting [1]. This cryo-EM map can be further exploited by fitting the structure at atomic resolution of the different subunits. However, only the experimental structures of the catalytic domain of the γ subunit (two thirds of its polypeptidic chain), and of the δ subunit (intrinsic calmodulin) have been solved. Experimental data are not yet available for the α and β subunits (two thirds of the PhK total mass), which probably arose from gene duplication, and play a key role in the regulation of the enzyme.

We choose sensitive approaches of sequence analysis, such as Hydrophobic Cluster Analysis (HCA) to detect unsuspected relationships between these regulatory subunits and other proteins, which could help to understand the molecular mechanisms by which they might regulate the enzyme activity and furnish valuable templates for homology modelling of their 3D structures.

A comparative analysis of the PhK α and β subunits from different species allowed the prediction of the probable limits of four domains (labelled A, B, C and D). Domain A shares significant similarities with members of the Glycoside Hydrolase family 15 (GH15), as previously reported by Pallen [2].The alignment of domain A with GH15 of known 3D structures was refined to build an accurate model of its 3D structure (Figure 2). Domains C and D are related to each other, and share significant similarities with Calcineurin B-like (CBL) proteins, which are members of the EF-hand family involved in the regulation of plant specific kinases [3]. The 3D model of domain D suggests the conservation of a regulatory switch in the C terminal extension of the domain, which in the uncomplexed CBL structure occupies the crevice involved in the interaction with the regulatory extension of the kinase subunit (Figure 2). Sequence similarities of domain B with other proteins are currently investigated, in order to provide valuable templates for homology modelling of the only domain for which no 3D structure can yet be ascribed. Fitting of these modelled 3D-structures in the cryo-EM map is in progress.

A. Aretz, B. Hermanns-Sachweh, J. Mayer (Eds.): EMC 2008, Vol. 3: Life Science, pp. 17–18,
DOI: 10.1007/978-3-540-85228-5_9, © Springer-Verlag Berlin Heidelberg 2008

1. S. Jonic, C. Carrière, E. Larquet, V. Skamnaki, L. Johnson, C. Venien-Bryan, J.-P. Mornon, I. Callebaut and N. Boisset (submitted to the 14th European Microscopy Congress, September 1 - 5, 2008, Aachen, Germany).
2. Mark J. Pallen, Protein Science (2003), p. 1804.
3. C. Carriere, J.P. Mornon, C. Venien-Bryan, N. Boisset and I. Callebaut, Proteins (2008) in press.

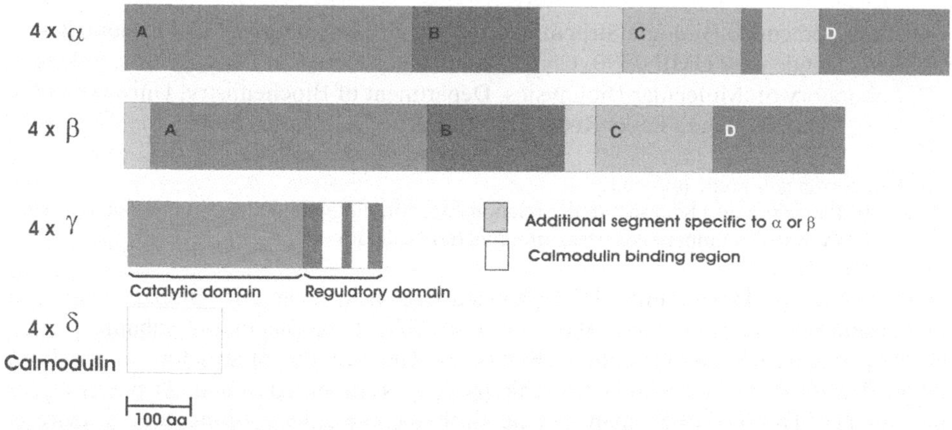

Figure 1. Schematic architecture of the PhK α, β, γ and δ subunits. The four distinct domains of the PhK α and β subunits are labelled A to D.

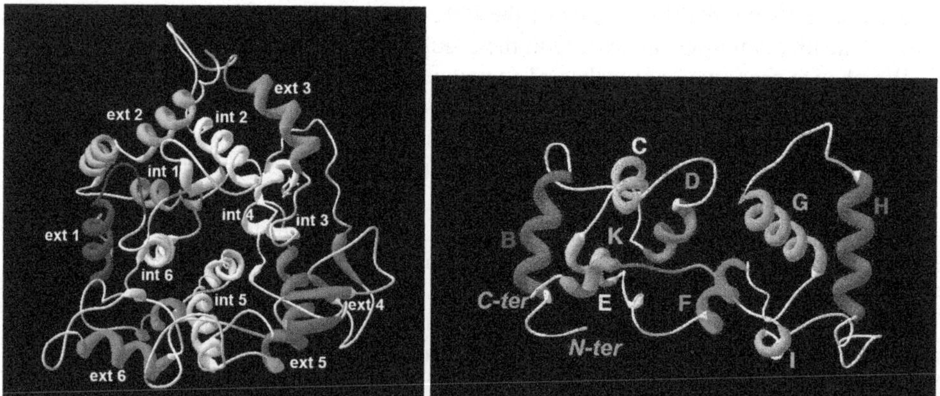

Figure 2. Models of the three-dimensional structure of domains of the α and β subunits (ribbon representation). Left panel, Domain A of PhK α, which is related to the GH15 family. The 3D structure consists in a $(\alpha/\alpha)_6$ toroid fold with an alternation of external/internal helices. Right panel, Domain D of PhK β, which is related to calcineurin B-like (CBL) proteins. The 3D structure consists in a two-domain structure (B to E and F to I) connected by a linker; each domain being formed by two EF-hand motifs. A C-terminal extension including a 3_{10} helix occupies a hydrophobic crevice displayed at the surface of the two EF-hand motifs.

Cellular Uptake of Polymer Nanoparticles Imaged by Electron Microscopy

M. Dass[1,2], A. Musyanovych[2], K. Landfester[2] and P. Walther[1]

1. Central Electron Microscopy Laboratory, University of Ulm, 89069 Ulm, Germany
2. Institute of Organic Chemistry III, and macromolecular Chemistry, University of Ulm, 89069 Ulm, Germany

martin.dass@uni-ulm.de

Keywords: Poly-L-lactide nanoparticles, cellular uptake, high pressure freezing

Polymer nanoparticles are used for technical, biological and medical applications. We are interested in the cellular pathway of polymer nanoparticles because they could be used in drug delivery and diagnostic fields. For this purpose biocompatible and degradable polymers are of special interest.

We studied the cellular uptake of poly-L-lactide (PLLA) in comparison to other polymer nanoparticles in a size range from 100 to 200 nm. These polymer nanoparticles were synthesised by the emulsion/solvent evaporation method combined with the miniemulsion technique[1]. As fluorescence dye perylene monoimide (PMI) was added during the formulation process. The cellular uptake was analysed by confocal laser scanning microscopy (cLSM), fluorescence activated cell sorting (FACS) measurements, and with electron microscopy for a more detailed view of the uptake behaviour.

HeLa cells were cultivated on carbon coated sapphire discs, treated with polymer suspension during 4h and 24h and fixed with a high pressure freezer Compact 01 (Engineering Office M. Wohlwend GmbH, CH-9466 Sennwald, Switzerland). Afterwards all samples were cryo substituted in a substitution medium of acetone, osmiumtetroxide, uranylacetate and 5% of water for 16 h[2]. The samples were embedded in EPON and ultrathin sectioned with a Leica microtome. For electron microscopical analysis a Zeiss EM 10 was used. Tomography studies on 500 nm thick semi thin sections were performed with a 300 kV STEM.

Figure 1A shows intracellular vesicles filled with PLLA-nanoparticles after 24 h incubation. The particles did not stain with heavy metals, so they appear bright compared to the stained cellular structures. Figure 1B shows a group of extracellular particles.

1. Musyanovych, A., et al.,. Macromolecular Bioscience, **8** (2008), p. 127-139.
2. Walther, P. and A. Ziegler, . Journal of Microscopy, **208** (2002), p. 3-10.

A. Aretz, B. Hermanns-Sachweh, J. Mayer (Eds.): EMC 2008, Vol. 3: Life Science, pp. 19–20,
DOI: 10.1007/978-3-540-85228-5_10, © Springer-Verlag Berlin Heidelberg 2008

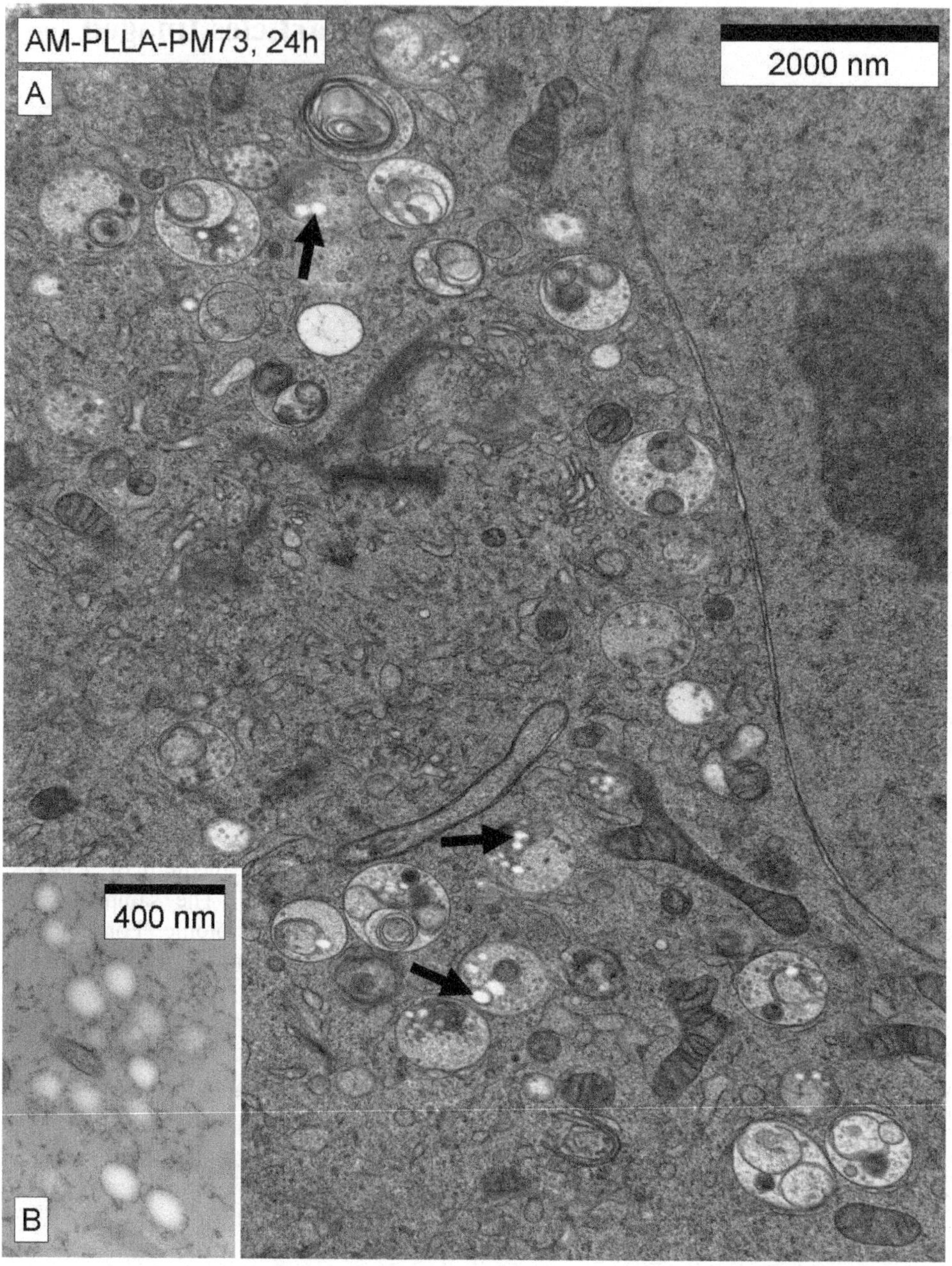

Figure 1. HeLa-cells have been incubated for 24 h with Poly-L-lactide nanoparticles. Cells have incorporated particles (arrows) in endosomes. The endosomes are not filled completely with PLLA and are distributed over the whole cell. Figure B. Group of extracellular PLLA nanoparticles.

9 Å cryo-EM structure and molecular model of a gastropod hemocyanin didecamer (KLH1) reveals the architecture of the asymmetric collar

C. Gatsogiannis, K. Büchler, F. Depoix & J. Markl

Institute of Zoology, Johannes Gutenberg University, D-55099 Mainz, Germany

gatsogia@uni-mainz.de
Keywords: hemocyanin, three-dimensional cryo-electron microscopy, molecular model

Hemocyanins are blue copper proteins that transport oxygen in the hemolymph of many arthropods and molluscs. Molluscan hemocyanins are decamers, didecamers or multidecamers of a 350-400 kDa polypeptide subunit that is subdivided into seven or eight different functional units (FUs, each with a single copper active site). The quaternary structure is a semi-hollow cylinder consisting of a wall and a collar. Recently, we published a 9 Å cryo-EM structure and molecular model of a cephalopod hemocyanin decamer (NpH, from *Nautilus pompilius*) that answered many hitherto unsolved questions concerning the quaternary structure of molluscan hemocyanin. Notably, it revealed the twisted pathway of the 10 subunit copies (each with seven FUs) and the various molecular interfaces between the 70 FUs [1]. Six FU types (termed NpH-a through NpH-f) constitute the cylinder wall, whereas the seventh type (NpH-g) forms the internal collar.

Now we present the first cryo-EM structure of a gastropod hemocyanin (KLH1) that allowed molecular fitting of the modelled FUs. In contrast to NpH, gastropod hemocyanins are didecamers, with two decamers assembled face-to-face, and the collar contains an additional FU type termed FU-h. Compared to the other seven FU types, FU-h is enlarged by a tail of ca. 100 amino acids [2]. KLH (keyhole limpet hemocyanin) is obtained from the Californian keyhole limpet *Megathura crenulata* and intensively applied, in immunological research and clinics, as an immunoactivator and tumour vaccine carrier; it occurs in two isoforms termed KLH1 and KLH2 [3]. Our present 9 Å cryo-EM structure of KLH1 shows significantly more detail than a previous 3D reconstruction of this protein [4]. It exhibits a very similar cylinder wall as NpH and reveals the potential attachment sites for N-linked glycans that might be primarily responsible for the observed immunological effects. In addition it shows, for the first time, the contact zones between the two decamers (containing potential Ca^{++}/Mg^{++} binding sites). And ultimately, by incorporating a recently obtained 4 Å crystal structure of KLH1-h [Büchler, K., Jaenicke E. et al., to be published], it completely solves the old mystery of the gastropod hemocyanin collar.

A. Aretz, B. Hermanns-Sachweh, J. Mayer (Eds.): EMC 2008, Vol. 3: Life Science, pp. 21–22,
DOI: 10.1007/978-3-540-85228-5_11, © Springer-Verlag Berlin Heidelberg 2008

1. C. Gatsogiannis, A. Moeller, F. Depoix, U. Meissner, J. Markl, JMB 374 (2007), p. 465-86.
2. B. Lieb, B. Altenhein, J. Markl, J Biol Chem 275 (2000), p. 5675-5681.
3. J.R. Harris, J. Markl, Eur Urol 37 (2000), p. 25-31
4. F. Mouche, Y. Zhu, J. Pulokas, C.S. Potter, B. Carragher, J Struct Biol. 144 (2003), p. 301-12.

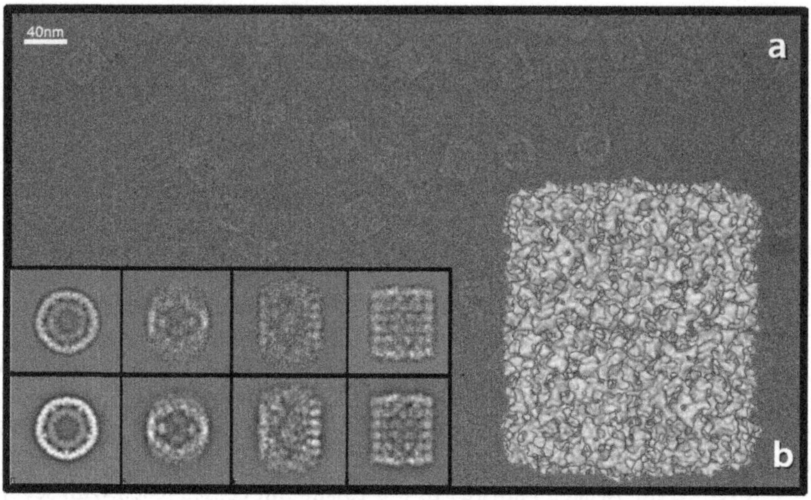

Figure 1. (a) Cryo-electron micrograph of keyhole limpet hemocyanin type 1 (KLH1). (b) 9 Å structure of KLH1 shown in side-view. The inset shows four different class average images (upper row) and the corresponding reprojections (lower row).

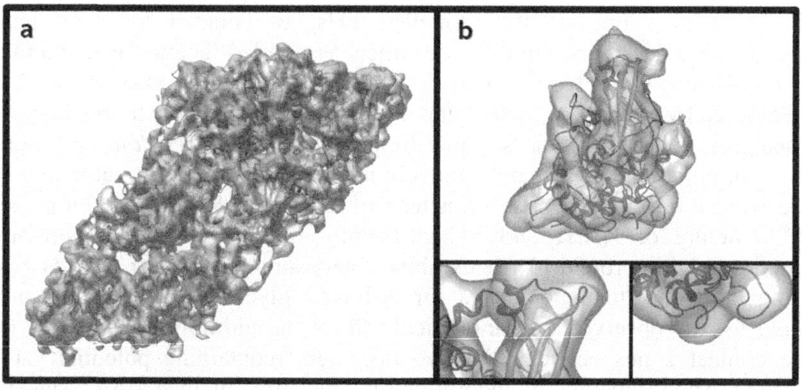

Figure 2. (a) Extracted cryo-EM structure and molecular model of the subunit dimmer, with the two subunits shown in different colours (b) Extracted cryo-EM structure and molecular model of a single FU (upper image). Details of the rigid body fitting (lower images)

Microstructure of model systems for sauces based in polysacharides observed by Cryo-SEM

I. Hernando, L.M. Guardeño, E. Llorca, A. Quiles, I. Pérez-Munuera

Departamento de Tecnología de Alimentos, Universidad Politécnica de Valencia, Camino de Vera s/n 46022 Valencia, Spain.

mihernan@tal.upv.es
Keywords: Cryo-SEM, polysaccharides, sauces.

The demand for foodstuffs which are easily and fast prepared has increased considerably in recent years. Sauces are often used in this kind of products and their formulation usually includes polysaccharides (starch and/or gums) as thickening agents [1]. Sauce microstructure gives information about the components arrangement and how it affects physical and chemical properties such as viscosity. The aim of this work is to study the effect of different thickeners and salt concentrations on the microstructure of model systems.

Four types of thickeners were used: native and modified corn starch, ι-carrageenan and λ-carrageenan. Samples were prepared by mixing different thickener concentrations (0.3-6% w/w) with different salt concentrations (1-7% w/w) in distilled water. The total mixture was heated and stirred in a food processor at 2000 rpm during 5 min in order to reach 90 ºC and this temperature was then maintained for 6 min.

The microstructure was studied by Low Temperature Scanning Electron Microscopy (Cryo-SEM). In the Cryo-SEM technique a JSM-5410 SEM microscope (Jeol, Tokyo, Japan) was used with a Cryo CT-1500C unit (Oxford Instruments, Witney, UK). The sample was frozen in slush nitrogen ($T \leq -210$ºC), fractured, etched (-90ºC) and gold-coated (2mbar, 2 mA) for imaging, working at 15 kV and -130ºC. Image analysis was performed with analySYS AUTO software (Soft Imaging Systems GmbH, Münster, Germany).

Native corn starch components exit the granule during gelatinization and disperse forming a continuous matrix of cells (figure 1A); these cells are formed during the etching of the sample for Cryo-SEM observation and are called eutectic artefact or solutes aggregation phenomena. Figure 1B shows some modified corn starch granules that resist heating during the sample preparation; a similar structure was observed in starch based custards [2]. ι-carrageenan (figure 1C) and λ-carrageenan (figure 1D) disperse in the water forming a continuous matrix of cells as in native starch.

Using image analysis, the area of the cells formed during the etching of the different samples has been measured. Figure 2 shows, by way of example, the images obtained when native starch was used as thickener at different concentrations: 1% w/w (figure 2A), 3.5% w/w (figure 2B) and 6% w/w (figure 2C). As it can be noticed, the higher concentration the smaller size of cells (from 82.67 ± 17.14 μm^2 in sample with 1% of native starch to 9.06 ± 2.83 μm^2 in sample with 6%). The decrease in size is related to a

A. Aretz, B. Hermanns-Sachweh, J. Mayer (Eds.): EMC 2008, Vol. 3: Life Science, pp. 23–24, DOI: 10.1007/978-3-540-85228-5_12,

high interaction between system components. The other polysaccharides used as thickening agents showed similar behaviour to native starch.

Variation of salt concentration caused different results depending on the thickener used in the system. Some polysaccharides, such as native starch and ι-carrageenan, decrease their cell size when salt concentration rose. However, salt concentration did not have significant influence in the cell size when λ-carrageenan was used.

1. I.G. Mandala, T.P. Savras, A. E. Kostaropoulos, Journal of Food Engineering **64** (2004), pp. 335-342.
2. I. Hernando, I. Pérez-Munuera, V. Larrea, E. Llorca, A. Quiles, M.A. Lluch, Proceedings of COST Action 921 (2006), pp.82-87.
3. The authors are indebted to the Project AGL-2006-11653-C02-02/ALI for financial support and to the Universidad Politécnica de Valencia for the grant awarded to L.M. Guardeño.

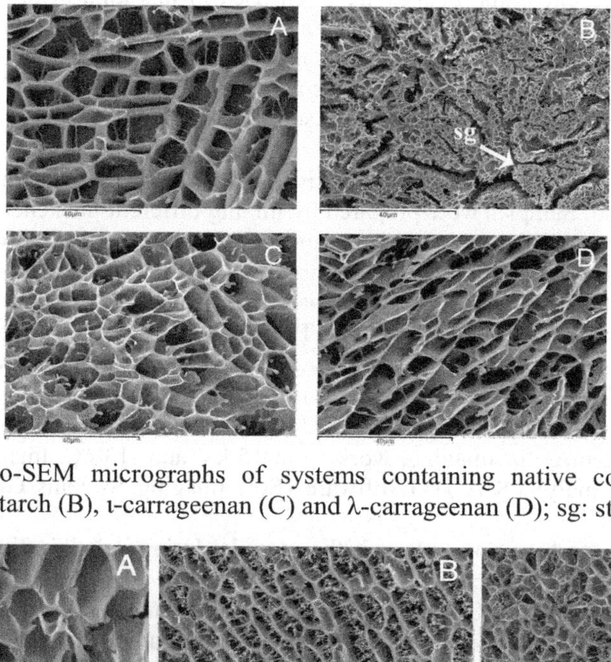

Figure 1. Cryo-SEM micrographs of systems containing native corn starch (A), modified corn starch (B), ι-carrageenan (C) and λ-carrageenan (D); sg: starch granule.

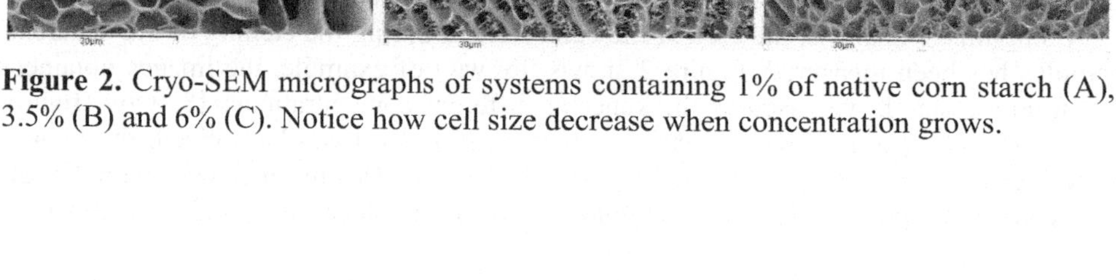

Figure 2. Cryo-SEM micrographs of systems containing 1% of native corn starch (A), 3.5% (B) and 6% (C). Notice how cell size decrease when concentration grows.

Three-dimensional architecture of outer- and inner-dynein arms in flagella revealed by cryo-electron tomography and single particle analysis

T.Ishikawa[1], K.H. Bui[1], T. Movassagh[1], H. Sakakibara[2], K. Oiwa[2]

1. Department of Biochemistry, ETH Zurich, HPK F7, ETH Hönggerberg, CH8049, Zurich, Switzerland
2. KARC, NIICT, 588-2 Iwaoka, Nishi-ku, Kobe, 651-2492, Japan

takashi.ishikawa@mol.biol.ethz.ch
Keywords: cryo-electron microscopy, tomography, flagella

Flagella are the bending organelles to drive cells and have "9+2" structure with nine microtubule doublets connected by dynein arms, which are ATP-driven force generating apparatus. Inner dynein arms are localized on microtubule doublets with 96nm periodicity and regulate flagellar bending motion, whereas outer dynein arms, the accelerators for axonemal motion, form arrays on microtubule doublets with 24nm repeat. Our goal is to elucidate the mechanism of bending motion generated by dynein arms, which make only linear motion as a monomer on microtubules, and their regulation.

Here in this study, using cryo-electron tomography and single particle analysis as well as the newly developed algorithm of image analysis, we analysed 3D structure of dynein motor protein binding to microtubules in. Previously, we reconstructed 3D structure of outer-dynein/microtubule complexes by dealing them as independent single fragments [1]. By our new algorithm which take continuous distribution of dynein particles on one microtubule into account, small components of outer dynein complexes (for example, docking complexes), which had not been visible previously, were visualized for the first time. This means our resolution from tomography was improved from ~45Å resolution to ~35Å. This approach also enabled us to analyse other complexes like inner-dynein and radial-spoke (Bui et al., in preparation) and visualize how dynein heavy chains, which consist of a AAA-ring, N-terminus tail and coiled-coil stalk (totally 4500 amino acids) form network; vertical trimers in outer-dynein arms (Figure 1) and horizontal dimers in inner-dynein arms.

1. T. Ishikawa, H. Sakakibara and K. Oiwa, J. Mol. Biol. **368** (2007), p. 1249.
2. We thank EMEZ (Electron Microscopy Center, ETH Zurich) for technical support.

A. Aretz, B. Hermanns-Sachweh, J. Mayer (Eds.): EMC 2008, Vol. 3: Life Science, pp. 25–26,
DOI: 10.1007/978-3-540-85228-5_13, © Springer-Verlag Berlin Heidelberg 2008

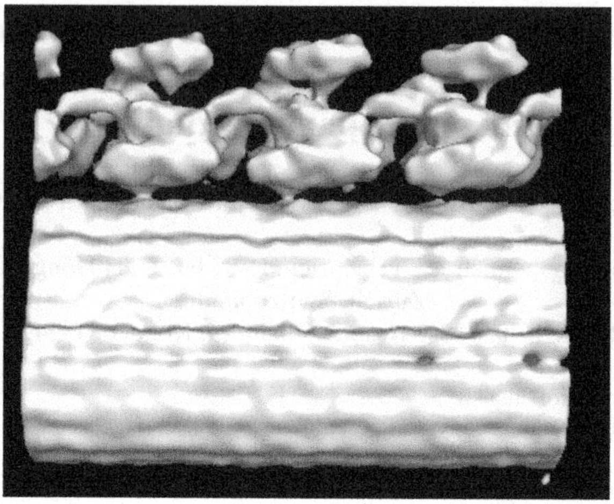

Figure 1. Surface rendering representation of outer-dynein arms and microtubule doublets from flagella. Intact flagella were purified, quickly frozen and continuously tilted in the microscope for tomographic image collection. After 3D reconstruction was carried out by back projection, dynein arms were extracted three-dimensionally, aligned and averaged.

Structural and functional studies of rabbit skeletal muscle Phosphorylase Kinase

S. Jonic[1], C. Carrière[1], E. Larquet[1], V. Skamnaki[2], L. Johnson[2], C. Venien-Bryan[2], J.-P. Mornon[1], I. Callebaut[1] and N. Boisset[1]

1. Département de Biologie Structurale, Institut de Minéralogie et de Physique des Milieux Condensés, UMR 7590, CNRS, Universités Paris 6 et Paris 7, Paris, France.
2. Laboratory of Molecular Biophysics, Department of Biochemistry, University of Oxford, South Parks Road, Oxford OX1 3QU, United Kingdom.

slavica.jonic@impmc.jussieu.fr
Keywords: Phosphorylase Kinase, cryo-electron microscopy, three-dimensional reconstruction, homology modeling, hydrophobic cluster analysis, fitting, structure, function

During a physical exercice, the activity of the Phosphorylase Kinase (PhK) is very high, promoting an optimal level of glucose in skeletal muscle. Its major role is to catalyze the conversion of the inactive Glycogene Phosphorylase (GPb) to an active form (GPa) by phosphorylation. The PhK is composed of 4 copies of the α, β, γ and δ subunits forming a hexadecameric structure. This structure is organised into 2 lobes with D2 symmetry. Each lobe comprises 2 protomers of α β γ δ, with bridges connecting the 2 lobes. The α and β subunits account for 2/3 of the protein mass ($1.3x10^6$ Da); the similarity of α and β sequences is probably due to gene duplication. The δ subunit is a typical calcium binding protein, a calmodulin, which is bound tighly to the complex. The γ subunit is the catalytic subunit, which functions to phosphorylate serine and threonine residues on PhK α and β, and GPb.

Despite its importance and several studies devoted to the analysis of its biochemistry and structure, very little is known about the intricate Ca^{2+}-sensitive communication network among the different subunits of the enzyme and the allosteric regulation of the holoenzyme activity in response to effector ligands. In addition, information is still lacking about the structures and functions of the α and β subunits of the enzyme.

In this work, we coupled several approaches to improve our understanding of the PhK. These are: 3D reconstruction from images acquired in a cryo-electron microscope (cryo-EM) (Figure 1), 3D structure prediction by hydrophobic cluster analysis (for the details on the prediction of the structures of α and β subunits, see the companion paper [2]), and fitting of atomic structures into the 3D cryo-EM density map (preliminary results of the fitting are shown in Figure 2).

1. C. Venien-Bryan, E.M. Lowe, N. Boisset, K.W. Traxler, L.N. Johnson and G.M. Carlson, Structure **10** (2002), p. 33.
2. C. Carrière, J.-P. Mornon, S. Jonic, E. Larquet, C. Venien-Bryan, N. Boisset and I. Callebaut. (submitted to the 14th European Microscopy Congress, September 1 - 5, 2008, Aachen, Germany)

A. Aretz, B. Hermanns-Sachweh, J. Mayer (Eds.): EMC 2008, Vol. 3: Life Science, pp. 27–28,
DOI: 10.1007/978-3-540-85228-5_14,

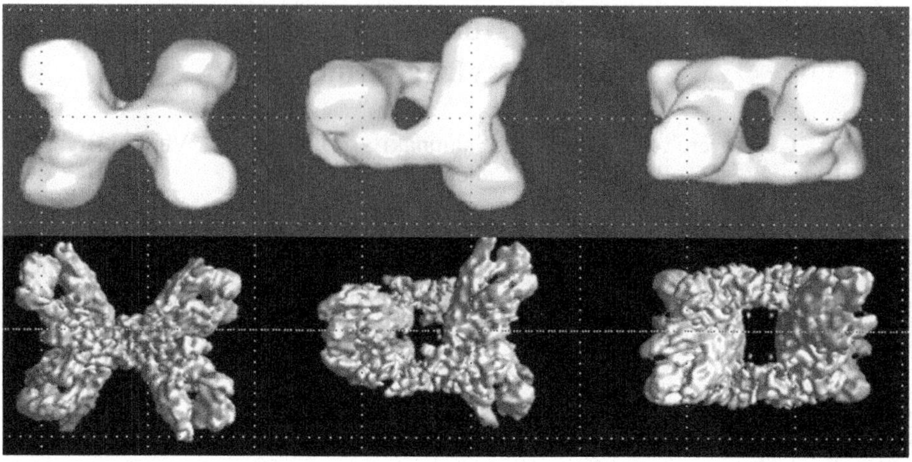

Figure 1. Three views of an isosurface of the 3D density map of the PhK. Top panel: negative-stain volume (resolution: 2.2 nm) [1]. Bottom panel: cryo-EM volume (resolution: 0.99 nm) obtained by refining the negative-stain volume.

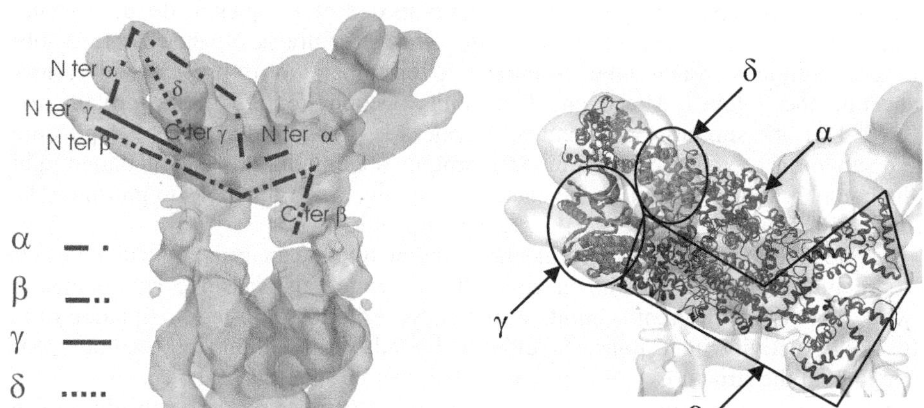

Figure 2. Preliminary fitting results. Global localisation of the PhK domains in an intermediate-resolution (1.4 nm) cryo-EM PhK density map. Left panel: model of interactions of the PhK domains. Right panel: manual fit of the structures solved or predicted at the atomic level.

Single-particle analysis of the Cdc6/Orc1, archaeal DNA replication initiator protein

N. Kajimura[1,a], H. Suzuki[1,b], K. Mayanagi[1,c], Y. Ishino[2], and K. Morikawa[1,d]

1. Department of Structural Biology, Biomolecular Engineering Research Institute, 6-2-3 Furuedai, Suita-shi, Osaka 565-0874, Japan
2. Laboratory of Protein Chemistry and Engineering, Faculty of Agriculture, Kyushu University, Hakozaki, Higashi-ku, Fukuoka-shi, Fukuoka 812-8581, Japan
a. Present address: Research Center for Ultra-High Voltage Electron Microscopy, Osaka University, 7-1 Mihogaoka, Ibaraki-shi, Osaka 567-0047, Japan
b. CREST, Japan Science and Technology Agency
c. Present address: Department of Bioscience, Faculty of Bioscience, Nagahama Institute of Bio-Science and Technology c/o Biomolecular Engineering Research Institute, 6-2-3 Furuedai, Suita-shi, Osaka 565-0874, Japan
d. Present address: Institute for Protein Research, Osaka University c/o Biomolecular Engineering Research Institute, 6-2-3 Furuedai, Suita-shi, Osaka 565-0874, Japan

kajimura@uhvem.osaka-u.ac.jp
Keywords: single particle analysis, DNA replication, AAA$^+$ ATPase

The formation of the pre-replication complex (pre-RC) is essential for DNA replication initiation in eukarya and archaea. Origin recognition complex (ORC) and Cdc6 are the key component of pre-RC [1]. There are six different ORC proteins in eukaryotes, and ORC complexes are known to recognize and bind to the specific DNA sequence called replication origin. Once bound, initiator proteins participate in other aspects of replication initiation; they facilitate origin unwinding and recruit other replication proteins, such as Cdc6 and mini-chromosome maintenance (MCM) helicase to the origin. However, the mechanism and structure of the ORC complexes are not known. An anaerobic hyperthermophile, *Pyrococcus friosus*, has only one gene showing a sequence similarity to ORC, and it also has a homology to Cdc6 [2]. Cdc6/Orc1 protein is the most likely candidates for a replication initiator in archaea, and this provides a simpler system for study of replication initiation.

ORC and Cdc6 proteins belong to the AAA$^+$ family ATPase, which is known to be class of chaperone-like ATPase involved in broad range of cellular processes [3]. The defining features of AAA$^+$ proteins are highly conserved motif in the ATPase domain called the second region of homology (SRH), and the formation of homo- or hetero-oligomers. The ATP-binding pockets lie in the subunit interface, and the hydrolysis of ATP is believed to be triggered by the conserved arginine residue protruding from the adjacent subunit. The other archeal Cdc6/Orc1s ever known exist as a monomer *in vitro,* like DnaA that is a bacterial initiator. Recently the crystal structures of DNA complex of initiator has been solved subsequently. DnaA was has solved in ATP-analogue bound form [4], and the Cdc6/Orc1 from other archaeal species have also been solved in ADP-form [5, 6]. However the mechanism of the hydrolysis of ATP and its relationship with DNA unfolding and are unclear.

A. Aretz, B. Hermanns-Sachweh, J. Mayer (Eds.): EMC 2008, Vol. 3: Life Science, pp. 29–30,
DOI: 10.1007/978-3-540-85228-5_15,

We have used electron microscopy and three-dimensional reconstruction to examine Cdc6/Orc1 protein from *P. friosus*, and showed they oligomerize into double-hexameric ring typical structure for AAA$^+$ family protein *in vitro*. The oligomerization did not need any nucleotide and DNA. The rings were arranged in tail-to-tail manner. The outer diameter of each ring is about 140 Å and the height is about 70 Å. A channel with 20 Å width that is sufficient to accommodate DNA is penetrating the center of the EM map. We tried fitting atomic structures of Cdc6/Orc1 from other archaeal species into the EM reconstruction map. Several structures of DNA-free Cdc6/Orc1 protein have reported to date [7,8], and the structures showed that a conformational change occurs in the position of domain-III, when they bound ADP. Within the structures, ADP-bound form of ORC2 from *Aeropyrum pernix* spieces fitted into our map with highest correlation score. The overall architecture of the ring is consistent with six Cdc6/Orc1 monomers arranged around a six-fold axis.

We also used a simultaneous multiple-model refinement method to search [9] for the presence the structural conformers in the arrangement of domain-III from the electron microscopic images of Cdc6/Orc1. We demonstrated the unexpected result that the resulting model varied not only in height, but also in diameter. This means that there are some variations in intra-molecule and in inter-molecule interaction in the hexameric oligomer. Individual domain fitting into the EM map also showed a variation in the distance between domain –I and domain –II. Both results explain a flexibility of Cdc6/Orc1 molecule.

1. S.P. Bell, and A. Dutta, Annu Rev Biochem. **71** (2002), p.333-374.
2. B. Grabowski, and Z. Kelman, Annu. Rev. Microbiol. **57** (2003), p. 487-516.
3. A.F. Neuwald, L. Aravind, J.L. Spouge, and E.V. Koonin, Genome Res. **9**(1999), p.27-43.
4. J.P. Erzberger, M.L. Mott, and J.M. Berger, Nat Struct Mol Biol. **13**(2006) p.676-683.
5. E.L. Dueber, J.E. Corn, S.D. Bell, and J.M. Berger, Science. **317**(2007), p.1210-1213.
6. M. Gaudier, B.S. Schuwirth, S.L. Westcott, and D.B. Wigley, Science. **317**(2007), p.1213-1216.
7. J. Liu, C.L. Smith, D. DeRyckere, K. DeAngelis, G.S. Martin, and J.M. Berger, Mol Cell. **6**(2000) p.637-648.
8. M.R. Singleto, R. Morales, I. Grainge, N. Cook, M.N. Isupov, and D.B. Wigley, J Mol Biol. **343**(2004) p.547-557.
9. J. Brink, S.J. Ludtke, Y. Kong, S.J. Wakil, J. Ma, and W. Chiu, Structure, **12**(2004), p.185–191.

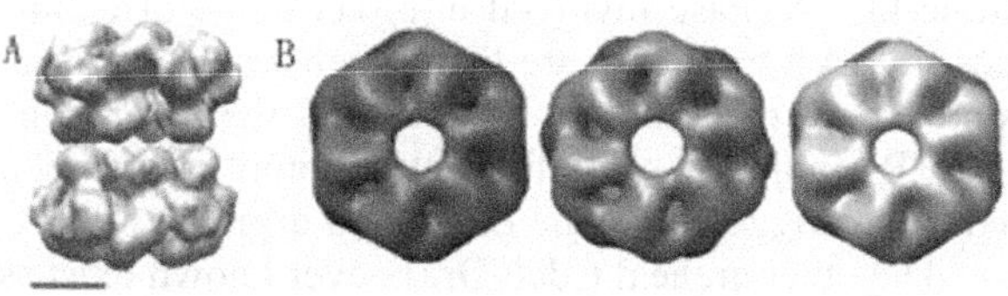

Figure 1. 3D Reconstruction of Cdc6/Orc1 oligomer. A) the original model with D6 symmetry. Scale bar is 50 Å. B) refined models obtained from *Multirefine* procedure with different starting models.

A high-throughput acquisition toolbox in Matlab for cryo-electron microscopy studies

A. Korinek[1], F. Beck[1], S. Nickell[1], W. Baumeister[1] and J. M. Plitzko[1]

1. Max-Planck-Institute of Biochemistry, Dept. of Molecular Structural Biology, Am Klopferspitz 18, 82152 Martinsried, Germany

korinek@biochem.mpg.de
Keywords: high-throughput cryo-electron microscopy, TOM toolbox, Titan Krios

High-throughput single particle analysis offers new perspectives in structural investigations of macromolecular complexes. Especially for complexes with a high amount of heterogeneity this technique can lead to structural insights at a reasonable resolution. The 'TOM software toolbox' designed for the acquisition and analysis in electron tomography was therefore extended to deal with the special needs of high-throughput single particle analysis. The system is designed in a flexible way to provide an easy-to-use platform for automated acquisition. Implemented as an object-oriented microscope driver framework, the software package can be easily adapted to different kinds of microscopes.

For a robust automated data collection, we have developed a three-step procedure comprising an initial scan of the central area of an EM grid at low magnification, the selection of positions suitable for imaging and the automated acquisition of 2D images at higher magnifications as illustrated in Figure 1 [1]. The system has been implemented at a Titan Krios (Figure 2) and a Tecnai Polara G2 (FEI Company, Eindhoven, The Netherlands).

The Polara G2 was the first generation of transmission electron microscopes (TEM) that enabled fully automated acquisition in a high-throughput fashion for single particles studies. The system is equipped with a stable specimen holder and a nitrogen cooling system which needs to be refilled only once per day. The Tecnai user interface is fully scriptable via ActiveX making the development of custom microscope control software in modern programming languages such as Matlab (The Mathworks, Natick, USA) feasible [2].

The Titan Krios TEM is a big step forward towards autonomous collection of large data sets. It is a dedicated cryo-EM instrument comprising the latest technological achievements in robotics and automation. The key features of this instrument are the automatically refilling nitrogen cooling system, which needs maintenance only once per week, the automated sample loading/transfer and a complete new holder design. Furthermore, it is equipped with an 8kx8k CCD camera (TVIPS, Gauting, Germany) based on CMOS technology offering an increased field-of-view (FOV) at a higher readout rate [3]. We will report about our first experiences and our recent findings with this instrument regarding the structural analysis of the 26S Proteasome.

A. Aretz, B. Hermanns-Sachweh, J. Mayer (Eds.): EMC 2008, Vol. 3: Life Science, pp. 31–32,
DOI: 10.1007/978-3-540-85228-5_16,

1. S. Nickell, F. Beck, A. Korinek, O. Mihalache, W. Baumeister and J.M. Plitzko, FEBS Letters 581 (15) (2007), 2751-2756.
2. S. Nickell, F. Foerster, A. Linaroudis, W. Del Net, F. Beck, R. Hegerl, W. Baumeister and J.M. Plitzko, Journal of Structural Biology 148 (2005), 227-234.
3. D. Tietz, S. Nickell, H. Tietz, W. Baumeister and J.M. Plitzko, EMC2008 abstract volume.
4. This work was supported by the 3DEM Network of Excellence, The Interaction Proteome and the High throughput-3DEM Grant all within the Research Framework Program 6 (FP6) of the European Commission.

Figure 1. The Titan Krios as installed at the MPI of Biochemistry (A and B). The control unit of the microscope is seated on the 'bridge' (B upper right and C).

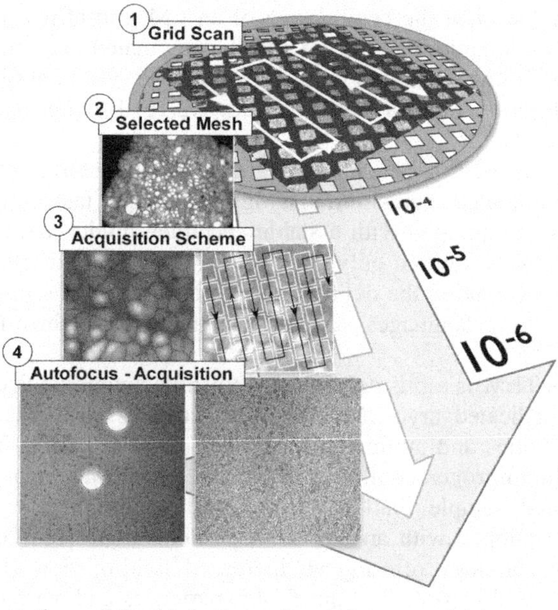

Figure 2. Representation of the high-throughput data acquisition mode. Initial scan (1, grid scan) of the central EM grid area. During the acquisition the stage is directed in a meandering fashion (1, white arrows and at higher magnifications 3, black arrows). Auto-focusing is supported by the generation of a pair of holes (4).

Architecture of bacterial glutamate synthase complexe using hybrid approaches

M. Cottevieille[1], E. Larquet[1], S. Jonic[1], I. Callebaut[1], D. Svergun[3,4], M.A. Vanoni[2], and N. Boisset[1]

1. Département de Biologie Structurale, Institut de Minéralogie et de Physique des Milieux Condensés, UMR 7590, CNRS, Universités Paris 6 et Paris 7, Paris, France.
2. Dipart. di Scienze Biomolecolari e Biotecnologie, Universita' degli Studi di Milano, Via Celoria 26, I-20133 Milano, Italy.
3. European Molecular Biology Lab. Hamburg Outstation, EMBL c/o DESY, Notkestrasse 85, D-22603 Hamburg, Germany.
4. Institute of Crystallography, Russian Academy of Sciences, Leninsky pr. 59, 117333 Moscow, Russia.

eric.larquet@impmc.jussieu.fr

Keywords: Glutamate Synthase, cryo-electron microscopy (cryo-EM), three-dimensional (3D) reconstruction, contrast transfer function (CTF), small-angle X-ray scattering (SAXS), homology modelling, structure, function

Glutamate synthase (GltS) is a mutienzymatic complex forming with glutamine synthetase the main ammonia assimilation pathway in micro-organisms and plants. GltS catalyses the synthesis of L-glutamate from 2-oxoglutarate (2-OG) and L-glutamine [1]. Bacteria contain a NADPH-dependent form of the enzyme (NADPH-GltS), composed of two types of subunits (α subunit, 162 kDa, and β subunit, 52.3 kDa). The two subunits form the catalytically active $\alpha\beta$ holoenzyme, which contains 1 FAD, 1 FMN, and 3 different Fe-S clusters. The β subunit acts as a FAD-dependent NADPH oxidoreductase, by transferring reducing equivalents to the site of glutamate synthesis on the α subunit through at least 2 of the 3 Fe-S clusters. The α subunit, whose structure has been recently determined by X-ray crystallography [2], contains 2 catalytic sites: one site where L-Gln is hydrolyzed to yield L-Glu and ammonia; the latter is transferred through an intramolecular ammonia tunnel to the other catalytic site where 2-OG is converted to the iminoglutarate (2-IG) intermediate, and reduced to L-Glu by receiving reducing equivalents from the reduced FMN cofactor.

We have applied cryo-electron microscopy (cryo-EM), small-angle X-ray scattering, homology modeling, site-directed mutagenesis and kinetic studies in a synergic fashion in order to gain more detailed information on the structure of the NADPH-GltS holoenzyme and on that of the $\alpha\beta$ protomer [3] (Figures 1 and 2).

1. M.A. Vanoni and B. Curti B , Cell Mol Life Sci **55** (1999), p. 617.
2. C. Binda, R. Bossi, S. Wakatsuki, S. Artz, A. Coda, B. Curti, M.A. Vanoni and A. Mavetti, Structure Fold Des **8** (2000), p. 1299.
3. M. Cottevieille, E. Larquet, S. Jonic, M. V. Petoukhov, G. Caprini, S. Paravisi, D. I. Svergun, M. A. Vanoni, N. Boisset, J. Biol. Chem. (2008) (in press).

A. Aretz, B. Hermanns-Sachweh, J. Mayer (Eds.): EMC 2008, Vol. 3: Life Science, pp. 33–34,
DOI: 10.1007/978-3-540-85228-5_17,

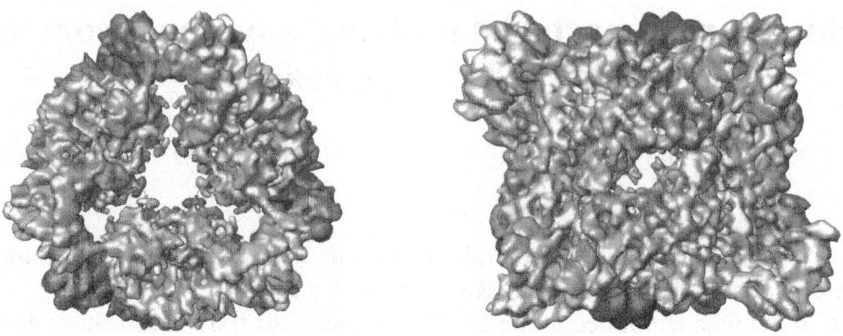

Figure 1. Two views of an isosurface of the 3D cryo-EM map of the GltS at 9.5 nm resolution. Left: top view. Right: side view.

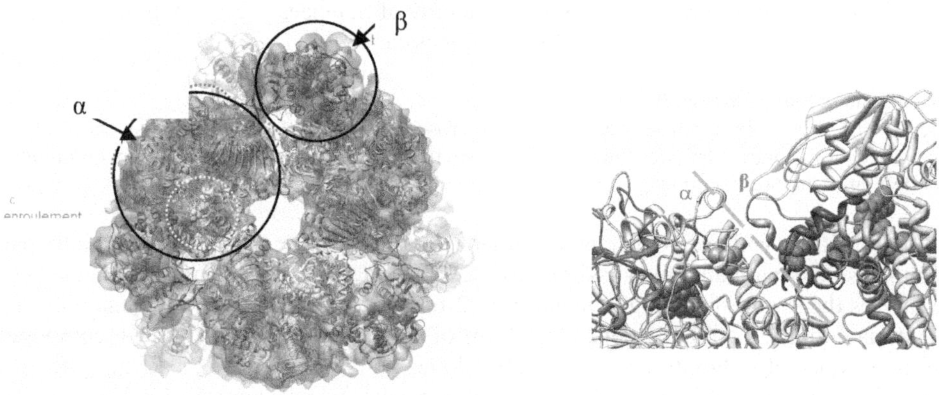

Figure 2. Left: Fitting of the subunit α (obtained by X-ray crystallography [2]) and of the subunit β (obtained by homology modelling [3]) into the 3D cryo-EM map from Figure 1. Right: Study of α-β contacts.

3D Structure of DNA Repair Macromolecular Complexes Participating in Non-Homologous End-Joining (NHEJ)

O. Llorca[1], A. Rivera-Calzada[1], L. Spagnolo[2], M.A. Recuero-Checa[1] and L.H. Pearl[2]

1. Centro de Investigaciones Biológicas. Consejo Superior de Investigaciones Científicas (CSIC). Ramiro de Maeztu, 9. Campus Universidad Complutense. 28040 Madrid. Spain
2. Section of Structural Biology and Cancer Research UK DNA Repair Enzyme Research Group, Institute of Cancer Research. Cancer Research UK. 237 Fulham Road, London SW3 6JB, U.K.

ollorca@cib.csic.es
Keywords: DNA repair, DNA-PK, NHEJ

Double-strand DNA breaks (DSBs) fracture the chromosomes, generating genomic instability and chromosomal translocations which favor tumor transformation [1]. Cells use two mechanisms to repair these DSBs, Homologous Recombination (HR) and Non-Homologous End-Joining (NHEJ) [2]. NHEJ is the most frequent DNA repair pathway in mammalian cells but this is restricted to G0 and G1 phases of the cell cycle. During NHEJ the broken ends are processed with potential loss of genetic information, and the backbones ligated without using a template. In addition, the NHEJ system repairs programmed DSBs generated as intermediates during V(D)J recombination in lymphocytes. Several recessive human syndromes and diseases can be related to mutations of proteins that participate in these DNA repair pathways.

Primary recognition of damaged DNA during NHEJ is performed by a highly abundant heterodimer of the homologous Ku70 and Ku80 proteins. The Ku70/Ku80 complex binds free DNA ends and recruits DNA-PKcs to the DNA lesion to form a high molecular weight complex [3]. DNA-PKcs (DNA-dependent protein kinase catalytic subunit) is a ~470 kDa enzyme belonging to the family of PI3-kinase related protein kinases (PIKK), which comprises other large kinases such as ATM, ATR or TOR. The Ku70/Ku80/DNA-PKcs complex acts as a scaffold for other proteins required to complete the repair reaction mainly the DNA Ligase IV/Xrcc4 complex containing a ligase activity specific for the NHEJ reactions that closes the phosphodiester backbone [4].

Although atomic structures of truncated versions of some of the proteins involved in NHEJ have been solved using X-ray crystallography [5-7], the size and complexity of this multi-component repair system presents severe challenges for structural analysis. We have purified individual components of the NHEJ reaction and analyzed them using single-particle analysis. We have performed three-dimensional reconstructions using data from both cryo-electron microscopy (cryo-EM) and negatively stained specimens, and the reconstructions have been interpreted after fitting atomic structures and models. We propose a structural model of DNA-PKcs based on a ~15 Å cryo-EM map and domain mapping [8] and of the full-length Ku70/Ku80 in isolation [9]. Moreover, we have reconstituted in vitro a human Ku70/Ku80/DNA-PKcs complex assembled on a

A. Aretz, B. Hermanns-Sachweh, J. Mayer (Eds.): EMC 2008, Vol. 3: Life Science, pp. 35–36,
DOI: 10.1007/978-3-540-85228-5_18,

DNA fragment mimicking damaged DNA, and the three dimensional structure of this 650 kDa complex has been solved at ~25 Å resolution using cryo-EM [10]. A synaptic complex between two Ku70/Ku80/DNA-PKcs complexes was visualized in the electron microscope and its structure revealed the mechanism by which two Ku70/Ku80/DNA-PKcs complexes interact and hold two broken DNA ends in proximity. More recently, we have purified several forms of the DNA ligase IV/Xrcc4 complex. 3D reconstructions of these complexes reveal likely structural mechanisms by which this complex performs the actual ligation of the broken DNA ends at the last steps in the repair reaction

Given the relative small size of some of these proteins for EM analysis, their asymmetry, their flexibility, and the difficulties in their purification and in the assembling of the multi-protein complexes, innovations in the image processing strategy were required to reconstruct these structures. We have applied several approaches to deal with (a) the model bias of the initial templates for refinement; (b) the heterogeneity of the sample, and (c) the possible distortions from staining agents. The collection of EM structures we have obtained provides relevant snapshots of these complex DNA repair processes [11].

1. K.K Khanna, and S.P. Jackson, Nat Genet **27** (2001), p. 247-54.
2. S.P. Jackson, Carcinogenesis **23** (2002), 687-96.
3. S.J. Collis et al., Oncogene **24** (2005), p. 949-61.
4. Y. Wang, , B.J. Lamarche, and M.D. Tsai. Biochemistry, **46** (2007), p. 4962-76.
5. M.S. Junop et al., Embo J. **19** (2000), p. 5962-70.
6. B.L. Sibanda et al., Nat Struct Biol **8** (2001), p. 1015-9.
7. J.R. Walker, R.A. Corpina, and J. Goldberg. Nature **412** (2001), p. 607-14.
8. Rivera-Calzada et al.. Structure **13** (2005), p. 243-55.
9. Rivera-Calzada et al. EMBO Rep **8** (2007), p. 56-62.
10. L. Spagnolo et al., Mol Cell **22** (2006), p. 511-9.
11. Llorca, Curr Opin Struct Biol **17**(2007), p. 215-20.

Quaternary structure of recombinant human meprin β, a zinc peptidase of the astacin family

Ulrich Meissner[1], Philipp Arnold[1], Arne Moeller[1], Frank Depoix[1], Christoph Becker-Pauly[2] & Walter Stöcker[2]

1. Department of Molecular Physiology, Institute of Zoology, Johannes von Muellerweg 6, 55128 Mainz, Germany
2. Department of Cell and Matrix Biology, Institute of Zoology, Johannes von Muellerweg 6, 55128 Mainz, Germany

meissner@uni-mainz.de
Keywords: meprin, metalloproteases, cryo-TEM, 3D-EM

Meprins are astacin-type zinc peptidases distantly related to matrix metalloproteinases (MMPs) [1]. They are expressed in various epithelia, intestinal leukocytes and cancer cells. They cleave basement membrane proteins, cytokines and adhesion molecules, suggesting a role in epithelial differentiation, cell migration and immune reactions [2-4]. There is evidence for detachment of cells from the basal membrane caused by meprin alpha, while meprin beta probably plays a role in terminal differentiation and cell migration [5, 6].

Although meprin alpha and beta have an amino acid sequence identity of 44%, they show marked differences in activation, substrate specificity and significantly, in their quaternary structure. From negatively stained electron micrographs it is known that meprin alpha forms ring- and chain shaped oligomers, up to mega Dalton size. This characterizes meprin alpha as the largest known secreted protease [7, 8]. In contrast, meprin beta, which is normally membrane bound, exists as homodimers, linked by an intermolecular disulfide bond.

Different quaternary structures have been discussed for meprin based on structure analysis: E.g. preliminary electron microscopy studies resulted in a low resolution model (resolution less than 20 Å) based on negatively stained samples as the data set for the 3D reconstruction [8]. The resolution acheived here was mainly limited by the technique of negative staining.

Here we present preliminary data of the quaternary structure of recombinant meprin beta homo dimers obtained by cryo-electron microscopy. The cryo-electron microscopy benefited from the high purity of the protein solution derived from insect cell (HiFive) supernatants [9]. Neverthless, the low molecular mass of the meprin beta dimer, remains a challenge for single particle analysis. We now show a more detailed model of the quaternary structure of the meprin β based on unstained cryo TEM data: At a molecular resolution of less than 12 Å (½-bit criterion) we performed a preliminary rigid docking of closely related atomic structures, which supports the biochemical data and gives a preliminary impression of the orientation of the single domains within the overall structure of the meprin β dimer in solution. This data presents a more detailed view of the complex quaternary structure (Figure 1) of this molecule, which offers a challenge

A. Aretz, B. Hermanns-Sachweh, J. Mayer (Eds.): EMC 2008, Vol. 3: Life Science, pp. 37–38,
DOI: 10.1007/978-3-540-85228-5_19, © Springer-Verlag Berlin Heidelberg 2008

to TEM-based 3D reconstruction due to its relatively small mass and asymmetric structure.

1. W. Stöcker and W. Bode, Curr Opin Struct Biol. **5** (1995), p. 383-390
2. C. Herzog, G. P. Kaushal and R. S. Haun, B. Cytokine. **31** (2005), p. 394-403
3. L. P. Norman, W. Jiang, X. Han, T. L. Saunders and J. S. Bond, Mol Cell Biol. **23** (2003), p. 1221-1230.
4. J. M. Crisman, B. Zhang, L. P. Norman, and J. S. Bond, J Immunol. **172** (2004), 4510-4519
5. S. Rösmann, D. Hahn, D. Lottaz, M. N. Kruse, W. Stöcker and E. E. Sterchi,. J Biol Chem. **277** (2002), p. 40650-40658
6. C. Becker-Pauly, M. Höwel, T. Walker, A. Vlad, K. Aufenvenne, V. Oji, D. Lottaz,, E. E. Sterchi, M. Debela, V. Magdolen, H. Traupe and W. Stöcker, J Invest Dermatol. **127** (2007), p. 1115-1125
7. C. Becker, M.-N. Kruse, K. A. Slotty, D. Köhler, J. R. Harris, S. Rösmann, E. E. Sterchi, and W. Stöcker, Biol. Chem. **384** (2003), p. 825-831
8. G. P. Bertenshaw, M. T. Norcum and J. S. Bond, J. Biol. Chem. **278** (2003), p. 2522-2532
9. M.-N. Kruse, C. Becker, D. Lottaz, D. Köhler, I. Yiallouros, H.-W. Krell, E. E. Sterchi, and W. Stöcker, Biochem J. **378** (2004), p. 383-389

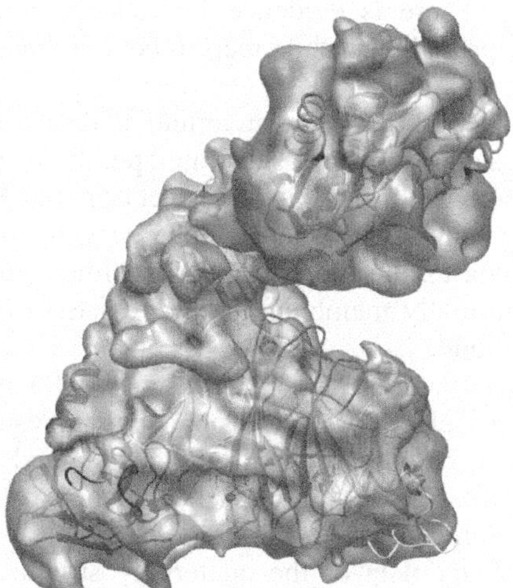

Figure 1. High-resolution 3D-reconstruction of the meprin beta.dimer: The fitted structures are close related X-ray data of the meprin domains: cyan and magenta: proteases domains (astacin); red and yellow: MAM domains; blue and green: the TRAF domains

Allosterism of *Nautilus pompilius* hemocyanin as deduced from 8 Å cryo-EM structures obtained under oxy and deoxy conditions

A. Moeller[1], C. Gatsogiannis[1], F. Depoix[1], U. Meissner[1] & J. Markl[1]

1. Institute of Zoology, Johannes Gutenberg University, 55099 Mainz, Germany

markl@uni-mainz.de , moellarn@uni-mainz.de
Keywords: single particle analysis, cryo electron microscopy, *Nautilus pompilius*, hemocyanin

Hemocyanins are the blue copper-containing respiratory proteins of many molluscs. *Nautilus pompilius* hemocyanin (NpH) is a cylindrical decamer composed of ten copies of a 350 kDa polypeptide subunit, in turn consisting of seven O_2-binding functional units (FUs, termed NpH-a to NpH-g). Ten copies of the subunit segment NpH-a to NpH-f form the cylinder wall (*ca.* 35 nm in diameter), whereas the ten copies of NpH-g build the internal collar. Recently we published a 9Å cryo-EM structure and molecular model of NpH that solved the principal architecture of this protein [1]. Hemocyanins are highly allosteric, and the cooperativity of oxygen binding should be transferred between functional units by conformational changes. In this context, we try to monitor structural changes caused by the reversible oxygen binding process. Our approach is to prepare the specimens in their fully oxygenated and deoxygenated state, respectively, and to perform independent 3D reconstructions of both states. To achieve this, we developed an automated plunge-freeze apparatus capable of flash freezing the specimen in liquid ethane under controlled atmospheric conditions. Prior to fixation, the molecules are either incubated in an atmosphere with 25 % O_2 + 75 % N_2 (oxy-state), or in an atmosphere with 100 % N_2 (deoxy-state). This led to two independent NpH datasets resulting in two 3D reconstructions with resolutions of 7.8 and 8.4 Å (**Figure 1**), respectively. Their correlative analysis shows significant structural differences especially concerning the spatial orientation of certain FU types (**Figure 2**). On the basis of the two resulting molecular models it is now the task to interpret these differences in terms of allosteric interaction mechanisms, but also to reveal possible pitfalls.

1. Gatsogiannis C, Moeller A, Depoix F, Meissner U, Markl J (2007). *Nautilus pompilius* hemocyanin: 9 A cryo-EM structure and molecular model reveal the subunit pathway and the interfaces between the 70 functional units. J Mol Biol. 374: 465-86.

A. Aretz, B. Hermanns-Sachweh, J. Mayer (Eds.): EMC 2008, Vol. 3: Life Science, pp. 39–40,
DOI: 10.1007/978-3-540-85228-5_20,

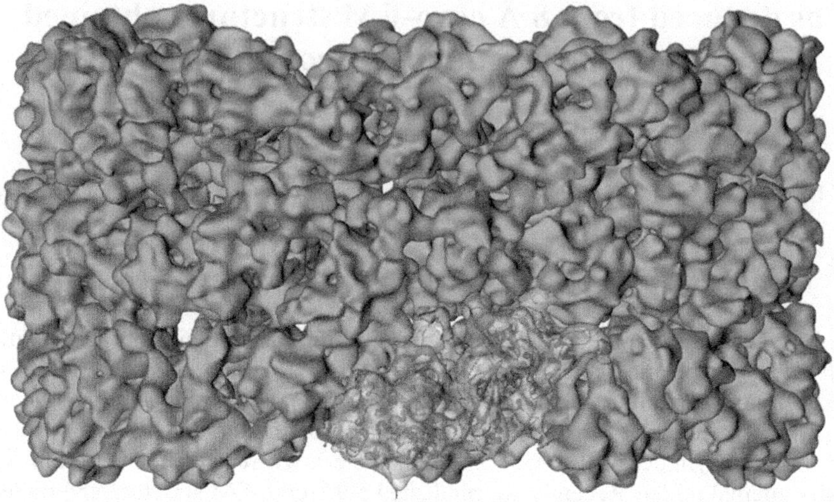

Figure 1. 8 Å density map deduced from single particle analysis of Nautilus hemocyanin (oxy-form). Fittings shown in FU a and b.

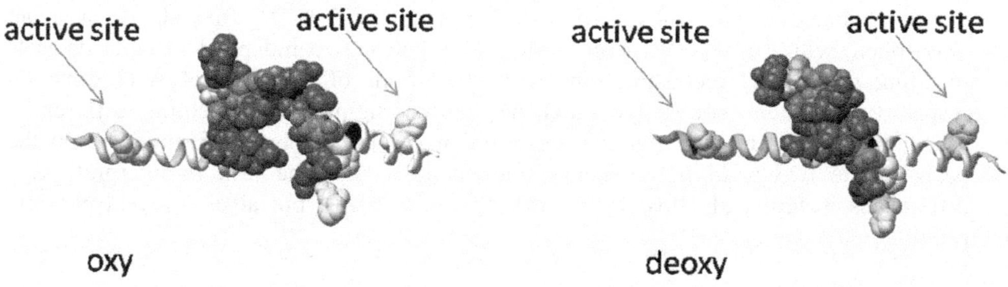

Figure 2. Changes of inter FU-contacts (FU-a and b) deduced from rigid body fitting within the density maps of the oxy- and deoxygenated models.

High resolution structure of a 6 MDa protease by xray-crystallography and cryo-EM

B. Rockel[1], C. K. Chuang[2], G. Seyit[1], P. J. Walian[2], J. Peters[1], W. Baumeister[1] and B. K. Jap[2]

1. Max-Planck-Institute of Biochemistry, Am Klopferspitz 18, 82152 Martinsried, Germany
2. Lawrence Berkeley National Laboratory, Life Sciences Division, 1 Cyclotron Road, Berkeley, CA 94720, U.S.A.

rockel@biochem.mpg.de
Keywords: hybrid structure, cryo-electron microscopy, tripeptidyl peptidase II

Cytosolic protein degradation proceeds largely via large protein complexes in which the active sites are located in secluded compartments. The paradigm for such a complex is the 26S proteasome, which degrades ubiquitinated proteins in an ATP-dependent manner. In the successive degradation of the resulting, relatively small products large complexes are also involved. Among them is Tripeptidyl Peptidase II (TPPII), a eukaryotic serine protease with a subtilisin-like active site, which acts mainly as an exopeptidase cleaving tripeptides from the free N-terminus of aminopeptides but also as an endopeptidase, albeit with a lower activity. Recently, TPPII has been attracting increasing attention owing to its extraordinary size, its apparent potential to substitute for some of the proteasome's functions as well as its implication in MHC-class I peptide trimming, in neuropeptide degradation, in apoptosis, and in sepsis [1].

Compared to ATP-dependent proteolytic systems like the 26S proteasome, where the linear assembly of the unfoldase- and protease-subcomplexes reflects the sequence of steps necessary for substrate degradation, the functional reason for the architecture of TPPII is not as obvious: As shown by cryo-electron microscopy, the 150 kDa subunits of *Drosophila* TPPII assemble into a spindle-shaped 6 MDa complex consisting of two twisted strands. Each of the strands comprises ten stacked and interdigitated dimers. The two strands enclose a central cavity system that is accessible via lateral openings arranged like an arcade [2]. When studying the assembly of heterologously expressed TPPII in E. coli cell extracts, we found, that in the course of strand elongation the specific proteolytic activity increases with strand length and correlates with the number of inter-dimer contacts formed [3].

In order to understand the structural basis of this activation, to see whether the active sites are sequestered in a secluded compartment and the route along which a substrate might be transported, a high resolution structure of the complex is indispensable. Since the system is highly dynamic and *in vitro* the length of TPPII-strands depends on protein concentration, crystallization of the holocomplex is not feasible. However, the building blocks of TPPII –the dimers- can be crystallized. We have obtained a crystal structure of TPPII dimers and are integrating our current cryo-EM map of the

A. Aretz, B. Hermanns-Sachweh, J. Mayer (Eds.): EMC 2008, Vol. 3: Life Science, pp. 41–42,
DOI: 10.1007/978-3-540-85228-5_21,

holocomplex with the high resolution structure of the dimers. We expect the resulting hybrid structure of the complex to provide initial clues on the functional mechanism of TPPII.

1. B. Tomkinson and A.C. Lindas, International Journal of Biochemistry & Cell Biology **37** (2005), p. 1933.
2. B. Rockel, J. Peters, S. A. Müller, G. Seyit, P. Ringler, R. Hegerl, R. M. Glaeser and W. Baumeister, Proceedings of the National Academy of Sciences of the United States of America **102** (2005), p. 10135.
3. G. Seyit, B. Rockel, W. Baumeister and J. Peters, Journal of Biological Chemistry **281** (2006), p. 25723.

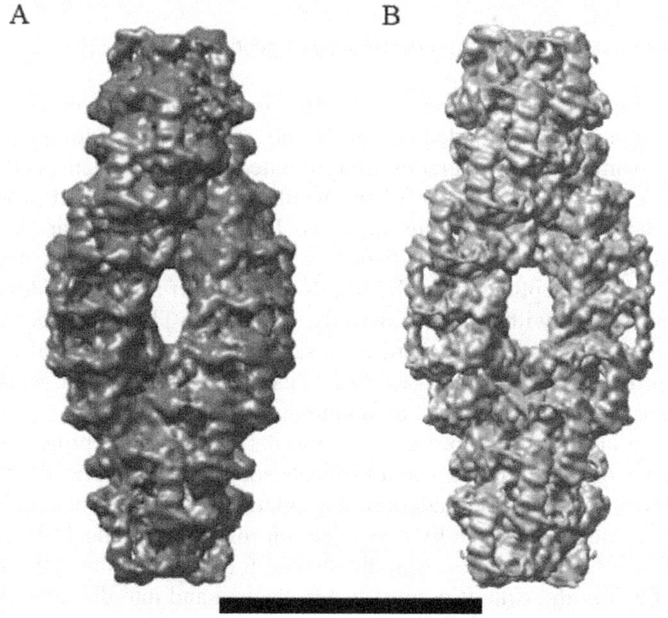

Figure 1. A. Cryo-EM reconstruction of *Drosophila* TPPII. B. High resolution structures of the TPPII dimers fitted into the EM-envelope and low-passed to a resolution of 1.5 nm. Scale bar: 30 nm.

Cryo-TEM of liquid crystalline particles: Application of materials science techniques to study liquid crystalline particle structures.

L. Sagalowicz

Nestlé Research Center, Vers-Chez-Les-Blanc, CH-1000 Lausanne 26

Laurent.Sagalowicz@rdls.nestle.com
Keywords: cryo-TEM, liquid crystals, monoglycerides.

Natural liquid crystalline phases, such as the ones formed with amphiphiles like monoglycerides and phospholipids, have generated large interest due to their applications in understanding membrane fusion, increasing efficiency of drugs, sustained release, controlling chemical reactions and crystallizing proteins[1]. The macroscopic properties of liquid crystal objects, such as molecular release, depend on structure and more precisely on space groups and interfaces between the particle and the surrounding matrix. Therefore, it is crucial to have methods adapted for determining those structural features.

Macroscopic methods, such as small angle X-ray scattering (SAXS), are very efficient for determining the space group. In real biological or food samples, though, concentration and homogeneity are often too low for these analytical techniques[2]. In addition, interfaces between particles and matrix cannot be determined in this way.

Cryo-TEM is an ideal method and is probably the most used technique to study liquid crystalline phases. However, it is often ambiguous and limited two-dimensional information is obtained. In addition, dispersed reversed cubic phases have only been observed with the electron beam (EB) aligned to the <100> or <111> axis, for which the motifs observed are in agreement with three different cubic structures and even in some cases (<111> axis of observation) with hexagonal structures[2].

In order to differentiate between the various space groups, we incline the sample in such a way that the contributing planes to the fast Fourier transform (FFT) enable differentiation between the various crystalline structures (Fig.1). In applying this technique, we show that we can unambiguously distinguish among particles having the reversed hexagonal, the inverted micellar cubic and the three types of inverted bicontinuous cubic structures.

In addition to the effects of internal structure and space group, the interface between the particle and the matrix also has an important influence on macroscopic properties. Here tilting experiments are also used to propose model of interfaces[1,3]. In addition we are currently simulating the TEM image contrast to obtain more information as this was done previously for vesicles.

In conclusion the use of methods developed for hard materials, such as controlled tilting or simulation of the image contrast, can bring unique information in short time (less than one hour for the controlled tilting method) and using a standard TEM machine (80kv with tungsten filament).

A. Aretz, B. Hermanns-Sachweh, J. Mayer (Eds.): EMC 2008, Vol. 3: Life Science, pp. 43–44,
DOI: 10.1007/978-3-540-85228-5_22,

1. L. Sagalowicz, R. Mezzenga and M. E. Leser. COCIS **11**(2006), p. 224.
2. L. Sagalowicz, S. Acquistapace, H. J. Watzke and M. Michel. Langmuir **23**(2007), p. 1203
3. L. Sagalowicz et al. Journal of Microscopy **221**(2006), p.110.
4. M. Johnsson, J. Barauskas, L. Yam and F. Tiberg. Langmuir **21**(2005), p. 5159.

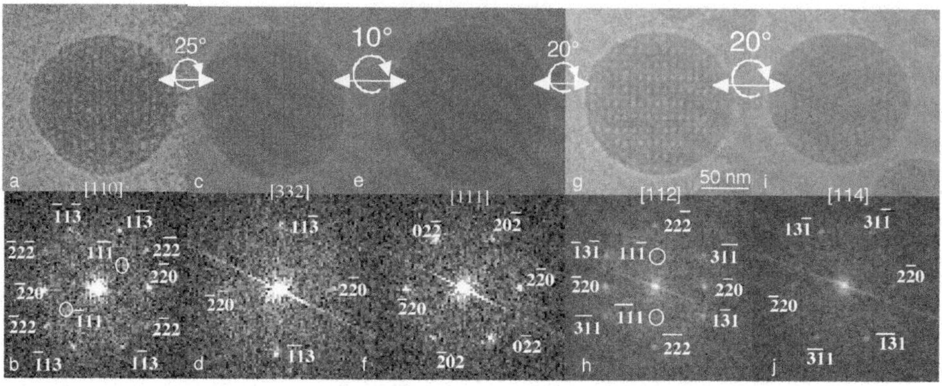

Figure 1. tilt series showing the same type of particle imaged under different directions. This analysis enables to differentiate between the different space groups and therefore allows an unambiguous determination of the internal structure (space group Fd-3m, inversed micellar cubic phase). Adapted from [2]

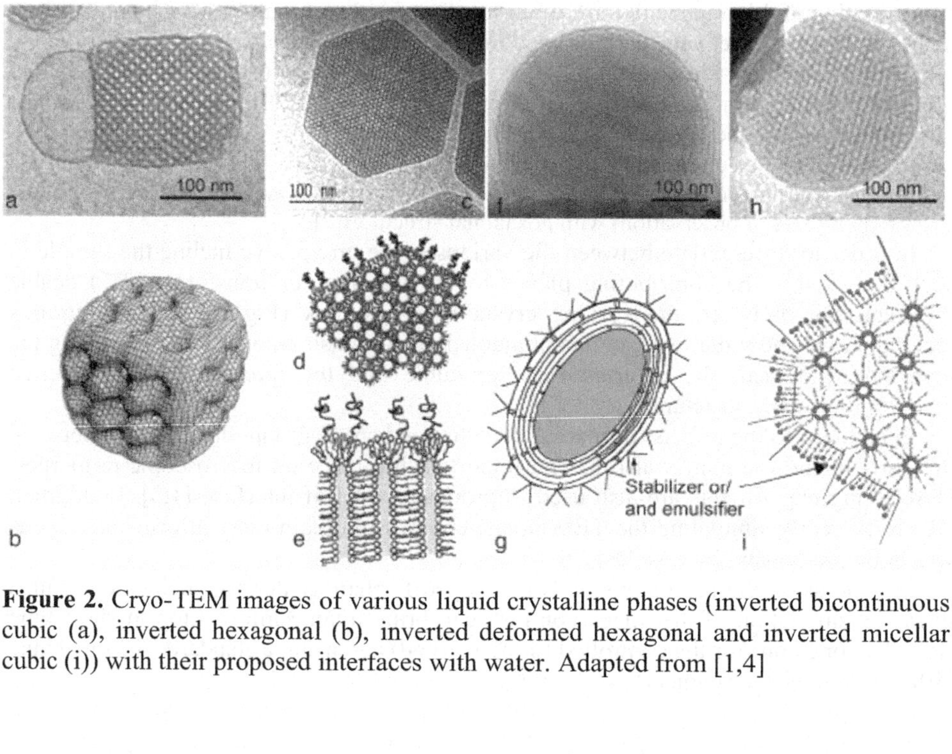

Figure 2. Cryo-TEM images of various liquid crystalline phases (inverted bicontinuous cubic (a), inverted hexagonal (b), inverted deformed hexagonal and inverted micellar cubic (i)) with their proposed interfaces with water. Adapted from [1,4]

Structural basis for the concerted integration of HIV-1 DNA in the human genome, role of the human cofactor LEDGF/p75.

P. Schultz [1], F. Michel [1], R. Benarous [2], M. Gottikh[3], D. Moras [1], and M. Ruff [1]

1. Département de Biologie et de Génomique Structurales, IGBMC, 1 rue Laurent Fries, 67404 Illkirch; France
2. Institut Cochin, Université Paris Descartes, Paris, France
3. Belozersky Institute of Physico-Chemical Biology, Moscow State University, 119992 Moscow, Russia

patrick.schultz@igbmc.u-strasbg.fr
Keywords: HIV, integrase, LEDGF, cryo-electron microscopy, DNA integration

Integration of the human immunodeficiency virus (HIV-1) cDNA into the human genome is catalyzed by integrase (IN) and requires additional cellular factors. The functional IN complex catalyses two endonucleolytic reactions. In the 3'-processing reaction, IN cleaves a dinucleotide from the U5 and U3 ends of viral cDNA thereby exposing a 3'-OH group. In the strand transfer reaction, IN generates a double strand break in the host DNA and joins the newly formed ends to the viral 3' ends by transesterification [1]. The complex can thus bind two viral DNA fragments and the host DNA in which the viral genome will be integrated.

Using a structural genomics approach, a stable and soluble complex between the wild type full length IN and the cellular cofactor LEDGF/p75 was produced. This complex is active in *in vitro* integration assays. Cryo-electron microscopy of negatively stained and hydrated samples [2] was used in combination with single particle analysis to determine a 14Å resolution map of the complex. This analysis revealed that the functional unit comprises two asymmetric IN dimers and two LEDGF/p75 molecules. Each IN dimer appears as a molecular clamp hold in a position suitable for DNA binding by the LEDGF molecules which act as molecular chaperones. Atomic structures of the IN domains were fitted into the 14Å resolution map and allowed the correct positioning of the highly flexible N- and C-terminal domains of the IN molecules which flack the catalytic core domain (Figure 1).

A 3-D structure of the IN-LEDGF complex was obtained in the presence of DNA and the DNA binding sites were revealed as stain excluding rods protruding out of the structure and as additional stain exclusion volumes. The structure obtained showed that each asymmetric IN dimer can bind two DNA molecules. The host DNA is bound between the clamps formed by the IN dimer, whereas the viral DNA is positioned in the active catalytic site for 3'OH processing. The binding of the DNA molecules is accompanied by small changes in the positions of the N- and C-terminal domains of IN. Particularly, the IN clamp closes around the host DNA molecule. The numerous cross-linking and mutational experiments that have been carried out [3,4] are all in excellent agreement with our model. The positions of the target and viral DNAs for the 3'

A. Aretz, B. Hermanns-Sachweh, J. Mayer (Eds.): EMC 2008, Vol. 3: Life Science, pp. 45–46,
DOI: 10.1007/978-3-540-85228-5_23,

processing and integration reaction shed light on the mechanism for concerted integration, a process with wide implications for the understanding of "viral induced" pathologies (Figure 2).

1. A. Engelman, F.D. Bushman and R. Craigie, EMBO J. **12** (1993), p. 3269.
2. M.M. Golas, B. Sander, C.L. Will, R. Luhrmann and H. Stark, Mol. Cell. **17** (2005), p. 869.
3. T.S. Heuer and P.O. Brown, Biochemistry **36** (1997), p. 10655.
4. T.S. Heuer and P.O. Brown, Biochemistry **37** (1998), p. 6667.

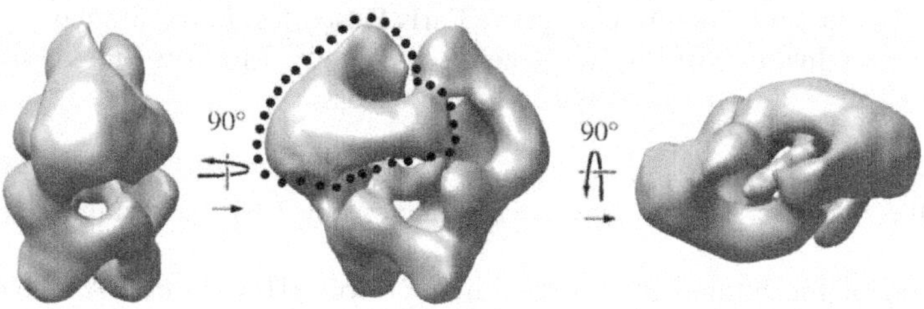

Figure 1. Three-dimensional model of the HIV-1 Integrase/human LEDGF complex. An integrase dimer is highlighted by a dotted line.

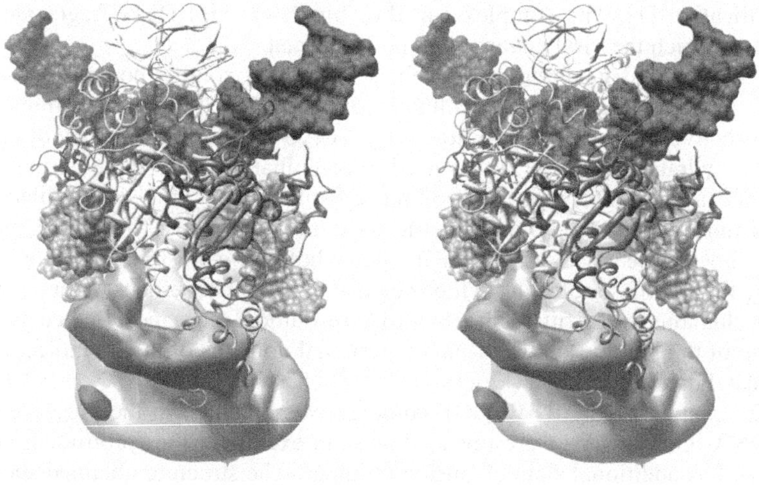

Figure 2. Stereo view of the proposed model of the integration complex.

Titan Krios: Automated 3D imaging at ambient and cryogenic conditions

Marc Storms, Raymond Wagner, Uwe Luecken, Werner Hax, Matt Harris

FEI Company, Achtseweg Noord 5, 5651 GG Eindhoven, The Netherlands

marc.storms@fei.com
Keywords: Automation, Cryo Microscopy, Vitrobot, 3D Imaging, Tomography

FEI's Titan Krios™ TEM allows investigating the biological world at the nanometer scale. Designed specifically to address the needs of structural and cellular biologists, Titan Krios allows them to visualize the intricate interactive mechanisms of individual proteins and molecular machines, and to localize that activity within the three dimensional architecture of the cell.

Of paramount importance in this context is specimen integrity. Titan Krios optimizes imaging performance on delicate biological samples. Life scientists face a fundamental dilemma: longer exposures improve the signal-to-noise ratio, but also risk damaging the sample. Titan Krios is specifically designed to obtain the maximum information with minimum exposure, preserving sample integrity and ultimately enabling investigation of increasingly smaller macromolecular detail.

- Robotic sample handling helps to guarantee a controlled and reproducible specimen environment with minimum exposure or delay from operator interaction
- The cold stage operates at liquid nitrogen or helium temperature, reducing radiation damage and other beam induced impact on samples.
- The availability of Low Dose and SpotScan operating modes, optimized for ease of use and speed, further reduce sample exposure.

Together these capabilities allow images to be recorded at higher doses or the number of images to be increased for better resolution or higher signal-to-noise ratios.

Titan Krios is the tool for obtaining fast, fully automated, high content three dimensional information about biological cells down to the scale resolving individual molecular complexes. It perfectly complements the cellular scale observations available from light microscopy and the atomic scale analysis provided by NMR and X-Ray diffraction. Advanced automation makes analysis fast and easy, increasing experimental productivity and enabling the high volume analysis needed for complex biological systems. The unique capabilities of Titan Krios open the gates to the investigational frontier of biological science, presenting unprecedented opportunities for ground-breaking results and breakthrough discoveries.

A. Aretz, B. Hermanns-Sachweh, J. Mayer (Eds.): EMC 2008, Vol. 3: Life Science, pp. 47–48,
DOI: 10.1007/978-3-540-85228-5_24,

The system is based on the world class modular Titan™ technology providing mechanical, electronic and thermal stability and is designed to deliver the ultimate performance in both Life and Materials Science applications.

In this presentation, sample preparation and transfer, data acquisition and processing of results, including a description of the critical steps and factors for obtaining 3D-cryo results at either liquid nitrogen and/or helium temperatures, will be discussed.

Self assembly and phase behaviour of new sugar based gemini amphiphiles

M. C. A. Stuart[1], M. Scarzello[2], J. E. Klijn[2], E. J. Boekema[1] and J. B. F. N Engberts[2]

1. Electron Microscopy Group, Groningen Biomolecular Sciences and Biotechnology Institute, University of Groningen, Nijenborgh 4, 9747 AG Groningen, The Netherlands
2. Stratingh Institute, University of Groningen, Nijenborgh 4, 9747 AG Groningen, The Netherlands.

m.c.a.stuart@rug.nl
Keywords: cryo-electron microscopy, phase behaviour, gemini surfactant.

Amphiphilic molecules possess parts which are distinctively hydrophobic and other parts which are distinctively hydrophilic. This ambivalence lies at the basis of the intriguing properties of these molecules, selected by nature as the building-blocks of architectures essential for life. Amphiphile self-assembly can lead to a variety of structures, differing in shape and size, depending on the molecular structure of the amphiphile and on the solution conditions such as temperature, ionic strength and pH. A relatively new and particularly interesting class of amphiphilic molecules, the gemini surfactants (GS), is obtained by connecting two single-tailed surfactants via a spacer at the level of the headgroups. The sugar-based gemini surfactant consists of several unique structural elements by which self-assembly properties can be made to fit the desired properties. The tertiary amino moieties in the headgroup can be protonated depending on the solution pH. The reduced sugars can be varied in length and stereochemistry. The spacer is either hydrophobic or hydrophilic and can be varied in length and the hydrophobic alkyl chain can be varied in length and degree of saturation. Most gemini surfactants form vesicles around neutral pH and exhibit a phase change towards micelles upon decreasing the pH, figure 1. When the nitrogens are protonated the cross-sectional headgroup area increases which gives the molecule a cone-shape appearance. The pH at which the phase changes from lamellar to micellar is dependent on the spacer length [1]. The shorter the spacer the more difficult it is to protonate both nitrogens which gives a lower pH of transition. At high pH the vesicles become neutral which leads to a decreased colloidal stability and the vesicles aggregated. Surprisingly the vesicles redispersed upon a further increase of the pH which could only be explained by OH^- binding to the vesicle surface giving the vesicle a negative ζ potential. The pH at which vesicle redispersion starts was found to be correlated with the sugar stereo chemistry [2].

1. J. E. Klijn, M. C. A. Stuart, M. Scarzello, A. Wagenaar, and J. B. F. N. Engberts. *J.Phys.Chem.B.* **110** (2006), 21694.
2. J. E. Klijn, M. C. A. Stuart, M. Scarzello, A. Wagenaar, and J. B. F. N. Engberts. *J.Phys.Chem.B.* **111** (2007), 5204.

A. Aretz, B. Hermanns-Sachweh, J. Mayer (Eds.): EMC 2008, Vol. 3: Life Science, pp. 49–50,
DOI: 10.1007/978-3-540-85228-5_25,

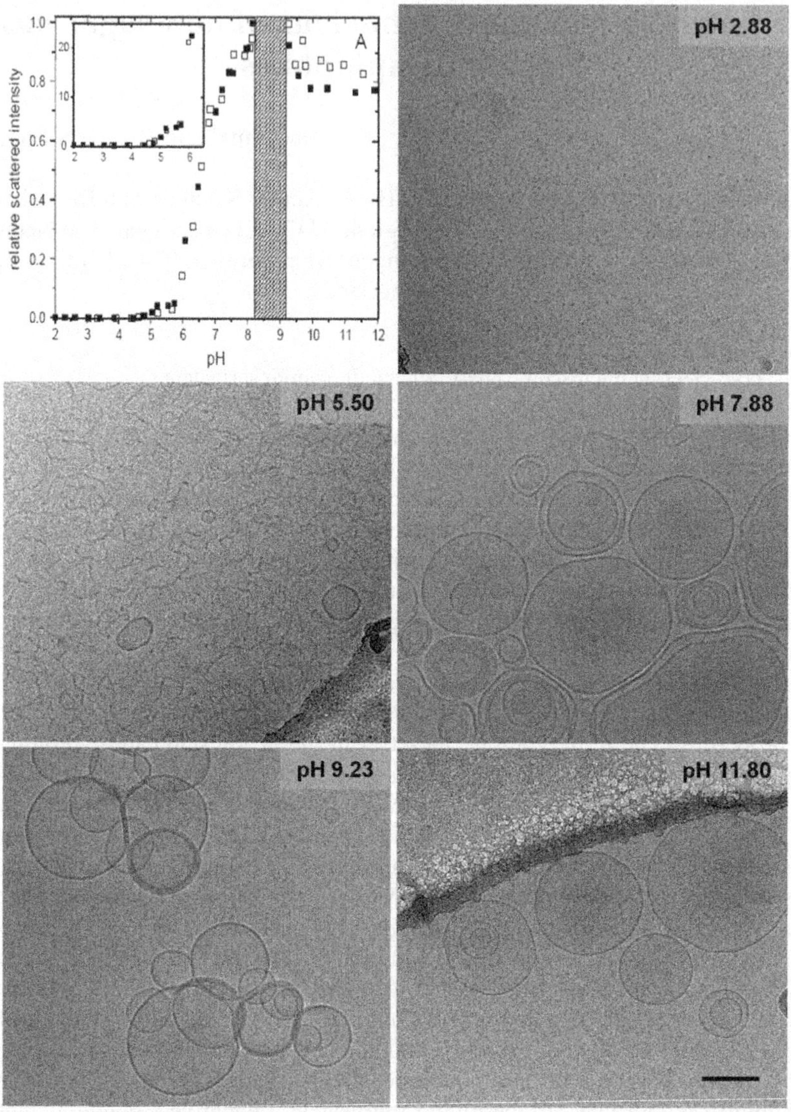

Figure 1. pH depended phase and aggregation behaviour of reduced glucose gemini surfactant with C6 spacer and oleoyl tails followed with dynamic light scattering and cryo-electron microscopy. At low pH micelles are formed. At somewhat higher pH the micelles aggregate into worm-like micelles. At neutral pH and above vesicles are formed. Between pH 8.2 and 9.3 the vesicles aggregate due to charge neutralization. At high pH the vesicles redisperse as the result of OH^- binding to the vesicle surface. Bar 100 nm.

Peptide-nanotube formation by lysine based lipoamino acids

L. Ziserman, A. Mor and D. Danino

Department of Biotechnology and Food Engineering, Technion, Haifa, Israel 32000

zlior@tx.technion.ac.il
Keywords: peptide nanotube, ribbon, chiral self-assembly, lysine, cryo-TEM

Lipid and peptide nanotubes are at the spotlight of nanoscience today due to their tremendous applicative potential in various medical and industrial disciplines. They were suggested to serve as templates for electrical nanowires, drug delivery systems, molecular bio-sensors, filtration and purification systems, and for many more applications.

We are investigating the unique self-assembly properties of a library of novel lipoamino acids (LAAs), composed of repeating sequences of acyl chains linked to positively charged amino-acids (mostly lysine), termed OAKs (Oligo Acyl Lysine). The nanostructural analysis involves few advanced microscopy techniques: cryogenic-transmission electron microscopy (cryo-TEM), negative-stain (NS)-TEM, high resolution scanning electron microscopy (HR-SEM), atomic force microscopy (AFM) and light microscopy. The microscopy work is supported by spectroscopic measurements such as light scattering and Fourier transform infra red spectroscopy (FTIR), and calorimetric methods such as differential scanning calorimetry (DSC).

The chemical structure of one of the simplest OAK molecules, N^{α}-dilauryllysine (DLL), is shown in the inset of Figure 1B. Cryo-TEM experiments show that already at low concentrations as 0.3 mM in aqueous solutions, at room temperature, DLL molecules gradually self-organize in a process known as *chiral self-assembly* [1] into hollow nanotubes (figure 1). This gradual aggregation process starts with the creation of nanofibers, which then transform into twisted ribbons (ribbons with Gaussian curvature). Gradually, the solution is being dominated by helical ribbons (ribbons with cylindrical curvature) that close upon themselves into nanotubes, 70-100 nm in diameter and a few microns long. This assembly path results from expansion of the width as more material is continually added to the ribbons edges. Complementary NS-TEM work suggests that the fibers and ribbons are constructed from tightly bound sub-fibers, a few nanometer in width (insert in Figure 1A). Thus, the fibers may, in fact, be very narrow, "young", ribbons composed from only few such sub-fibers.

The twisted ribbon-to-cylindrical ribbon transformation course we find here was suggested theoretically only recently [2] and as far as we know this is the first experimental evidence in peptide nanotube system.

1. J.H. Fuhrhop and W. Helfrich. Chemical Reviews. **93** (1993), p. 1565-1582.
2. R.L.B. Selinger, J.V. Selinger, A.P. Malanoski, and J.M. Schnur. Physical Review Letters. **93** (2004), p.158103 / 1-4.

A. Aretz, B. Hermanns-Sachweh, J. Mayer (Eds.): EMC 2008, Vol. 3: Life Science, pp. 51–52,
DOI: 10.1007/978-3-540-85228-5_26, © Springer-Verlag Berlin Heidelberg 2008

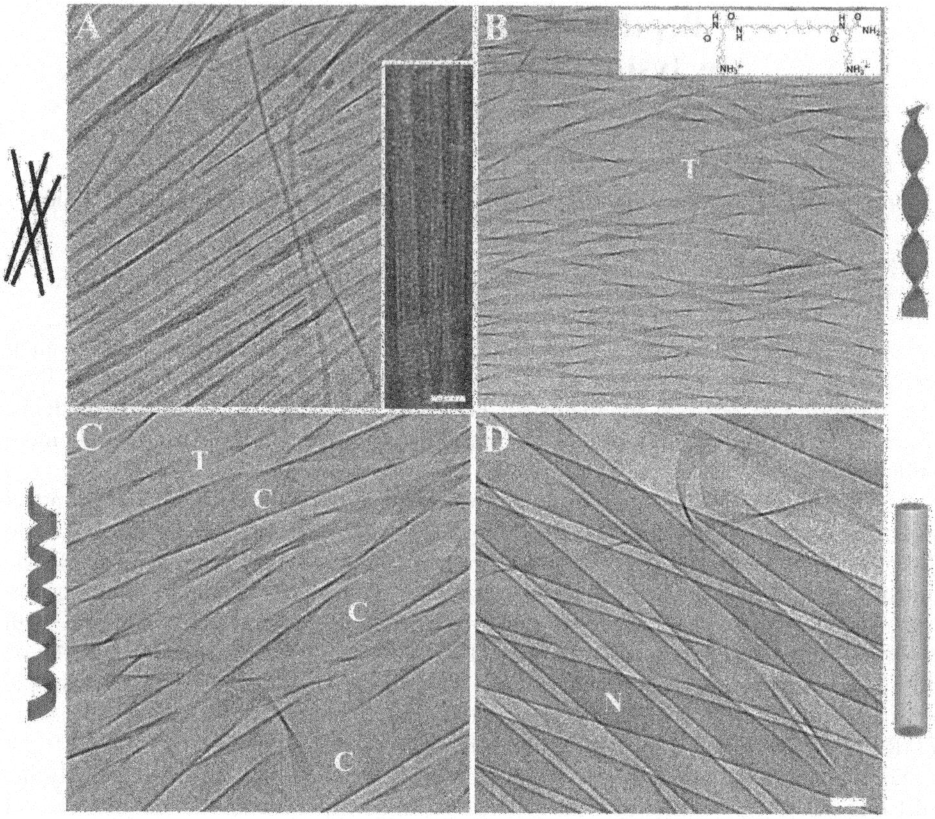

Figure 1. Cryo-TEM images showing the time-dependent evolution of DLL nanostructures in water at 3 mM: (A) Fibers and narrow ribbons in a fresh sample (few minutes after preparation). (B) Twisted ribbons after overnight incubation. (C) Helical-coiled ribbons after four weeks. (D) Nanotubes after four months. T, C, and N mark a twisted ribbon, a helical-coiled ribbon and a nanotube, respectively. Bar = 50 nm. Inset in (A): NS-TEM images of a fresh 3 mM solution showing enlargement of one of the fibers, suggesting it is constructed from joined, parallel "sub-fibers". Bar = 25 nm. Inset in (B): M$^{\text{olecular structure of N}\alpha}$-lauryl-lysyl-aminolauryl-lysine-amide, referred to as N$^{\alpha}$-dilauryllysine (DLL).

Advances in biological microanalysis using event streamed spectrum imaging and programmed beam acquisition

Peter Ingram,[1] Scott Davilla,[2] and Ann LeFurgey[1]

1. Duke University and Veterans Affairs Medical Centers, Durham, 27710, USA
2. 4pi Analysis Inc., 3500 Westgate Dr., Suite 403, Durham, 27707, USA

peter.ingram@duke.edu
Keywords: x-rays, imaging, event-streaming

The current trend of x-ray mapping in Electron Probe X-ray Microanalysis (EPXMA) is toward full spectrum collection, or spectrum imaging [1]. Spectrum imaging is the collection and spatial registration of all x-ray events, yielding a spectral data cube. Recent developments in spectral analytical techniques have greatly improved the efficiency and precision of data handling from quantitative chemical images; principal component and cluster analysis has significantly aided in unambiguous identification and assessment of localized data as well as the reduction of statistical noise in the images; fitting of all elemental peaks at each pixel to standards accomplishes a similar result as well as providing real-time or off-line quantitative images; "programmed beam acquisition" (PBA) enables the exciting beam to only dwell for sufficient times per pixel in regions of the specimen necessary to acquire a predetermined statistical precision (e.g.±2 standard deviations). PBA, while not a new concept [1,2], is rarely used. However it is of particular importance to biological samples, which are sensitive to beam damage and inherently produce low x-ray count rates.

Event Streamed Spectrum Imaging (ESSI) is a new method [3] that treats x-ray events as simply another signal source (Fig. 2). Instead of passing the current pixel x-y position to the x-ray pulse processor for registration and slowly building up the images, x-ray events are passed to the scan generator and assembled into a pixel event which contains all information (transmitted electrons, beam current, x-rays etc.) related to that pixel position. The pixel events are streamed as packets to a host computer, where they are buffered and stored to disk for subsequent processing. A high degree of interactive analysis is possible, including real-time alteration of acquisition conditions such as dynamic dwell modulation (i.e. programmed beam acquisition) where the x-ray dwell is altered on a frame-to-frame basis to increase (or decrease) dwell times over areas of interest (Fig. 3). Similarly real-time drift correction can readily be implemented. ESSI is an extremely efficient process for the collection of all events simultaneously registered to every x-y beam position over single or multiple frames with zero overhead.

1. P. Ingram et al. *Biomedical Applications of Microprobe Analysis,* Acad. Press (1999)
2. P. Ingram, D.A. Kopf, and A. LeFurgey. Microbeam Analysis, J. Friel, Ed., (1994) 111
3. S.D. Davilla, Scanning 28, 2 (2006) 120

A. Aretz, B. Hermanns-Sachweh, J. Mayer (Eds.): EMC 2008, Vol. 3: Life Science, pp. 53–54,
DOI: 10.1007/978-3-540-85228-5_27,

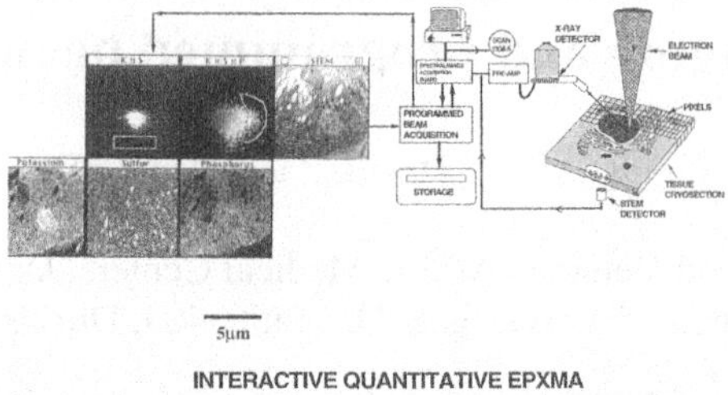

Figure 1. Slow scan conventional Programmed Beam Acquisition. Regions selected from scatterplots between various elemental maps (Potassium, Phosphorus and Sulfur from freeze-dried cryosections of mouse kidney) are used to generate masks that are assigned different pixel dwell times for acquisition of x-ray data [1]. Reproduced with author's permission.

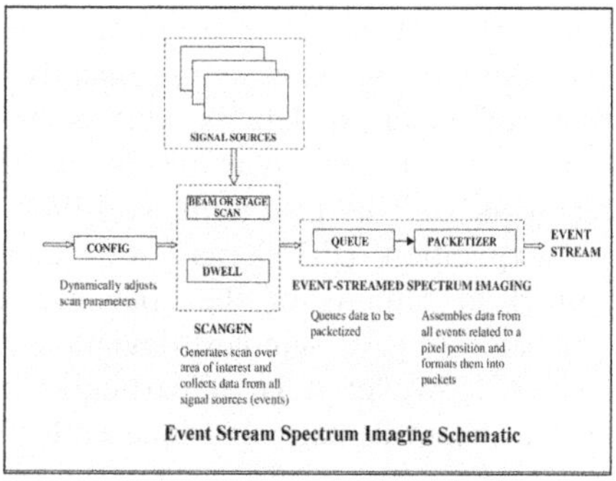

Figure 2. Event Stream Spectral Imaging (ESSI) schematic

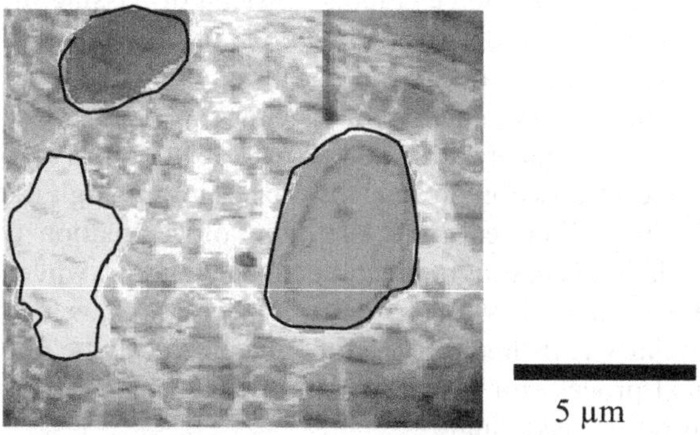

Figure 3. Dynamic Dwell Modulation. Chemically fixed, stained and embedded thin section of frog cardiac muscle. Masks are drawn around regions of the cells in which a dwell times (clockwise from top left) of 16, 40 and 80 msec/pixel were applied. A dwell time of 8 µsec/pixel was applied to the rest of the field. the image was frame locked to prevent drift.

3D-correlation-averaging for membrane-protein-crystals

P.J.B. Koeck, P. Purhonen, R. Alvang, B. Grundberg and H. Hebert

Karolinska Institutet and The Royal Institute of Technology, Department of Biosciences and Nutrition at Novum, S-141 57 Huddinge, Sweden

Philip.Koeck@ki.se
Keywords: electron crystallography, single particle reconstruction, membrane proteins, crystal distortion, shrink-wrap algorithm, missing cone, resolution extension

Few 2-dimensional protein crystals can be used to determine high-resolution structures, whereas most electron crystallography projects remain at a resolution around 10 Ångström. This might be partly due to lack of flatness of many two-dimensional crystals [1]. We have investigated this problem and suggest single particle projection matching (3D-correlation averaging) of locally averaged unit cells to improve the quality of three-dimensional maps. Theoretical considerations and tests on simulated data demonstrate the feasibility of this refinement method [2].

Applying the method to crystal images of melibiose permease from *Escherichia coli* we have calculated a three-dimensional map that is clearer and easier to interpret than the map derived using purely electron-crystallographic methods [2, 3] (figure 1 A-C).

By filling in the missing cone using a shrink-wrap algorithm [4] we generated an improved starting model that allowed single-particle refinement to a higher resolution than that achieved without shrink-wrapping (figure 1 D,E).

A practical advantage of this method is the improved success-rate particularly for images recorded at tilt-angles between 40 and 70 degrees, many of which cannot be used in traditional electron crystallography (probably due to the aforementioned crystal warping). We therefore propose 3D-correlation-averaging even in cases were near-atomic resolution can be reached by traditional electron crystallography using images and diffraction patterns since it would greatly improve the efficiency of data-collection.

1. M.B. Sherman, T. Soejima, W. Chiu and M. van Heel, Ultramicroscopy **74** (1998), p. 179-199.
2. P.J.B. Koeck, P. Purhonen, R. Alvang, B. Grundberg and H. Hebert, Journal of Structural Biology **160** (2007), p. 344-352.
3. P. Purhonen, A.-K. Lundbäck, R. Lemonnier, G. Leblanc and H. Hebert, Journal of Structural Biology **152** (2005), p. 76-83.
4. S. Marchesini, H. He, H.N. Chapman, S.P. Hau-Riege, A. Noy, M.R. Howells, U. Weierstall and J.C.H. Spence, Physical Review B **68** (2003), p. 140101.
5. This work was supported by the EU 6th framework (3DEM NoE)

A. Aretz, B. Hermanns-Sachweh, J. Mayer (Eds.): EMC 2008, Vol. 3: Life Science, pp. 55–56, DOI: 10.1007/978-3-540-85228-5_28,

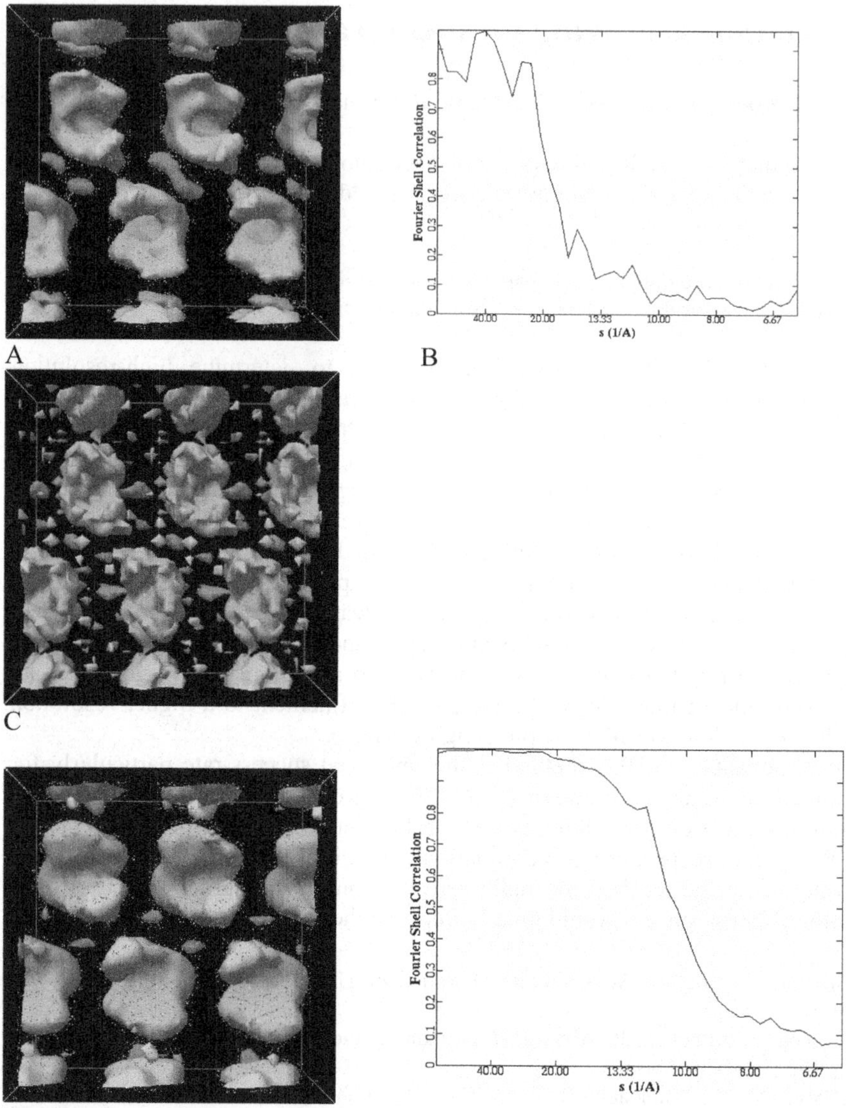

Figure 1. After refinement by 3D-correlation-averaging with the electron-crystallographic reconstruction shown in C as starting model the map, shown surface rendered in A, is consistent with the data to a resolution of about 19 Ångström according to the 0.5-criterion of the Fourier shell correlation plot shown in B. After filling in the missing cone in the map of A and continuing refinement based on this new starting model the map shown in D with a resolution of approximately 11 Ångström according to the same criterion, as shown in E, emerges.

8 Å cryo-EM structure of the giant hemoglobin from the planorbid snail *Biomphalaria glabrata*

J. Markl & F. Depoix, A. Moeller, W. Gebauer

Institute of Zoology, Johannes Gutenberg University, D-55099 Mainz, Germany

markl@uni-mainz.de
Keywords: hemoglobin, three-dimensional cryo-electron microscopy, Gastropoda

Until 2006, snail red hemoglobin remained a phylogenetic enigma because it occurs quite isolated in a single gastropod family, the Planorbidae, whereas all other gastropods use blue hemocyanin as a respiratory protein (for recent cryo-EM of hemocyanin, see [1,2]). Moreover, sequence data on this snail hemoglobin were completely lacking. In 2006, our group published the complete cDNA and predicted amino acid sequence of two *Biomphalaria glabrata* hemoglobin polypeptides, termed BgHb1 and BgHb2 [3]. (*Biomphalaria* is intermediate host of the human parasite *Schistosoma mansoni* that causes Bilharziosis.) Resembling pearl-chains, both polypeptide subunits encompass 13 different, cysteine-free globin domains (~17 kDa each; total subunit M_r ~238 kDa). In addition, there is a small N-terminal non-globin "plug" domain (~10 kDa) which contains three cysteines for subunit dimerization. Phylogenetic interpretation of the sequence data showed that *Biomphalaria* hemoglobin evolved from myoglobin which is present in the radula muscle of most gastropods. Possibly, this hemoglobin evolved to replace a less efficient hemocyanin which was detected in *Biomphalaria* hemolymph as a trace component [3].

But why did these snails change their respiratory protein, from hemocyanin to hemoglobin, when hemocyanin has successfully performed its task in countless mollusks for ~740 million years? Functional analysis has revealed that *Biomphalaria* hemoglobin exhibits a high oxygen affinity (p50 *ca.* 6 mmHg). This is interpreted as adaptation of the snails to their pond habitat, where they experience extreme variation in temperature and oxygen partial pressure [4]. It has been suggested that hemocyanin is probably incapable to evolve into high affinity forms, in contrast to hemoglobin [5].

In our previous study, in addition to cDNA sequencing, the native hemoglobin was purified and submitted to a preliminary 3D electron microscopy of negatively stained single particles. Although lacking any detail, the resulting 3D reconstruction (~30Å resolution) suggests a hexameric quaternary structure (calculated M_r ~1.43 million Da), and point-group D3 symmetry [3]. This contrasts to an earlier publication that propose an octameric quaternary structure based on D2 symmetry [6].

This controversy, and the surprisingly simple evolutionary mechanism to create a high molecular mass respiratory protein from 78 paralogous globin domains, stimulated us to analyse this protein more closely by 3D cryo-electron microscopy. From ~10.000 single particles we developed a cryo-EM structure that according to the 0.5 criterion has *ca.* 8 Å resolution (Fig. 1). It confirms the previously assumed D3 symmetry. However, even at this resolution, rigid-body fitting of the small globin domains was unsuccessful

A. Aretz, B. Hermanns-Sachweh, J. Mayer (Eds.): EMC 2008, Vol. 3: Life Science, pp. 57–58,
DOI: 10.1007/978-3-540-85228-5_29, © Springer-Verlag Berlin Heidelberg 2008

as yet. Consequently, a molecular model of BgHb is still lacking, and the twisted pathway of the "pearl-chain" subunit remains to be unravelled. Nevertheless, from the cryo-EM structure the general localization and shape of the six subunit copies has emerged, and for the subunit dimer, only two possibilities remain (see Figure 1).

1. C. Gatsogiannis, A. Moeller, F. Depoix, U. Meissner, J. Markl, JMB 374 (2007), p. 465-86.
2. A. Martin, F. Depoix, M. Stohr, U. Meissner, J. Markl et al. JMB 366, (2007), 1332-50.
3. B. Lieb, K. Dimitrova, H.-S. Kang, S. Braun, J. Markl et al., PNAS 103 (2006), p. 12011-16.
4. J. Bugge, R.E. Weber, Am J Physiol Regul Integr Comp Physiol 276 (1999), p. R347-56.
5. C.P. Mangum, Am J Physiol Regul Integr Comp Physiol 248 (1985), p. R505-17.
6. M.H.L. Arndt et al. Biopolymers 69 (2003), 470-79.

7. Supported by the DFG graduate school 1043 "Antigen specific immunotherapy", and the "Immunointervention" cluster of excellence (ICE) of Rheinland-Pfalz.

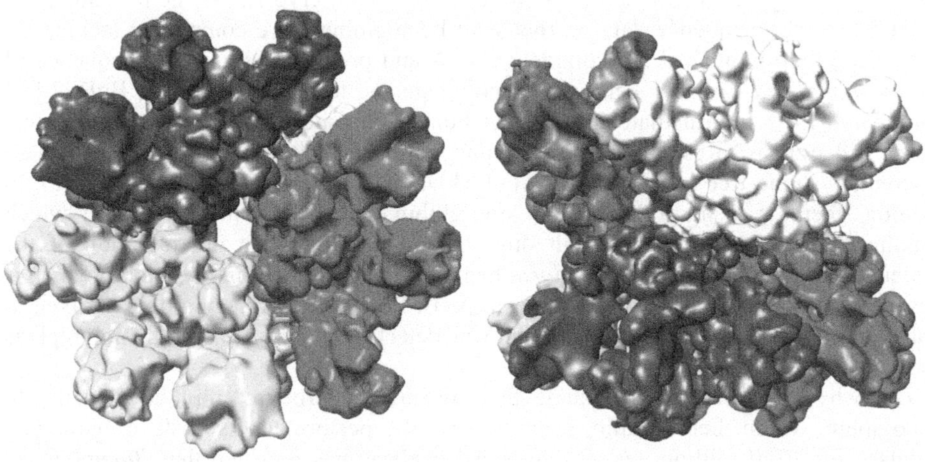

Figure 1. 8 Å cryo-EM structure of *Biomphalaria* hemoglobin, exhibiting point-group D3 symmetry (left, top view; right, side view). The molecule measures ca. 20 nm in diameter. The putative subunits are indicated by different colors. The three subunit dimers either form a right-handed (e.g. blue/yellow) or a left-handed helix (e.g. blue/purple). Note that in addition, the handiness of the structure is still unknown. Each subunit contains 13 globin domains and a cysteine-rich plug domain that is required for subunit dimerization via disulphide bridges.

Structure of Pex5p and Pex5-20 complexes in the yeast *Hansenula polymorpha*. Pex20p causes a conformational change upon binding to Pex5p tetramers involved in peroxisomal protein transport.

Kasia Moscicka[1], Sandra H. Klompmaker[2] ,Dongyuan Wang[2], Ida J. van der Klei[2] and Egbert J. Boekema[1]

1. Electron Microscopy, and [b]Eukaryotic Microbiology, Groningen Biomolecular Sciences and Biotechnology Institute (GBB), University of Groningen, [a]Nijenborgh 4, 9747 AG Groningen, The Netherlands
2. Kerklaan 30, 9751 NN Haren, The Netherlands

K.B.Moscicka@rug.nl
Keywords: HpPex5p, HpPex5p,-HpPex20p, single particle electron microscopy

Peroxisomal matrix proteins are synthesized on free polyribosomes and directed to the organelle by specific peroxisomal targeting signals (PTSs).

The *PEX5* gene encodes the PTS1 receptor, Pex5p, which interacts with the PTS1 signal via a series of tetratricopeptide repeats (TPRs) within its C terminus. A crystal structure has been determined of a 41 kDa fragment of human Pex5p that includes six TPR motifs in complex with a small peptide containing a PTS1 sequence [1,2] or the sterol carrier protein [3]. This structure reveals the molecular basis for PTS1 recognition which is mostly formed by two clusters of three TPRs almost completely surrounding the PTS1-peptide.

However, whether or not Pex5p functions as an oligomer, is still a matter of debate. Gel filtration chromatography and electron microscopy studies indicated that human Pex5p (HsPex5p) is a homotetramer [4]. Fluorescence spectroscopy studies on Pex5p of the yeast *Hansenula polymorpha* (HpPex5p) indicated that HpPex5p also forms oligomers [5].

In this study, the projection structures of HpPex5p and HpPex5p-HpPex20p complexes were investigated by single particle electron microscopy. The analysis shows that HpPex5p is a tetramer and that HpPex20p is able to induce a major conformational change leading to a rather open space in the centre of the HpPex5p tetramer. In a successive set of experiments we show that HpPex5p-HpPex20p complexes are able to bind folded copies of tetrameric catalase at the periphery. Since catalase is one of the major peroxisomal proteins this indicates that such HpPex5p-HpPex20p-catalase complexes are functional as receptor complex.

1. Gatto, G.J. et al., 2000, Nature Str. Biol., 7 (12), 1091-1095
2. Gatto, G.J. et al., 2003, Biochem., 42 (6), 1660-1666
3. Stanley, W. A. et al., 2006, Mol. Cell., 24 (5), 653-663

A. Aretz, B. Hermanns-Sachweh, J. Mayer (Eds.): EMC 2008, Vol. 3: Life Science, pp. 59–60,
DOI: 10.1007/978-3-540-85228-5_30,

4. Schliebs, W. et al., 1999,J. Biol. Chem., 274 (9), 5666–5673
5. Boteva, R. et al., 2003, Eur. J. Biochem., 270 (21), 4332-4338

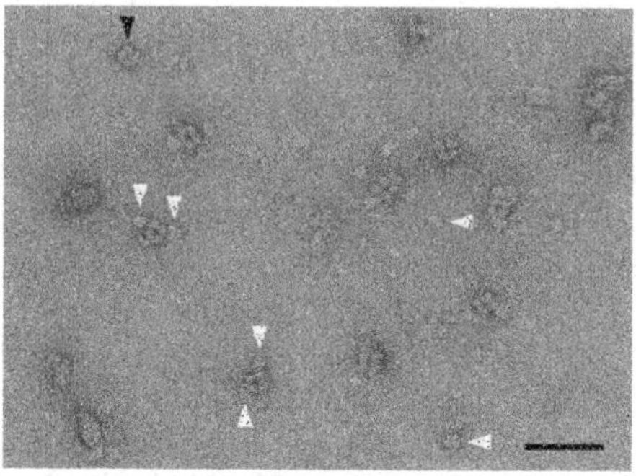

Figure 1. Electron micrograph of negatively stained Pex5p-20p complexes. The black arrowhead points to a Pex5p tetramer in the closed conformation; the white arrowheads indicate Pex20, either attached to Pex5p in the open conformation or as single complexes. The space bar indicates 50 nm.

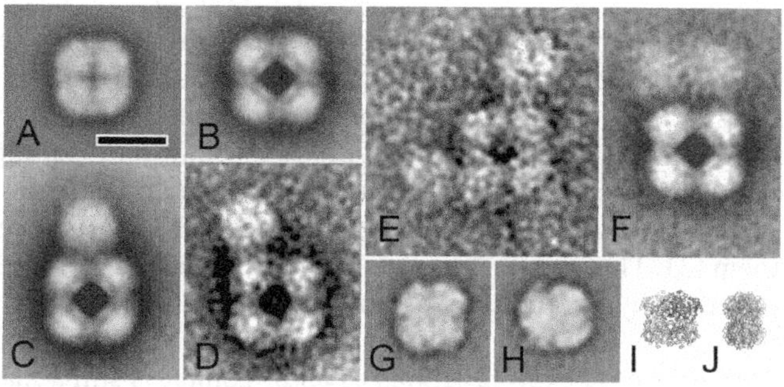

Figure 2. Single particle image analysis of Pex5p, Pex5-20 and Pex5-20-catalase complexes. (A) Average projection map of purified Pex5p in the closed conformation. (B) Average projection map of purified Pex5p in the open conformation. (C) main view of the Pex5-20 complex. (D) averge map of a small class of Pex5-20 complexes in which the upper Pex20 multimer is displaced. (E). Another class of Pex5-20 complexes in which a Pex20 multimer second multimer is binding to the left side of the Pex5p tetramer. (F) main class of Pex5-20-catalase complexes (G,H) main views of purified catalase tetramers. (I) High-resolution catalase X-ray model in a position similar to the EM projection of image G, in which it is slightly tilted out of its 4-fould symmetrical view. (H) side-view of the X-ray model in which two monomers are almost in overlap with two others. Four-fold symmetry was imposed on images of (A) and (B) after completion of analysis. The space bar equals 10 nm.

Structure of OprM-MexA interacting complex revealed by cryo electron tomography

S. Trépout[1], JC. Taveau[1], H. Benabdelhak[2], A. Ducruix[2], A.S. Frangakis[3], O. Lambert[1]

1. Architecture of Molecular Complexes and Cellular Processes, CBMN, University Bordeaux 1, Avenue des Facultés, 33405 Talence, France
2. Laboratoire de Cristallographie et RMN Biologiques, Faculté de Pharmacie, 75270 Paris Cedex 06, France
3.EMBL, Meyerhofstrasse, 69117 Heidelberg, Germany

s.trepout@iecb.u-bordeaux.fr
Keywords: cryo-electron tomography, multidrug resistance, membrane reconstitution

Pseudomonas Aeruginosa is a gram negative bacterium responsible for nosocomial diseases in hospitals, infecting immune-compromised patients. The resistance of this bacterium to a broad spectrum of antibiotics is mainly due to the presence of the MexAB-OprM multidrug resistance efflux pumps constitutively expressed, capable of extruding most of the antibiotics. These protein assemblies are tripartite complexes comprising an internal membrane transporter exchanging antibiotics against proton motive force (MexB), a periplasmic protein attached to the inner membrane by its fatty acid anchor (MexA) and an external membrane channel facilitating the exit of drugs (OprM). MexA is assumed to play a role in bridging the gap between the two others. Although atomic structures of all three proteins have been determined, the structure of the whole complex still remains unknown.

Here we report the architecture of OprM/MexA interacting complex, both proteins reconstituted into two different membranes, in Cryo-Electron Microscopy (CryoEM) and Cryo-Electron Tomography (CryoET). CryoEM studies reveal repetitive structures composed of a layer of OprM proteins facing a layer of MexA, constituting the layer of the interacting complex whose thickness is 20 nm "Figure 1A". This extraordinary assembly of OprM/ MexA complex has been more profoundly studied in investigated in CryoET. From large calculated 3D volumes of the assembly, sub-volumes containing OprM/MexA interacting complex were aligned and averaged "Figure 1B and 1C" providing new highlights on the topology and stoechiometry, taking in to account our previous results on OprM and MexA reconstituted into lipid bilayers on solid supports [1, 2]. A new model of the OprM/MexA interacting complex and its stoechiometry will be discussed in regards to X-ray-based models.

1. S. Trépout, S. Mornet, H. Benabdelhak, A. Ducruix, A.R. Brisson and O. Lambert, Langmuir **23** (2007), p. 2647.
2. S. Trépout, JC. Taveau, S. Mornet, H. Benabdelhak, A. Ducruix and O. Lambert, European Biophysical Journal **36** (2007) p. 1029.

A. Aretz, B. Hermanns-Sachweh, J. Mayer (Eds.): EMC 2008, Vol. 3: Life Science, pp. 61–62,
DOI: 10.1007/978-3-540-85228-5_31,

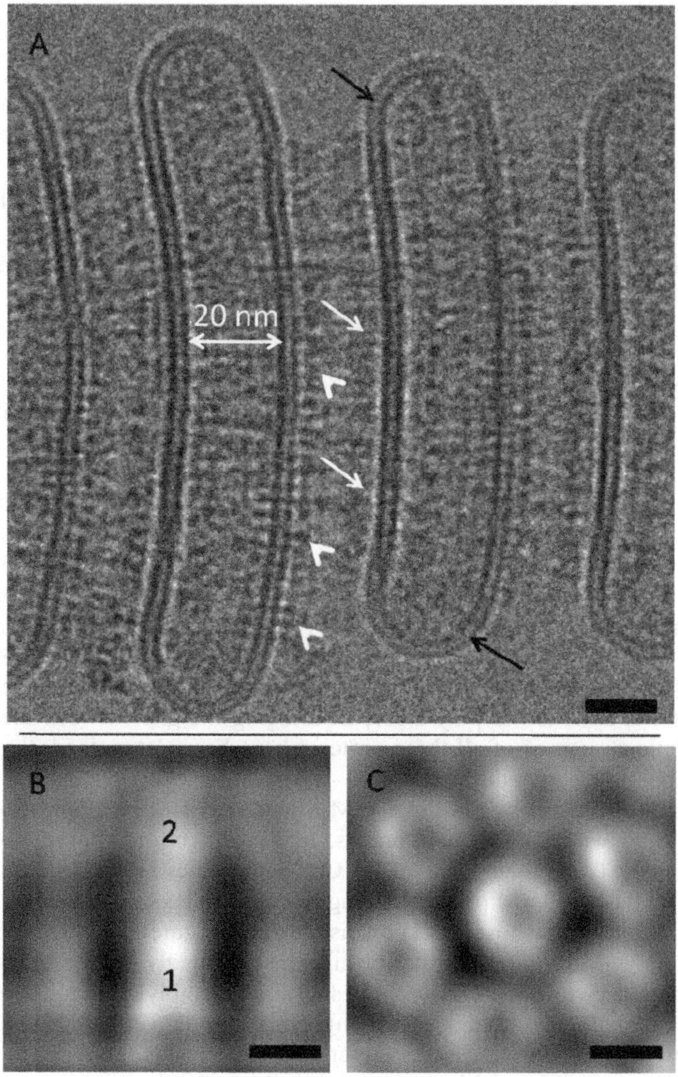

Figure 1. OprM/MexA interacting complex revealed by CryoEM and CryoET. A) CryoEM image of both OprM and MexA reconstituted in lipid membranes (black arrow) showing the regular repetitive assembly of proteins of 20 nm. The layer of OprM (white arrowhead) facing the one of MexA (white arrow). B) 4 nm projection of the average of 600 sub-tomograms, revealing OprM (1) and MexA (2). C) Projection of the OprM part revealing the hexagonal pattern of the complex and the channel shape of OprM. Scale bar are 20 nm (A) and 8 nm (B and C).

Pleiomorphic viruses revealed by cryo tomography: the structure of coronaviruses

M. Bárcena[1], W. Baterlink[2], G.T. Oostergetel[3], A. Verklijk[4], P.J. M. Rottier[2], A.J. Koster[1] and B.J. Bosch[2]

1. EM Section, Molecular Cell Biology, Leiden University Medical Center, P.O.Box 9600, 2300 RC, Leiden, The Netherlands
2. Virology Division, Infectious Diseases and Immunology, Faculty of Veterinary, Utrecht University, 3584 CL Utrecht, The Netherlands
3. Biophysical Chemistry, Groningen Biomolecular Sciences and Biotechnology Institute, University of Groningen, Nijenborgh 4, 9497 AG, Groningen, The Netherlands
4. Molecular Cell Biology, Utrecht University, Padualaan 8, 3584 CH Utrecht, The Netherlands

m.barcena@lumc.nl
Keywords: cryo-electron tomography, coronavirus, murine hepatitis virus.

Cryo-electron microscopy has provided in the last decades a wealth of three-dimensional information on viral structures. However, most of this knowledge stems from single particle methods, which rely on averaging and are therefore restricted to viral homogenous structures such as icosahedral capsids [1]. This type of analysis excludes pleiomorphic viruses, that is, viruses that, having defined general architectures, arrange their components into different shapes or sizes. Cryo-electron tomography, which provides three-dimensional reconstructions of unique specimens, is now starting to shed light into the structure of this type of viruses [2].

Coronaviruses (CoVs) are a good example of the difficulties posed by pleiomorphic viruses. CoVs are enveloped viruses containing the largest reported RNA genome [3,4]. A lot of attention has been drawn to these viruses since the outbreak in 2002-2003 of SARS (Severe Acute Respiratory Syndrome), whose causative agent was found to be a member of this group. In spite of the interest in CoVs, our structural understanding of them is rather limited, with basic question such as their internal arrangement still unclear. Previous works have been based on 2D electron (cryo)microscopy of complete/disrupted virions and were hampered by their pleiomorphic nature.

To directly investigate the global 3D structural organization of intact coronaviruses, we have applied cryo-electron tomography on purified samples of murine hepatitis virus (MHV), the prototype for CoVs. Seven tilt series were collected in a 300 kV FEI Polara microscope equipped with a post-column energy filter. Each tilt series covered an angular range of ~130° in 2° increments. The nominal underfocus value used was 4 µm, which ensured no contrast inversions due to the contrast transfer function up to about 3 nm. The pixel size was 0.58 nm at the specimen level. In the tomograms, a total of about 140 viruses were reconstructed and independently examined in 3D.

A. Aretz, B. Hermanns-Sachweh, J. Mayer (Eds.): EMC 2008, Vol. 3: Life Science, pp. 63–64, DOI: 10.1007/978-3-540-85228-5_32,

The reconstructed MHVs show a distinct spherical shape and a relatively homogenous size (85 nm envelope diameter, SD ± 6 nm), although viruses from 65% to 200% the volume of the average MHV were measured. Coiled structures and tubular fragments are observed in the interior of the virus in agreement with a helical RNP nucleocapsid model. The reconstructions provide no evidence of a shelled core. Instead, the RNP seems to be extensively folded into itself in compact structure that tends to closely follow the envelope. The viral envelope was found to be significantly thicker than a lipid bilayer and focal pairs revealed the existence of an extra layer, most likely formed by the C-terminal domains of the major envelope protein. Focal contact points and thread-like densities connecting the envelope and the RNP were revealed in the tomograms. Our results suggest a model for the structure of coronaviruses based upon local structural motifs that might serve as the basic framework for the global architecture of the virion.

1. T.S. Baker, N.H. Olson, and S.D. Fuller, Microbiol. Mol. Biol. Rev. **63** (1999) 862.
2. K. Grunewald and M. Cyrklaff, Curr. Opin. Microbiol. **9** (2006), 437.
3. C. A. M. de Haan and P. Rottier, Adv. Vir. Res. **64** (2005), 165.
4. P. S. Masters, Adv. Vir. Res. **66** (2006), 193.
5. Acknowledgments. This work was partially supported by a Marie Curie Intra-European fellowship to M.B. (MEIF-CT-2004-501540)

Combined structural and chemical analysis of unique anammox bacteria that contain a prokaryotic organelle

Laura van Niftrik[1], Willie J.C. Geerts[2], Elly G. van Donselaar[2], Bruno M. Humbel[2], Arie J. Verkleij[2], Marc Strous[1] and Mike S.M. Jetten[1]

1. Department of Microbiology, Radboud University Nijmegen, Heyendaalseweg 135, 6525 AJ Nijmegen, The Netherlands
2. Department of Cell Biology, Electron Microscopy group, Utrecht University, Padualaan 8, 3584 CH Utrecht, The Netherlands

L.vanNiftrik@science.ru.nl
Keywords: anammox bacteria, compartmentalization, electron tomography

The anaerobic oxidation of ammonium to dinitrogen gas (anammox) is a recently discovered pathway of the biological nitrogen cycle [1] and is currently estimated to be a major source of gaseous nitrogen on Earth [2]. Anammox is performed by a deep-branching clade of planctomycete bacteria, which have been applied in wastewater treatment for the removal of ammonium. Anammox bacteria divide only once per two weeks at maximum speed and possess a prokaryotic organelle, the "anammoxosome", an intracytoplasmic compartment surrounded by a bilayer membrane. This major compartment is dedicated to anammox catabolism and is the locus of respiration by these bacteria, possibly analogous to the mitochondria in Eukaryotes. The anammoxosome membrane consists mainly of so-called ladderane lipids [3], which make this membrane more impermeable to protons and the valuable intermediates of anammox catabolism.

Here we used complementary electron microscopy methods to investigate anammox cell biology and to compile a three-dimensional image of the anammox cell and its anammoxosome (Figure 1). With their small size (< 1 μm) and intriguing cell biology, these prokaryotes are an ideal target to be investigated with electron tomography. Different stages of the cell cycle were investigated and compared, to resolve how the anammoxosome divides and is passed on to daughter cells. Additional structures were observed in the anammox cell, such as glycogen storage particles, iron-containing particles, membrane foldings and fibrous protein structures [4,5]. Electron tomography was combined with energy dispersive x-ray analysis, immunogold localization of anammox enzymes, and histochemical staining methods to provide a link to the functional significance of these visualized structures. The recently completed genome of the anammox bacterium "*Candidatus* Kuenenia stuttgartiensis" [6] provided us with many candidate genes that may have important functions in anammox cell biology.

1. M. Strous, J.A. Fuerst, E.H.M. Kramer, S. Logemann, G. Muyzer, K.T. van de Pas-Schoonen, R. Webb, J.G. Kuenen and M.S.M. Jetten, Nature **400** (1999), p. 446-449.
2. M.M.M. Kuypers, A.O. Sliekers, G. Lavik, M. Schmid, B. Barker Jørgensen, J.G. Kuenen,
3. J.S. Sinninghe Damsté, M. Strous and M.S.M. Jetten, Nature **422** (2003), p. 608-611.
4. J.S. Sinninghe Damsté, M. Strous, W.I.C. Rijpstra, E.C. Hopmans, J.A.J. Geenevasen, A.C.T. van Duin, L.A. van Niftrik and M.S.M. Jetten, Nature **419** (2002), p. 708-712.

A. Aretz, B. Hermanns-Sachweh, J. Mayer (Eds.): EMC 2008, Vol. 3: Life Science, pp. 65–66,
DOI: 10.1007/978-3-540-85228-5_33, © Springer-Verlag Berlin Heidelberg 2008

5. L. van Niftrik, W.J.C. Geerts, E.G. van Donselaar, B.M. Humbel, A. Yakushevska, A.J. Verkleij, M.S.M. Jetten and M. Strous, J. Struct. Biol. **161** (2008), p. 401-410.
6. L. van Niftrik, W.J.C. Geerts, E.G. van Donselaar, B.M. Humbel, R.I. Webb, J.A. Fuerst, A.J. Verkleij, M.S.M. Jetten and M. Strous, J. Bacteriol. **190** (2008), p. 708-717.
7. M. Strous, E. Pelletier, S. Mangenot, T. Rattei, A. Lehner, M.W. Taylor, M. Horn, H. Daims, D. Bartol-Mavel, P. Wincker, V. Barbe, N. Fonknechten, D. Vallenet, B. Segurens, C. Schenowitz-Truong, C. Médigue, A. Collingro, B. Snel, B.E. Dutilh, H.J.M. Op den Camp, C. van der Drift, I. Cirpus, K.T. van de Pas-Schoonen, H.R. Harhangi, L. van Niftrik, M. Schmid, J. Keltjens, J. van de Vossenberg, B. Kartal, H. Meier, D. Frishman, M.A. Huynen, H-W. Mewes, J. Weissenbach, M.S.M. Jetten, M. Wagner and D. Le Paslier, Nature **440** (2006), p. 790-794.

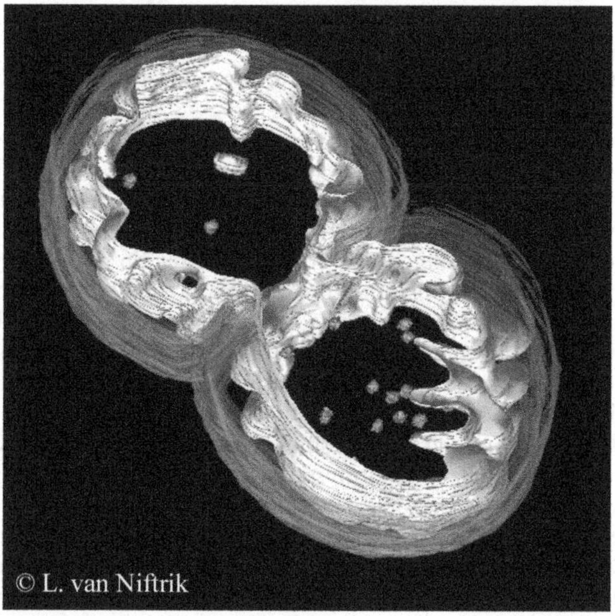

Figure 1. Snapshot of an electron tomography model showing a 300 nm section of a dividing "*Candidatus* Kuenenia stuttgartiensis" cell [4].

Structural and functional considerations on the 3-D organization of the fenestrated cytoplasm of hepatic endothelial cells

F. Braet[1], W. Geerts[1,2], J. Riches[3,4], S. Satchell[5], E. Wisse[1,6], P. Frederik[6]

1. Australian Key Centre for Microscopy & Microanalysis, The University of Sydney, NSW 2006, Australia
2. Department of Molecular Cell Biology, Utrecht University, The Netherlands
3. Centre for Microscopy & Microanalysis, The University of Queensland, Australia
4. EMBL Heidelberg, Meyerhofstraße 1, 69117 Heidelberg, Germany
5. Academic Renal Unit, Southmead Hospital, University of Bristol, UK
6. Electron Microscope Unit, University of Maastricht, P.O. Box 616, The Netherlands

f.braet@usyd.edu.au
Keywords: cell transport, correlative microscopy, cytoskeleton, fenestra, liver sieve, tomography

Hepatic endothelial cell (HEC) fenestrae are membrane-bound pores that are grouped in sieve plates and act as a bidirectional sieve in regulating transendothelial liver transport. The high permeability of the endothelial lining is explained by the presence of numerous fenestrae, and by various membrane-bound transport vesicles [1,2]. The question of whether fenestrae relate with other transport compartments, such as pinocytotic vesicles, remains unclear and has been debated since their discovery almost forty years ago [3]. In this study, novel insights concerning the 3-D organization of the fenestrated cytoplasm were built on transmission electron tomographic observations on isolated and cultured whole-mount HECs. Classical TEM and AFM imaging was performed to accumulate cross-correlative structural evidence [4].

Dual-axis transmission electron tomographic (ET) analysis showed two types of organization of fenestrae; i.e., complexes of multifolded fenestrae situated in the cytoplasm around the nucleus (Fig. 1A-B), and membrane-bound open fenestrae in the peripheral cytoplasm (Fig. 1C-D). A high degree of 3-D complexity in case of the multifolded fenestrae complexes could be discerned when studying the models under different angles, - disclosing that these complexes resemble a labyrinth-like organization and that the openings at the luminal face are diametrically opposed to the ones situated at the basal cell membrane. In contrast, ET modelling of cytoplasmic fenestrae confirms that these fenestrae are open pores lacking any spatial hindrance by superimposed intracytoplasmic fenestrae labyrinths. Porosity measurements showed that these open connections take up about two thirds of the total surface available for fenestrae-mediated exchange (Fig. 1E). The 3-D data also revealed that small membrane-bound vesicles (Ø 18 ± 4 nm) which span along the cell body (175 ± 34 nm), together with coated pits, are exclusively interspersed in the non-fenestrated cytoplasmic arms.

The data presented here indicate that mechanisms of transendothelial transport through HEC fenestrae have to be reconsidered: i.e., free passage of substances throughout open fenestrae mainly located in the ultrathin peripheral cytoplasm and, the

A. Aretz, B. Hermanns-Sachweh, J. Mayer (Eds.): EMC 2008, Vol. 3: Life Science, pp. 67–68,
DOI: 10.1007/978-3-540-85228-5_34,

trapping or delaying of substances in fenestrae-labyrinths which are found close to the proximity of the perinuclear area. Fenestrae in labyrinths constitute about one third of the total HEC porosity, indicating that the surface available for unobstructed transport is considerably reduced.

1. E. Wisse, F. Braet, D. Luo, R. De Zanger, D. Jans, E. Crabbé, A. Vermoesen. Toxicol Pathol **24** (1996), p. 100-111.
2. F. Braet, E. Wisse. Comp Hepatology **1** (2002), p. 1.
3. E. Wisse. J Ultrastruct Res **31** (1970), p. 125-150.
4. F. Braet, E. Wisse, P. Bomans, P. Frederik, W. Geerts, A. Koster, L. Soon, S. Ringer. Microsc Res Techniq **70** (2007), p. 230-242.

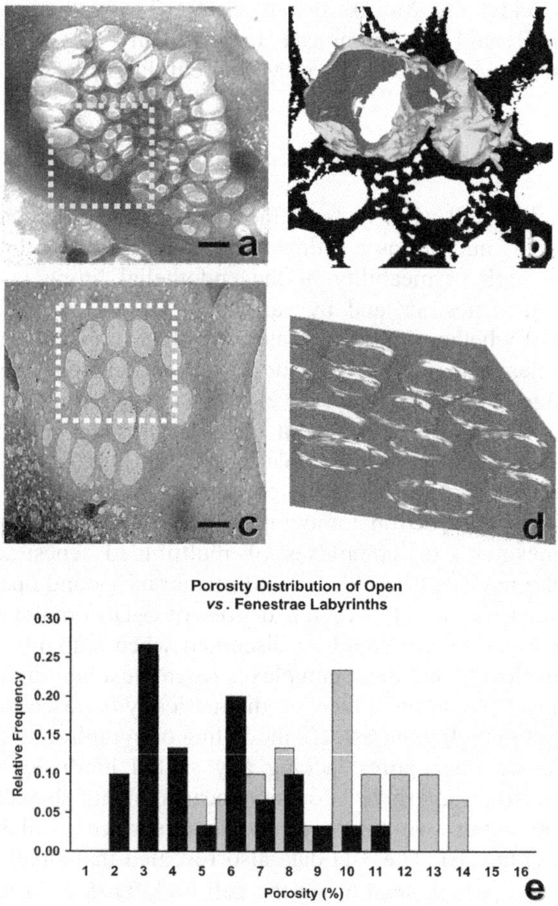

Figure 1. ET data of perinuclear (**a-b**) *vs*. cytoplasmic located fenestrae (**c-d**). Dotted line boxes in a & c correspond with the models depicted in c & d. Scale bar, 200 nm. (**e**) Porosity data of cytoplasmic (open) *vs*. perinuclear (labyrinth) located fenestrae.

Molecular architecture of the presynaptic compartment studied by cryo-electron tomography

R. Fernández-Busnadiego[1], B. Zuber[2], U. Maurer[1], G.W.K. Kong[1], M. Cyrklaff[1], J. Dubochet[2], W. Baumeister[1] and V. Lučić[1]

1. Max Planck Institute of Biochemistry, Am Klopferspitz 18a, D-82152 Martinsried, Germany
2. Laboratoire d'Analyse Ultrastructurale, Université de Lausanne, Biophore building, CH-1015 Lausanne, Switzerland

ruben@biochem.mpg.de
Keywords: cryo-ET, synapse, automated data analysis

Cryo-electron tomography allows the three dimensional visualization of cells and other biological material in a frozen, close-to-life state. In order to achieve the best preservation of the biological structures, the samples must be frozen in a manner that only vitreous ice is formed. High-pressure freezing is the method of choice for vitrification of tissue samples, whereas dissociated cells can simply undergo plunge freezing. Once frozen, the samples are inserted directly into the electron microscope, therefore avoiding dehydration and use of chemical fixatives, necessary for conventional electron microscopy (EM).

Presynaptic ultrastructure has been analyzed both in plunge-frozen nerve terminals (synaptosomes) and in cryosections obtained from high-pressure-frozen hippocampal tissue [1]. Hippocampal cryosections offer a unique window into the brain under the most physiological conditions, whereas synaptosomes can be easily isolated and allow application of different pharmacological treatments.

All the canonical synaptic features (presynaptic terminal containing numerous synaptic vesicles, a very dense protein network in the synaptic cleft and a postsynaptic counterpart) can be observed in both kinds of samples, with a resolution not achievable for other EM techniques. Our work focuses on the small filamentous strands that connect synaptic vesicles to each other and to the cell membrane. We have used an semi-automatic segmentation algorithm, previously developed in our laboratory [2], to study the architecture and connectivity of this network.

Figure 1 and 2 show, respectively, a tomogram of hippocampal cryosections and its corresponding molecular connectors detected in a semi-automated manner.

1. B. Zuber, I. Nikonenko, P. Klauser, D. Muller and J. Dubochet, PNAS **102** (2005), p. 19192-19197.
2. V. Lučić, T. Yang, G. Schweikert, F. Forster, W. Baumeister, Journal of Structural Biology **13** (2005), 423-434

A. Aretz, B. Hermanns-Sachweh, J. Mayer (Eds.): EMC 2008, Vol. 3: Life Science, pp. 69–70,
DOI: 10.1007/978-3-540-85228-5_35, © Springer-Verlag Berlin Heidelberg 2008

Figure 1. Tomographic z-slice of a synapse in a hippocampal cryosections. SV: Synaptic vesicles, SC: synaptic cleft, PSD: postsynaptic density.

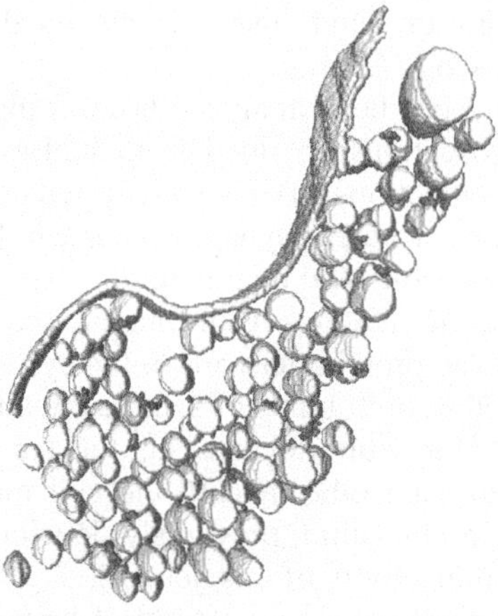

Figure 2. Corresponding surface rendering of the SVs and part of the cell membrane (yellow), together with the molecular connectors linking SVs to each other and to the cell membrane (red).

Synapses in high pressure frozen *Caenorhabditis elegans*

J. Hegermann[1], S. Eimer[1]

1. European Neursciene Insitute, Göttingen; DFG Research Center for Molecular Physiology of the Brain (CMPB)

jhegerm1@gwdg.de
Keywords: high pressure freezing, electron microscopy, *Caenorhabdits elegans*

C. elegans is a very well and extensively described model organism concerning its biochemistry, genetics and ultrastructure. Its organs and different tissues have been subject to numerous electron microscopic investigations [1]. Even though the resolution in light microscopy is improving significantly at present, electron microscopy is still the method of choice to obtain insights into subcellular details including single vesicles, individual cisternae of golgi stacks, or the architecture of synapses. Cellular details of a few nanometers can be resolved. However, in biological specimen the bottleneck in achieving a life-like portray of the living cell is still the specimen-preparation. This procedure includes harsh dehydration and plastic embedding and makes stabilisation of the tissue indispensable. Conventional preparation techniques include chemical fixation, during which the tissue gets infiltrated by crosslinking agents. However, this process takes much more time than the duration of certain cellular events. Chemical fixation causes more or less a slow cell death and afterward mostly leads to images of the beginning decomposition of tissues.

High Pressure Freezing (HPF) has overcome these drawbacks of the conventional preparation of biological material for electron microscopy. During HPF the tissue is not chemically fixed but frozen rapidly within milliseconds. Afterwards it can be investigated either in frozen-hydrated state [2] or can be embedded in plastic without further alteration. The process of freezing is fast enough to capture highly dynamic cellular processes like endo- and exocytosis, the real-life state of golgi stacks and the distribution of synaptic vesicles at the synapse.

Using HPF and three-dimensional reconstruction we are investigating the overall ultrastucture of synapses in *C. elegans*. At the synapse, synaptic vesicles gather around an electron-dense specialisation, called "presynaptic density" (PD) which is the site of vesicle docking and release. Compared to results obtained by chemical fixation we show that the structure of the PD is much more complex than described before. We compare different mutants of Rab-GTPases as to the distribution of synaptic vesicles around the PD using a standardised analysis method adopted for *C. elegans* synapses. Additionally we are investigating neuropeptide-containing dense core vesicles (DCV) in neurons. Some of these appear also near to synapses, however, we find them to be distributed all along the axons. We are adressing the question where DCV go, where they are secreted and if they derive from the Golgi-complex in the neuronal cell body.

A. Aretz, B. Hermanns-Sachweh, J. Mayer (Eds.): EMC 2008, Vol. 3: Life Science, pp. 71–72,
DOI: 10.1007/978-3-540-85228-5_36,

We are studying the appearance of the Golgi-complex also in other tissues in *C. elegans*. Being very dynamic membrane systems, golgi stacks are predominantly suffering from chemical fixation, which makes it worthwile to redescribe this organelle in high-pressure-frozen samples by tomography and 3D-reconstruction.

1. http://www.wormatlas.org/index.htm
2. Tokuyasu, K. T. (1973). A technique for ultracryotomy of cell suspensions and tissues. J Cell Biol **57**, 551-65.

High resolution EM-tomography of a whole vitrified cell

H. Heijnen[1,2], H. van Nispen tot Pannerden[1,2], F. de Haas[3]

1. Department of Hematology, University Hospital, 3508 GA Utrecht, The Netherlands
2. Department of Cell Biology, Medical School, University of Utrecht, 3584 CX Utrecht, The Netherlands
3. FEI Company, Achtseweg Noord 5, 5600 KA Eindhoven, The Netherlands

H.F.G.Heijnen@umcutrecht.nl
Keywords: Tomography, cryo-EM, vWF, Collagen, electron microscopy, microtubules, granules

Conventional transmission electron microscopy (EM) studies have contributed much to our present understanding of the platelet ultrastructure and the membrane dynamics during activation. However, most of these studies rely on chemical fixation procedures and have the disadvantage that conclusions are drawn from two-dimensional images.

Ultrastructural analysis of cellular volumes and simultaneous visualization of macromolecular complexes in their native state requires rapid freezing of the cell combined with high resolution EM-tomography of the whole vitrified cell. Due to cell size limitations and high-angle tilting axis, whole cell EM tomography is restricted to small cells and thin cell borders. Blood platelets are a-nucleated discoid cells, 1-3 µm in diameter, that circulate in the bloodstream. Upon vascular injury platelets are tethered to the damaged vessel wall and interact rapidly with exposed components of the subendothelium (vWF, collagen). During this process platelets undergo dynamic shape changes and release the content of their storage granules, thereby enabling optimal interaction with exposed adhesive molecules and formation of a stable thrombus.

We have taken advantage of the small size of blood platelet and performed EM tomography 3D reconstruction on cryo-fixed human blood platelets. Platelet-rich plasma was isolated from healthy individuals and allowed to interact for 15 min with a fibrinogen-coated electron microscopic grid. 10 nm colloidal gold particles were briefly added as fiducial markers. The cells were vitrified in liquid ethane using the Vitrobot™ (FEI Company) plunge freezing unit, and observed in a Tecnai 12 BioTWIN (FEI Company) at 120 kV using a GATAN 626 cryo-holder. Tilt series were collected at low dose conditions and recorded on a 4k Eagle CCD camera (FEI company) from -65° to 65° with 2° increments at -5 µm defocus. Tomograms were aligned and reconstructed using Inspect3D (FEI Company).

The thin size of the adherent platelets allowed optimal vitrification of the sample. Three-dimensional reconstructions showed filamentous actin in protruding filopodia and irregular arrays of severed microtubules at the focal adhesion plane. Direct visualization of secretory alpha granules (Figure 1) showed intra tubular substructure (Figure 2) and luminal membranes. Extensive cryo electron tomography studies on vitrified platelets will lead to novel structural details of this cell.

A. Aretz, B. Hermanns-Sachweh, J. Mayer (Eds.): EMC 2008, Vol. 3: Life Science, pp. 73–74,
DOI: 10.1007/978-3-540-85228-5_37,

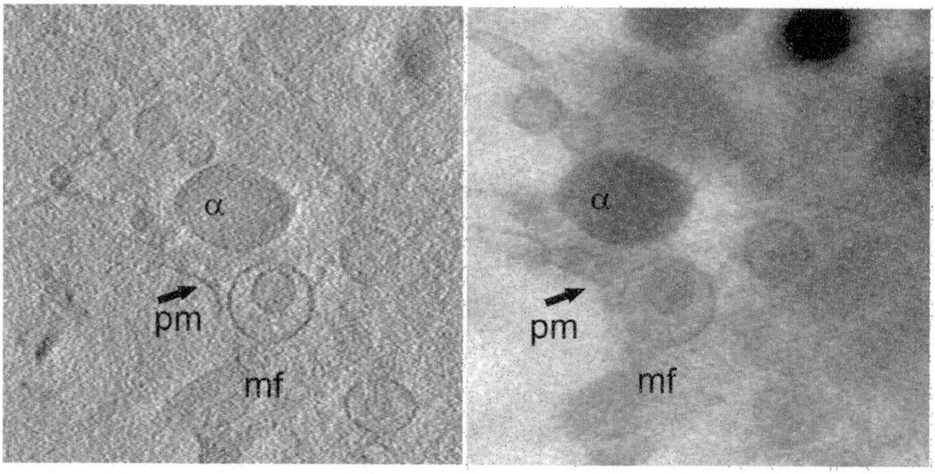

Figure 1. From left to right; Reconstruction using Weighted Backprojection, SIRT.

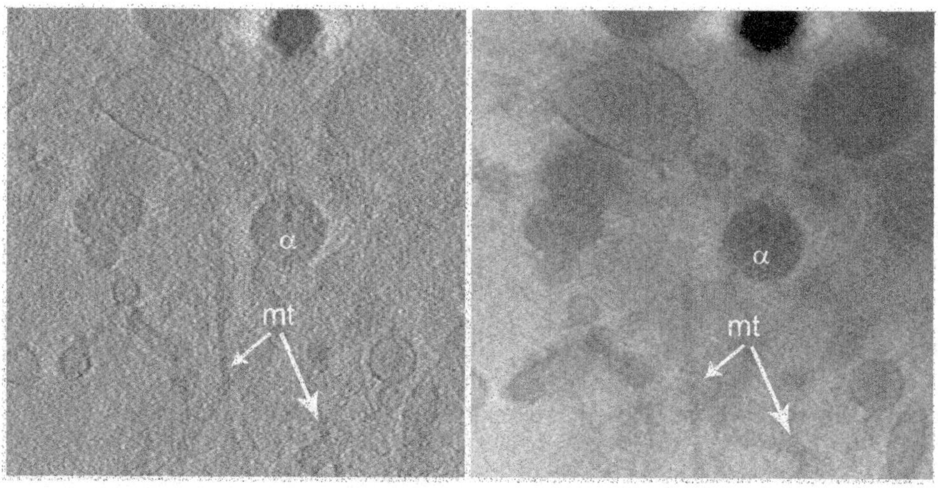

Figure 2. Direct visualization of microtubules in the presence of alpha granules.(left; BP, right; SIRT)

Evaluation of beam damage in catalase crystals observed in vitrified sections

Hong-Mei Han[1,2], Bryant Gibson[3] and Jacques Dubochet[2]

1. CMB, Karolinska Institutet, Nobels väg 3, S-17177 Stockholm, Sweden
2. LAU, Université de Lausanne, CH-1015 Lausanne, Switzerland
3. MCB, University of California, Shields Ave. Davis, CA 95616, USA

hongmei.han@ki.se
Keywords: beam damage, CEMOVIS, catalase crystals

It was a major breakthrough that Gleaser [1] quantitatively determined beam damage by measuring the fading of electron diffraction spots in organic crystal. The method was soon extended to catalase crystal, which became a kind of working horse for quantitative evaluation of specimen preparation methods [2,3]. For such purpose, catalse crystals are well suited because they can be grown slowly into very thin platelet. The crystal lattice of the platelet is tetrahedral with lattice constant of 6.9 nm and 17.4 nm in the length and width respectively and 20.6 nm in the thickness. Cryo-electron microscopy of vitreous sections (CEMOVIS) is a technique which has recently reached maturity after decades-long development [4,5]. What is shown with CEMOVIS is surprizingly different from what is used to be seen with conventional electron microscopy. And it seems that a better resolution is generally obtained. However, the quality of the recorded images and the plethora of significant information contained over the thickness of a section call for 3D reconstruction of the section. This can be achieved with the remarkable method of computerized electron tomography (CET). It requires that a large number of images of the same region are recorded in different orientations. Therefore, beam damage is the major limitation of the method.

In the present work, it was intended to find out what is the maximum electron dose that can be given to a biological specimen according to a given resolution of its structure. And a quantitative evaluation of beam damage was needed to gain an estimation of the most important parameter needed for optimizing the recorded information. For this purpose, catalase crystal [6] was used as a test object and the fading of diffraction spots was measured as Glaeser initiated 35 years ago [1]. In case of CEMOVIS, a bulky solution of randomly oriented thick crystals was vitrefied in EMPACT high-pressure freezer (Leica Vienna). Thin vitreous sections led to random sections through randomly oriented crystals. As expected, image and optical diffractogram compare favorably with any previous images obtained from thin sections of protein crystals [3]. Fig.1a and b show a vitreous section of catalase crystal in low (2000 X) and high (25 5000 X) magnification. The diffractogram of Fig.1b, recorded at low dose (<200 electrons/nm^2), extends to less than 2 nm (Fig.1c). Resolution decreases with the electron dose increasing from 400 to 7000 electrons/nm^2 (Fig.2). The fading of high resolution spots is more rapid than that of low resolution spots (Fig.3).The critical dose for the fading of diffraction spot is roughly 3000 e/nm^2 at 2.5nm and 5000 e/nm^2 at

A. Aretz, B. Hermanns-Sachweh, J. Mayer (Eds.): EMC 2008, Vol. 3: Life Science, pp. 75–76,
DOI: 10.1007/978-3-540-85228-5_38,

5nm. However, these values are only general trends with considerable variations from one spot to the other, even at comparable dimensions. This variability may have two causes. Firstly, the orientation of the crystal section may be changed during electron exposure. In the depicted series, this effect probably explain the apparent rapid fading in the vertical direction of the diffractogram. Secondly, it is probable that some structures are more beam resistant than other even at comparable dimension.

1. Glaeser, R. M. (1971) Limitations to significant information in biological electron microscopy as a result of radiation damage. J. Ultrastructure Research. 36:466-482.
2. Unwin P.N. (1975) Beef liver catalase structure: interpretation of electron micrographs. J. Mol. Biol. 98(1):235-42.
3. Jesior, J. C.(1982) The grid sectioning technique: a study of catalase platelets. Embo. J. 1(11):1423-8.
4. Reid, T. J., Murthy, M. R.Sicignano, A.Tanaka, N.Musick, W. D.Rossmann, M. G. (1981) Structure and heme environment of beef liver catalase at 2.5 A resolution. PNAS. 78(8):4767-71.
5. Al-Amoudi, A., Chang, J. J., Leforestier, A., McDowall, A., Salamin, L.A., Norlén, L. P. O., Richter, K., Sartori Blanc, N., Studer, D. & Dubochet, J. (2004) Cryo-electron microscopy of vitreous sections. EMBO J. 23: 3583 - 3588.
6. Dubochet, J., Zuber, B., Eltsov, M., Bouchet-Marquis, C., Al-Amoudi, A. & Livolant, F. (2007) How to "Read" a Vitreous Section. Cellular Electron Microscopy. McIntosh, J. R., Elsevier. 79: 385-406.
7. The authors gratefully acknowledge the discussions with Daniel Studer, Benoît Zuber and Henning Stahlberg. This work was financed by 3D-EM european network of excellence.

Figure 1.

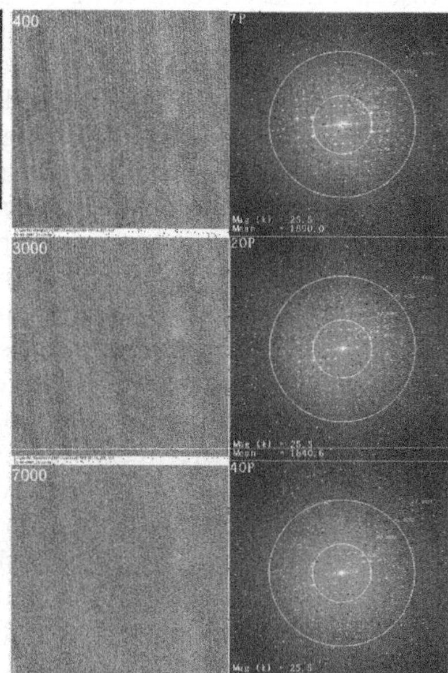

Figure 2.

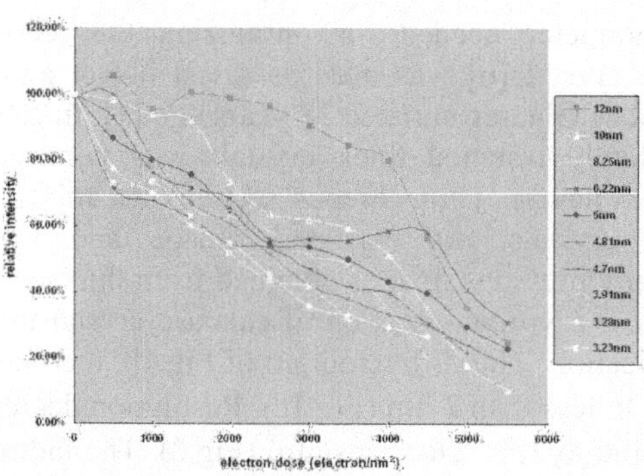

Figure 3.

Urothelial fusiform vesicles are formed in post-Golgi compartment

S. Hudoklin[1], R. Romih[1], J. Neumüller[2], M. Pavelka[2], K. Jezernik[1]

1. Institute of Cell Biology, Medical Faculty, Lipiceva 2, 1000 Ljubljana, Slovenia
2. Department of Cell Biology and Ultrastructure Research, Center for Anatomy and Cell Biology, Medical University Vienna, Schwarzspanierstraße 17, A-1090 Vienna, Austria

samo.hudoklin@mf.uni-lj.si
Keywords: urothelium, fusiform vesicles, Golgi apparatus, EM tomography, immunocytochemistry

Introduction: Asymmetric unit membrane (AUM) is a hallmark of terminally differentiated urothelial superficial cells. This specialized membrane covers the apical surface and forms a pool of flattened cytoplasm vesicles, called fusiform vesicles. AUM contains four major proteins, uroplakins (UPIa, UPIb, UPII, UPIII). UPIa and UPIb form heterodimer with UPII and UPIII, respectively in ER. Heterodimers are organized in tetramers and further into 16 nm particles in the Golgi apparatus (GA). Uroplakin particles assemble into crystalline, rigid-looking plaques of AUM [1]. Fusiform vesicles function as highly specialized transportation carriers that can fuse with apical membrane during bladder filling with urine, hence they help to increase luminal surface. Since the formation of fusiform vesicles is poorly understood, we aimed to determine cellular compartments involved in this process.

Methods: Male mice bladders were used. For morphological studies, urothelium was rapidly frozen, freeze-substituted, and embedded in Epon. Three dimensional tomography was performed on 300 nm thick sections with FEI Tecnai 20 TEM microscope, running under 200kv Modeling was done with Imod software (http://bio3d.colorado.edu/imod/). For immunocytochemical studies, cryo sections were prepared by modified Tokuyashu method and labeled antibodies against all four uroplakins.

Results: Golgi consisted of multiple cisterns and numerous vesicles. AUM was never observed inside the Golgi apparatus. Three dimensional reconstructions revealed prominent tubulo-vesicular *trans*-Golgi network. AUM was not seen on small vesicles that were detached from *trans*-Golgi. These vesicles fuse to form larger vesicles with patches of AUM. With increasing length of post-Golgi vesicles the amount of AUM membranes increased and their shape changed from rounded to disk-shaped. In post-Golgi area, fusiform vesicles were organized into stacks containing 5-10 vesicles. Direct detachment of mature fusiform vesicles from Golgi apparatus was never seen. Antibodies against uroplakins did not label Golgi apparatus membranes. Small vesicles close to Golgi apparatus were weakly labeled and mature fusiform vesicles were densely labeled.

A. Aretz, B. Hermanns-Sachweh, J. Mayer (Eds.): EMC 2008, Vol. 3: Life Science, pp. 77–78,
DOI: 10.1007/978-3-540-85228-5_39,

Discussion: Using morphological and immunocytochemical methods for TEM, we showed that AUM plaques are not formed in the Golgi apparatus, but modifications in post-Golgi are necessary. Plaques of AUM are gradually formed by fusion of small, uroplakin positive vesicles, not by direct detachment of mature fusiform vesicles from *trans*-Golgi [2, 3].

1. C.C. Hu, F.X. *et al*, Assembly of urothelial plaques: tetraspanin function in membrane protein trafficking, Mol. Biol. Cell., volume 6-9(2005), p. 3937-3950.
2. R.M. Hicks, The function of the golgi complex in transitional epithelium. Synthesis of the thick cell membrane. J Cell Biol, volume 30-3 (1966), p. 623-643.
3. N.S. Severs, *et al,* Analysis of membrane structure in the transitional epithelium of rat urinary bladder. J Ultrastruct Res, volume 69-2 (1979), p. 279-296.

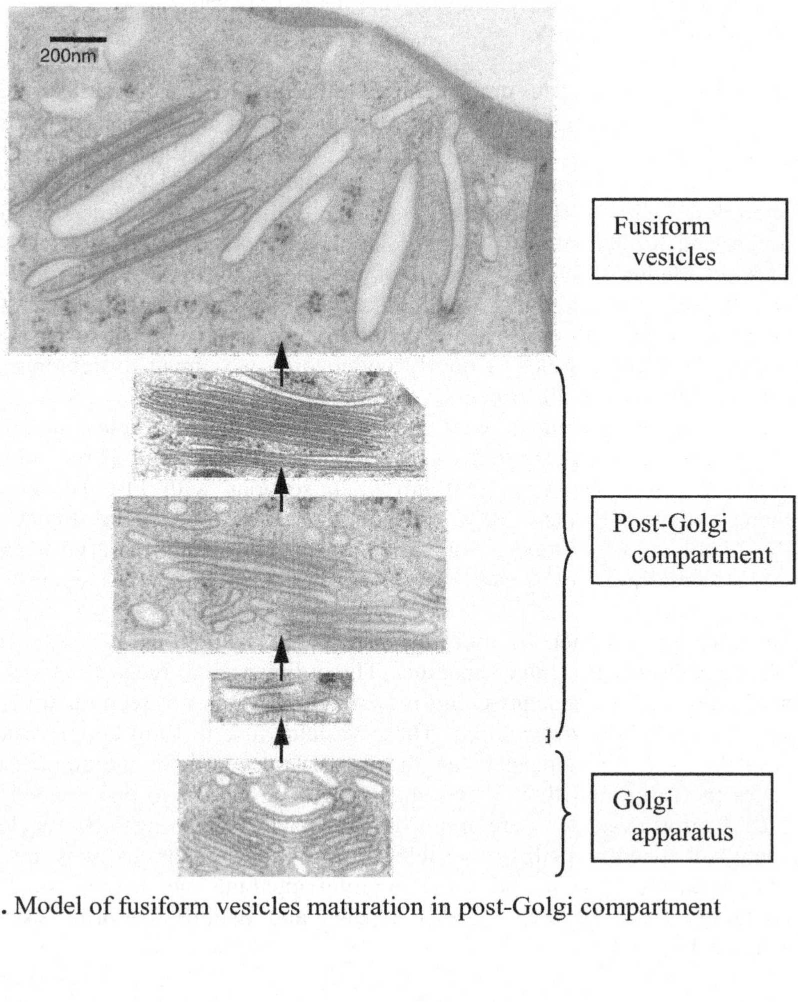

Figure 1. Model of fusiform vesicles maturation in post-Golgi compartment

Localisation of GFP and GFP tagged PKD2 in cultivated pancreatic cancer cells using high-pressure freezing

H. Krisp[1], G. von Wichert[2], T. Seufferlein[2] and P. Walther[1]

1. Zentrale Einrichtung Elektronenmikroskopie, Universität Ulm, 89069 Ulm
2. Klinik für Innere Medizin I, Universität Ulm, 89071 Ulm

Holger.krisp@uni-ulm.de
Keywords: Pancreas, GFP, high-pressure freezing, fluorescence microscopy, TEM, STEM

Protein kinase D (PKD) is an important regulator of intracellular signalling pathways [1]. Members of the PKD family, especially PKD2, have been detected in pancreatic cancer. Its functional characterisation in these tumours is under investigation. The PKD enzymes have been implicated in the organisation of the Golgi apparatus, regulating the fission of vesicles from the trans-Golgi network [2]. The aim of this study is to combine the localisation of PKD2 with good structural preservation. This should be achieved by high-pressure freezing fixation of cultivated adherent pancreatic cells.

In this study, PKD2 was overexpressed by transfection of pancreatic cancer cells (BON) using GFP as a marker. For the first part of the study we applied GFP-(pmaxGFP, Amaxa, Cologne, Germany) and PKD2-vector-DNA for the nucleofection, for a second set of experiments we used GFP tagged PKD2-vector-DNA (pEGFP, Clontech, Saint-Germain, France). The cells were grown as a monolayer on special 3 mm sapphire discs that fit into 3 mm aluminium platelets used for high-pressure freezing in a Compact 01 (Wohlwend Engineering, Sennwald, CH). Afterwards the samples were freeze-substituted in acetone containing 5% of water [3] and uranyl acetate. All samples were substituted without osmium and embedded in LR-Gold. Ultrathin sections (70 nm) were cut and collected on formvar-coated copper grids. The same grids were analysed in the fluorescent light microscope (LM) and in the transmission electron microscope (TEM). Ultrathin sections of pmaxGFP-DNA transfected cells clearly show a fluorescence signal. The same ultrathin section was then analyzed in the TEM. The combination of LM and TEM provides an accurate ultrastructural localisation of the GFP signal. With the help of this method six anti-GFP antibodies were tested on the thin sections (Table I). On thin sections of pEGFP tagged PKD2 expressing cells, unfortunatelly, the GFP-fluorescence signal was too weak. Therefore we did GFP immunolabelling with a Cy3 fluorescence antibody. The sections (70 nm and 500 nm) were analysed in the fluorescence microscope, and in the conventional TEM or in a scanning TEM (STEM) for tomography. Figure 1A shows areas with a high pEGFP tagged PKD2 expression rate. In the STEM we made tomographic analysis of the Golgi to find differences between transfected and non transfected cells. The reconstructed Golgi shown in Figure 1B was found in a non transfected cell.

1. A. Rykx et al., FEBS Letters (2003) 81-86.
2. Ch. Yeaman et al., Nature cell biology Vol. 6 No. 2 (2004) 106-112.
3. P. Walther & A. Ziegler, J. Microsc. 208 (2002) 3-10.

A. Aretz, B. Hermanns-Sachweh, J. Mayer (Eds.): EMC 2008, Vol. 3: Life Science, pp. 79–80,
DOI: 10.1007/978-3-540-85228-5_40,

Table I. pmaxGFP expressing cells. (A) Labelling intensity in nucleus and cytoplasm of six GFP-antibodies . (B) More detailed statistical analysis of the both positive tested antibodies showed that the Rockland anti-GFP antibody gave the most specific signal on LR-gold sections.

A

Primary antibodies	Secondary antibodies	Intensity of immunogold labelling	
		nucleus	cytoplasm
Anti-GFP Abcam AB 6556-25	GAR	+	-
Anti-GFP BD Biosciences Clontech Lot 632377	GAR	++	-
Anti-GFP Molecular Probes A 6455	GAR	+	++
Anti-GFP Roche	GAM	++	+
Anti-GFP Rockland Lot 17549	RAG	-	+++
Anti-GFP Torrey pines biolabs TP-401	GAR	++	-

B

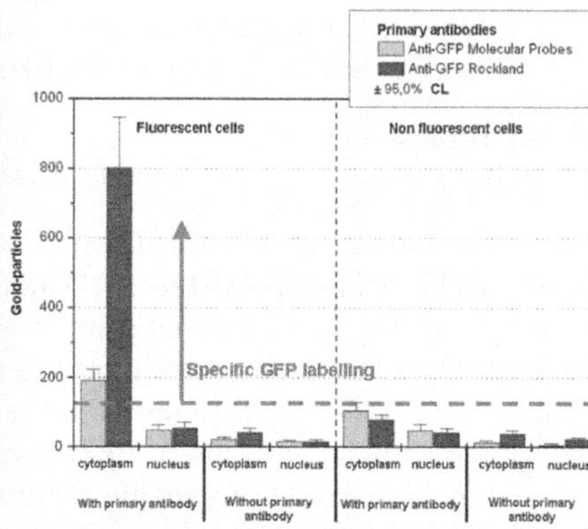

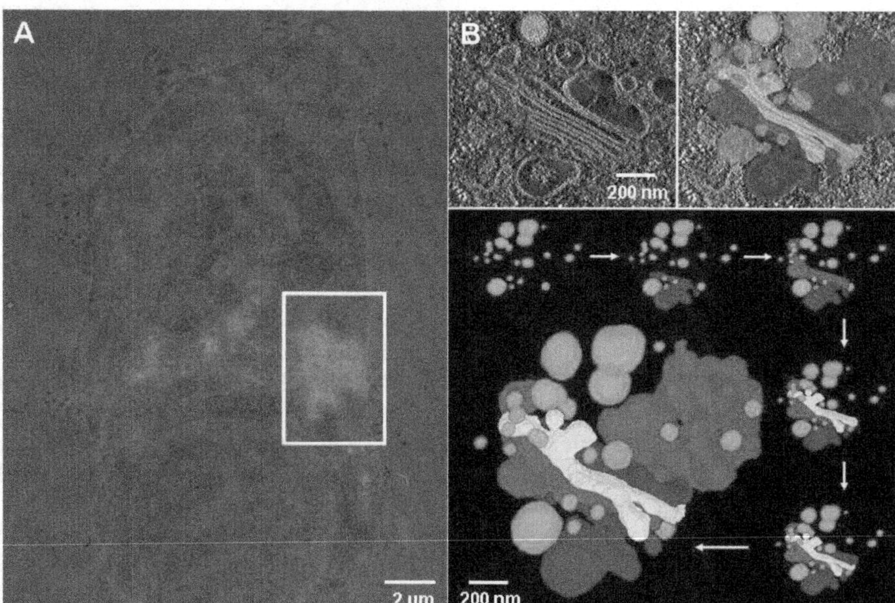

Figure 1. pEGFP tagged PKD2 expressing cells. (A) Overlay of a fluorescence microscope and a TEM image of a Cy3 GFP immunolabelling. Areas of high expression rate close to the Golgi were clearly visible (rectangular marking). (B) First results of a STEM-Tomogram of a Golgi apparatus of a non transfected cell.

Cytomegalovirus membrane envelopment revisited – a STEM-tomography study based on high-pressure freezing and freeze substitution.

S. Landwehr[1,2], J. von Einem[1], T. Mertens[1] and P. Walther[2]

1. Institute for Virology, University Hospital Ulm, 89081 Ulm, Germany
2. Electron Microscopy Facility, University of Ulm, 89069 Ulm, Germany

sandra.landwehr@uniklinik-ulm.de
Keywords: HCMV, envelopment, high-pressure freezing, STEM

Membrane envelopment of viruses, such as secondary envelopment of human cytomegalovirus (HCMV) has been investigated in the electron microscope since many years. In this study we have investigated this process with novel electron microscopical methods like high-pressure freezing (which allows for arrest of dynamic events) freeze-substitution and scanning transmission electron microscopical (STEM) tomography [1] (which enables three-dimensional data collection of relatively thick sections up to 500 nm). The method is ideal to investigate membrane envelopment of viruses, since this event is fast and an enveloped cytomegalovirus (150 nm) fits well into these sections.

Human fibroblast cells were cultivated and infected with HCMV on carbon coated sapphire discs [2], and high pressure frozen with a Compact 01 (Wohlwend GmbH, Sennwald, Switzerland). Afterwards, samples were freeze substituted in acetone, osmiumtetroxide, uranylacetate and 5% of water [3] and embedded in Epon. Tilt series (-72° to +72°) on 350 to 500 nm thick sections were recorded in a 300 kV STEM (FEI, Eindhoven) with a high angle annular dark field detector. Tomograms were reconstructed with iMod by standard weighted back projection (Figure 1) and for visualization the software Amira was used (Figure 2).

Figure 1 is an artificial (computer generated) section through the reconstructed tomogram and Fig. 2 is a visualization of the same volume. Membranes occur with high contrast and the two membrane leaflets can be well discerned (Figure 1). This allows for observation of the different stages of envelopment. Capsids (red) with attached tegument (yellow) that are not yet enveloped with a membrane (1a, b and c) can be seen in Figure 2 (but not in Figure 1, since they are not in the section plane). An envelope that is almost closed (2) and two capsids (3) that share an envelope and a vesicle membrane are visible in both figures. Some virions have completed their secondary envelopment and sit in vesicles (4a, b and c) as a consequence of the budding event. We often observed virions in vesicles that are cracked (5). It is not clear yet, whether this is an effect of viral envelopment or represents a preparation artifact. [4]

1. A. E. Yakushevska et al., J Struct Biol. **159** (2007), p. 381-391.
2. Ch. Buser et al., J Virol. **81** (2007), p. 3042-3048.
3. P. Walther, and A. Ziegler, J Microsc. **208** (2002), p. 3-10.
4. This work is supported by the ***DFG Schwerpunktprogramm 1175***.

A. Aretz, B. Hermanns-Sachweh, J. Mayer (Eds.): EMC 2008, Vol. 3: Life Science, pp. 81–82,
DOI: 10.1007/978-3-540-85228-5_41,

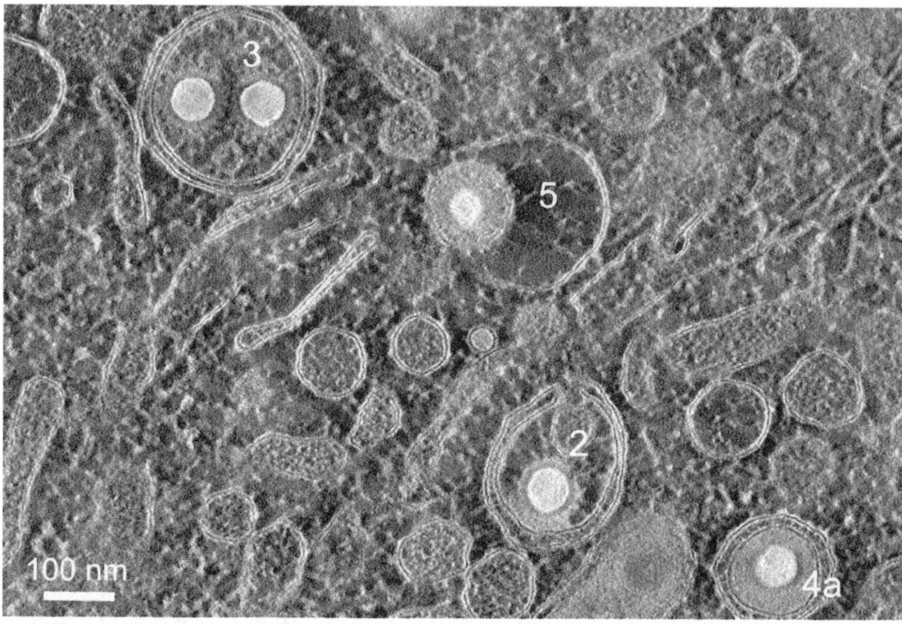

Figure 1. Section through the reconstructed tomogram of a HCMV infected cell, five days post infection.

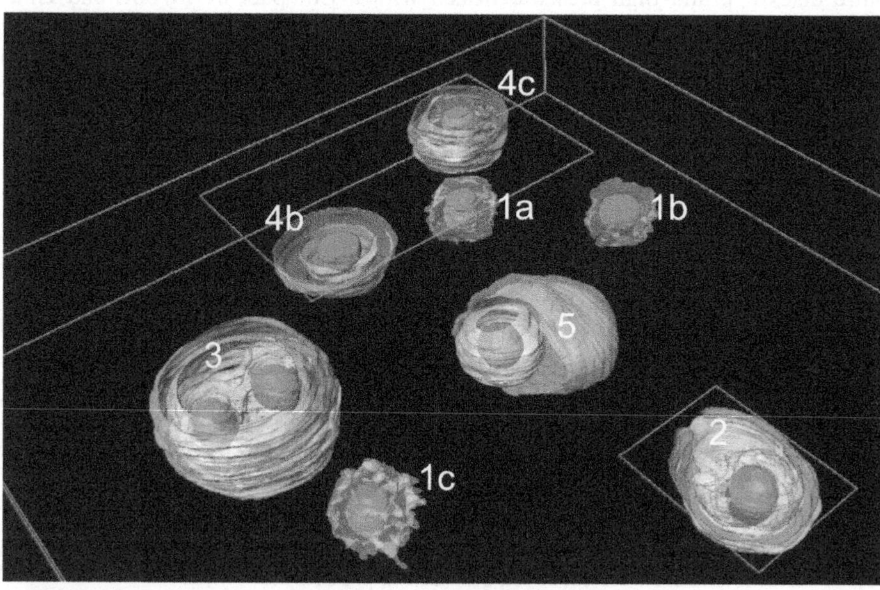

Figure 2. Visualization of capsids (red), teguments (yellow) and some membranes (blue) of the tomogram of Fig. 1. See text for details.

Electron tomography & template matching of biological membranes

M.N. Lebbink[1], W.J. Geerts[1], E. van Donselaar[1], B.M. Humbel[1], J.A. Post[1], L.O. Hertzberger[2], A.J. Koster[3], A.J. Verkleij[1]

1. Cellular Architecture & Dynamics, Utrecht University
2. Department of Computer Science, University of Amsterdam
3. MCB Electron Microscopy, Leiden University Medical Center

M.N.Lebbink@UU.nl
Keywords: electron tomography, template matching, membrane domains

In recent years, electron tomography has proven to be a valuable addition to existing EM techniques, providing new and exiting insights into cellular architecture in three dimensions. Constant technical developments have played a crucial role in establishing the role of the technique in modern electron microscopic studies. The next technical bottleneck to overcome is that of a reliable, objective, and (semi)-automatic analysation of the acquired data. One such approaches is three dimensional template matching based on normalized cross-correlation analysis (figure 1) [1].

We investigated the feasibility of this approach in biology, and specifically in membrane-research. Conventional electron tomography has provided a wealth of information on membrane morphology, and organellar interaction; but observations about structural differences throughout the membrane (domains) were limited to some electron dense coats.

Based on our previous work [2], we used temperature induced domains in the *E.coli* outer-membrane as a model system. Since the inner- (phospholid) and outer- (LPS) leaflet are composed differently, they react fundamentally different to temperature shifts (figure 2). While in the former lipid phase separation is induced, the latter is largely unaffected. We were able to visualize this effect in three dimensions in our tomographic micrographs by matching the leaflets - both at a temperature higher than the phase transition point and a temperature below – to two artificial cuboid-shaped templates of different thickness. Due to the lipid phase separation the inner leaflet becomes largely heterogeneous at lower temperatures, while the heterogeneity of the outer leaflet is largely unaffected (figure 3) [3].

By combining electron tomography and template matching we present a state of the art approach for objective investigation of membrane domain morphology at a high resolution in three dimensions. The technique is strong enough to localize and visualize domains in tomographic data that are invisible to the naked eye (but known to exist from other techniques). Such cannot be observed on an electron tomographic level using any other technique to date [4].

A. Aretz, B. Hermanns-Sachweh, J. Mayer (Eds.): EMC 2008, Vol. 3: Life Science, pp. 83–84,
DOI: 10.1007/978-3-540-85228-5_42,

1. Frangakis AS, Bohm J, Forster F, Nickell S, Nicastro D, Typke D, Hegerl R, Baumeister W. *Proc Natl Acad Sci U S A.* **99(22)** (2002) : 14153-8
2. Lebbink MN, Geerts WJ, van der Krift TP, Bouwhuis M, Hertzberger LO, Verkleij AJ, Koster AJ. *J Struct Biol.* **158(3)** (2007) : 327-35.
3. Lebbink MN, Geerts WJ, van Donselaar E, Humbel BM, Post JA, Hertzberger LO, Koster AJ, Verkleij AJ. [In peparation.]
4. This research was supported by a grant from Senter Novum, IOP Genomics (IGE03012)

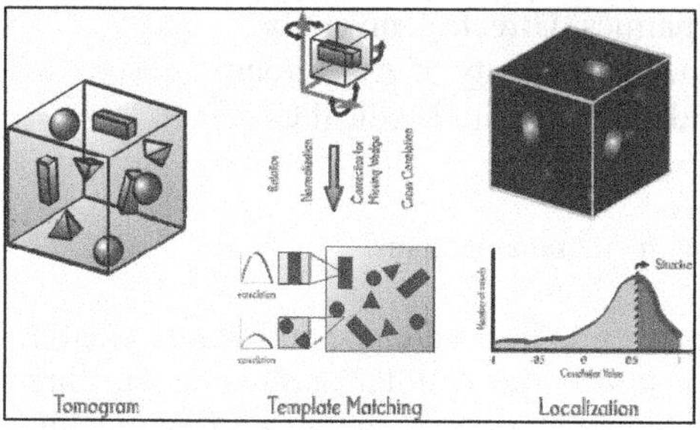

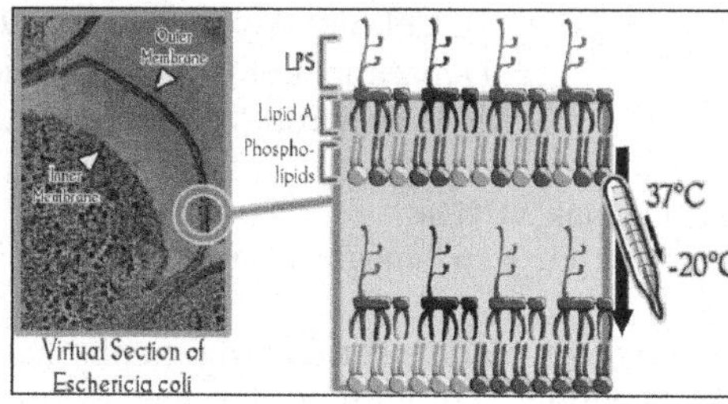

Figure 1 (left). Schematic representation of the template matching procedure for a single template (adapted from Frangakis 2002). Specific structures are localized within a tomographic volume, and assigned a correlation value (value of similarity).
Figure 2 (right). Schematic drawing of composition of the *E.coli* outer membrane, and the effect of temperature shift to lipid organization and domains formation.

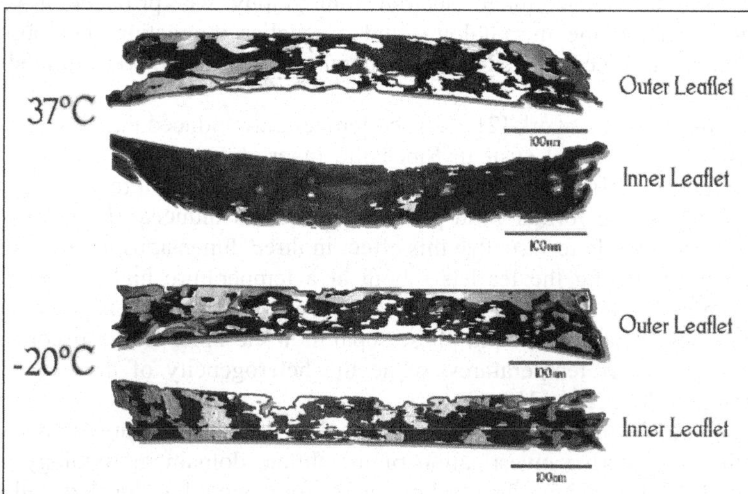

Figure 3. Three dimensional visualization of the differences in membrane organization above and below the lipid phase transition point. These domains were visualized by matching the membranes with two templates of different thickness (4 voxels (light) and 6 voxels (dark) respectively).

Correlated Microscopy: From Dynamics to Structure

S. Lepper[1], M. Kudryashev[1], A. Sartori[2], S. Bohn[3], M. Cyrklaff[1] and F. Frischknecht[1]

1. Department of Parasitology, Hygiene Institute, University of Heidelberg Medical School, INF 324, 69120 Heidelberg, Germany
2. Institut Pasteur, Imagopole, Dynamic Imaging and Ultrastructural Microscopy Platform, 25-28 rue du Dr. Roux, 75015 Paris, France
3. Department of Molecular Structural Biology, Max Planck Institute for Biochemistry, Am Klopferspitz 18, 82152 Martinsried, Germany

Simone.Lepper@med.uni-heidelberg.de
Keywords: Correlation, Cryo-Electron Tomography, Malaria

It has always been the aim of functional studies in biology to establish structure-function relationships of the molecular machineries involved in the process of interest. A powerful tool within this field has always been electron microscopy continuously adapting to answer latest questions [1, 2] and therefore, principally shaping our view on cell architecture. Correlative approaches now allow to bridge the gap of information by combining dynamic data obtained by fluorescent light microscopy (LM) with the structural information of cryo-electron microscopy (CEM) or cryo-electron tomography (CET) and *vice versa. Plasmodium berghei* sporozoites are an ideal system for such studies. They are the reactive agents transmitting malaria from mosquitoes to rodents, and due to their close relation to the human pathogen *Plasmodium falciparum* they are widely used as a model organism. Since they are only 1 μm thick they can be viewed *in toto* with ET. The sporozoites represent both, a motile, highly polarized eukaryotic cell and a devastating pathogen.

A correlative approach is opening new doors for the understanding of pathogenesis and dynamic cell architecture. Not only the spatial organisation of the cell organelles remains barely understood, but so does also one crucial factor for pathogenesis: the motility of the parasite. Unlike the immobile sporozoites from the mosquito midgut, only the motile ones isolated from the salivary glands of the mosquito are infectious [3]. During a mosquito bite a very variable number of parasites is transmitted [4]. Any regulation of this process is unrecognized, but might be triggered by the spatial organisation within the salivary glands. Once inoculated into the dermis the parasite uses his highly adapted gliding machinery to start his destructive journey through the host. Some key players of the gliding machinery are characterized, but their dynamic interaction stays largely unclear. Various drugs such as cytochalasin D and jasplakinolide inhibit actin dynamics and modulate the motility of the parasite [5]. The distribution of actin is reorganized under the influence of jasplakinolide (Figure 1A, B), a drug that stabilizes actin in the filamentous form. However, the better characterization of the role of actin within the gliding machinery is still necessary. In order to answer this question the behaviour of sporozoites expressing fluorescently tagged actin was monitored by LM (Figure 1A, B) and the different, drug-concentration dependent phenotypes were quantified. Further structural investigations have been hampered so far

A. Aretz, B. Hermanns-Sachweh, J. Mayer (Eds.): EMC 2008, Vol. 3: Life Science, pp. 85–86,
DOI: 10.1007/978-3-540-85228-5_43,

due to the known shortness of the actin filaments and because only a minor part of the total actin pool of the cell occurs as filaments [5].

Our aim now is to complete the description of the different phenotypes with structural information on the EM level. *Plasmodium* sporozoites are drug treated and filmed on EM grids in LM followed by plunge freezing in liquid ethane within a few seconds. This preserves even sensitive structures like the gliding machinery (Figure 1C), thus making higher resolution data accessible (Figure 1D, E, F). To investigate the spatial organisation within the salivary glands and of certain organelles, vitreous cryosections are obtained from intact salivary glands and fluorescently tagged organelles are retrieved in cryo-LM and CEMs are taken. This leads to a new understanding of host-vector interaction and gives new insights into the highly adapted, but still unravelled, architecture of *Plasmodium* sporozoites.

1. W. Baumeister, FEBS **579** (2005) 933.
2. A. Al-Amoudi, J.J. Chang, A. Leforestier, A. McDowall, L. M. Salamin, L . P. O. Norlén, K. Richter, B. N. Sartori, D. Studer and J. Dubochet, EMBO Journal **23** (2004) 3583.
3. K. Matuschewski, Journal of Cellular Microbiology **8** (2006) 1547.
4. F. Frischknecht. P. Baldacci, B. Martin, C. Zimmer, S. Thiberge, J. C. Olivo-Marin, S.L. Shorte and R. Ménard, Cell Microbiol. **6** (2004) 687.
5. D. M. Wetzel, S. Håkansson, K. Hu, D. Roos, L. D. Sibley, Molecular Biology of the Cell **14** (2003) 396.
6. H. Schüler, A.-K. Müller, K. Matuschewski, FEBS Letters **579** (2005) 655.
7. Funding by the German Ministry of Education and Research (BMBF: BioFuture) and the German Research Foundation (DFG: SPP 1128) are acknowledged.

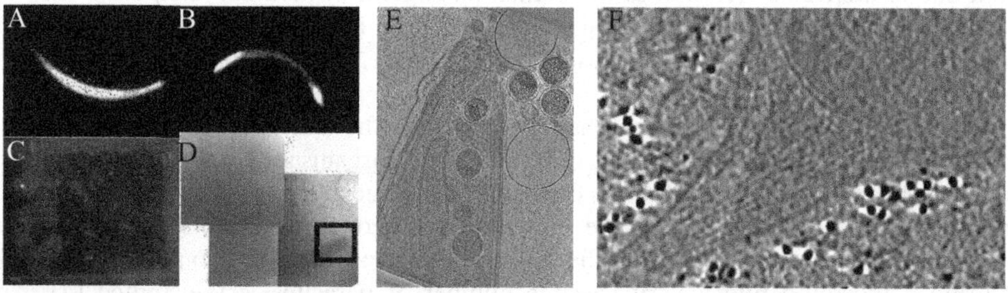

Figure 1. Massive rearrangement of the spatial actin distribution within the sporozoite occurs under jasplakinolide (A and B, fluorescent LM of a sporozoite expressing mCherry tagged actin treated with jasplakinolide (B) and without jasplakinolide (A)). The fluorescent signal can be preserved in cryo-LM (C) allowing the access to higher resolution data (D, low magnification CEM). Parts of the sporozoite of special interest can be further examined and CETs are taken (E and F, cross-sections from a tomogram of the apical (E) and back (F) end of a jasplakinolide treated sporozoite; arrowheads indicate the actin filaments).

3DEM-analyses of Golgi apparatus subunits during physiologic and pathologic reorganizations

J. Neumüller, M. Vetterlein, C. Meißlitzer-Ruppitsch, C. Weiss, A. Ellinger and M. Pavelka

Department of Cell Biology and Ultrastructure Research, Center for Anatomy and Cell Biology, Medical University Vienna, Schwarzspanierstraße 17, A-1090 Vienna, Austria

margit.pavelka@meduniwien.ac.at
Keywords: Golgi apparatus, electron tomography, cell dynamics

The Golgi apparatus is a dynamic organelle that rapidly reorganizes upon changes of functional states, and environmental conditions. Physiologic reorganizations take place continuously with secretory and endocytic traffic into and across the stacks of Golgi cisternae [1, 2]. We show here 3D-electron microscopic (3DEM) analyses of newly formed and transformed subcompartments of the Golgi apparatus during endocytosis of Wheat germ agglutinin (WGA 33μg/100μl, labelled with peroxidase and visualized electron microscopically by means of oxidation of diaminobenzidine) in HepG2 hepatoma cells within internalization periods of 10 to 60 min. WGA is an N-acetyl-glucosamine and sialic acid-specific lectin, which is known for many years to be transported to the Golgi apparatus in retrograde direction [3], and also is used for developments of drug delivery systems [4]. Following internalization, WGA is rapidly transported to a cell region close to the Golgi apparatus. Early globular endosomes accumulate at the *trans*-Golgi side, and an endocytic *trans*-Golgi network (endocytic TGN) is formed. Portions of the endocytic TGN attach to *trans*-Golgi cisternae, thus becoming integrated parts of the Golgi apparatus stacks. The *trans*-Golgi attachment of the endocytic TGN leads to interconnections of the small Golgi stacks, and causes the formation of Golgi apparatus ribbons. Three-dimensional reconstructions of the endocytic TGN following different preparation methods show that this compartment consists of globular and cylindrical pieces, which in dimensions and shapes resemble the earlier globular endosomes, and which are interconnected by cristae and fine bridges.

The 3DEM analyses of Golgi subunits revealed that pathologic Golgi apparatus transformations due to depletion of the cytoplasmic ATP pool [5] in part show similarities with physiologic Golgi apparatus reorganizations. For ATP-depletion, HepG2 hepatoma cells were washed in glucose-pyruvat-free Dulbecco's medium containing 1% dialysed FBS (GPF medium), and incubated at 37°C in the same medium containing 50mM desoxy-glucose (DOC medium) for different times up to 60min. To restore ATP levels, the cultures were washed again 3 times in GPF medium to remove remnants of DOC and subsequently, were incubated for 15, 30, 45, 60 or 90 minutes at 37°C in GPF medium with 1M glucose and 100mM pyruvat added (GLUC medium). Following chemical fixation, or high pressure freezing and freeze substitution, the cultures were embedded in Epon. Upon depletion of the cytoplasmic ATP pool, the

A. Aretz, B. Hermanns-Sachweh, J. Mayer (Eds.): EMC 2008, Vol. 3: Life Science, pp. 87–88, DOI: 10.1007/978-3-540-85228-5_44,

Golgi apparatus looses the regular stacked organization of the cisternae, and transforms in glomerular-like organelles. At early stages of transformation, the cisternae reorganize; they detach from each other, form branches, arches, and helical pieces, which twist through cisternal windows, thus connecting different parts of the stacks. Figure 1a-d shows 4 different prospects of a 3D-model of a Golgi apparatus stack at an early transformation stage. Regularly organized Golgi apparatus stacks are reconstituted during replenishment of the ATP-pools.

1. M. Vetterlein, A. Ellinger, J. Neumüller and M. Pavelka, Histochem Cell Biol **117** (2002), p.143.
2. M. Pavelka, J. Neumüller and A. Ellinger, Histochem Cell Biol **129** (2008), p.277.
3. N.K. Gonatas, S.U. Kim, A. Stieber, and S. Avrameas, J Cell Biol **73** (1977), p.1.
4. A. Weissenböck, E. Bogner, M. Wirth and F. Gabor, Pharma Res **21** (2004), p. 1917.
5. M. del Valle, Y. Robledo, and I.V. Sandoval, J Cell Sci **112** (1999), p. 4017.
6. The authors gratefully acknowledge the valuable technical assistance of Mrs.Elfriede Scherzer, Mrs.Beatrix Mallinger, Mr.Peter Auinger, and Mr.Ulrich Kaindl.

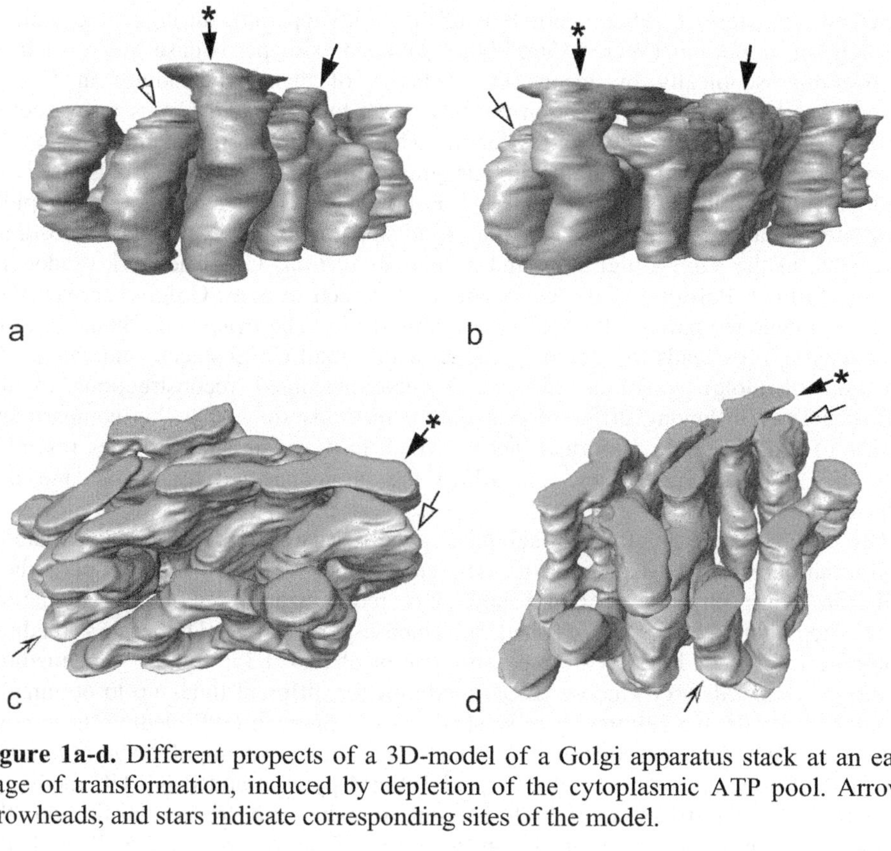

Figure 1a-d. Different propects of a 3D-model of a Golgi apparatus stack at an early stage of transformation, induced by depletion of the cytoplasmic ATP pool. Arrows, arrowheads, and stars indicate corresponding sites of the model.

Visualization of the 80s ribosome *in situ* using cryo electron tomography of vitreous sections

J. Pierson[1], J.J. Fernández[2,3], E. Bos[1], H. Gnäegi[6], J.L. Carrascosa[2], M. Vos[1] and P.J. Peters[1]

1. Division of Tumor Biology, The Netherlands Cancer Institute, Plesmanlaan 121, 1066 CX Amsterdam, The Netherlands
2. Centro Nacional de Biotechnologia-Conejo Superior de Investigaciones Cientificas, Campus Universidad Autonoma de Madrid, Cantoblanco, 28049 Madrid, Spain.
3. Department of Computer Architecture and Electronics. University of Almeria. 04120 Almeria. Spain.

J.pierson@nki.nl
Keywords: Cryo Electron Microscopy, Tomography, Volumetric Averaging, 80s Ribosome, TOVIS, CEMOVIS.

The ribosome is a structurally complex biological machine that is responsible for protein biosynthesis. Elucidation of the ultra-structure has been a formidable challenge in order to provide valuable insights into the mechanism of translation. Often, imaging has been associated with isolation procedures using detergents [1], which can possibly introduce structural rearrangement of ribosomal components and/or associated artifacts.

Here we have used cryo-electron tomography (Cryo-ET) [2] of vitreous sections [3] to explore the close-to-native structure of the 80s eukaryotic ribosome *in situ,* not as an isolated entity. Novel vitreous sectioning procedures allowed superior alignment of tilted images using fiduical markers applied to the carbon film resulting in high-resolution tomographic reconstructions.

Furthermore, using novel volumetric averaging procedures we have constructed a ~35Å resolution density map of the 80s ribosome, which was utilized for template matching with single particle and X-ray crystal structures. Finally, we have placed the density map back into the original reconstructed volume to create a "pseudo" atomic atlas of the 80s ribosome within a vitreous section.

Interestngly, ribosome pockets were observed that assumed a circular orientation, or putative polysomes. Furthermore, a dense structure, absent of a clear membrane, was seen in the center of many of the putative polysomes.

1. Gabashvili, I.S. et al. Cell. **100(5),** (2000), p. 537-49.
2. Baumeister, W. FEBS Lett. **579(4),** (2000), p. 933-7.
3. J. Dubochet, J.J. Adrian, J.C. Chang et al., Q. Rev. Biophys. **21** (1988), p. 129-228.
4. We kindly acknowledge Victoria Menéndez Benito, Bram Koster, Nico Sommerdijk, and Bert de With

A. Aretz, B. Hermanns-Sachweh, J. Mayer (Eds.): EMC 2008, Vol. 3: Life Science, pp. 89–90,
DOI: 10.1007/978-3-540-85228-5_45,

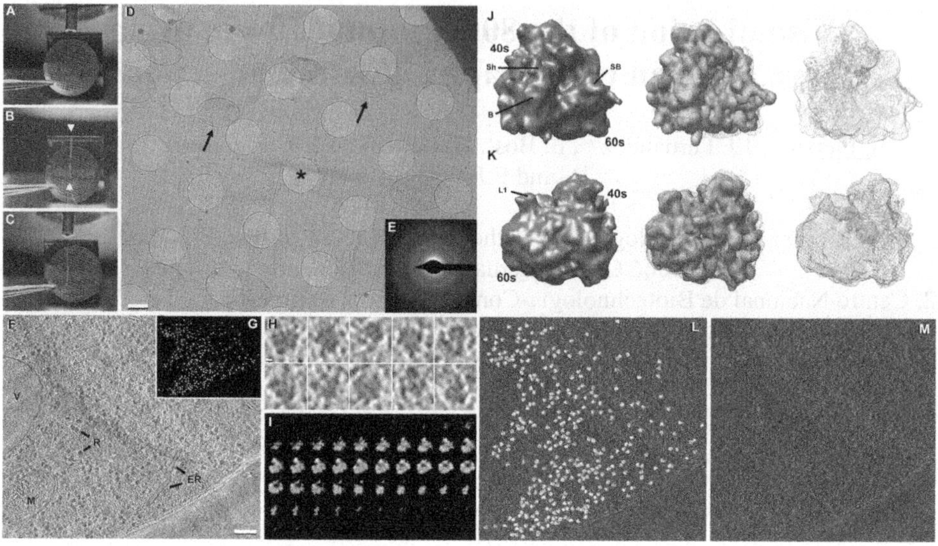

Figure 1. Cryo Electron Tomography of Vitreous Sections. A) The copper tube containing the vitreous sample of *S.cerevisiae* can be seen at the top of the picture. **B)** A ribbon of vitreous sections is guided over the EM grid, the ionizer is switched off and the ribbon is charged onto the EM grid **(C)**. **D)** *S.cerevisiae* are scattered throughout the vitreous section (Black arrows). Bar = 2μm. A hole in the C-Flat carbon film **(*)** can be seen in the MMM. **F)** ~5nm thick central slice from a tomographic reconstruction (defocus -5μm) showing the sub-cellular organization. On the left side of the micrograph a vacuole (V) and a mitochondrion (M) can be seen, ribosomes are deposited throughout the cytosol (R), and Endoplasmic Reticulum profiles are also present (ER). Bar = 100nm. **G)** Semi automatic particle selection of the 80s ribosome, each round contour represents a ribosome with the tomogram **(F)**. **H)** A small gallery of individual ribosome sub-volumes. Bar = 10nm. **I)** A gallery of slices through the un-biased averaged ribosome density, the resolution was estimated at 34Å according to the FSC. **J)** An isosurface representation of the ribosome density map (Purple colored density map). Both subunits, 40s and 60s, can be discerned and also important ribosomal landmarks, such as shoulder (Sh) and the base (B) of the 40s and the stalk base (SB) and L1 protuberance of the 60s in **(K)**. Single particle data was docked into the (Purple) TOVIS density map (Aqua Blue), along with a theoretical high-resolution model (Blue and Green x-ray structure into a transparent density map). **L) 80s Ribosome Molecular Atlas**. The tomogram from **(F)** with the averaged ribosome located in the positions from where they were picked up, and in the orientations found by the alignment. **M)** Weighted ribosomes according to the cross-correlation coefficient obtained with the alignment, the lighter the ribosome the higher the correlation to the averaged density.

Three-dimensional analysis of the intermediate filament network using SEM-tomography

M. Sailer[1], S. Lück[3], V. Schmidt[3], M. Beil[2], G. Adler[2] and P. Walther[1]

1. Electron Microscopy Facility, University of Ulm, D-89069 Ulm, Germany
2. Department of Internal Medicine I, University Hospital Ulm, D-89070 Ulm,
3. Department of Stochastics, University of Ulm, D-89069 Ulm, Germany

michaela.sailer@uni-ulm.de
Keywords: cytoskeleton, intermediate filaments, SEM-tomography

Pancreatic cancer is characterized by its early infiltration of adjacent tissue. There is a correlation between the keratin filament network structure and the migration ability of cells [1]. In earlier papers we quantitatively investigated two-dimensional images of the three-dimensional keratin network [2]. It is the goal of our ongoing research to expand these studies to three-dimensional datasets. We investigated, therefore, intermediate filaments from detergent-extracted cells by SEM-tomography.

The term *SEM-tomography* sounds contradictory, since the SEM maps physical *surfaces*, whereas tomography stands for the analyzation of a *volume*. The clue is the extraction protocol that is widely applied in cytoskeletal research. Most of the cellular structures are, thereby, removed and only the finely woven cytoskeleton remains, with the filaments surrounded by vacuum. (Possible artifacts of this preparation protocol have also been discussed [3].) This network can be imaged in an SEM using the standard secondary electron signal and, as can be seen in Fig. 1, also filaments in some depth of the cell are clearly visible. A tomographical dataset is recorded by tilting the sample, similar to TEM-tomography. The tilt series is then back-projected into a three-dimensional model [4] (Fig. 2).

Methods. Panc 1 cells were grown as a monolayer on gold EM grids coated with formvar and carbon. The cells were then extracted with 1% Triton X-100. The samples were afterwards chemically fixed and contrasted with OsO_4 and uranyl acetate, critical point dryed and coated with a thin (5 nm) layer of carbon. Tomograms of the intermediate filament network were recorded in a S-5200 in-lens SEM (Hitachi, Tokyo) at 30 kV using the secondary electron signal [5] at tilt angels of -40° to +40° with 2° step size and reconstructed using the iMod software. [6]

In future work we plan to increase the tilting angle to reduce the missing wedge problem, in order to make the data amenable to mathematical modeling.

1. M. Beil et al., Nature Cell Biology **5** (2003), 803-811.
2. M. Beil et al., J Microsc **220** (2005), 84-95.
3. G. P. Resch, et al., Cell Science **115** (2002), 1877-1882.
4. M.S. Ladinsky, D. Mastronarde et al., J Cell Biology **144** (1999), 1135-1149.
5. P. Walther in: Biological Low-Voltage Scanning Electron Microscopy. Schatten H, Pawley J. B. eds. (Springer New York) (2008), 245-261.
6. This work is supported by the *DFG Sonderforschungsbereich 518*, project B21.

A. Aretz, B. Hermanns-Sachweh, J. Mayer (Eds.): EMC 2008, Vol. 3: Life Science, pp. 91–92,
DOI: 10.1007/978-3-540-85228-5_46,

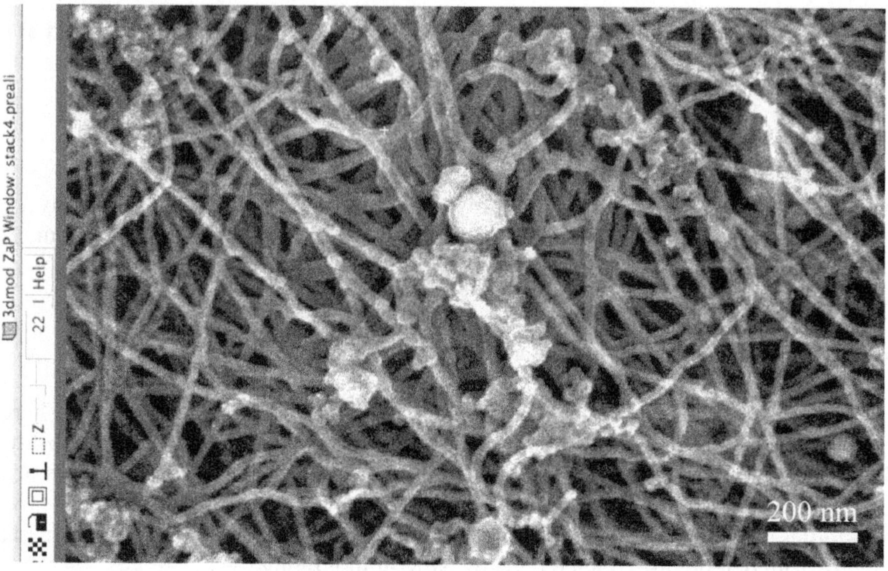

Figure 1. One SEM image of an aligned +- 40° tilt series of the intermediate filament network. The thin filaments occur with high contrast in the secondary electron image. Filaments in the depth of the sample are imaged too, although with less contrast than filaments at the periphery.

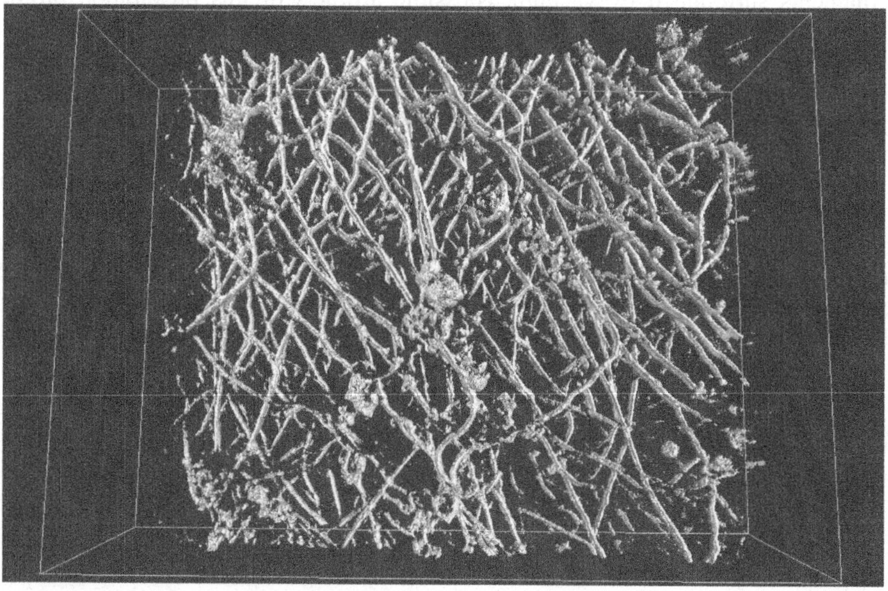

Figure 2. Reconstructed volume from the above tilt series (by weighted back projection using iMod) visualized with an isosurface, using amira.

The apical vesicles of the tuft cells in the main excretory duct of the rat submandibular gland by EFTEM-TEM tomography

A. Sato[1], J. Kodama[1], Y. Inoue[2], Y. Sawa[1], T. Oikawa[3]

1.Biological Morphology (Functional structure)
2.Oral growth & Development (Orthodontics), Fukuoka Dental College
3.France JEOL Ltd.

sato@college.fdcnet.ac.jp
Keywords: tuft cells, apical vesicles, EFTEM-TEM tomography

Tuft cells exist in the hallow organs such as gastro-intestinal tract including the ducts of the rat salivary gland. They are characterized by an abundance of vesicles in apical cytoplasm and the prominent microvilli [1, 2]. The number of apical vesicles changes according to the length of the microvilli. The precise morphology of the apical vesicles and their function is not yet entirely clear. This study investigated the three-dimensional structure of the apical vesicles of tuft cells in the main excretory duct (MED) of the rat submandibular gland using a combination of transmission electron tomography and computer modeling.

MEDs were obtained from 9-week-old starvation- and isoproterenol (IPR) (30 mg/kg body weight)-treated male Wistar rats. Semi-thin sections (0.3 μ m) were prepared by standard TEM techniques. Sections were examined with JEM-3200FS-TEM tomography (JEOL). The 3-D images were analyzed using Tri-3D Bon software (Ratoc System Engineering Co., Ltd., Tokyo).

Treatment for two days with IPR resulted in elongation of the microvilli and a decrease in vesicles. In contrast, the length of microvilli shortened in the rats starved for two days, and the number of vesicles increased. Three-dimensional reconstruction of the apical vesicles demonstrated that adjacent vesicles were partially connected and appeared to be coherent vesicles [2, 3]. The 3-D images of these coherent vesicles analyzed using Tri-3D Bon software, which express the continuous area in same color, revealed that the coherent vesicles connected by very narrow tubules formed a network (Figure 1). In cases of shortened microvilli separated vesicles increased and there were few coherent vesicles (Figure 2).

The network of apical vesicles seems to supply the membranous constituent, which elongates the microvilli.

1. A. Sato, S. Miyoshi, Anat Rec 248 (1997), 325-331.
2. A. Sato, Anat Sci Inter 82 (2007), 187-199.
3. A. Sato, Y. Hisanaga, Y. Inoue, T. Nagato, H. Toh, Eur J Morphol 40 (2002), 235-239.
4. This study was supported by a Grant-in-Aid for Scientific Research (No. 18592024) from Japan Society for the Promotion of Science and Nanotech project of Kyushu University.

A. Aretz, B. Hermanns-Sachweh, J. Mayer (Eds.): EMC 2008, Vol. 3: Life Science, pp. 93–94,
DOI: 10.1007/978-3-540-85228-5_47,

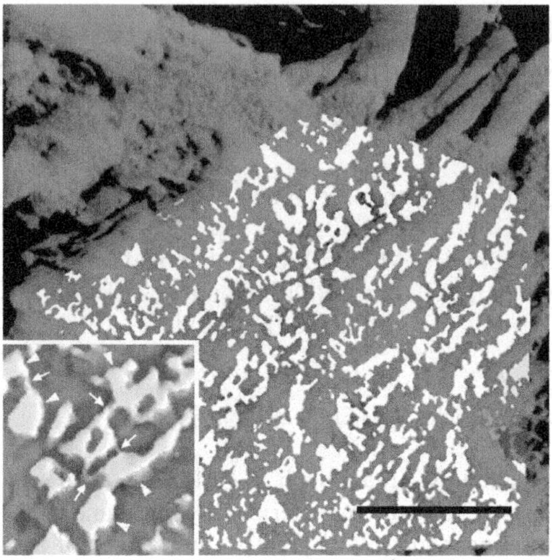

Figure 1. Apical vesicles of the tuft cell show a tubulovesicular system in case of the elongated microvilli. All most vesicles are continuous, indicated in the same color (yellow). Coherent vesicles (arrowheads) are connected by narrow tubules (arrows). Very few vesicles are dependent, indicated in the different colors. Bar, 1 μm.

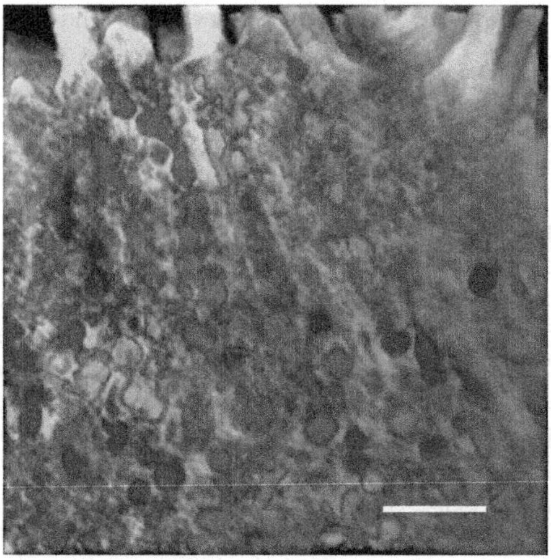

Figure 2. Apical vesicles do not show a tubulovesicular system in case of shortened microvilli. Only adjacent vesicles are continuous, indicated in the different colors (green, red, blue, pink, yellow, dark blue, etc.). Bar, 1 μm.

(S)TEM Dual Axis Tomography

Andreas Voigt, Alevtyna Yakushevska, Felix de Haas, Wim Voorhout, Remco Schoenmakers

FEI Company, Achtseweg Noord 5, 5651 GG Eindhoven, The Netherlands

andreas.voigt@fei.com
Keywords: Tomography, Xplore3D, TEM

Single axis (S)TEM Tomography data suffers from the existence of a "missing wedge" due to the geometry of samples and stage, images can usually only be acquired for tilt angles in a range of ±70°. In 3D-reconstructions this gives rise to artifacts and non-isotropic resolution and contrast. To reduce these shortcomings, samples are often rotated by about 90° in the holder and a second tilt series is acquired (= Dual Axis Tomography). The "missing wedge" is thus reduced to a "missing pyramid". FEI's Xplore3D tomography package now supports both the acquisition and the reconstruction of dual axis tomography data sets.

Data Acquisition:

The process of acquiring "dual axis tomography" data is quite tedious, the main complication being the task of finding back the sample features for which tomography data has already been acquired before the rotation. The features may be small (~100nm), have low contrast and the sample may contain many features that resemble each other. The task may be even more complicated, if the features are dose sensitive, like in cryo-tomography.

The new version of Xplore3D supports an accurate determination of specimen rotation angle and translation vector and the storage of tomography feature positions (including reference images). The features can then be relocated after specimen rotation with an accuracy of several microns. The solution integrates with the existing batch tomography solution and the TEM low dose server. The current approach consists of a series of semi manual tasks. We expect that in the future more of these tasks can be fully automated.

Volume Reconstruction:

FEI's reconstruction package Inspect3D supports alignments using cross-correlations or bead tracking. Subsequent reconstruction can be done (on the GPU if needed) using weighted back projection or SIRT. The dual axis reconstruction routines need to support all combinations of alignment routines and reconstruction algorithms. This is achieved by aligning and reconstructing the volumes separately and then finding the best match (3D shift, rotation, magnification and sheer) between the volumes before adding them in Fourier space (Mastronarde, 1997) or direct space. Finding this match is fully automatic in most cases, but can be guided by the user if needed.

In this presentation, the complete workflow for dual axis tomography will be discussed and examples will be shown that demonstrate the power of dual axis tomography.

A. Aretz, B. Hermanns-Sachweh, J. Mayer (Eds.): EMC 2008, Vol. 3: Life Science, pp. 95–95,
DOI: 10.1007/978-3-540-85228-5_48,

STEM tomography of high-pressure frozen cell monolayers

P. Walther[1], S. Landwehr[2], H. Krisp[1], T. Seufferlein[3], T. Mertens[2] and G. Adler[3]

1. Electron Microscopy Facility, University of Ulm, D-89069 Ulm, Germany
2. Institute for Virology, University Hospital Ulm, 89081 Ulm, Germany
3. Department of Internal Medicine I, University Hospital Ulm, 89070 Ulm, Germany

paul.walther@uni-ulm.de
Keywords: High-pressure freezing, freeze substitution, STEM tomography

We are investigating structures of the Golgi apparatus in relation to protein kinase D2 expression in pancreatic cells (see also Krisp et al, this issue) and membrane envelopment of capsids during viral morphogenesis in the cytoplasm of the host cells (see also Landwehr et al., this issue). Both processes are highly dynamic and for a better understanding, a three-dimensional description is preferable. For this purpose we have optimized the existing preparation and imaging protocols for electron tomography with the following goals: -Fast immobilization of a physiologically defined state. -Good preservation and good visibility of cellular components, especially of membranes. -Imaging of sections as thick as possible, so that a good three-dimensional impression of the dynamic processes can be achieved.

The cell monolayers were cultivated on carbon coated sapphire discs which can be used for light microscopical analysis as well and which fit into a Compact 01 high-pressure freezer (Wohlwend GmbH, Sennwald, CH). The frozen samples were then freeze substituted in acetone, uranylacetate, (with or without osmium) and 5% of water to enhance membrane visibility [1] and embedded in Epon or LR-gold. 350 to 520 nm thick sections were placed on naked EM grids; 10 nm colloidal gold as fiducial marker was brought onto both surfaces with the help of poly-lysine. Tilt series (-72° to +72°) were recorded in a 300 kV field emission STEM [2] (FEI, Eindhoven) with a high angle annular dark field detector (Fischione, Export, PA, USA). Tomograms were reconstructed with iMod by weighted back projection (Fig. 1 and 2) and for visualization the software Amira was used (Fig. 3). The figures represent portions of cells infected with human cytomegalovirus (HCMV). Fig. 1 is a Golgi apparatus in a 400 nm thick section, reconstructed with iMod, Fig. 2 shows a higher magnification of a similar tomogram of the same sample that demonstrates the good visibility of membranes due to the freeze substitution protocol and the dark-field STEM imaging. In Fig. 3 some capsids, teguments and some membranes have been visualized with the help of the software Amira.

1. A. E. Yakushevska et al., J Struct Biol. **159** (2007), p. 381-391.
2. P. Walther and A. Ziegler, J Microsc. **208** (2002), p. 3-10.
3. This work is supported by the *DFG Sonderforschungsbereich 518*, project *A15* and the *Schwerpunktprogramm 1175*.

A. Aretz, B. Hermanns-Sachweh, J. Mayer (Eds.): EMC 2008, Vol. 3: Life Science, pp. 97–98,
DOI: 10.1007/978-3-540-85228-5_49,

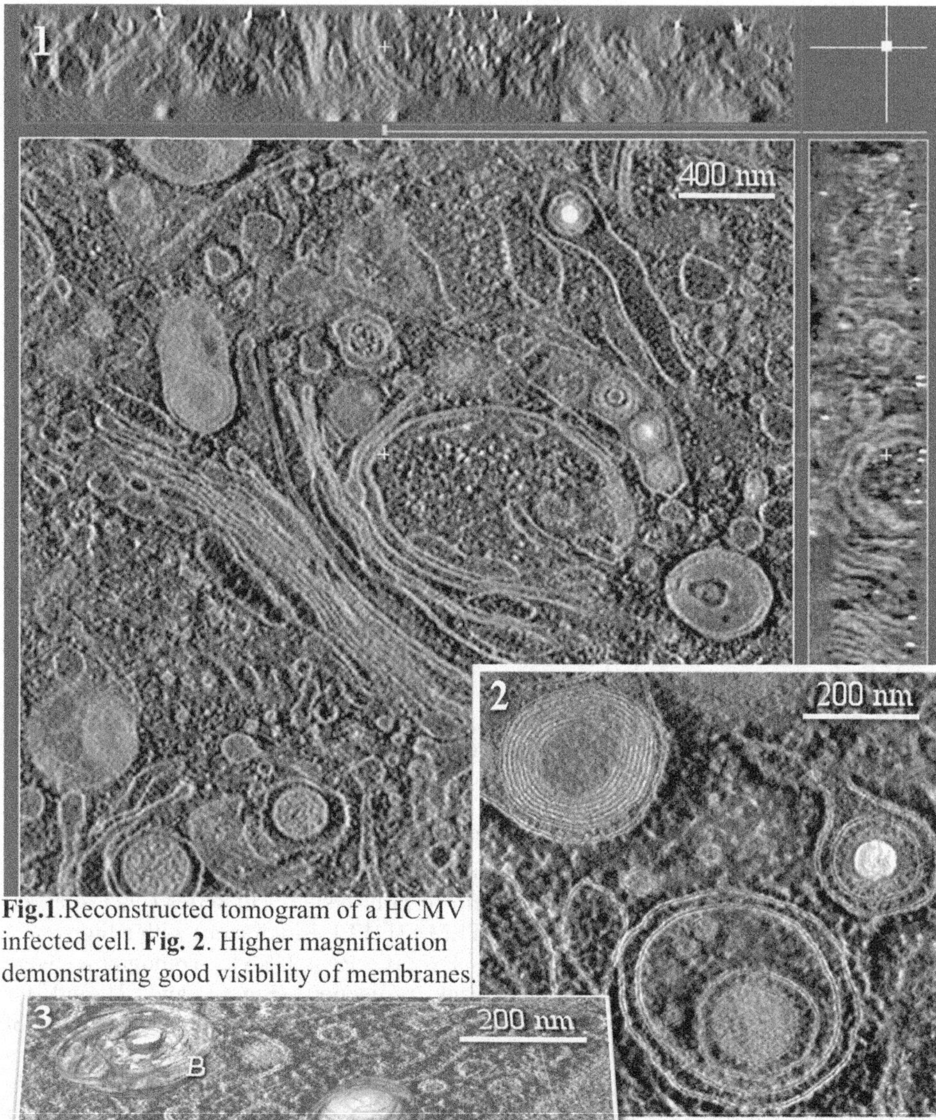

Fig.1.Reconstructed tomogram of a HCMV infected cell. **Fig. 2**. Higher magnification demonstrating good visibility of membranes.

Fig. 3. Visualization of not yet enveloped capsids with teguments (A), a partially enveloped capsid (B), and fully enveloped virions in vesicles (C1 to C3).

Multimodality of pericellular proteolysis in cancer cell invasion

Peter Friedl

Department of Cell Biology, Nijmegen Center for Molecular Life Science, Radboud University Nijmegen, The Netherlands and
Rudolf Virchow Center, DFG Research Center for Experimental Biomedicine, University of Würzburg, Germany

Peter.friedl@gmx.de
Keywords: proteolysis, cancer

For cancer progression and metastatic dissemination, cancer cells migrate and penetrate through extracellular tissues. Cancer invasion is frequently facilitated by proteolytic processing of components of the extracellular matrix (ECM)[1]. The cellular regions executing surface-localized proteolysis of tissue components are diverse and depend upon the physical structure, composition, and dimensionality of the ECM. Multimodal life- and fixed cell imaging, i.e. combining confocal fluorescence with reflection or multiphoton-excited fluorescence with second harmonic detection now permits to reconstruct proteolytic cell-matrix interactions in time and space at subcellular resolution[2]. Proteolytic invasion through 3D fibrillar ECM, as monitored for HT1080 fibrosarcoma and MDA-MB-231 breast carcinoma cells is mediated by several types of lytic structures rich in filamentous actin, ß1 integrin, and MT1-MMP with distinct location and function. These comprise (i) zones of local cell compression by constraining collagen fibers, i.e. between anterior branching pseudopodia and nearby the forward-moving nucleus mediating step-wise ECM fiber cleavage and realignment; (ii) lateral small spikes that protrude into the ECM and cause small spot-like lytic zones; and (iii) a strongly proteolytic cell rear sliding along reorganized ECM fibers [3,4]. Through their combined action these proteolytic surface structures are suited to remove ECM barriers, support rear end retraction, generate tube-like matrix defects and laterally widen existing tracks during 3D tissue invasion (Figure 1).

1. Wolf, K. & Friedl, P. Tube travel: protease functions in individual and collective cancer invasion. *Cancer Res.* **(in press)**(2008).
2. Friedl, P., Wolf, K., von Andrian, U.H. & Harms, G. Biological second and third harmonic generation microscopy. *Curr. Prot. Cell Biol.* **4.15**, 1-21 (2007).
3. Wolf, K. et al. Multi-step pericellular proteolysis controls the transition from individual to collective cancer cell invasion. *Nat Cell Biol* **9**, 893-904 (2007).
4. Wolf, K. & Friedl, P. Mapping proteolytic cancer cell-extracellular matrix interfaces. *Clin. Exp. Metast.* **(in press)**(2008).

A. Aretz, B. Hermanns-Sachweh, J. Mayer (Eds.): EMC 2008, Vol. 3: Life Science, pp. 99–100,
DOI: 10.1007/978-3-540-85228-5_50, © Springer-Verlag Berlin Heidelberg 2008

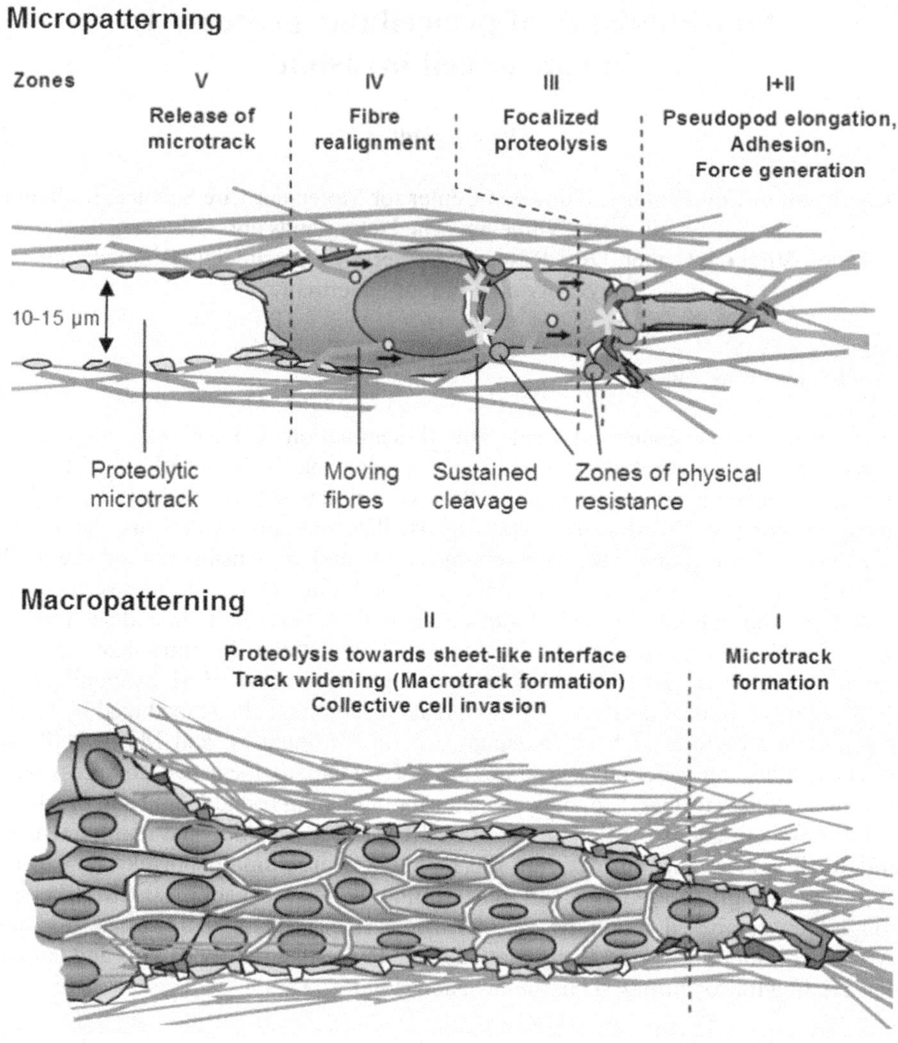

Figure 1. Individual and collective tissue invasion are mediated by two distinct types of pericellular proteolysis. (A) Single-moving proteolytic cells cause ECM micropatterning. After protrusion of an anterior pseudopod (step I) and traction force generation (step II), focal cleavage of individual ECM fibres is executed slightly backward to leading adhesion sites (step III). Transport of loose fibre ends (step IV) results in a small microtrack detected upon forward gliding of the cell rear (step V). (B) ECM marcopatterning is executed by multiple cells that collectively fill a small pre-existing tissue gap while remaining connected. By focalizing proteolytic activity towards the cell-ECM interface, a near-continuous ECM layer is generated and further cleaved. Colors and symbols: proteases and MT1-MMP (yellow), degraded collagen (green), ß1 integrin (blue), F-actin (red).

Photonic crystal type nanoarchitectures in butterfly wing scales

[1]Z. Vértesy, [1]K. Kertész, [2]Zs. Bálint, [1]G. Márk, [1]L.Tapasztó, [3]J.P. Vigneron, [1]L.P. Biró

1. Research Institute for Technical Physics and Materials Science, H-1525 Budapest, P.O.Box 49, Hungary, (http://www.nanotechnology.hu/)
2. Hungarian Natural History Museum, H-1088 Budapest, Baross str. 13, Hungary
3. Facultés Universitaires Notre-Dame de la Paix, LPS, 61 rue de Bruxelles, B-5000 Namur Belgium

vertesyz@mfa.kfki.hu
Keywords: natural photonic crystals, wing scales, structural color

Photonic crystals also called Photonic Band Gap (PBG) materials – are periodic dielectric composite structures, which affect the propagation of the light. These materials are intensively studied since two decades [1, 2], because of perspectives of their applications in optical devices, textiles and colorants. The diversity of photonic crystal type nanoarchitectures found in Nature is also in the focus of researchers' attention. Living organisms, during millennia of evolution, achieved very complex and highly optimized micro- and nanostructures not only for interaction with visible light but also for other specified functions as for example UV protection, thermal regulation or adhesion. A very remarkable attribute is that these nanostructures are produced by a limited range of materials, which offers only a moderate refractive index contrast (1/1.6). The investigation of natural nanostructures leads to the knowledge that can be applied in the design of artificial biomimetic materials with photonic properties and can reduce the engineering effort in designing these materials and costs of their fabrication.

Butterflies wing scales are excellent examples for natural PBG composites that consist of chitin and air i.e. materials with very moderate refractive index contrasts. Scales cover dorsal and ventral surfaces of wings and their coloration is generated in two ways: by pigmentation (chemical color) or/and by structure (structural or physical color). Blue and green colors of scales originate from 3D nanoarchitectures found in the volume of scales, as the green and blue pigments do not exist in butterfly wings.

The aim of our study is to find and understand the dependence between structural colors and nanoarchitectures of wing scales. In order to acquire the highest possible amount of information we perform complex investigations: characterization by means of the light, scanning (SEM) and transmission electron microscopy (TEM), spectral measurement followed by numerical simulations.

Wing scales display a huge diversity of nanostructures; here only some examples will be discussed.

1. The role of photonic crystal type structures in the thermal regulation [3] will be demonstrated with an example of Lycaenid butterfly sister species pair *Polyommatus daphnis* and *Polyommatus marcidus* living in different habitats at low and high altitudes.

A. Aretz, B. Hermanns-Sachweh, J. Mayer (Eds.): EMC 2008, Vol. 3: Life Science, pp. 101–102,
DOI: 10.1007/978-3-540-85228-5_51,

2. The Brazilian butterfly *Cyanophrys remus* is a peculiar case when from only two materials in the same species poly- and single-crystalline photonic structures are obtained [4]. This species displays different photonic-crystal-type structures on the ventral and dorsal surfaces of their wings. Matt green color on the ventral surface of wing is generated by photonic polycrystalline nanoarchitectures (shown in Figure 1 a and 1c), while metallic blue color on the dorsal surface by photonic single crystals (shown in Figure 1 b and 1d).

3. Many species belonging to the family Lycaenidae have blue coloration of different hues. Correlations between blue color and nanostructure of the scales will be demonstrated with an example of 4 species belonging to the subfamily Lycaeninae [5].

1. E. Yablonovitch, Phys. Rev. Lett. **58**, (1987) p. 2059-
2. J. D. Joannopoulos, R. D. Meade, and J. N. Winn, Molding the Flow of Light Princeton University Press, Princeton, 1995
3. L. P. Biró, Zs. Bálint, K. Kertész, Z. Vértesy, G. I. Márk, Z. E. Horváth, J. Balázs, D. Méhn, I. Kiricsi, V. Lousse, and J.-P. Vigneron., Phys. Rev. E 67, (2003) p. 021907.
4. K. Kertész, Zs. Bálint, Z. Vértesy, G. I. Márk, V. Lousse, J.-P. Vigneron and L. P. Biró, Phys. Rev. E 74, (2006) p.021922.
5. Z. Vértesy, Zs. Bálint, K. Kertész, J.-P. Vigneron, V. Lousse and L. P. Biró, Journal of Microscopy, Vol. 224, (2006), pp. 108–110
6. The work was supported by Hungarian Scientific Research Fund OTKA-NKTH K 67793 and EU FP6 NEST/PATHFINDER/BIOPHOT- 012915 project.

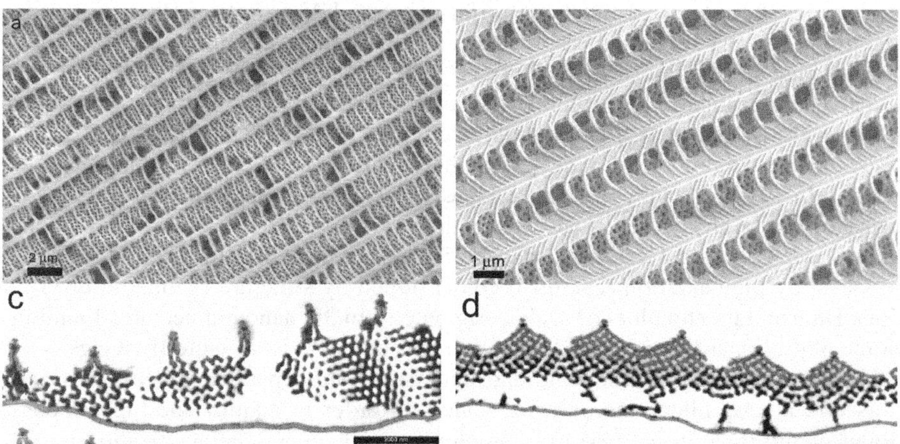

Figure 1. Electron microscopic images of the wing scales of the *Cyanophrys remus*. SEM micrograph of the photonic crystal grains with random orientation (a) and cross-sectional TEM micrograph (c) of scales covering ventral surface of wing. SEM micrograph of the photonic crystal ordered structure (b) and cross-sectional TEM micrograph (d) of scales covering dorsal surface of wing.

Iridovirus-like viruses of *Lacerta monticola* from Serra da Estrela, Portugal

A.P. Alves de Matos[1], M.F. Caeiro[2,3], Guilherme Costa[2,3], Teresa Granja[3], Tibor Papp[4], Rachel Marschang[4]

1. Dental Medical School, Lisbon University and Anatomic Pathology Department, Curry Cabral Hospital, 1069-166 Lisbon, Portugal.
2. University of Lisbon, Faculty of Sciences, Dept. of Plant Biology, Portugal.
3. University of Lisbon, Faculty of Sciences, Environmental Biology Centre, Portugal.
4. Institut für Umwelt- und Tierhygiene, Hohenheim University, Garbenstr. 30, 70599 Stuttgart, Germany.

apamatos@kanguru.pt
Keywords: Iridoviruses, Lizards, Portugal

Lizard erythrocytic viruses (LEVs) were previously found with high prevalence in mountain areas of Portugal in three different species of lizards. Similar viruses are known to infect other reptiles, amphibians and fish [1]. A previous attempt to adapt these viruses to cell cultures resulted in the isolation of a morphologically different *Ranavirus* of uncertain significance [2]. In this work we report molecular data obtained from blood samples of *Lacerta monticola* captured in Serra da Estrela, Portugal, that confirms the presence of potentially pathogenic members of the genus *Ranavirus*, family *Iridoviridae* in this lizard population.

Infected lizards were detected by light microscope examination of Giemsa-stained smears of blood from clipped tips of tails. A drop of blood from animals displaying red inclusions in the cytoplasm were processed for transmission electron microscopy using previously described procedures [1] for detection of virus particles. DNA extracted from infected blood was used for PCR using primers for the conserved regions of the major capsid protein (MCP) of the genera *Ranavirus* and *Iridovirus*. The PCR product obtained with the PCR reaction for *Ranavirus* was purified and sequenced using the same primers.

LEV infected erythrocytes displayed large cytoplasmic inclusions containing Iridovirus-like particles 220nm in diameter. The virions had a complex nucleoid consisting of a network of dense strands (Fig. 1). DNA extracted from the erythrocytes of one LEV infected and one uninfected lizard were positive for Iridoviruses of the genus *Ranavirus* and negative for the genus *Iridovirus* (Fig. 2). The sequence of the PCR product differed from the published sequence of the type species FV3, in only five base pairs. This sequence was identical to the sequence of a *Ranavirus* previously isolated from a different lizard of the same species and provenience (fig. 3).

Our results confirm the presence of a *Ranavirus* infecting the *Lacerta monticola* population of Serra da Estrela. This virus was designated as LMV. It is highly unlikely that these viruses are related to the larger LEVs detected in the same animals by light and electron microscopy. Further studies are under way to characterize the LEVs.

A. Aretz, B. Hermanns-Sachweh, J. Mayer (Eds.): EMC 2008, Vol. 3: Life Science, pp. 103–104, DOI: 10.1007/978-3-540-85228-5_52,

1. A.P. Alves de Matos, I. Paperna and E. Crespo, Intervirology **45** (2002) p. 150.
2. F. Caeiro, T. Papp, D. Correia *et al.* (2006) XVIth IMC. Sapporo, Japan.

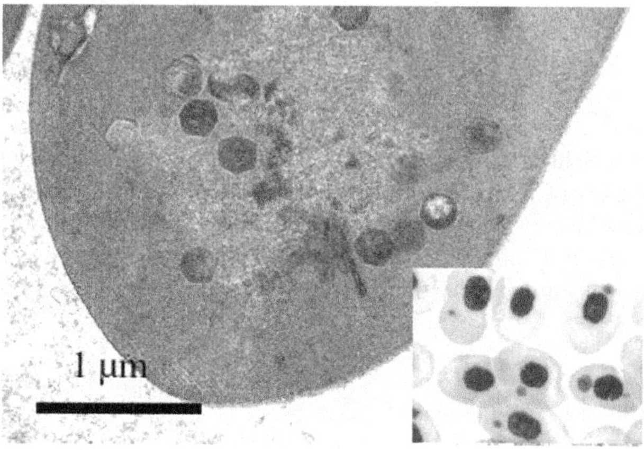

Figure 1. Erythrocytes infected with LEVs. Inset – inclusions in Giemsa-stained blood smears.

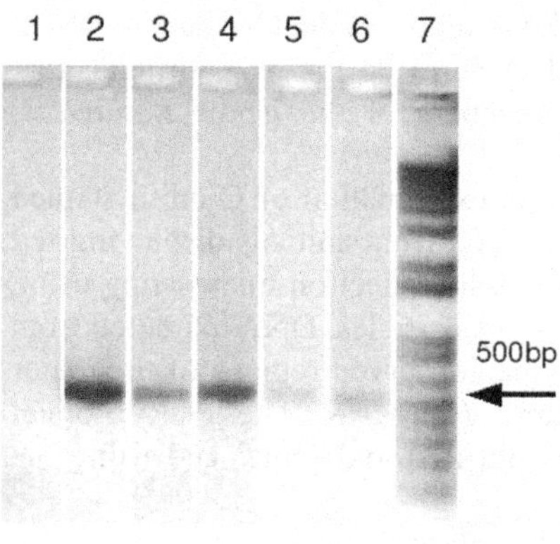

Figure 2. PCR products: 1- H2O; 2 – FV3 DNA; 3, 4 - LEV infected; 5, 6 - LEV free. 7 – MW marker.

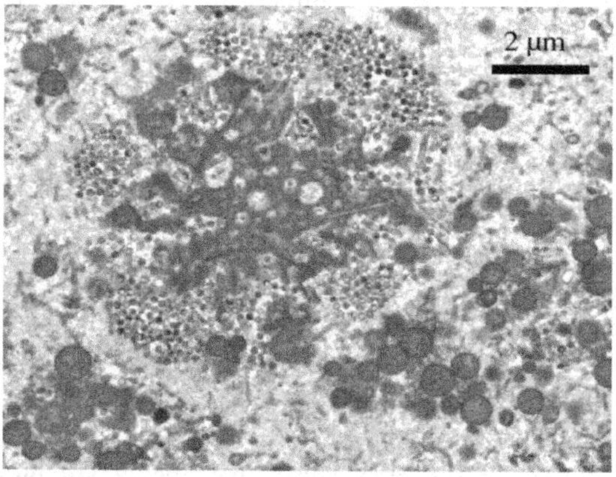

Figure 3. LMV infected IgH-2 cells

Structural and ultrastructural investigations on the *in vitro* effects of plant extracts in chicken and rabbit whole blood cultures

L. Barbu-Tudoran[1], M. Spanu[2], C. Craciun[1] and C.V. Mihali[1]

1. Electron Microscopy Center, Babes-Bolyai University, 400006 Cluj-Napoca;
2. USAMV Cluj-Napoca, Romania

lbarbu@biolog.ubbcluj.ro
Keywords: ultrastructure, blood cells

Recently published data underline the beneficial effects of plant extracts, in correlation with those of associated classical treatments, when immune intervention is required. Such information strongly support researches on the possible immune-modulating activities of vegetal extracts with an increased bioavailability, pronounced therapeutic activity and lacking side effects, when compared to synthetic drugs. Still, the interactions between the plant extractions and the immune mechanisms are far from being solved. Thus, after the initial investigations on anti-tumor and anti-microbial effects, the researches switched to characterize the influence of plant extracts on different components of the immune system.

The *in vitro* effects of alcoholic extracts of *Calendula officinalis* and *Echinacea angustifolia* were tested in whole blood cultures, from healthy chickens and rabbits, in amounts of 1.5 and 6.5 μl, compared to the 70° alcohol, which served as solvent. The cultures were incubated for 48 hours in a 5% CO_2 atmosphere and the cells were harvested by centrifugation. Cellular pellets of the cultures were processed for electron microscopy studies, as follows: fixation in glutaraldehyde (2.7%) and osmium acid (2%), dehydration in acetone and infiltration with Epon 812. Semi-thin (0.5 μm) and ultra-thin (50-60 nm) sections were obtained with a diamond knife (Diatome) on a Leica UC6 ultramicrotome. Images from the semi-thin sections were recorded on an Olympus BX51 LM. From the ultra-thin sections stained with uranyl acetate and lead citrate images were recorded on a Jeol JEM1010 TEM with MegaViewIII CCDcamera.

When compared to the untreated control cultures (Figure 1 – chicken and Figure 2 – rabbit), revealing mostly physiological changes, the 6.5% alcohol treated cultures (Figure 3 – chicken and Figure 4 – rabbit) showed cytoplasm vacuolization in most of the erythrocytes, fragments of destroyed erythrocytes, and large numbers of macrophages. The alcoholic plant extractions' treated cultures (*Echinacea* – Figures 5 and 6; *Calendula* – Figures 7 and 8) indicated the protective effect of active vegetal compounds, compared to the single solvent treated cultures, by reducing cytoplasm vacuolization, decreasing numbers of macrophages, and the presence of only a few spherical erythrocytes and some cellular debris. We concluded that the active principles from the plants protected the leucocytes and red blood cells, during the incubation period, against the harmful effects of the alcohol solvent, in a dose dependent manner.

A. Aretz, B. Hermanns-Sachweh, J. Mayer (Eds.): EMC 2008, Vol. 3: Life Science, pp. 105–106,
DOI: 10.1007/978-3-540-85228-5_53,

Surprisingly, the *Echinacea* extract, renowned as an immunomodulating one, had milder protective effects than the *Calendula* extract.

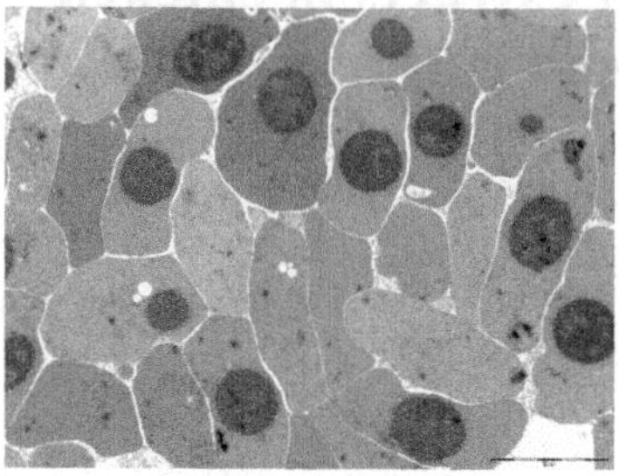

Figure 1. Chicken control group.

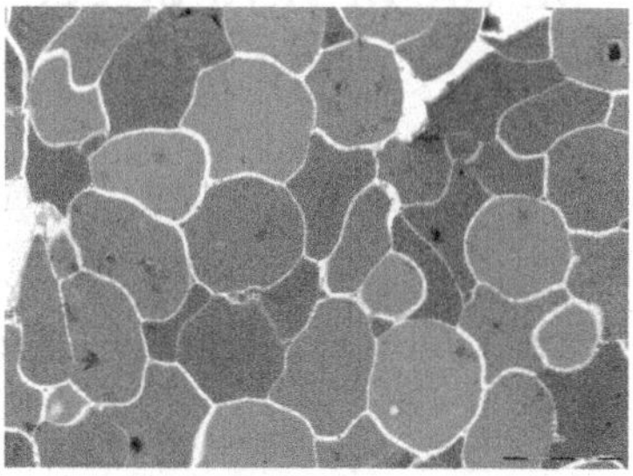

Figure 2. Rabbit control group.

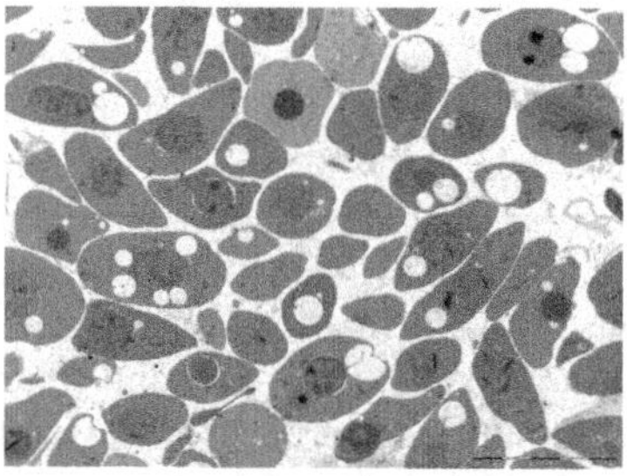

Figure 3. Chicken alcohol treated group.

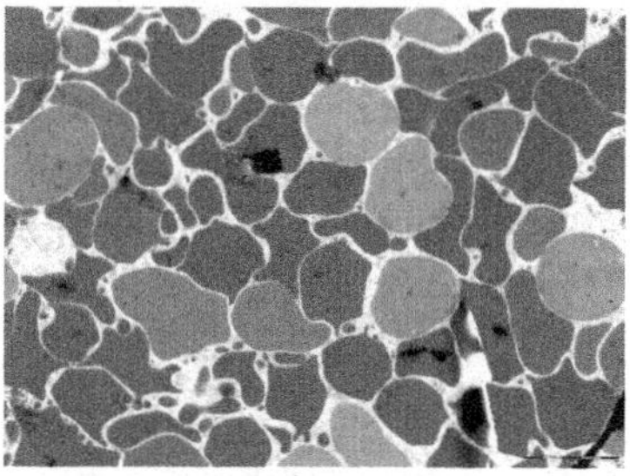

Figure 4. Rabbit alcohol treated group.

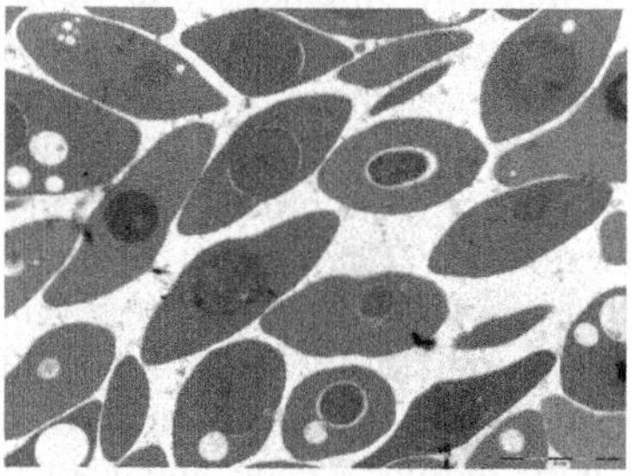

Figure 5. Chicken *Echinacea* treated culture.

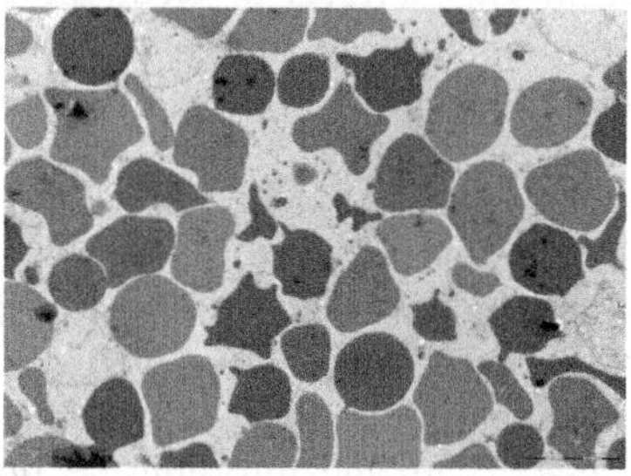

Figure 6. Rabbit *Echinacea* treated culture.

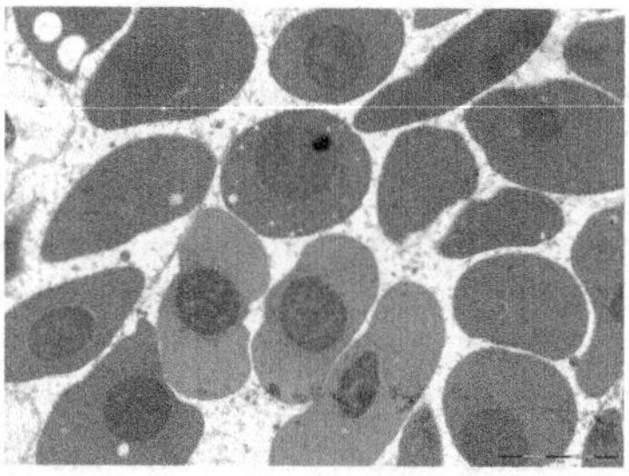

Figure 7. Chicken *Calendula* treated culture.

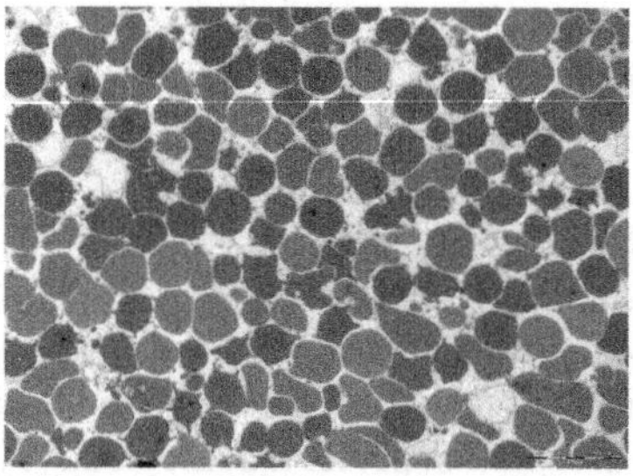

Figure 8. Rabbit *Calendula* treated culture.

Fine structural characterization of cloned human endogenous retrovirus HERV-K113

Klaus Boller, Kurt Schönfeld, Ralf R. Tönjes

Paul-Ehrlich-Institut, Paul-Ehrlich-Str. 51 – 59, D-63225 Langen, Germany

bolkl@pei.de
Keywords: retrovirus, morphology, immunoelectron microscopy

Endogenous retroviruses are retroviruses that have entered the germline of their host and integrated into its genome. They are transmitted vertically from parents to offspring just as mendelian genes. In humans, 8% of the genome derives from endogenous retroviruses. While most endogenous retroviruses were silenced by mutations and deletions, some are still transcribed and some produce intact proteins. The only endogenous retrovirus of man that has been shown to produce viral particles is HERV-K (human endogenous retrovirus type K). While the oldest of the roughly 30 HERV-K sequences integrated into the human genome more than 25 million years ago, evidence is accumulating that HERV-K was active until quite recently, the youngest sequence found being HERV-K113. It is however unclear, which HERV-K sequence encodes for the viral particles, that are produced by human germ cell tumors or melanomas.

Here we show that the genome of HERV-K113, cloned in an baculovirus expression vector, is capable of producing intact particles of retroviral morphology. Particles are structurally identical to those which were characterized in cell lines derived from human germ cell tumors. Thus, the HERV-K113 sequence is a candidate for particle production in vivo and therefore a candidate for a still infectious human retrovirus of today.

1. Boller, K., Schönfeld, K., Lischer, S., Fischer, N., Hofmann, A., Kurth, R., Tönjes. R.R. (2008). Human endogenous retrovirus HERV-K113 is capable to produce intact viral particles. J. Gen. Virol. 89, 567-572.

A. Aretz, B. Hermanns-Sachweh, J. Mayer (Eds.): EMC 2008, Vol. 3: Life Science, pp. 107–108,
DOI: 10.1007/978-3-540-85228-5_54,

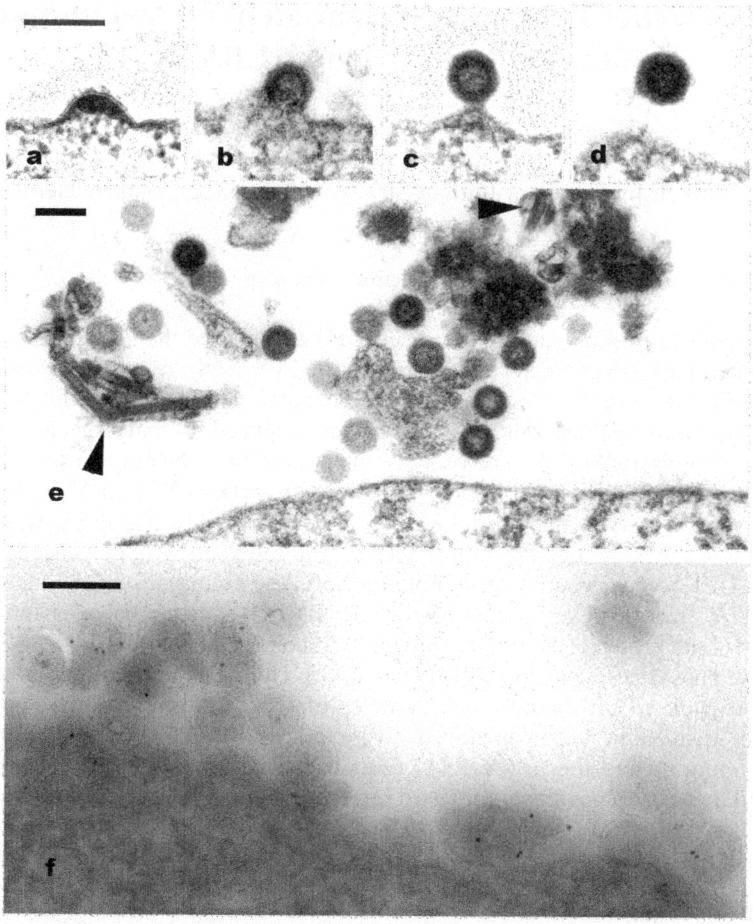

Figure 1. a – d) budding stages of HERV-K113, the particles clearly show retroviral morphology, but free mature particles with condensed cores were never observed; **e)** a group of HERV-K113 virus particles (note the baculovirus particles indicated by arrowheads); **f)** immunoelectron microscopical labelling of HERV-K113 particles with a mixture of the monoclonal anti HERV-K/Gag antibodies HERMA-1,-4,-6,-7,-8. Bars represent 150 nm

The procaine hydrochlorate effect onto the corpuscular anthocyans from the vacuolar sap of different plant cells

D. Cachita[1], A. Ardelean[1], C. Craciun[2], V. Turcus[1], L. Barbu-Tudoran[2]

1. Vasile Goldis West University, Arad, Romania
2. Electron Microscopy Center, Babes-Bolyai University, Cluj-Napoca, Romania

georgepribac@gmail.com
Keywords: procaine, anthocyans, liposomes

In diluted solutions of 1-10mg/l, the procaine hydrochlorate induce stimulating effects over the plant growth [1].

The concentrated solution of procaine hydrochlorate (1g/l), determines the production of liposomes from lecithin like phospholipoid compounds [1-3]. In time it will induce, mostly in the cell vacuoles of different plant species (peony, azalea, water lily etc.), liposomes production. Over 24-72 hours of procaine concentrate solution action the liposomes are working, this process is getting finally stable in vacuoles by forming 1-2 big corpuscles, strongly refracting the light. If the vacuoles contain anthocyans, the liposomes will increase it in their structures [1-3], and they become coloured. In the case of leucoanthocyans, the liposome corpuscles remain uncoloured. In both cases the liposomes remain refractive.

Researches made on monocotyledonous species, on similar epidermal tissues, revealed the absence of this reaction in their vacuolar sap. The significance of this phenomenon may be explained by the absence of the phospholipoid compounds from the vacuolar juice of their cells.

Observing the process described in the vacuoles from epidermal cells of the *Cyclamen* petals or petioles, after 24 hours of procaine action, we saw that 2-3 big liposomes were formed, together with several small violaceous corpuscles (Figures 1A and B). At high pH, in the presence of a diluted sodium bicarbonate solution, the corpuscles colour will turn red – violaceous bluish, as we can see in the Figures 1A and B. In the same experimental conditions, in the case of *Gynura* leaf epidermis (Figures 1C and D), although the pluricellular hairs are present, they reveal a red pigment in their vacuolar sap, the treatment with procaine did not start the forming of liposome corpuscles (Figures 1C and E). Because on the external walls (Figure 1D) of *Gynura* hairs, certain ruggedness could be observed, we asked ourselves if these structures could get the cell membrane impervious, after it a secondary change, making difficult for the procaine to penetrate in the cell. The scanning electron microscopy (Figure 1D), revealed the fact that *Gynura* epidermal hairs cell walls are easily deformed by dehydration, phenomena caused by the absence of secondary structures. The absence of the corpuscular process in the vacuolar sap, under the procaine action, at Gynura hairs, may be explained by the lack of anthocyans, respectively, the lack of phospholipoids from their vacuolar juice. In this case, vacuolar sap colour is produced by betacyanine, and not by anthocyans.

A. Aretz, B. Hermanns-Sachweh, J. Mayer (Eds.): EMC 2008, Vol. 3: Life Science, pp. 109–110,
DOI: 10.1007/978-3-540-85228-5_55,

1. D. Cachiţă Cosma, A. Ardelean, Procaina, bioefector la om, plante şi animale, ed. Mirton Timişoara, (1996), p.167
2. D. Cachiţă, M. Keul, Gh. Popovici, Bemerkungen über die Wirkung von Procain auf die Blutenepidermiszellen von *Convolvulus persicus* L., Rev. Roum. Biol., ser. Bot., **16, 2**, (1971), pp. 197-200
3. D. Cachiţă, C. Crăciun, The ultrastructure of the epidermal cells of the peony petals (Paeonia officinalis) under the action of procain, Cytologia, **44**, (1979), pp. 71-75

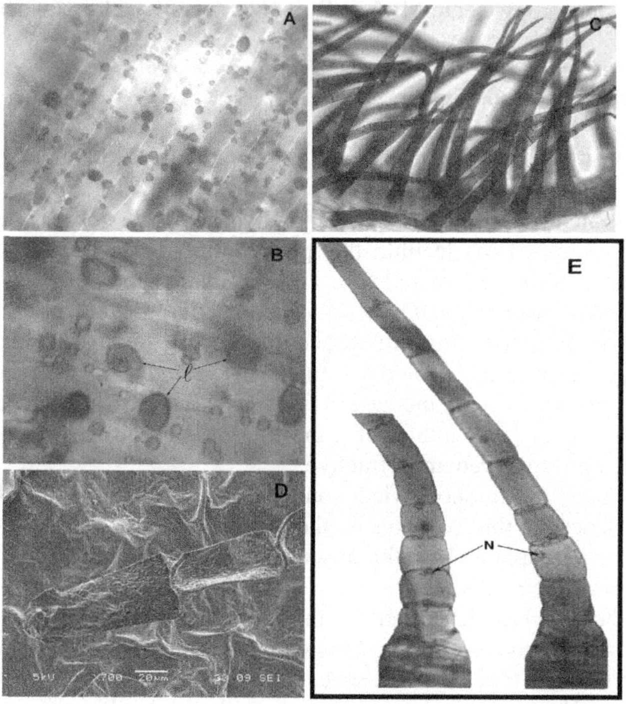

Figure 1. A-B: liposomes (l) formation in the vacuolar sap of epidermal petals (A) or petioles (B) after 24 h treatment with procaine solution in 1g/l concentration; **C-E**: cytological aspects about Gynura foliar, epidermal hairs: C- normal (without procaine treatment); D- SEM image; (h=hair); E- hairs after treatment with 1g/l procaine solution (N-nucleus)

Comparative anatomical and ultrastructural investigations on normal and vitroplantlets leaves from *Pistia stratiotes* L.

Dorina Cachiță-Cosma[1], Daniela Beles[2], C. Craciun[3], L. Barbu-Tudoran[3]

1. V. Goldis West University, Arad;
2. Oradea University, Oradea; 3. Electron Microscopy Center, Babes-Bolyai University, Cluj-Napoca, Romania

v.cachita@gmail.com
Keywords: *Pistia*, anatomy, SEM

Our studies, that made the object of these communication had as purpose the comparative examination of the structural and ultrastructural aspects of *Pistia* plants, generated *in vitro* or growth in the green house.

Pistia stationes L. (waterlettuce), is a natant vegetal specie growing in pantropical zones. In nature, the vegetative multiplication of *Pistia* is realized spontaneously by budding. We initiated *Pistia* vitroculture [1], on a liquid *Murashige – Skoog* aseptic medium [2], from 2-3 mm buddy explants. From these inocully, leaf minirosettes are regenerated, which – in contrast to plants grown in septic conditions – as they grow, its lose their floatability and, gradually, sink into the liquid growth medium. The examination of the leaf epidermis (especially of the ventral side) with a scanning electron microscope showed that the tectorial hairs from the inferior side of the leaf limbs (from the base of the leaf minirosettes of the vitroplantlets) are bigger, longer, bat les rigids as the naturale one, things that cause their flattening, and leads to the impossibility of forming an air layer thick enough to support the floatability of these leaf rosettes. *Pistia* bud anatomy, which is the starting point for a new leaves ring, respectively a new plant, wasn't studied so far.

Our investigations permited to identify the meristematic tissues, integrated in anatomic structures of the explant. These investigations were important for obtaining small sized inocula, without decreasing their regenerative capacity. The transversal sections, made at different levels on these explant types, were examined with an optical microscope, and with a transmission electron microscope, as well.

The electron microscopic examinations have revealed – in the central zone of the explants – the location of the primary meristematic tissue. Next to it, centrifugally, the foliar primordia was found, and outside these, around the buddy explant, 3-4 foliar eboches could be observed. Both the anatomic study and the ultrastructural one permitted us to distinguish – excepting the meristem – the presence in the potential phytoinocul, of a structural heterogeneity of cells found in the microscopic field.

The vitrocultured buddy explants from the aseptic medium maintained cells, revealed the presence of an early specialization tendency of proplastides into amyloplasts, in whose stroma the starch resulted after sucrose metabolisation is stored. Sucrose was added to vitroculture medium as the energetic source, needed to supply the cells, witch were detached from the leaves rosettes they belonged to.

A. Aretz, B. Hermanns-Sachweh, J. Mayer (Eds.): EMC 2008, Vol. 3: Life Science, pp. 111–112,
DOI: 10.1007/978-3-540-85228-5_56,

The ultrastructural aspects found in different cell types, which belong to the buddy explant, are very various. In the regenerative processes and in vitroplantlet neogenesis, these cells manifest different proliferative capacities, and the new formed cells can express themselves morphogenically, or not.

1. D. Cachita-Cosma, D. Beles., L. Barbu-Tudoran. Studiu privind aspectul epidermei frunzelor de *Pistia stratiotes* L., normale sau provenite din vitrocultura, examinata la microscopul electronic cu baleiaj, Analele SNBC, vol **XI**, Cap VI-Biologie Vegetala, C. Craciun., A. Ardelean (Ed. coord.), Ed. RISOPRINT Cluj-Napoca (2006), pp. 462-472.
2. Murashige, T., Skoog, F., 1962, *A revised medium for growth bioassays with tobacco tissue cultures*, Physiol. Plant., 15, pp. 473 – 497.

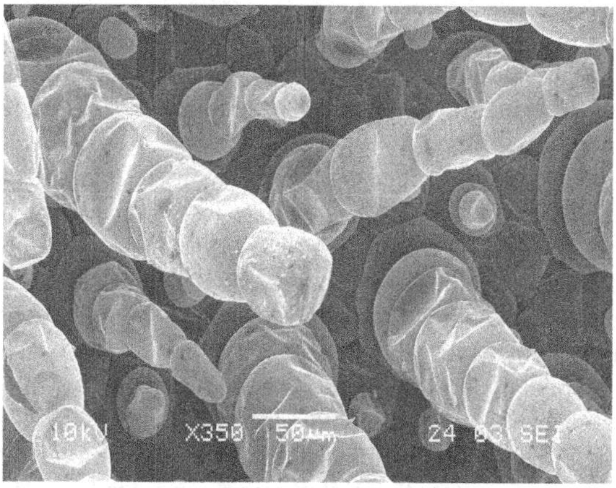

Figure 1. Tector hairs from the epidermal surface (0.8 mm^2) of *Pistia* leaf from green house.

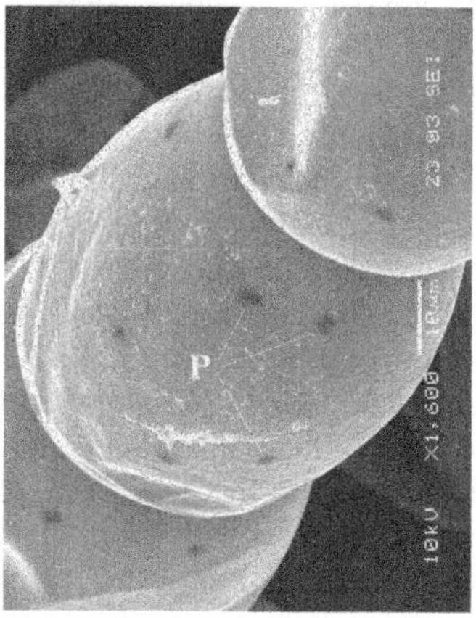

Figure 2. Detail of pores (P) from tector hair.

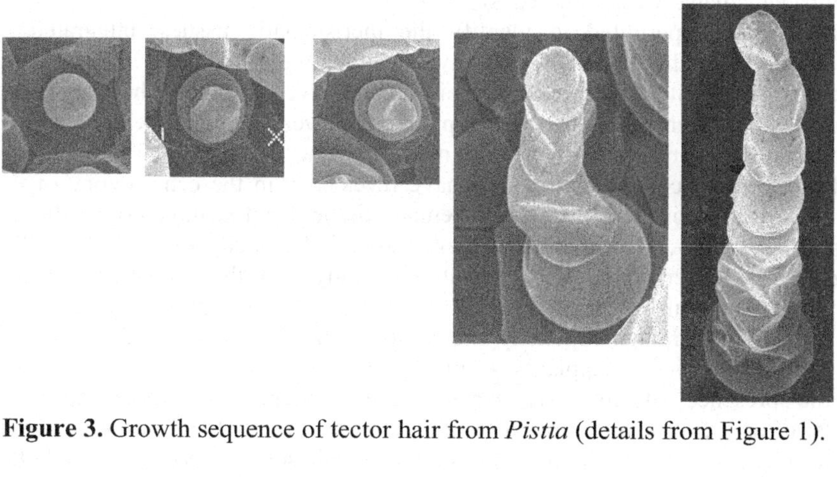

Figure 3. Growth sequence of tector hair from *Pistia* (details from Figure 1).

Autophagy of mitochondria in brown adipocytes of chemically thyroidectomised rats

M. Čakić-Milošević, A. Korać and M. Ukropina

Institute of Zoology, Faculty of Biology, University of Belgrade, Studentski trg 16, 11000 Belgrade, Serbia

maja@bf.bio.bg.ac.yu
Keywords: brown adipocytes, mitochondria, autophagy

Thyroid hormones (TH) have an important role in modulation of brown adipose tissue (BAT) thermogenic function. In hypothyroid animals, iodothyronine-5'-deiodinase II expressed in BAT, elevates local triiodothyronine level, protecting tissue thermogenic activity [1]. However, during adaptation to such serious endocrine disbalance as hypothyroidism is, brown adipocytes undertake remarkable ultrastructural alterations and the main objective of this work was to investigate some of them.

Male Wistar rats maintained under standard laboratory conditions were used in this study. The animals from the experimental group (n=8) were made hypothyroid with 0.02% methimazole (Sigma, St. Louis, MO, USA) in drinking water for three weeks. The other animals (n=6) were untreated controls. At the end of the experiment, interscapular BAT was isolated and routinely prepared for electron microscopy. Ultrathin sections, contrasted with uranyl-acetate and lead-citrate were examined with a Philips CM 12 electron microscope.

One of the most conspicuous findings obtained by electron microscopic observations was an increased number of mitochondria subjected to autophagy. Presented electron micrographs show progression of this process. At early stages, isolation membrane forms a cup-shaped structure that gradually envelops targeted mitochondrion (Figure 1A-C). Then, isolation membrane seals, closing into double-membrane autophagosome containing mitochondrion inside (Figure 1, D-E). Upon this, autophagosome fuses with lysosome to form autolysosome in which sequestered mitochondrion is degraded (Figure 1F). The appearance of quite numerous residual bodies in brown adipocytes from hypothyroid animals (Figure 1G), might be, although not exclusively, related to mitochondrial degradation. In some instances, autophagosome formation comprises merging of two or three isolation membranes around mitochondrion determined for elimination (Figure 1, H). Mitochondria subjected to degradation usually are closely apposed to the surface of lipid droplets. They have well preserved ultrastructure, with numerous long cristae, without any visible sign of dysfunction (Figure 1, A-C).

In normal physiology, cell uses autophagy to remove dysfunctional or unneeded organelles in order to adjust its internal organisation to altered physiological demands [2]. Since mitochondria involved in autophagy are structurally well preserved, we ruled out damage and/or regressive alterations as reasons for their elimination. On the other hand, as we reported previously, hypothyroidism significantly increases catalase activity and peroxisome proliferation in brown adipocytes [3, 4]. Based upon these facts, we

A. Aretz, B. Hermanns-Sachweh, J. Mayer (Eds.): EMC 2008, Vol. 3: Life Science, pp. 113–114,
DOI: 10.1007/978-3-540-85228-5_57,

suggest that peroxisomes, as organelles capable for fatty-acid oxidation, become involved in oxidative processes, thus decreasing substarate delivery to mitochondria. Accordingly, mechanisms that regulate autophagy recognize "unemployed" mitochondria as superfluous and assigned them for degradation.

1. J. E. Silva, Physiol. Rev. **86** (2006), p. 435.
2. I. Kim, S. Rodriguez-Enriquez and J. J. Lemasters, Arch. Biochem. Biophys. **462** (2007), p. 245.
3. N. Petrović, G. Cvijić and V. Davidović, Physiol. Res. **50** (2001), p. 289.
4. M. Čakić-Milošević, A. Korać and M. Ukropina, Proceedings of the 13th EMC, Antwerp, Belgium, Aug. 22-27, 2004, p. 479.

5. The Ministry of Science and Environmental Protection of the Republic of Serbia (Grant No.143050) supported this work.

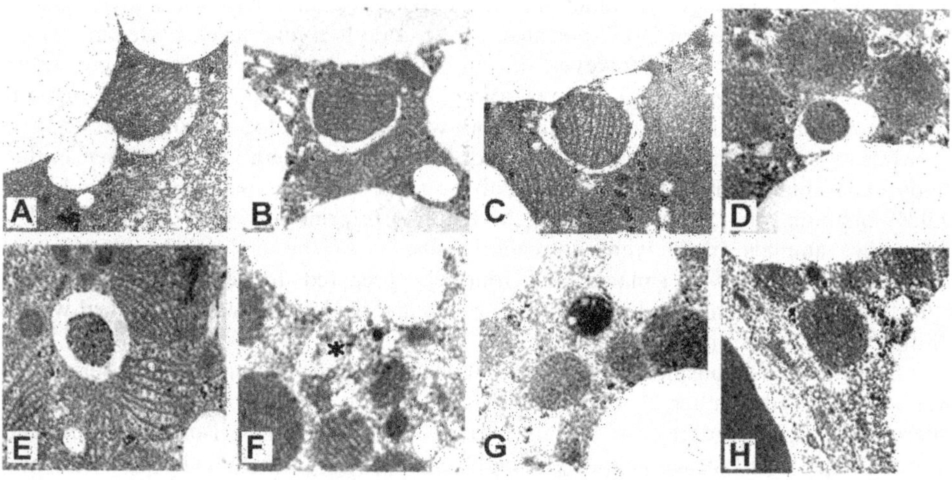

Figure 1. Sequences of autophagy of mitochondria in brown adipocytes of hypothyroid rats (A-G); asterisk on F indicates mitochondrion in process of degradation. H – merging of few isolation membranes around mitochondrion as an alternate mode of autophagosome formation. Magnification x 6300 orig.

Ultrastructural studies of immune organs in antigen primed chickens treated with vegetal extractions

C. Craciun[1], M. Spanu[2], and A. Ardelean[3]

1. Electron Microscopy Center, Babes-Bolyai University, 400006 Cluj-Napoca;
2. USAMV Cluj-Napoca
3. Western University Vasile Goldis, Arad, Romania

ccraciun@biolog.ubbcluj.ro
Keywords: ultrastructure, plant extracts, chicken organs

The researches were carried out to establish the immune stimulating influence of alcoholic extracts from *Calendula officinalis* and *Echinacea angustifolia* on chickens' immune organs.

The experiment was carried out on 28 Rock x Cornish chickens, divided into four equal groups (n=7), subjected each, for seven days (days 0 to 7), to a differentiated oral treatment, using the same pattern (0.5 ml/bird/day): group I - alcoholic *Calendula officinalis* extraction; group II –alcoholic *Echinacea angustifolia* extraction; group III – 70° alcohol, solvent control, and group IV – saline treated, environmental control. Samples from thymus, spleen, adrenal glands, and liver were processed for ultrastructural studies on day 14 of the experiment.

The results in the alcohol treated group indicated alterations in all studied organs. Thus, in thymus, decrease of the thymocytes number, irregular outline of nuclei, and increase of the intracellular spaces were present (Figure 1). Fragmentation of some lymphocytes and accumulation of amorphous material in edematous spaces were observed in the spleen (Figure 2). The adrenal glands showed irregular shaped cells and altered steroid component of CSH at the fasciculate zone (Figure 3). Hepatocytes were filled with lipids, the nuclei having irregular outlines, while RER was proliferated. Similarly, an increased amount of collagen fibrils were observed in the Disse space (Figure 4). The alcoholic extraction treatment showed a protective effect in all organs against alterations induced by alcohol. No side effects were noticed for any of the extractions. The images from thymus (Figure 5), spleen (Figure 6), adrenal glands (Figure 7), and liver (Figure 8) of the alcoholic extractions' treated birds show an ultrastructure similar to the environment control group. It was concluded that both vegetal extractions, fed to chickens under the experimental conditions, showed beneficial effects at cellular and sub-cellular levels in all studied organs, with a slightly better effect of *Calendula officinalis* extraction. Therefore, we considered that both extractions could be used as immune adjuvants in birds.

A. Aretz, B. Hermanns-Sachweh, J. Mayer (Eds.): EMC 2008, Vol. 3: Life Science, pp. 115–116,
DOI: 10.1007/978-3-540-85228-5_58,

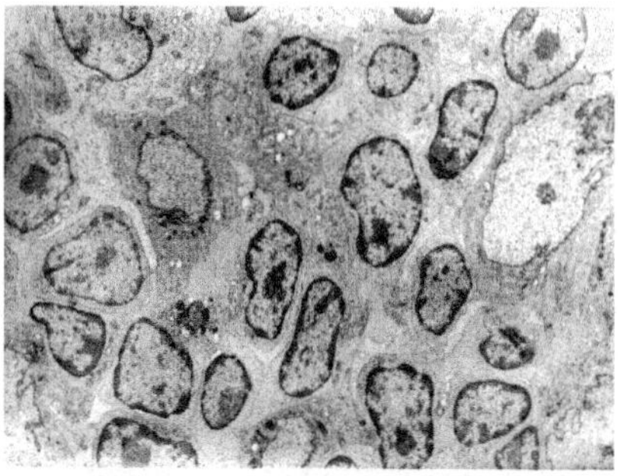

Figure 1. Thymus - alcohol

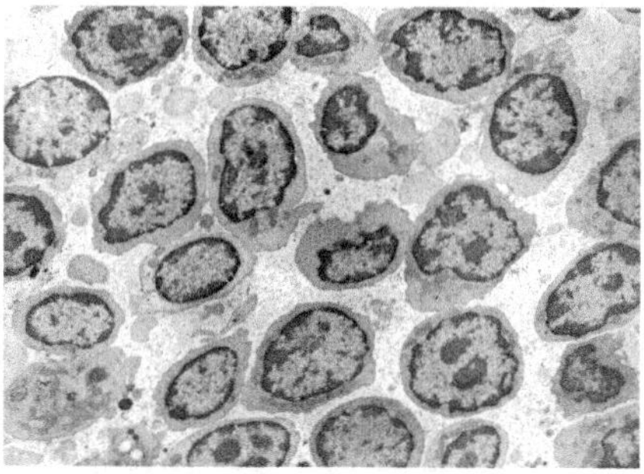

Figure 2. Spleen - alcohol

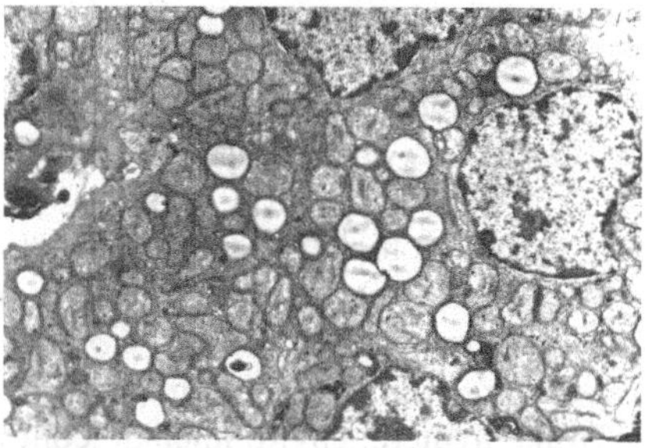

Figure 3. Adrenal glands - alcohol

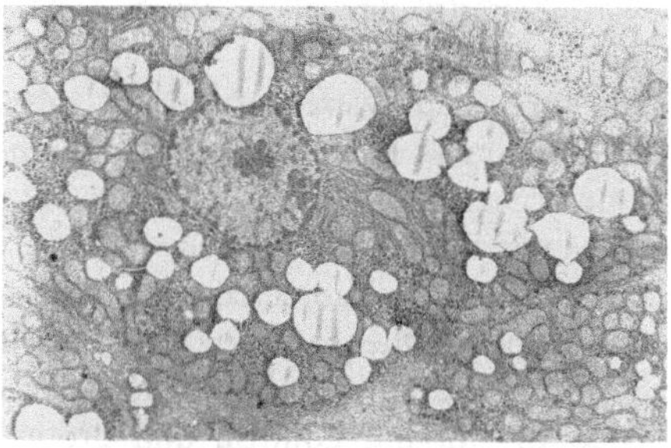

Figure 4. Liver - alcohol

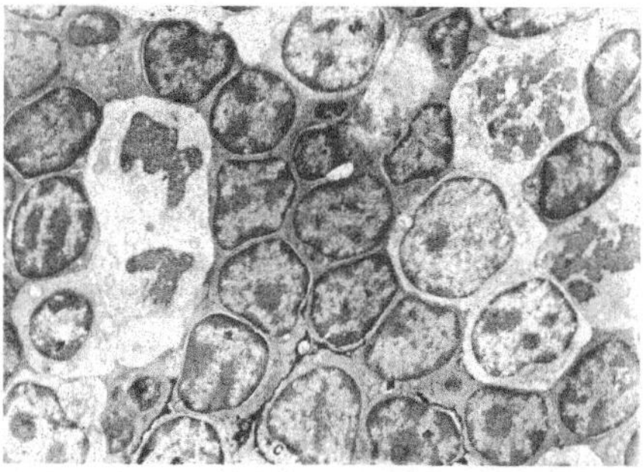

Figure 5. Thymus – alcohol + *Calendula*

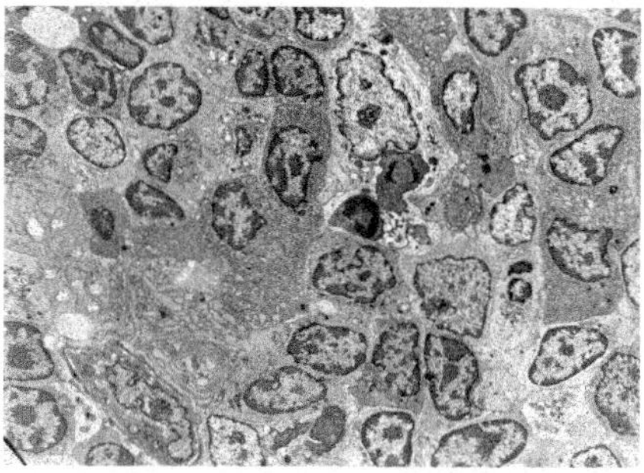

Figure 6. Spleen – alcohol + *Echinacea*

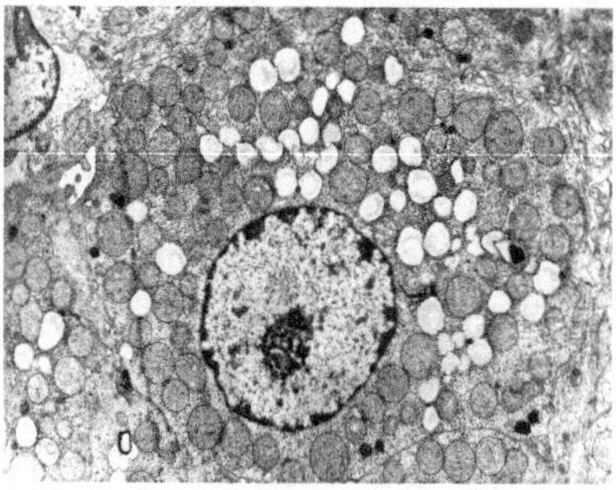

Figure 7. Adrenal glands – alcohol + *Calendula*

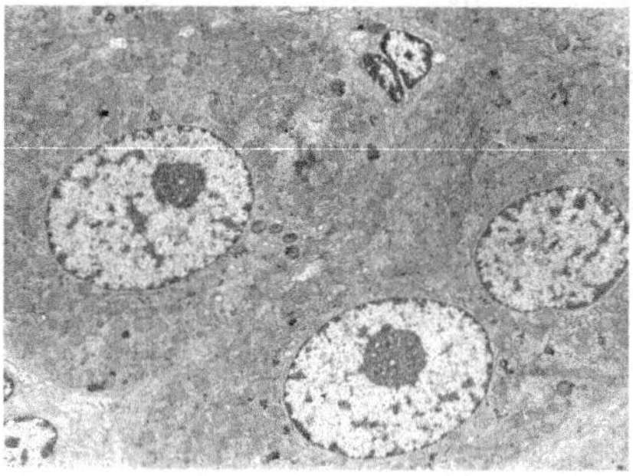

Figure 8. Liver – alcohol + *Echinacea*

Structural features of HIV-1 immunologic activation

A. Foglia

via Bosia 4, CH-6900 Lugano-Paradiso

dralberto.foglia@bluewin.ch
Keywords: HIV-1, erythrocytes binding, T-cell syncytia, filopodia

Immunologic activation by HIV-1 virus presents several structural features. In this case the HIV-1 infection is characterized by a high level of viremia allowing scanning electron microscopic visualization [1] of the virus on the surface of a red blood cell “Figure 1” [2,3]. However a recent paper has denied such observation [4].

On isolated T-cells we observe the presence of podosomes to form a T-cell syncytium “Figure 2”, described as a potentially destructive structure [5].

Podia formation and in particular filopodia in immunologic activated lymphocytic cells has been currently scanned [6], here we see an HIV-1 activated T-cell with a protruding filopodium “Figure 3”.

Methods: high viremic HIV-1 EDTA-blood sample is fixed in 2,5% glutaraldehyd pH 7,2 and stored at room temperature for 24h in Sörensen solution. After one short washing procedure (centrifugation in bidest. H2O), the solution is dehydrated in increasing concentrations of acetone (20, 40, 60, 80, 95 and 100%; 10 min each), placed on non coated 6 mm coverslips and air dried for 2h. Platin coating was performed with a Balzers SCD 004 sputter coater and visualized with a Scanning Electron Microscope Jeol JSM 840, with 15,0 kV accelerating voltage, magnification 10,000x “Figure 1”. Lymphocytic cells isolation was performed with centrifugation of the fixed solution with Leucosep (Assaf Scientific Development Co.), magnification 14,000x “Figure 2”. A clearer picture was possible after 3 repeated washing procedures, magnification 5,500x “Figure 3”.

1. Withman AG., Dyson OF., Lambert PJ., Oxendine TL., Ford PW., Akula SM., Changes occurring on the cell surface during KSHV reactivation, *J. Electron Microsc* (2007) **56**(1), pp. 27-36.
2. Montefiori DC., Graham BS., Zhou JY, Ahearn JM., Binding of human immunodeficiency virus type 1 to the C3b/C4b receptor (CD35) and red blood cells in the presence of envelope-specific antibodies and complement, *J. Infect Dis* (1994) **170**, pp. 429-32.
3. Hess C., Klimkait T., Schlapbach L., et al., Association of a pool of HIV-1 with erythrocytes in vivo: a cohort study, *Lancet* (2002) **359**, pp. 2230-4.
4. Seth Fierer D., Vargas J., Patel N., Clover G., Absence of erythrocyte-associated HIV-1 in vivo, *J. Infect Dis* (2007) **196**(4), pp. 587-90.
5. Sylwester A., Daniels K., Soll DR., The invasive and destructive behavior of HIV-induced T cell syncytia on collagen and endothelium, *J. Leukoc Biol* (1998) **63**, pp. 233-44.
6. Kondo K., Yoshitake J., Scanning Electron Microscopy of Leukocytes in Motion, *J. Electron Microsc* (1976) **25**(2), pp. 99-102.

A. Aretz, B. Hermanns-Sachweh, J. Mayer (Eds.): EMC 2008, Vol. 3: Life Science, pp. 117–118,
DOI: 10.1007/978-3-540-85228-5_59,

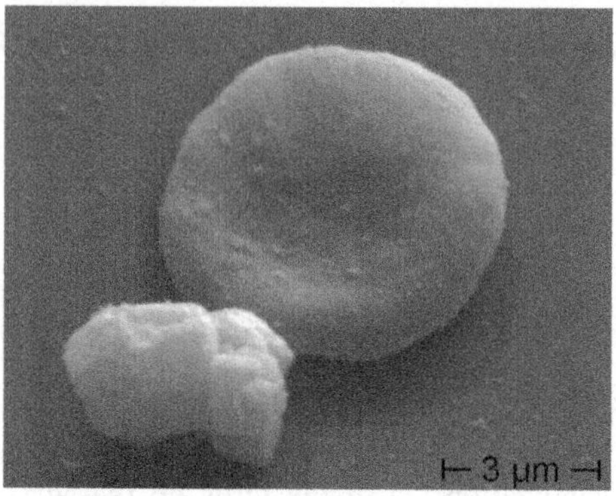

Figure 1. HIV-1 virus on the surface of a red blood cell (near an unidentified crystal).

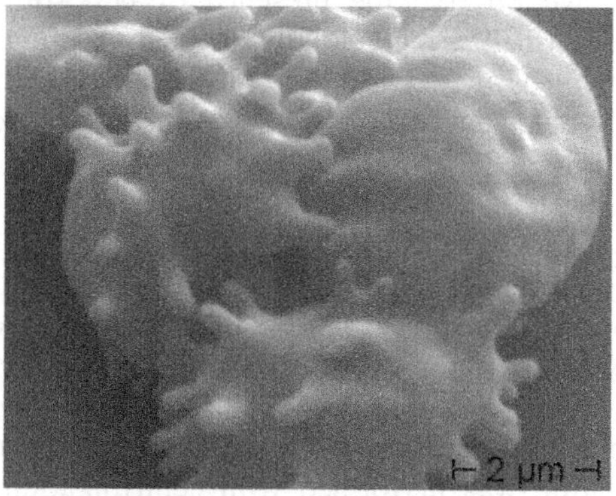

Figure 2. T-cell syncytium formation in HIV-1 acti-vation.

Figure 3. Filopodium in HIV-1 activated T cell.

Functions of actin-binding proteins in the cell nucleus

J. Hofmannová[1], M. Maninová[1], V. Filimonenko[2] and P. Hozák[2]

Institute of Molecular Genetics, Dept. of Biology of the Cell Nucleus, Academy of Sciences of the Czech Republic, Prague, Czech Republic

hofmannova@img.cas.cz, maninova@img.cas.cz
Keywords: actin-binding proteins, actinin, filamin, spectrin, vinculin

Actin is not only a component of the cytoplasm of eukaryotic cells but is abundant also in the nucleus. However, the functional significance of occurrence of actin in the nucleus is still largely unknown, although several important roles of the actin in nuclear processes have been suggested. Nuclear actin participates in regulation of gene transcription, nucleocytoplasmic transport, and chromatin and nuclear structure.

Numerous actin-binding proteins (ABPs) regulate the dynamics of actin filaments in the cytoplasm. Some ABPs have an ability to regulate actin polymerization and depolymerization, other ABPs cross-link or bundle actin filaments, and still other ABPs breakdown actin filaments by "severing" mechanismus. The presence of some ABPs in the nucleus have been reported (α-actinin, CapG, emerin, filamin A, gelsolin, profilin, protein 4.1, spectrin, thymosin β4) but little is known about their nuclear functions.

We address the questions of nuclear localization and of possible nuclear functions of actinin, filamin, spectrin and vinculin. Spectrin was first characterized as a cytoskeletal protein underlying the plasma membrane of the erythrocyte and linking actin filaments to an integral membrane protein. Recent studies has provided evidence that spectrin is present in the nucleus of mammalian cells. Nonerythroid αspectrin (αSpII) is part of nuclear protein complex involved in DNA repair [1, 2]. Spectrin βIV (βSpIVΣ5) associates with the nuclear matrix and PML nuclear bodies [3] and spectrin βII interacts with nuclear proteins [4].

Filamin A was the first actin filament cross-linking protein identified in non-muscle cells. During last few years filamin was observed in the cell nucleus too. Human filamin A (hsFLNa) interacts with BRCA2 in the nucleus and the lack of hsFLNa results in increased cellular sensitivity to DNA damaging agents [5]. C-terminal 100kDa fragment of FLNa colocalized with androgen receptor to the nucleus where it regulates androgen receptor activity [6].

Member of spectrin protein family and focal adhesion and stress fiber-associated proteins, α-actinin 1 and 4 were identified as class IIa histon deacetylases-interacting proteins and in addition a novel splice variant of α-actinin 4 which is predominantly localized in the nucleus was isolated [7]. Vinculin interacts with numerous other junctional proteins and links integral membrane proteins to actin filaments.

We performed a study on nuclear localization of these actin-binding proteins by means of electron microscopy. Immunogold labeling on ultrathin sections of HeLa cells was evaluated using spatial statistics with previously developed plugins to the Ellipse image processiong program. We demonstrate that several actin-binding proteins reveal

A. Aretz, B. Hermanns-Sachweh, J. Mayer (Eds.): EMC 2008, Vol. 3: Life Science, pp. 119–120, DOI: 10.1007/978-3-540-85228-5_60,

characteristic distribution within the cell nucleus implying important roles in nuclear functioning. The data on colocalization of ABPs with the key proteins of important nuclear processes, such as transcription, replication, splicing and cell cycle regulation will be discussed.

1. L. W. McMahon, C. E. Walsh, M. W. Lambert (1999). Human αspectrin II and Fanconi anemia proteins FANCA and FANCC interact to form a nuclear complex. J. Biol. Chem. 274, 32904-32908
2. D. Sridharan, M. Brown, W.C. Lambert, L. W. McMahon, M. W. Lambert (2002). Nonerythroid αII spectrin is required for recruitment of FANCA and XPF to nuclear foci induced by DNA interstrand cross-link. J. Cell Sci. 116, 823-835
3. W. T. Tse, J. Tang, O. Jin, C. Korsgren, K. M. John, A. L. Kung, B. Gwynn, L. L. Peters, S. E. Lux (2000). A new spectrin, bIV, has a major truncated isoform that associates with promyelocytic leukemia protein nuclear bodies and nuclear matrix. J. Biol. Chem 276, 23974-23985.
4. Y. Tang, V. Katuri, A. Dillner, B. Mishra, C. X. Deng (2003). Distribution of transforming growth factor-beta signaling in ELF beta-spectrin-deficient mice. Science 299, 574-577.
5. Y. Yuan and Z. Shen (2001): interaction with BRCA2 suggests a role for filamin-1 (hsFLNa) in DNA damage response. J. Biol.Chem. 276, 48318-48324.
6. C. J. Loy, K. S. Sim, E. L. Yong (2003): Filamin-A fragment localizes to the nucleus to regulate androgen receptor and coactivator functions. PNAS 100, 4562-4567.
7. S. Chakraborty, E. L. Reineke, M. Lam, X. Li, Y. Liu, Ch. Gao, S. Khurana, H.-Y. Kao (2006): α – Actinin 4 potentiates myocyte enhancer factor – 2 transcription activity by antagonizing histon deacetylase 7. J. Biol. Chem. 281, 35070-35080.
8. www.nucleus.img.cas.cz
9. This study was supported by the Grant Agency of the Czech Republic (Reg. No. 204/05H023), by the Ministry of Education, Youth and Sports of the Czech Republic (Reg. No. 2B06063 and LC545) and by Institutional Grant (Reg. No. AV0Z50520514).

Ultrastructural mechanisms of entamoebae movement system

Hovnanyan Margarita, Hovnanyan Karlen, Asatryan Rosa

hovkarl@mail.ru
Keywords: entamoeba, ultrastructura, movement cryoultramicrotomy

The aim of this work is to find out ultrastructural aspects of movement function of the cell of Ent. histolytica vegetative forms.

We used in this work Ent. histolytica cultures as well as bioptates of large intestine mucous membrane of patients with amebiasis, containing entamoebae hematophages. Electronicmicroscopic preparations are prepared according to generally established methodic. Together with classical method we also used cryoultramicrotomy.

Electronicmicroscopic analysis at the part of entamoebae hematophages specimen, together with specific for entamoebae elements of cellular formations (ribonucleic proteid aggregates, glycogen zones, virus-like particles), structural complex was also noticed at the back part of cells, which took part in the process of entamoebae-uroid movement (Fig. 1), which are the plasmatic membrane microexcrescences, pinocytar vacuoles, actinolike microfilaments. In formation of pseudopodia the attractive matter is the first revealed by us strange tubular formations 200nm diameter, with inner diameter 100nm and wall thickness 50nm, situated in different parts of cytoplasm both on longitudinal and crosspiece sections (Fig.2). In paranuclear area of entamoebae cysts there were revealed vacuoles containing filaments. Such differentiation of the separate parts of cytoplasm witness the variations of dynamic functional purposes performed by them, that allow qualifying Ent.histolytica cytoplasm according to its structural and functional indicators as high developed. Cryoultramicrotomy method usage [2] for entamoebae ultrastructure study allowed to reveal the canceling microfibrillae in protists cytoplasm, correlating with the results, received with the help of classical method in entamoebae in culture and tissual forms of Ent. histolytica.

Thus, received results of electronicmicroscopic study of tubular formations and cytoplasm actinolike microfibrillae, trophozoits uroid and entamoebae hematophages tell about their participation in cytoplasmatic flow movement process and formation of amoeba cell pseudopodia.

1. W.Bernhard, E.H.Leduc J.of CellBiology, v.34,(1967), p.757.

A. Aretz, B. Hermanns-Sachweh, J. Mayer (Eds.): EMC 2008, Vol. 3: Life Science, pp. 121–122,
DOI: 10.1007/978-3-540-85228-5_61,

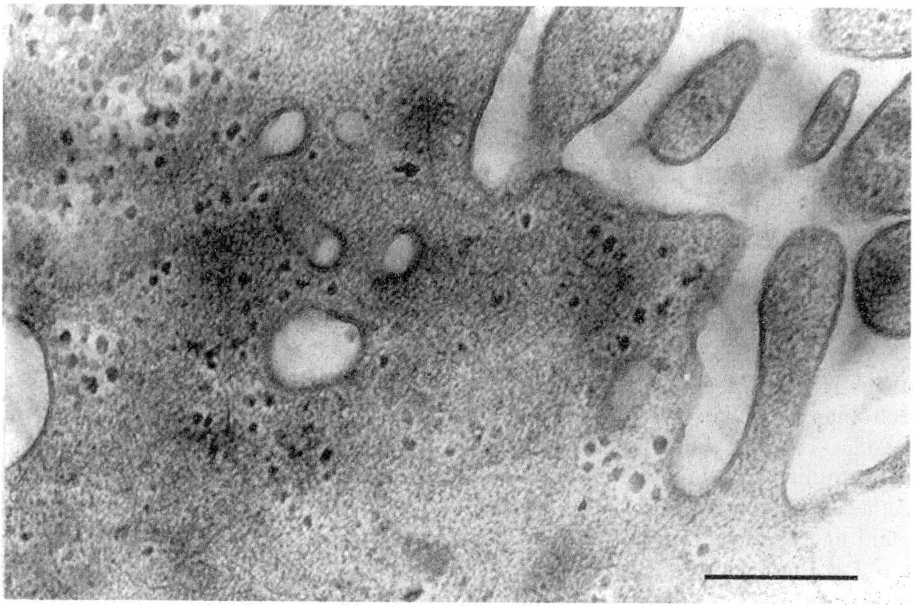

Figure 1. Hematophag of Ent.histolytica in patient with amoebiases in large intestine mucous membrane. Uroide, microfilaments. Scale=300nm.

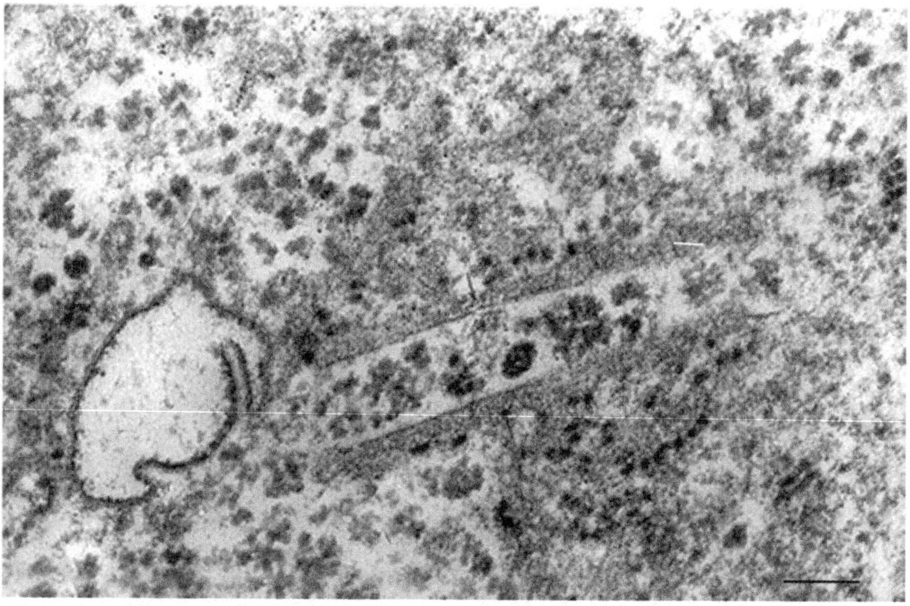

Figure 2. Hematophag of Ent.histolytica in patient with amoebiases in large intestine mucous membrane.Cytoplasmatic tubular formations.Scale=100nm

White adipocytes transdifferentiation into brown adipocytes induced by triiodothyronine

A. Korac[1], M. Cakic-Milosevic[1], M. Ukropina[1], M. Grubic[1], K. Micunovic[1], V. Petrovic[2], B. Buzadzic[2], A. Jankovic[2], A. Vasilijevic[2], B. Korac[2]

1. Institute of Zoology, Faculty of Biology, University of Belgrade, Studentski trg 16, 11000 Belgrade, Serbia
2. Institute for Biological Research »Sinisa Stankovic«, University of Belgrade, Despota Stefana 142, 11000 Belgrade, Serbia

sankor@bf.bio.bg.ac.yu
Keywords: white adipocyte, brown adipocytes, triiodothyronine, UCP1

The uncontrolled expansion of white adipose tissue (WAT) seen in obesity predisposes affected individuals to health complications. Another type of fat, brown adipose tissue (BAT) has an opposing physiological function because it allows dissipation instead of storage of energy. Recent studies on the transcriptional control of adipocyte differentiation offer a new perspective on conversion between brown and white adipocytes. The ability of the adipose organ to interconvert its main cytotypes in order to meet changing metabolic needs is highly pertinent to the physiopathology of obesity and related to therapeutic strategies [1]. Thyroid hormones (TH) have an important role in modulation of brown adipose tissue (BAT) thermogenic function. The main objective of this work was to investigate capability of triiodothyronine to induce structural and functional changes of rat white adipocytes in retroperitoneal white adipose tissue (RWAT).

Male Wistar rats maintained under standard laboratory conditions were used in this study. The animals from the experimental group (n=6) were treated with triiodothyronine (Sigma, St. Louis, MO, USA) for a five days. The other animals (n=6) were untreated controls. At the end of the experiment, interscapular BAT was isolated and routinely prepared either for light and electron microscopy. Immunohistochemistry was carried out on paraffin-embedded tissues with the avidin–biotin–peroxidase (ABC) method by using the UCP1 primary antibody. Ultrathin sections, contrasted with uranyl-acetate and lead citrate were examined with a Philips CM12 electron microscope.

Light microscopic examination of the retroperitoneal white adipose tissue after 5 days treatment with triiodothyronine showed numerous multilocular as well as unilocular adipocyte positive for UCP1, a marker of brown adipocytes (Figure 1A). Under electron microscope examination, however, the complement of cytoplasmic organelles, especially nucleus and mitochondria of these cells differed from that typical for the white adipocyte (Figure 1B, C). They contained large active nucleus and elongate, cristae-rich mitochondria with dense matrix (Figure 1C), quite dissimilar from the roundish mitochondria of typical white adipocytes. Although the one large lipid droplet was present in the cytoplasm of the white adipocyte, a numerous smaller once

A. Aretz, B. Hermanns-Sachweh, J. Mayer (Eds.): EMC 2008, Vol. 3: Life Science, pp. 123–124, DOI: 10.1007/978-3-540-85228-5_62, © Springer-Verlag Berlin Heidelberg 2008

were also notable. Furthermore, all lipid droplets were surrounded by mitochondria suggesting intensive lypolysis (Figure 1B, C).

Our results, presented here showed that the white adipocytes acquire features of brown adipocytes, i.e. an induction of UCP1 and respiratory chain gene expression, and an increased capacity to oxidize fatty acid. UCP1 biosynthesis is mainly controlled at the transcriptional level. During cold exposure, sympathetic nervous system stimulation of BAT is the primary signal that activates UCP1 gene expression. Retinoic acid (RA) and thyroid hormones are other positive regulators [2]. Based upon these facts, we suggest that triiodothyronine induced conversion of white adipocyte into brown one, e.g. their transdifferentiation. Pharmacotherapy targeted at molecular pathways that regulate adaptive thermogenesis provides a plausible and safe means of increasing energy expenditure. Reactivation of brown adipocytes is therefore an important goal.

1. S. Cinti. J. Endocrinol. Invest. **25** (2002), p. 823.
2. J. Markader, J. Ribot, I. Murano, F. Felipe, S. Cinti, M.L. Bonet and A. Palou. Endocrinology. **147** (2006), p. 5325.
3. The Ministry of Science and Environmental Protection of the Republic of Serbia (Grant No.143050) supported this work.

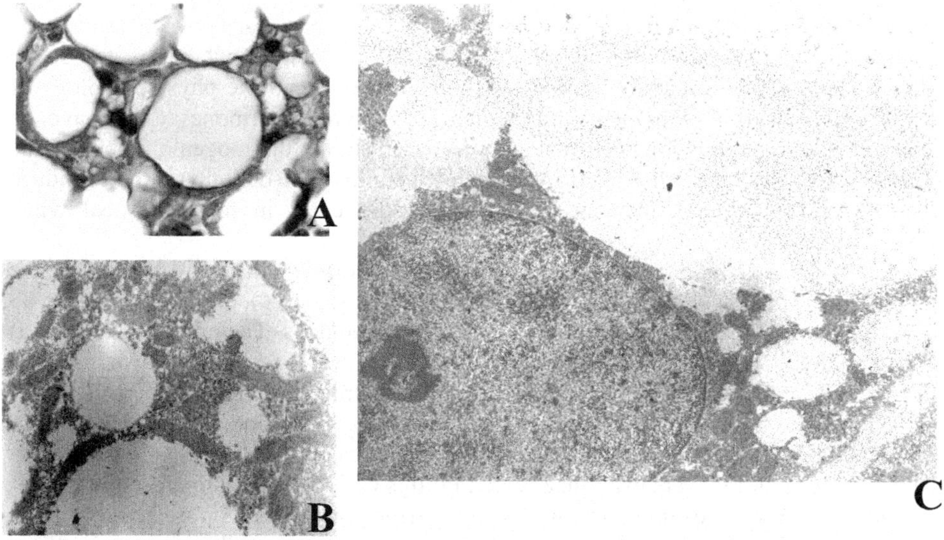

Figure 1. Retroperitoneal white adipose tissue of triiodothyronine-treated rats. (A-C). UCP1-positive white adipocytes (A) and their ultrastucture (B). Nucleus is active, numerous small lipid droplets appeared, and large, elongated mitochondria surrounds lipid droplets. Magnification: (A) x100, orig; (B and C) x 6300 orig.

Ultrastructural studies of the tegument of cestodes (Platyhelminthes): phylogenetic implications

C. Levron

Institute of Parasitology, Biology Centre of the Academy of Sciences of the Czech Republic, Branišovská 31, 37005 České Budějovice, Czech Republic

levron@paru.cas.cz

Keywords: tegument, microtriches, Cestoda, phylogeny

Electron microscopy has been used for a long time to study different organs and structure of parasitic flatworms (Platyhelminthes: Neodermata). Special attention has been paid to the fine structure of the tegument, which is a unique surface coverage of trematodes, cestodes and monogeneans [1, 2]. In 1985, Ehlers proposed the term "neodermis" to characterize the structure that replaces epidermis during ontogenetic development of parasitic flatworms. The neodermis (tegument) consists of a syncytium that covers the body in a continuous layer connected to perikarya that lie beneath the body-wall musculature [3]. The tegument is highly adaptated for a parasitic life-style. Neodermatan taxa can be distinguished from each others by differences in the structures covering the tegument (microvillli, microtriches).

Most distinctive is the tegument of tapeworms (Cestoda) which bears unique structures, microtriches. They are specialized surface extensions of the neodermis, characterized by possession of an electron-dense cap composed of numerous microtubules separated from distal cytoplasm by a base-plate [4]. They have multiple functions such as amplification of the surface area for digestion and absorption, excretion, movement, attachment to the intestinal mucosa of their hosts and external protection [1, 5]. The morphology of microtriches, mainly their shape, has been found to vary among species, life-cycle stages and body regions [6]. Two main types of microtriches have been described in cestodes, i.e. filiform and spiniform ones, each including several subtypes [5, 7]. Filiform microtriches (Figures 1A, B) are considered to be an ancestral type [8]. Spiniform microtriches (Figure 1C) have a relatively broad base and contain a long electron-dense cap [7].

Microtriches of cestodes supposed to be the most basal orders (Gyrocotylidea, Spathebothriidea, Diphyllobothriidea, Caryophyllidea) are mainly filiform ones (Figures 1A, B). However, their total length can vary among species. Spiniform microtriches (Figure 1C) first appear in more derived cestode groups (Haplobothriidea and Bothriocephalidea). They possess a great variety of shape, especially in parasites of elasmobranchs (Diphyllidea and Trypanorhyncha) [5, 9]

Type and distribution of microtriches are considered to be of systematic and phylogenetic importance. However, the existing knowledge is fragmentary and descriptions of microtriches need to be done using scanning and transmission electron microscopy.

A. Aretz, B. Hermanns-Sachweh, J. Mayer (Eds.): EMC 2008, Vol. 3: Life Science, pp. 125–126, DOI: 10.1007/978-3-540-85228-5_63,

1. M.K. Jones, International Journal for Parasitology **28** (1998), p. 913-923.
2. D.H. Halton, Micron **35** (2004), p. 361-390.
3. S. Tyler and M. Hooge, Canadian Journal of Zoology **82** (2004), p. 194-210.
4. J. Caira and D.T.J. Littlewood in "Encyclopedia of Biodiversity", ed. S.A. Levin (Academic Press, San Diego) (2001), p. 863-899.
5. H.W. Palm in "The Trypanorhyncha Diesing, 1863", ed. PKSPL-IPB Press (Bogor) (2004), p. 35-47.
6. R.C.A. Thompson, A.R. Hougton and L.P. Jue Sue, Zeitschrift für Parasitenkunde **64** (1980), p. 95-111.
7. J. Caira, K. Jensen and C.J. Healy, Systematic Parasitology **42** (1999), p. 77-151.
8. J.M. Holy and J.A. Oaks, Cell Tissue and Research **244** (1986), p. 457-466.
9. G.A. Tyler, Bulletin of the University of Nebraska State Museum **20** (2006), p. 1-142.
10. We are grateful to the staff of the Laboratory of Electron Microscopy, Institute of Parasitology, České Budějovice, Czech Republic. This study was supported by the Grant Agency of the Czech Republic (projects nos. 524/07/P039 and 524/08/0885), Research Centre of Ichthyoparasitology (LC 522) and the Institute of Parasitology (Z60220518).

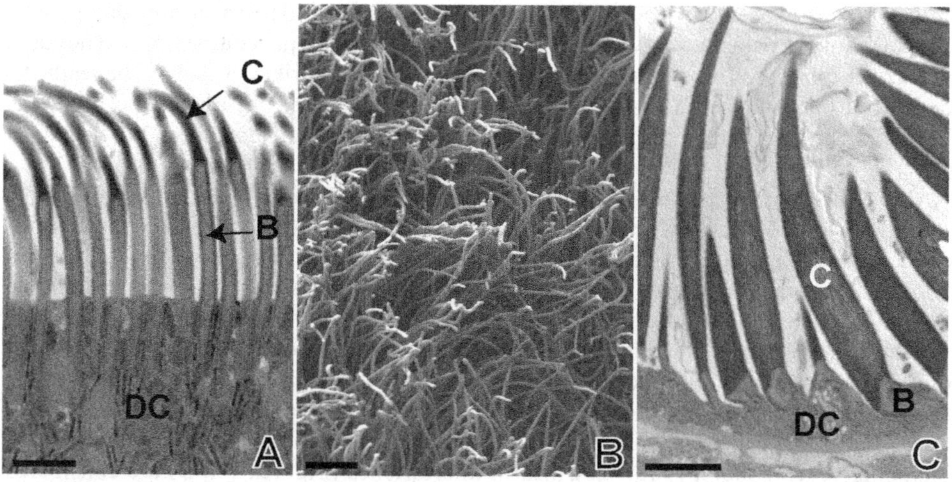

Figure 1. Ultrastructure of the tegument of fish tapeworms. **A.** TEM photomicrograph of filiform microtriches of *Caryophyllaeides fennica* (Caryophyllidea). Bar = 0.5 µm. **B.** SEM photomicrograph of long filiform microtriches of *Cyathocephalus truncatus* (Spathebothriidea). Bar = 2 µm. **C.** TEM photomicrograph of spiniform microtriches of *Paraechinophallus japonicus* (Bothriocephalidea). Bar = 1 µm.

Crystallization stages of the $CaCO_3$ deposits in the earthworm's calciferous gland

J. Méndez[1], J. B. Rodríguez-González[1], R. Alvarez-Otero[2], M.J.I. Briones[3], L. Gago-Duport[4]

1. Servicio de Microscopía Electrónica, CACTI. Universidad de Vigo, 36310 Vigo, España.
2. Departamento de Biología Funcional y Ciencias de la Salud. Universidad de Vigo, 36310 Vigo, España.
3. Departamento de Ecología y Biología Animal. Universidad de Vigo, 36310 Vigo, España
4. Departamento de Geociencias Marinas. Universidad de Vigo, 36310 Vigo, España

susomen@uvigo.es
Keywords: ACC, biomineralization, earthworms .

Calcium carbonate is the most widespread mineral phase involved in the biomineralization processes occurring in invertebrates. Besides the usual $CaCO_3$ crystalline modifications, i.e. vaterite, calcite and aragonite, an increasing number of investigations have demonstrated the presence of amorphous calcium carbonate (ACC) as a transient phase that plays an important role in the initiation of the biomineralization process.

In the earthworm's calciferous gland, $CaCO_3$ can be found in two different aggregation states [1,2]: solid concretions and as a suspension (known as 'milky fluid' in the literature). In this work we have analysed the morphological and microstructural aspects of both materials separately by means of electron microscopy techniques.

We performed Cryo-SEM analyses (Figure 1) of 20μm sections of the calciferous gland to characterise the morphological and compositional aspects of the freshly formed carbonate deposits. Further microstructural analyses allowed the identification of the transitional sequences towards more stable carbonates (Figure 2). Results showed that the spherulithic deposits merely consisted of ACC partially transformed to vaterite. Transformations of ACC to calcite takes place preferably on the surfaces of these amorphous globular aggregates, as revealed by the fact of their characteristic smooth characteristic surface becoming rougher with time, and suggest that transformations are solvent-induced following an Ostwald ripening mechanism. This transitional path was not unique and the presence of aragonite, as an intermediate phase, was also found. In this particular case, the transition process followed a completely different pathway with the crystallisation starting at the centre of the sphere, then progressively extending towards the periphery and finally leading to the formation of radial aggregates.

1. Briones M.J.I., Ostle N.J. and Piearce T.G. (2008). Stable isotopes reveal that the calciferous gland of earthworms is a CO_2-fixing organ. Soil Biology and Biochemistry 40, 554-557.
2. Gago-Duport L., Briones M.J.I., Rodríguez J.B. and Covelo B. (in press). Amorphous calcium carbonate biomineralization in the earthworm's calciferous glands. Journal of Structural Biology. doi:10.1016/j.jsb.2008.02.007

A. Aretz, B. Hermanns-Sachweh, J. Mayer (Eds.): EMC 2008, Vol. 3: Life Science, pp. 127–128, DOI: 10.1007/978-3-540-85228-5_64,

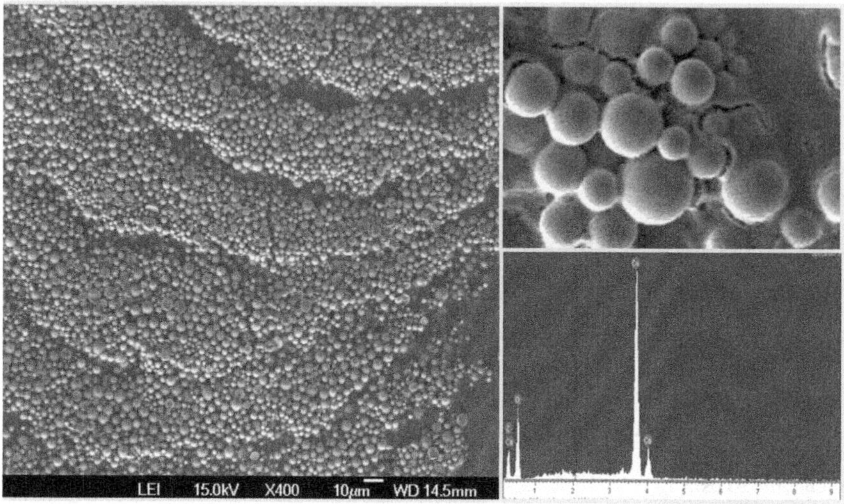

Figure 1. Cryo-SEM image of $CaCO_3$ deposits formed in the earthworm's calciferous gland. The initially formed spherulites are ACC embedded by an organic matrix.

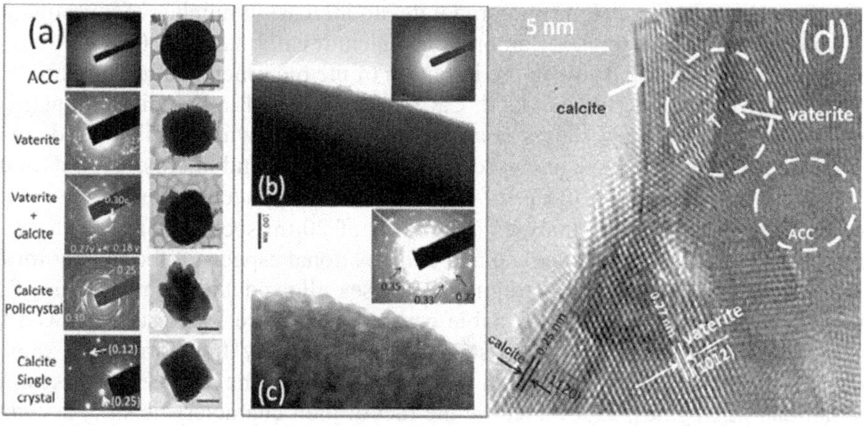

Figure 2. (a, b,c) TEM image sequence showing the morphological and structural evolution of the ACC spherulites. (d) HREM lattice fringes image which evidences the vaterite to calcite transformation occurring at the spherulite surfaces.

Scanning electron microscopic analysis of heavy metal resistant microorganisms

S. Nietzsche[1], A. Schmidt[2], E. Kothe[2], M. Westermann[1] and W. Richter[1]

1. Elektronenmikroskopisches Zentrum, Universitätsklinikum Jena, Ziegelmühlenweg 1, 07743 Jena, Germany
2. Institut für Mikrobiologie, Biologisch-Pharmazeutische Fakultät, Friedrich-Schiller-Universität, Neugasse 25, 07743 Jena, Germany

sandor.nietzsche@uni-jena.de
Keywords: heavy metal resistance, nickel, Streptomyces

Industrial or mining activities can cause serious environmental problems by the contamination of large areas. In these polluted habitats selection pressure have led to adaptation in microorganisms now containing special resistance mechanisms as a result of the permanent exposure to exceedingly high concentrations of heavy metals. Recently van Nostrand et al. [1] isolated four actinobacterial strains, among them two of the genus Streptomyces, from contaminated riparian sediments. One of the Streptomyces strains was able to grow on 85.2 mmol/l nickel which is to the best of our knowledge the highest detected nickel resistance so far.

In this work, a unique Streptomyces mirabilis strain that shows a very high resistance to nickel was analyzed. This strain is able to grow on concentrations higher than 100 mmol/l nickel [2,3]. Additional experiments revealed that it is also able to cope with 100 mmol/l zinc. The survival at the contaminated site could be mimicked using media prepared with soil that present the composition of mobile and adsorbed heavy metals in axenic culture that are present in the environment and determine the soil microflora composition. This analyzed strain was found at the banks of a creek in a former uranium mining area near Ronneburg, Germany. Soil samples of these banks of the creek Gessenbach contain high amounts of a range of heavy metals as a consequence of mining procedures over several decades. Therefore, the creek Gessenbach appears to be a good location for the search for heavy metal resistant microorganisms.

The scope of this contribution was to show the deposition of nickel and zinc in the Streptomyces strains by means of backscattered electron images using a LEO-1530 scanning electron microscope. After grey level calibration with element standards the material contrast was used to sense metals. Different electron energies and spot sizes were used to localize metal inclusions and estimate the penetration depth. Also strains of several growth times were tested to analyze the deposition dynamics.

This work aims to increase the understanding of the mechanism of heavy metal resistance of microorganisms. This knowledge could lead for example to microbiologic methods for decontamination of polluted areas in the future.

A. Aretz, B. Hermanns-Sachweh, J. Mayer (Eds.): EMC 2008, Vol. 3: Life Science, pp. 129–130,
DOI: 10.1007/978-3-540-85228-5_65,

1. J.D. Van Nostrand, T.V. Khijniak, T.J. Gentry, M.T. Novak, A.G. Sowder, J.Z. Zhou, P.M. Bertsch and P.J. Morris, Microb. Ecol. **53** (2007), pp. 670-682.
2. A. Schmidt, G. Haferburg, A. Schmidt, U. Lischke, D. Merten, F. Ghergel, G. Büchel and E. Kothe, Chemie der Erde – Geochemistry (2008), corr. proof in press.
3. G. Haferburg, D. Merten, G. Büchel and E. Kothe, J Basic Microbiol. **47**(2007), pp. 474-84.

Characterization of mouse embryoid bodies by Scanning Electron Microscopy (SEM) and Confocal Laser Scanning Microscopy (CLSM)

C. Nogués, R. Arroyo and N. Gaztelumendi

Departament Biologia Cel·lular, Fisiologia i Immunologia, Facultat Biociències, Universitat Autònoma de Barcelona, 08193-Bellaterra. Spain.

carme.nogues@uab.cat
Keywords: SEM, CLSM, Embryoid bodies

When embryonic stem cells (ESCs) are cultured in suspension, they spontaneously differentiate in vitro and form three-dimensional multicellular aggregates called embryoid bodies (EBs). EBs are composed of ectodermal, mesodermal, and endodermal tissues which reproduce many characteristics of cell differentiation during early mammalian embryogenesis. The formation of embryoid bodies is the principal step in a differentiation process of embryonic stem cells.

To maintain their pluripotency and self-renewal capacity mouse ESCs are cultured on feeder layers where they form very well delimited colonies. Colonies could be dissociated into single cell suspension and seeded into bacterial-grade dishes. Isolated ES cells cultured in non-adherent surfaces could not be attached and aggregate to form spherical three dimensional structures resembling an early embryo, the EBs. EB formation takes between 2 and 6 days, generating heterogeneous EBs in size and shape. When harvested and transferred to culture dishes, they attach and start to differentiate into a variety of cell types.

When ES cells are subcultured for a long period, their capacity to form well delimited colonies decrease. The number of EBs formed is lower and their potential to attach to culture adherent dishes and to differentiate diminishes. The pluripotentiality decrease with passage number.

We have analysed the capacity of two different mouse ESC lines (E14 and J1) to differentiate into embryoid bodies. To maintain their pluripotentiality, ESCs have been cultured on feeder layers for different period of time, up to 35 passages for J1 and 40 passages for E14. The evolution of both cell lines have been characterized (Fig. 1) through number, size and shape of the formed EBs. After chemical fixation, EBs have been further analysed by scanning electron microscopy (SEM) (Fig. 2) and confocal scanning laser microscopy (CSLM) (Fig. 3).

J1-cell line have kept their capacity to form delimited colonies and to produce EBs up to 35 passages while E14 line have loose their capacity to form delimited colonies and have decreased the production of EBs before passage 35.

A. Aretz, B. Hermanns-Sachweh, J. Mayer (Eds.): EMC 2008, Vol. 3: Life Science, pp. 131–132, DOI: 10.1007/978-3-540-85228-5_66, © Springer-Verlag Berlin Heidelberg 2008

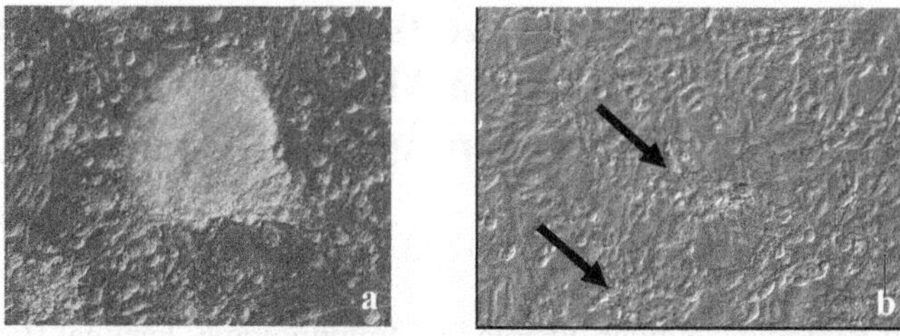

Figure 1. Phase-contrast photomicrographs of mouse ESC colonies. a) J1-ESC line with a well delimited colony and b) E14-ESC line with two diffuse colonies

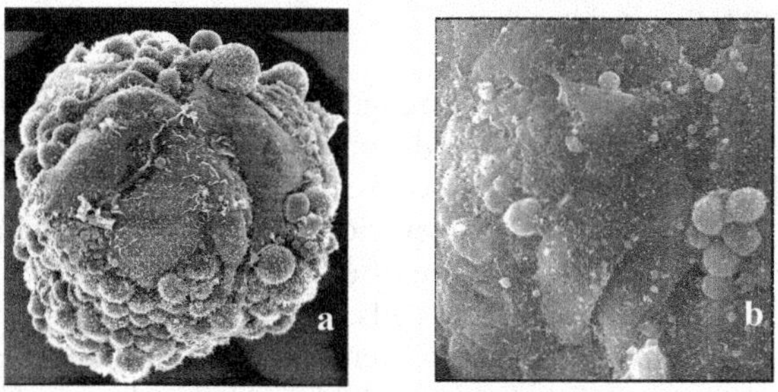

Figure 2. SEM (scanning electron microscope) images of a) mouse EBs from E14 ESC line and b) detailed morphology

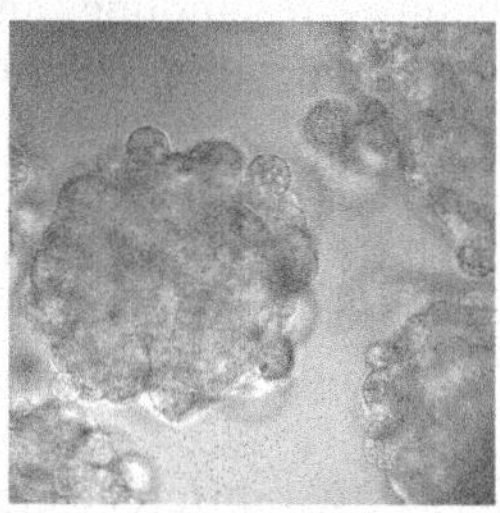

Figure 3. CLSM (confocal laser scanning microscope) image of a mouse EB.

Stage-dependent localization of mitochondrial DNA during the cell cycle of Euglena gracilis Z by immunogold electron microscopy

T. Osafune, N. Kiyohara, I. Watanabe and T. Ehara

Department of Life Science, Nippon Sport Science University, Yokohama 227-0033, and Department of Microbiology, Tokyo Medical University, Tokyo 160-8402, Japan

osafunet@aol.com
Keywords: Euglena, mitochondria, anti-DNA mitochondrial antibody

Examination of *Chlamydomonas* and *Euglena* mitochondrial morphology in synchronized cells by high resolution fluorescence microscopy, serial ultra-thin section-electron microscopy, and computer graphic analysis has found dynamic morphological changes such as fusion, fission, and branching during the cell cycle. Giant *Euglena* and *Chlamydomonas* mitochondria are formed by fusion of smaller reticular mitochondria at specific stages of the cell cycle. Giant mitochondria formation is associated with a temporary reduction of respiratory function suggesting a role in the cell cycle for giant mitochondria other than their primary respiratory function[1,2,3]. We examined the localization of mitochondrial DNA in synchronized cells of *Euglena gracilis* Z [2]. Mitochondrial DNA was localized by fixation of cells, staining with an anti-DNA monoclonal antibody, immuno-electron microscopy, and computer graphic analysis. Cross sections of normal size reticulate mitochondria in cells sampled 4 hours after the onset of light showed only a few colloidal gold particles confined to specific regions of the matrix while in other mitochondria, DNA molecules were not observed. Serial ultra-thin sectioning and immuno-electron microscopy found colloidal gold particles homogeneously distributed throughout the matrix of giant mitochondria. This demonstrates for the first time the distribution of mitochondrial DNA throughout the matrix of giant mitochondria suggesting that individual DNA molecules are not spatially separated as found in reticulate mitochondria. If mitochondria differ in their genetic composition, the temporary formation of giant mitochondrion through mitochondrial fusion would provide an opportunity during the cell cycle for gene exchange between individual mitochondrial DNA molecules and for mitochondrial DNA replication prior to mitochondrial division.

1. T.Osafune et al., Plant Cell Physiol.13 (1972) p.211.
2. T.Osafune, J.Electron Microsc.22 (1973) p.51.
3. T.Ehara, T.Osafune and E.Hase, Exp.Cell Res., 193 (1991) p.320.

A. Aretz, B. Hermanns-Sachweh, J. Mayer (Eds.): EMC 2008, Vol. 3: Life Science, pp. 133–134,
DOI: 10.1007/978-3-540-85228-5_67,

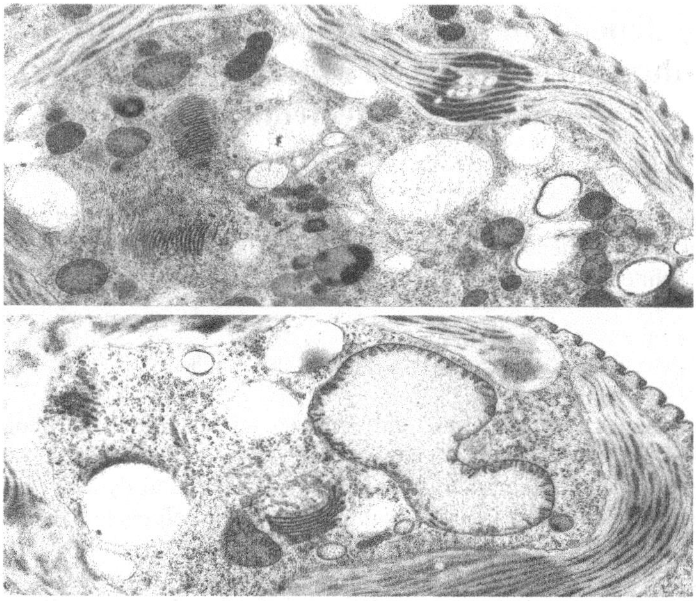

Figure 1,2. Cells ware fixed with freeze substitution method and the cross section of reticular mitochondria in cells at 4 h of synchronous culture (Fig.1) and giant mitochondria at 9 h (Fig.2) was observed by electron microscopy.

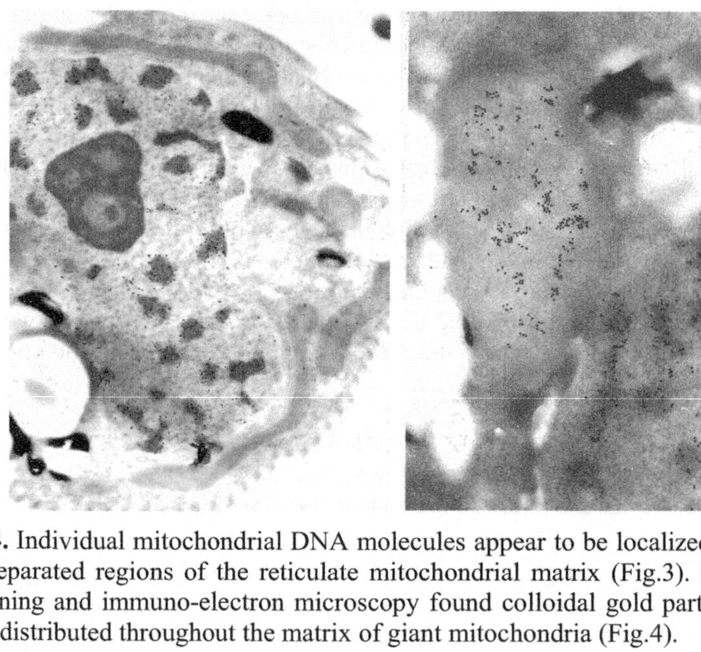

Figure 3,4. Individual mitochondrial DNA molecules appear to be localized to discrete spatially separated regions of the reticulate mitochondrial matrix (Fig.3). Serial ultrathin sectioning and immuno-electron microscopy found colloidal gold particles homogeneously distributed throughout the matrix of giant mitochondria (Fig.4).

The virus-inducted structures in RNA virus infected macrophages

Plekhova N.G.[1], Somova L.M.[1], and E.V. Pustovalov[2]

1. Lab. of Electron Microscopy and Pathomorphology, Institute of Epidemiology and Microbiology, Vladivostok, Russia
2. Lab. of Electron Microscopy, Far East National University, Vladivostok, Russia

pl_nat@hotmail.com
Keywords: virus, macrophages, ultrastructure

Beside nonspecific character pathology of cells infected by virus, within its the typical for viral infection changes can be defined as virus-inducted structures – the viroplasts, polyribosomal filaments, microfibrils and others. The system of mononuclear phagocytes phylogenically belongs to the most ancient system of immunity. Experimental evidences suggest that macrophages play the important role in resistance to viral infection [1]. In this study, we tested the ultrastructure of infected RNA enveloped (Hantavirus, caused hemorrhagic fever with renal syndrome, and Tick Borne encephalitis virus, TBEV) and nonenveloped (poliovirus, enteroviruses ECHO11, Coxsackie B1 and 71) viruses primary macrophages of mice.

The formation of virus-inducted structures in cytoplasm of cells infected by these viruses, aside from poliovirus, was fixed. The appearance of virus-inducted structures was indicated on the viral replication in determinate region of cell cytoplasm – fabrics of viral synthesis. The most intensive formation of viroplasts, polyribosomal filaments and microfibrils were identified after 9 hours post-infection, and near its the virions formation were determined "Figure 1". Moreover, in the event infection caused by nonenveloped virus the activity of synthesis of virus ECHO11 components in macrophages was more expressed, than in cells infected by Coxsackie B1 and 71 viruses. The RNA of these viruses was bound with ribosomes that expressed in form of polyribosomal filaments.

The morphology of viroplasts in Hantavirus infected macrophages differed from similar structures formed in cells infected by another viruses "Figure 2". The formation three types of viroplasts in cytoplasm of resident macrophages infected by Hantavirus was demonstrated: 1) the compactness viroplasts; 2) viroplasts forming from the first and surrounding by two-layer membrane and including structure like of mitochondria cristeas, 3) lamellar and tube structures. The synthesis of virus nucleoproteins was localized on the surface of first and second type viroplasts, then the envelope protein formed in lamellar and tube structures. The result of ultrastructure study were indicated that the places of hantavirus components synthesis had different localization. In infected by other viruses macrophages we did not observed the similar delimitations of places of the virus component synthesis.

1. V. Maréchal, M.-C. Prevost, Petit C., Perret E., Heard J.-M., Schwart O. J. Virology, 75, 2001, p. 11166-11177.

A. Aretz, B. Hermanns-Sachweh, J. Mayer (Eds.): EMC 2008, Vol. 3: Life Science, pp. 135–136,
DOI: 10.1007/978-3-540-85228-5_68,

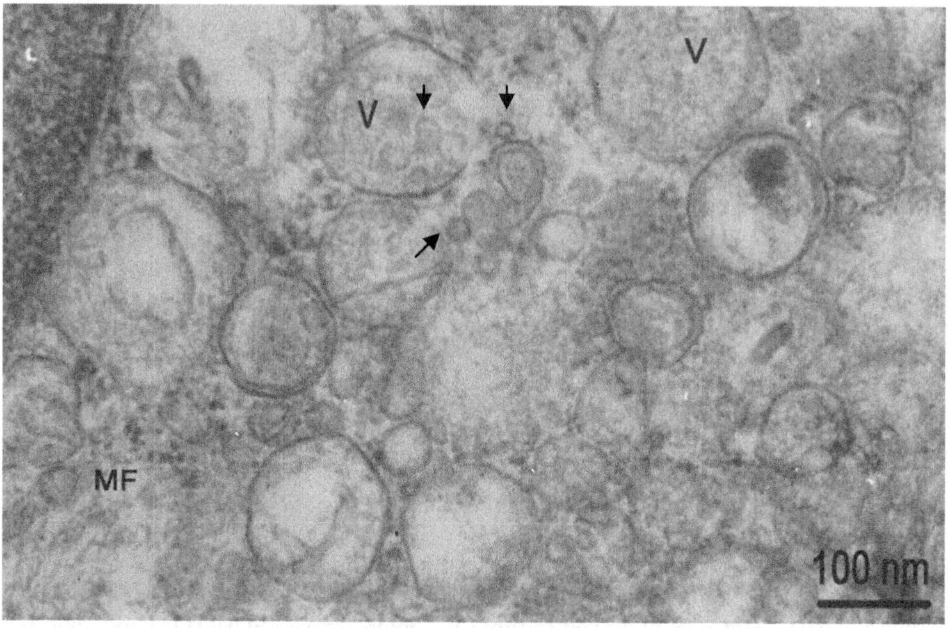

Figure 1. Virus-inducted structures in enterovirus ECHO11 (→) infected macrophages. V – viroplasts, MF – microfibrils; 9 hours post infection.

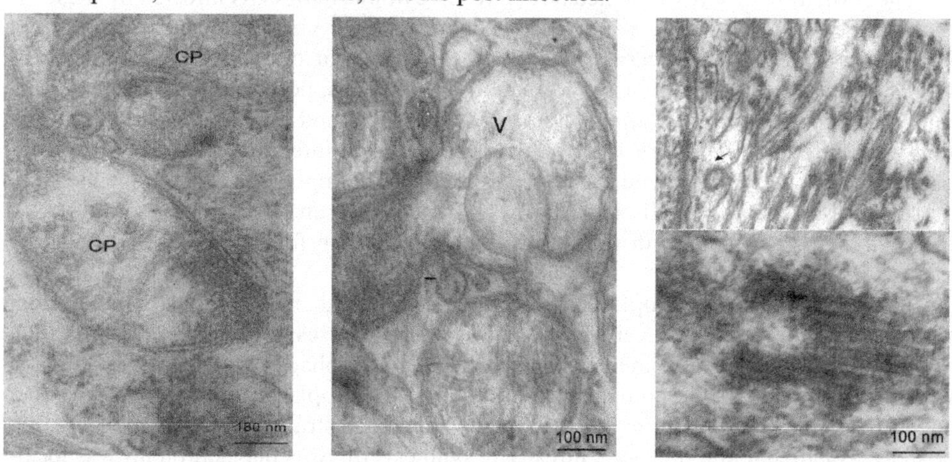

Figure 2. Virus-inducted structures in Hantavirus (→) infected macrophages: a) the compactness viroplasts; b) viroplasts forming from the first and surrounding by two-layer membrane and including structure like of mitochondria cristea, c) lamellar and tube structures.

Surface layers of ore-leaching Bacteria and Archaea

R. Rachel[1], A. Klingl[1], C. Neuner[1], W. Depmeier[2], G. Schmalz[3], M. Thomm[1], H. Huber[1]

1. Dept of Microbiology / Centre for Electron Microscopy, University of Regensburg, Universitaetsstr 31, D-93053 Regensburg, Germany
2. Inst of Geoscience, University of Kiel, Olshausenstr. 40, D-24098 Kiel, Germany
3. Dept of Conservative Dentistry and Periodontology, University Hospital of Regensburg, Franz-Josef-Strauß-Allee 11, D-93053 Regensburg, Germany

reinhard.rachel@biologie.uni-r.de
Keywords: Surface layers, cell-surface interaction, cryo-preparation, image processing

Etching experiments on pyrite [1] were performed with four different strains of microorganisms. Two are mesophilic Bacteria, *Acidithiobacillus ferrooxidans* SP5/1 and *A. caldus* HV2/2, and two are thermophilic Archaea, *Metallosphaera sedula* TH2 and *Sulfolobus metallicus* Kra23[2, 3]. They are all characterized by their ability to perform ore-leaching, as one (or the only) way to gain metabolic energy. The mechanism of this process is not understood yet.

For these microorganisms, different modes of ore-leaching activities are discussed, 'contact' mechanism vs. 'non-contact' mechanism [4], depending on whether the cells are in contact or not in contact with the pyrite crystal surface. In order to understand the bio-molecules involved, we currently investigate the cell surface molecules of the microbes, using a variety of transmission electron microscopy techniques, in particular freeze-etching and shadowing, and sections following high-pressure freezing, freezing substitution, and resin embedding. So far, we found pili on cells of both bacterial strains, and S-layers on all four strains (Figure 1). The S-layers formed well-ordered arrays and fully covered the surface of all strains; they appeared to be in direct contact with the mineral surface.

S-layer sheets were detached from the cells by detergent-treatment, and further characterized as negatively-stained samples, incl. image processing. We found characteristic differences between the S-layers of the Bacteria and of the Archaea: both strains of *Acidithiobacillus* exhibited an S-layer with an almost identical, unusual arrangement of protein subunits on a p2 lattice (Figure 2); they were composed of a single glycoprotein with a molecular mass of about 90 kDa. The S-layers of the two Archaea exhibited p3 symmetry, as already found for other *Sulfolobus* strains [5], and were each composed of at least one glycoprotein with a mass at about 150 to 200 kDa.

At present, we aim to extent our studies by determining the 3D structure of the novel bacterial S-layers. In addition, specific antibodies are raised, in order to visualize the distribution of the main protein and to test whether antibodies are sufficient to detach the cells from the pyrite surface.

1. K.J. Edwards et al., Environm Microbiol **2** (2000), 324.
2. G. Huber et al, Syst Appl Microbiol **12** (1989), 38.

A. Aretz, B. Hermanns-Sachweh, J. Mayer (Eds.): EMC 2008, Vol. 3: Life Science, pp. 137–138,
DOI: 10.1007/978-3-540-85228-5_69, © Springer-Verlag Berlin Heidelberg 2008

3. G. Huber et al, Syst Appl Microbiol **14** (1991), 372.
4. W. Sand and T. Gehrke, Res. Microbiol. **157** (2006), 49.
5. G. Lembcke and W. Baumeister, J. Bioenerg. Biomembr. **24** (1992), 567.
6. This work is supported by a grant of the DFG to WD, GS, MT, HH, and RR.

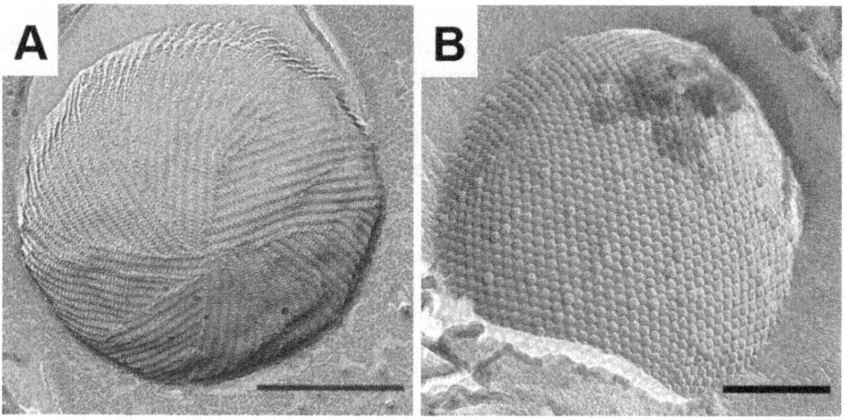

Figure 1. Transmission electron micrographs of freeze-etched cells of *Acidithiobacillus ferrooxidans* (A) and *Sulfolobus metallicus* (B). Bar, 0.2 μm

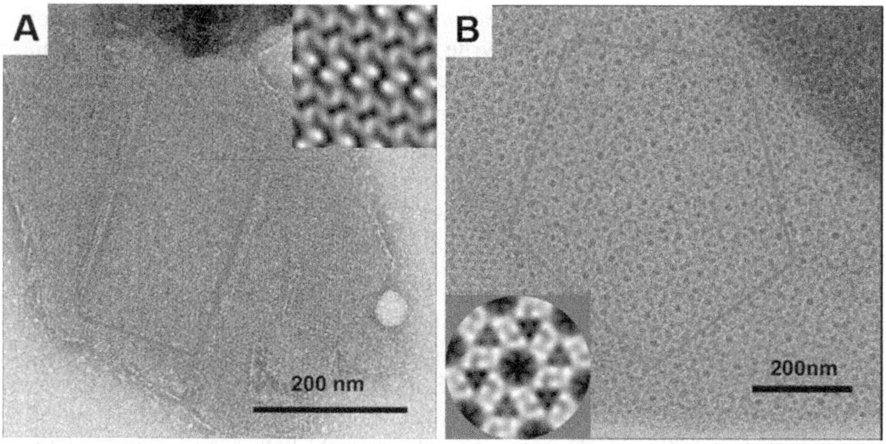

Figure 2. Examples of transmission electron micrographs of isolated S-layer sheets of ore-leaching Bacteria and Archaea: *Acidithiobacillus ferrooxidans* (A) and *Sulfolobus metallicus* (B). Bar, 0.2 μm. Insets, results of image processing of small S-layer patches using correlation averaging.

Structure of *Bacillus anthracis* spores analyzed by CEMOVIS

M. Sachse[1], G. Pehau-Arnaudet[1], E. Couture-Tosi[2], J.-L. Ranck[2], P. Sylvestre[2], M. Mock[2]

1. Plateforme de Microscopie ultrastructurale, Institut Pasteur, Paris
2. Unité des Toxines et Pathogénie Bactérienne, Institut Pasteur, Pasteur

msachse@pasteur.fr
Keywords: *Bacillus anthracis*, CEMOVIS, spore

Bacillus anthracis is a Gram+, aero-anaerobic, endospore-forming bacterium and the causative agent of anthrax disease. Within the life cycle of the bacteria the endospores represent the disseminating agent that is taken up from the environment by mammals. The transformation of the dormant spore into a vegetative cell is an important step in the pathogenesis of the bacterium [1]. We analyzed the structure of the spore by Cryo-ElectronMicroscopy Of Vitrified Sections (CEMOVIS), which allows observation of the sample as close to the native state as currently possible. After high pressure freezing the sample is sectioned in the frozen, hydrated state and the sections are observed at -180°C.

With aldehyde fixation, dehydration and embedding in EPON the mature spore consists of an electron dense core successively surrounded by the inner membrane, the cortex, the outer membrane, followed by the coat and finally the outermost structure, the exosporium. On CEMOVIS sections the core of the spore provides a surface with fractures, likely to be caused by the dense structure of the material. As observed after aldehyde fixation, the coat consists of several layers. However, on the CEMOVIS sections we identified new a repetitive structure representing the terminal layer of the coat. The distance from one element to the other was calculated to be 4 nm. We only observed this structure on frozen hydrated sections and not with any other preparation technique for electron microscopy. We are currently studying by CEMOVIS the morphological changes that occur during the transformation of the spore into a vegetative cell, which involves disappearance of the coat and the formation of the capsule of the vegetative bacteria. [2].

1. M. Mock, and A. Fouet, Anthrax. Annu. Rev. Microbiol. **55** (2001), p. 647-671.
2. We kindly acknowledge the support of M.C. Prevost (Institut Pasteur) for this study.

A. Aretz, B. Hermanns-Sachweh, J. Mayer (Eds.): EMC 2008, Vol. 3: Life Science, pp. 139–140,
DOI: 10.1007/978-3-540-85228-5_70,

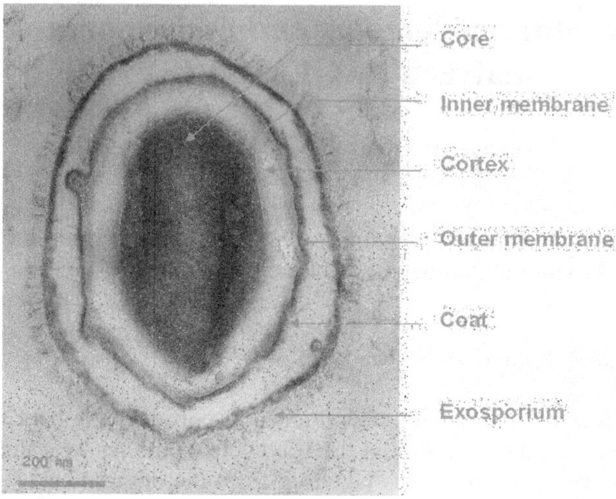

Figure 1. Spore of *Bacillus anthracis* after chemical fixation, dehydration and embedding in EPON. Bar: 200 nm

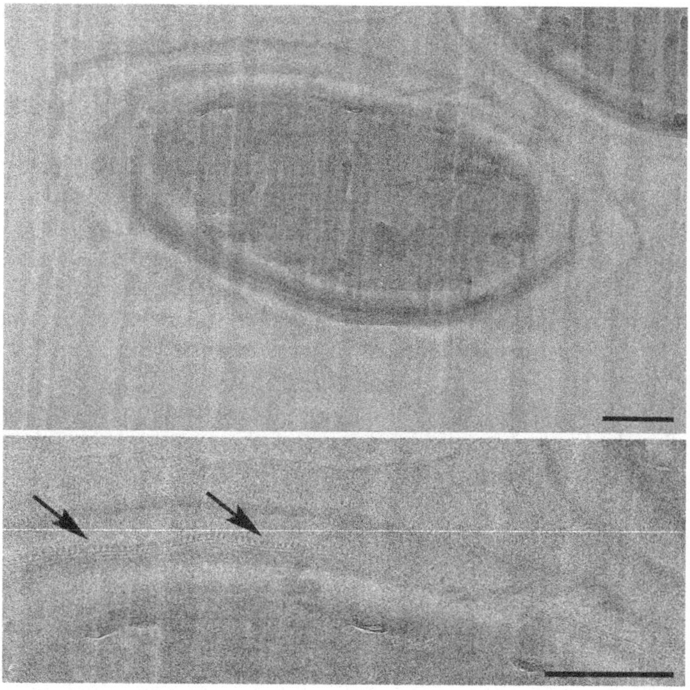

Figure 2. Spore after high pressure freezing and frozen hydrated sectioning. The upper picture shows an overview. A detailed view of the coat is shown in the lower picture. The terminal repetitive structure of the coat is highlighted by arrows. Bars: 200 nm

Insights into the role of pUL71 during HCMV morphogenesis

M. Schauflinger[1,2], S. Landwehr[1,2], P. Walther[1], Th. Mertens[2], and J. von Einem[2]

1. Zentrale Einrichtung Elektronenmikroskopie, Universität Ulm, 89081 Ulm, Germany
2. Abteilung Virologie, Institut für Virologie, Universitätsklinikum Ulm, 89081 Ulm, Germany

martin.schauflinger@uni-ulm.de
Keywords: HCMV, tegument protein pUL71, high pressure freezing

Human cytomegalovirus (HCMV) is a ubiquitous, double-stranded DNA virus that is a major cause of serious diseases in newborns and immunocompromised patients. In the HCMV virion, the DNA is packed in an icosahedral capsid, which is surrounded by a protein layer called the tegument, and a lipid envelope. The envelope is obtained by budding of tegumented capsids into vesicles, which are presumably trans-Golgi derived.

To explore the biological significance of the HCMV tegument protein pUL71, a UL71-stop mutant was generated by markerless BAC mutagenesis and reconstituted on human MRC-5 fibroblast cells. For electron microscopic investigation, MRC-5 cells were cultivated on carbon coated sapphire discs [1]. The cells were fixed with a high pressure freezer Compact 01 (Wohlwend, CH) 5 days after infection with wild-type and UL71-stop mutant virus. The samples were cryo substituted in a substitution medium containing acetone, osmiumtetroxide, uranyl acetate and 5% of water for 18 h [2]. Embedding in Epon was followed by sectioning of 70 nm slices. A Zeiss EM 10 was used on 80 kV to analyse the samples.

In contrast to wild-type virus infected cells, only few extracellular virus particles were observed in the sections of cells infected with the UL71-stop mutant. This goes along our observation that UL71-stop mutant infected cells contained significantly lower numbers of enveloped virus particles compared to wild-type virus. Interestingly, many UL71-stop mutant particles were found in the process of secondary envelopment (Fig. 1, arrows), suggesting an envelopment defect of this mutant, because those budding processes can only rarely be seen in wild-type infected cells.

Moreover, unenveloped, tegumented UL71-stop virus particles were found in contact with membranes of large multivesicular bodies (MVBs) and also budding into MVBs was observed (Fig. 1, arrowheads). In mutant virus infected cells MVBs are unusual large and seem to be increased in number, which could be a direct or indirect effect caused by the lack of pUL71.

1. Ch. Buser et al., J Virol. **81** (2007), p. 3042-3048.
2. P. Walther, and A. Ziegler, J Microsc. **208** (2002), p. 3-10.
3. This work is supported by the ***DFG Schwerpunktprogramm 1175***.

A. Aretz, B. Hermanns-Sachweh, J. Mayer (Eds.): EMC 2008, Vol. 3: Life Science, pp. 141–142,
DOI: 10.1007/978-3-540-85228-5_71,

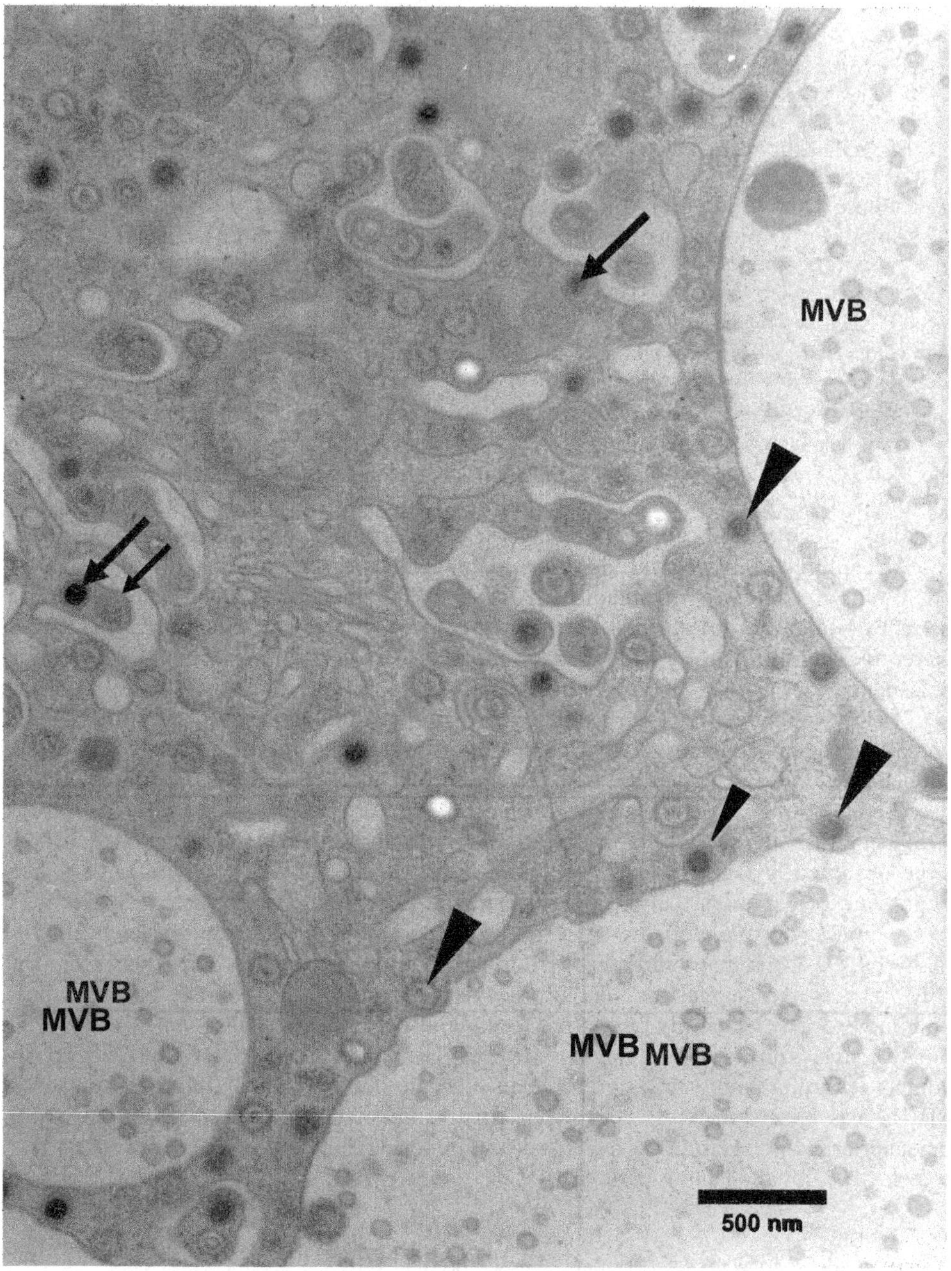

Figure 1. HCMV particles in the cytoplasm of a MRC-5 cell infected with UL71-stop mutant. Unenveloped, tegumented capsids were found budding in vesicles (arrows) and in close contact to large MVBs (arrowheads).

Fine structure of the genital chamber of female *Chrysomya megacephala* (F.) (Diptera: Calliphoridae)

Sirisuda Siriwattanarungsee[1], Tarinee Chaiwong[1], Kom Sukontason[1], Urai Chaisri[2], and Kabkaew L. Sukontason[1]

1. Department of Parasitology, Faculty of Medicine, Chiang Mai University, Chiang Mai 50200, Thailand
2. Department of Tropical Pathology, Faculty of Tropical Medicine, Mahidol University, Bangkok 10400, Thailand

kikcmu@hotmail.com
Keyword: *Chrysomya megacephala*, genital chamber, scanning electron microscopy, transmission electron microscopy

As an efficient mechanical vector, nuisance pest and myiasis causing fly *Chrysomya megacephala* (F.) (Diptera: Calliphoridae) is the most medically important blow fly species in Thailand [1]. In this study we investigated the ultrastructure of the genital chamber of female *C. megcephala*, which has not been previously described, by using scanning electron microscopy (SEM) and transmission electron microscopy (TEM) [2]. SEM imaging showed that the genital chamber is located in the posterior region of the abdomen where it is occasionally penetrated by tracheoles before opening to the exterior. Cross-sectional TEM imaging of the genital chamber revealed the chamber consists of a central lumen surrounded by five distinct cell layers. The first-layer, located adjacent to the lumen, consists of the cuticle- and basement membrane of the cuticle (Figure 1 and Figure 2). In the second-layer, the tracheoles and mitochondria are distinguishable (10-15 mitochondria/cell). The third-layer also consists of tracheoles but mitochondria exist in much higher densities (75-85 mitochondria/cell), and the endoplasmic recticulum are observable. The fourth-layer consists of multi-layer cells, which each are elongate and contain approximately 100 mitochondria/cell. The fifth layer consists of cells with two different types of nuclei, characterized by either compact chromatin in the nucleus or dispersed chromatin in the nucleus. Also in this fifth layer few mitochondria are visible. These morphological characteristics may be useful in understanding the function of the genital chamber of *C. megacephala*. Moreover, these features suggest that each cell layer may have potential to be used as targets in fly control strategies attempt to disrupt reproduction in this blow fly species.

1. S. Sucharit and W. Tumrasvin, Japanese Journal of Sanitary Zoology 32 (1981), p. 281-285.
2. A. Bansal and H. Murad, Japanese Journal of Sanitary Zoology 38 (1987), p. 233-238.
3. We thank the Thailand Research Fund through the Royal Golden Jubilee Ph.D. Program and Faculty of Medicine, Chiang Mai University.

A. Aretz, B. Hermanns-Sachweh, J. Mayer (Eds.): EMC 2008, Vol. 3: Life Science, pp. 143–144, DOI: 10.1007/978-3-540-85228-5_72,

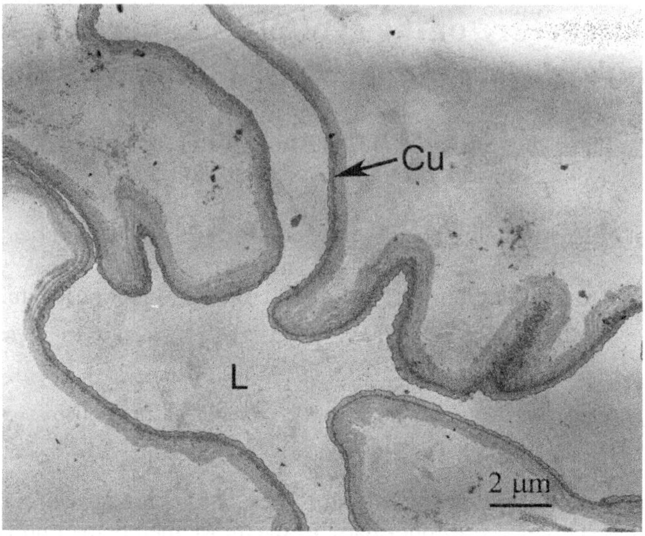

Figure 1. Transmission electron micrograph of the tubular genital chamber of a 3 day-old *Chrysomya megacephala* showing the first-layer located adjacent to the lumen (L) consists of thin cuticle (Cu).

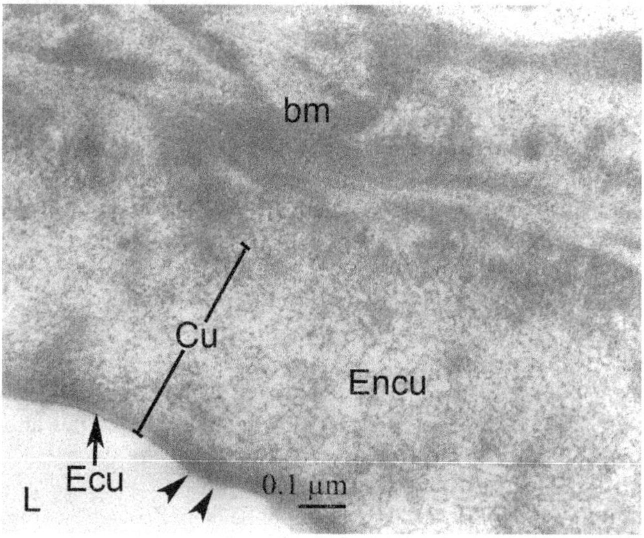

Figure 2. Transmission electron micrograph of the tubular genital chamber of a 3 day-old *C. megacephala* at the high magnification of the first-layer illustrating the cuticular layers (Cu) consisting of an epicuticle (Ecu) and the overlying endocuticle (Encu). A closer view of the first-layer showing the double-membrane (arrowheads) of the epicuticle (Ecu) and basement membrane (bm). The secretion material perforates to lumen (L) through the epicuticle layer (arrowheads).

Ultrastructure of the testes and spermatozoa of the blow fly *Chrysomya megacephala* (F.) (Diptera:Calliphoridae)

Tarinee Chaiwong[1], Kom Sukontason[1], Urai Chaisri[4], Hiromu Kurahashi[3], Jimmy K. Olson[2], Udom Chaithong[1], and Kabkaew L. Sukontason[1]

1. Department of Parasitology, Faculty of Medicine, Chiang Mai University, Chiang Mai 50200, Thailand
2. Department of Entomology, Texas A&M University, College Station, TX 77843-2475, USA
3. Department of Medical Entomology, National Institute of Infectious Diseases, Tokyo 162-8640, Japan
4. Department of Tropical Pathology, Faculty of Tropical Medicine, Mahidol University, Bangkok 10400, Thailand

tarinee_cw@hotmail.com
Keyword: *Chrysomya megacephala*, testes, scanning electron microscopy, transmission electron microscopy

Chrysomya megacephala (F.) (Diptera: Calliphoridae) is the most medically important blow fly species in Thailand as the adults are a mechanical vector of various pathogens and a nuisance pest, while the larvae can cause myiasis. Biological information on this fly species is limited, thus this study presents the ultrastructure of the testes using light microscopy (LM), scanning electron microscopy (SEM) and transmission electron microscope (TEM). This information may be particularly important for future systematic study of this species. Thirty adult *C. megacephala* males aged three- and seven- days-old were used in this study. Light microscopic observations revealed that the color of the testes changes from pale orange at emergence to a reddish orange color in 1 old-day flies, and remains this color throughout the entire adult stage. The shape of the testes correlates closely with the age of the fly. SEM imaging showed that the testes are a paired structure, which has a smooth surface and is occasionally penetrated by tracheoles. TEM analysis revealed that the testes of 3 day-old are characterized by three stages of developing spermatids. The first stage is the beginning of the first division of maturation, and the spermatid has a pear-shape with a large nucleus (the ratio of nucleus and cytoplasm is 1:1), that is profuse with dense chromatin material dispersed inside the nucleus (Figure 1). The second stage has a nucleus with elongated clusters of chromatin in the globular cell. The ultrastructure of the tail or flagellum includes a spherical nucleus, but the axonemal model is indistinguishable. In the testes of 7 old-day males, the spermatozoa are regularly arranged. The spermatozoa consist of two regions: the head, formed by an acrosome and nucleus, and the tail consisting of two mitochondrial derivatives, an axoneme and an accessory body (Figure 2). The axoneme pattern is similar to the 9+9+2 axonemal model [1, 2].

A. Aretz, B. Hermanns-Sachweh, J. Mayer (Eds.): EMC 2008, Vol. 3: Life Science, pp. 145–146, DOI: 10.1007/978-3-540-85228-5_73,

1. G. Alberti and V.B. Meyer-Rochow, Arthropod Structructure and Development 30 (2002), p. 315-328.
2. K.O. Owusu-Daaku, R.D. Butler and R.J. Wood, Arthropod Structructure and Development 36 (2007), p. 23-43.
3. We thank the Thailand Research Fund through the Royal Golden Jubilee Ph.D. Program and Faculty of Medicine, Chiang Mai University.

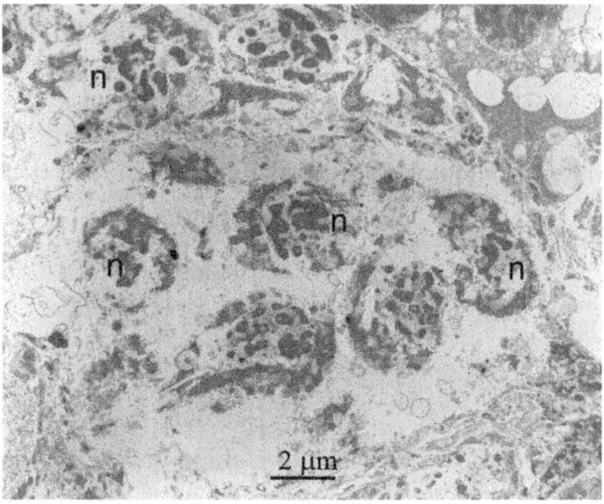

Figure 1. Transmission electron micrograph of the 3 old-day testes of *Chysomya megacephala* showing the first division of maturation, and the spermatocyte has a pear-shape with a large nucleus (n).

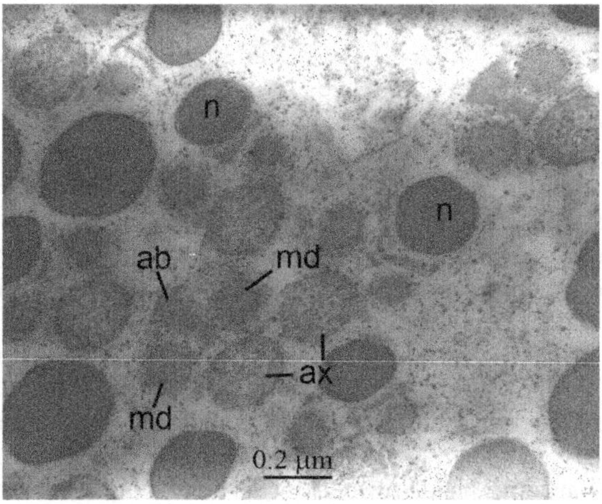

Figure 2. Transmission electron micrograph of the 7 old-day testes of *C. megacephala* showing axoneme pattern is similar to the 9+9+2 axonemal model; mitochondrial derivatives (md); an axoneme (ax); an accessory body (ab).

Biosynthesis of cellulose in red algae

I. Tsekos

Institute of Botany, School of Biology, Aristotle University of Thessaloniki, GR-54124 Thessaloniki, Macedonia, Hellas

tsekos@bio.auth.gr
Keywords: Cellulose microfibrils, freeze fracture, terminal complexes (TCs), red algae

There is now ample evidence that cellulose biosynthesis occurs at the plasma membrane-bound cellulose synthase [1,2]. Freeze-fracture studies of the supramolecular organization of the plasma membrane support the view that the rosettes (a six-subunit complex) in higher plants and both rosettes and the linear terminal complexes (TCs) in algae are the structures that synthesize cellulose and secrete cellulose microfibrils [1, 3]. One typical feature of the linear terminal complexes in red algae is the periodic arrangement of the particle rows transverse to the longitudinal axis of the TCs "Figures 1, 2". In bangiophyte red algae cellulose microfibrils are thin, ribbon-shaped structures, 1-1.5 nm thick and 5-70 nm wide "Figure 3". Terminal complexes appear to be made in the endoplasmic reticulum and are transferred as large globular particles (globules, precursor units) to Golgi cisternae, where the cellulose synthases are activated and may be transported to the plasma membrane"Figure 4" [1, 3]. The globules are composed of a central hole (the core) with small subunits forming a peripheral ridge and seem to represent zymogenic precursors "Figure 5". In the plasma membrane the globules (consisting of more subunits) aggregate, swell and unfold, and become closely arranged contributing to TC assembly "Figure 5" [3]. Longitudinal elongation of the TC occurs by the unfolding of globules attached to both ends of the TC nucleation unit until the TC is completed (double row-formation centers) [3]. In the red algae with linear TCs, deposition of microfibrils follows a precise pattern directed by the movement and the orientation of the TCs (membrane flow). A principal underlying theme is that the architecture of cellulose microfibrils (size, shape, crystallinity, and intramicrofibrillar associations) is directly related to the geometry of TCs.

1. I. Tsekos. J. Phycol. **35** (1999), 635.
2. S. Kimura, W. Laosinchai, T. Itoh, X. Cui, C.R. Linder and R.M. Brown, Jr. The Plant Cell **11** (1999), 2075.
3. I. Tsekos, K. Okuda and R.M. Brown, Jr. Protoplasma **193** (1996), 33.

A. Aretz, B. Hermanns-Sachweh, J. Mayer (Eds.): EMC 2008, Vol. 3: Life Science, pp. 147–148,
DOI: 10.1007/978-3-540-85228-5_74,

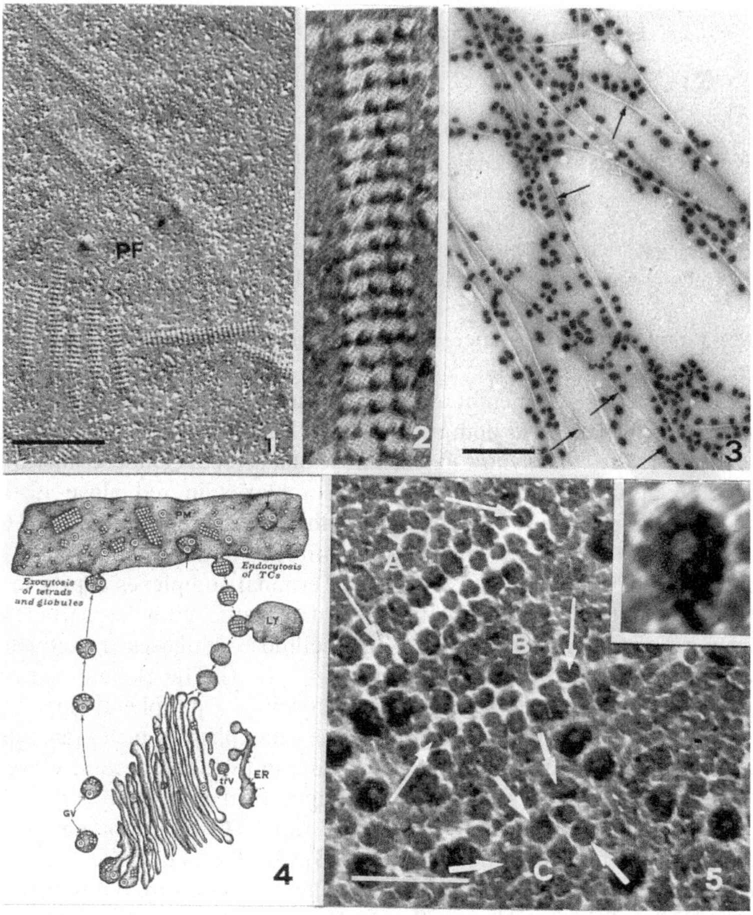

Figure 1. *Erythrocladia subintegra.* Fracture through a freeze-etched cell, showing the protoplasmic face (PF) of the plasma membrane and many linear terminal complexes (TCs). Bar: 200 nm.

Figure 2. *Erythrocladia subintegra.* A TC at higher magnification. X 375.000.

Figure 3. *Erythrocladia subintegra.* Microfibrils labeled by cellobiohydrolase I-gold complexes, twisted in some places (arrows). Bar 150 nm.

Figure 4. Scheme showing various possible routes for membrane flow in developing cells of *Erythrocladia*, which could lead to TC formation. D = dictyosome, GV = Golgi vesicles, LY = lysosome, PM = plasma membrane, trV = transitional vesicle. Explanation is provided in the text.

Figure 5. *Erythrocladia subintegra.* Rotary shadowed PF face of the plasma membrane showing the assembly of TCs at different stages. Nascent TCs with double row-formation centres (A and B, small arrows), and attachment and unfolding of globules to form nascent TC rows (C, big arrows). Bar = 50 nm. **Inset:** Details of a large globule consisting of multiple subunits. X 880.000.

Light and electron microscopy aspects of the glandular sessile hairs from the vitroplantlet leave of *Drosera rotundifolia*

V. Turcuş[1], D. Cachiţă[1], C. Crăciun[2], A. Ardelean[1], L. Barbu-Tudoran[2] and C. Mihali[2]

1. Vasile Goldis West University, Arad, Romania
2. Electron Microscopy Center, Babes Bolyai University, Cluj-Napoca, Romania

violeta.turcus@gmail.com
Keywords: Drosera, *in vitro* cultivation, hairs

Carnivore plants, named by *Darwin* (1835) insectivores [2], show morpho-anatomical characteristics which allow them to capture – through different types and structures of traps – small animals, and they compensate in this way the lack of nitrogen compounds in their growth substrate. One of the carnivore plant species is *Drosera rotundifolia*. On the leaf of this plant we can describe a diversity of hairs. Among these, the most numerous are the tentacular, bludgeon hairs, with a very complex role. In nature, the apex of *Drosera*'s tentacular hairs produces some dew-like drops. These drops attract and immobilize small insects, by curving the tentacular hairs over the "prey". In this case, the leaf's lamina is also curving. The external cells of the bludgeon secrete proteolytic enzymes that produce protein hydrolysis. Afterwards, the amino acids that result from the proteolysis of captured insect are absorbed in the leaf's laminar tissues. When the digestion products are depleted, the leaf and the tentacular hairs will retract in the normal position. The leaves are capable to repeat this process only for three or four times, and after that, they become senescent and die. The new leaves, after tissue differentiation, will develop this type of nutrition, in addition to photoautotrophic assimilation.

Recently, the *D. rotundifolia* plants are used in modern phytotherapy. This is the reason why the species is cultivated in vitro, and the resulting biomass is used to prepare valuable phyto-pharmaceutical extracts.

From the hair types that exist on *Drosera* leaves, we have studied with light and electron microscope the structure of the sessile hairs that are present on the leaves of *D. rotundifolia* in vitro plantlets. These plantlets were produced in vitro on *Murashige-Skoog* (1962) mineral basal medium [3]. To this medium we added various organic compounds, usually used in similar experiments, as FeEDTA, growth regulators, myo-inositol, vitamins, saccharose, agar-agar etc. On this kind of medium, the cultivated *Drosera* grew luxuriant, and after a longer cultivation period (about 3-4 months), it was necessary to sub-cultivate the newly formed rosettes on the same type of fresh medium.

The cytological investigations permitted the evidentiation of several types of pluri-cellular hairs on the leaves of *Drosera* vitroplantlets. Among these, the sessile hairs (Figure 1A) were less studied. They are formed of 6 cells (Figure 1B), arranged in two columns, each consisting of three cells, partially connected to each other. These types of hairs are distributed with high frequency, especially on the bottom side of the leaf's

A. Aretz, B. Hermanns-Sachweh, J. Mayer (Eds.): EMC 2008, Vol. 3: Life Science, pp. 149–150,
DOI: 10.1007/978-3-540-85228-5_75, © Springer-Verlag Berlin Heidelberg 2008

lamina, and rarely found on the upper side of it. As a novelty, we identified the same kind of structures on the tract of tentacular hairs, on petiole and on the inflorescences' rachis, developed on the vitro-plantlets rosettes. Like the tentacular hairs, the sessile hairs are considered to be glandular. The function of the sessile hairs is just partially known. From an ultrastructural point of view, our studies revealed that from the 3 cells that form a half of the sessile hair, the first 2 apical cells are the largest. They have a spherical form (Figure 1A) and show a very thin cellular wall. By dehydrating the external wall of these cells will lead to their implosion (Figure 1C). Using the transmission electron microscope, it can be observed that the cytoplasm is disorganised and their vacuoles presents massive, electrondense deposits (one of these vacuoles is larger and it's visible on light microscope – Figure 1D). We previously observed these types of structures in the vacuolar sap of the plant cells rich in anthocyans, after exposure to procaine solution [1]. In the vacuoles of senescent cells that contain anthocyans, there is a tendency of conglomeration for these compounds, due to modifications in chemical and physical equilibrium of the vacuolar sap. In conclusion, the vacuolar sap of the apical cells of the sessile glandular hairs, probably contain anthocyans.

1. D. Cachiță-Cosma and C. Crăciun, The ultrastructure of the epidermal cells of the Peony petals (Paeonia officinalis) under the action of the Procain, Cytologia **44** (1979), pp. 71-75.
2. C. Darwin in "Insectivorus Plants", ed. John Murray, London (1875), p. 462.
3. T. Murashige and F. Skoog, A revised medium for rapid growth and bioassays with tabaco tissue culture, Phisiol. Plant **15** (1962), pp. 473-497.

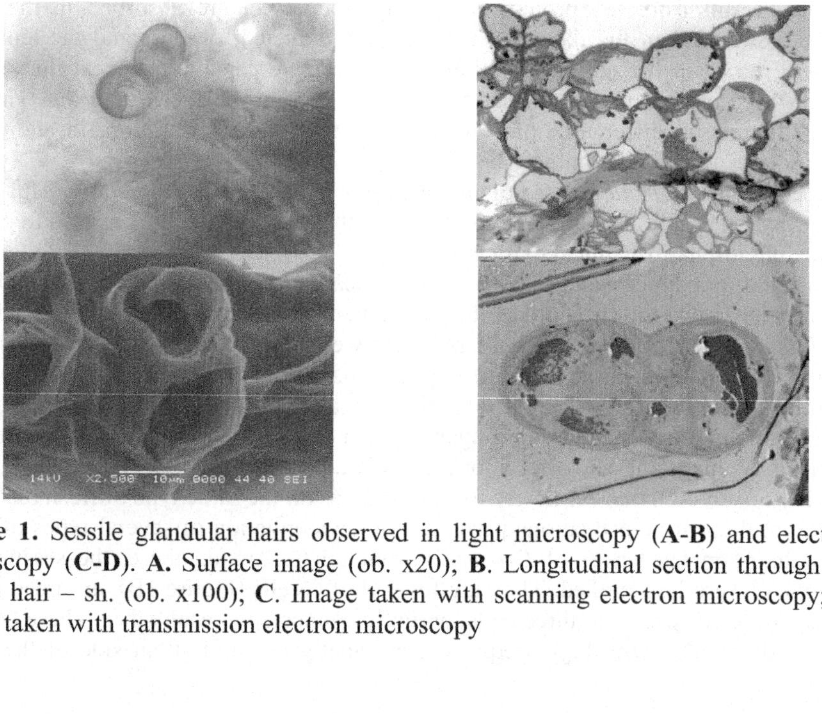

Figure 1. Sessile glandular hairs observed in light microscopy (**A-B**) and electron microscopy (**C-D**). **A.** Surface image (ob. x20); **B**. Longitudinal section through the sessile hair – sh. (ob. x100); **C**. Image taken with scanning electron microscopy; **D.** Image taken with transmission electron microscopy

Ultrastructure of selected *Euglena* species in relation to their taxonomy

M. Weidinger[2] and E. Kusel-Fetzmann[1]

1.Centre of Ecology, University of Vienna, Althanstrasse 14, A-1090 Vienna, Austria
2.Cell Imaging and Ultrastructure Research, University of Vienna, Althanstrasse 14, A-1090 Vienna, Austria

marieluise.weidinger@univie.ac.at
Keywords: Euglenophyceae, ultrastructure, species characteristics

The taxonomy of the Euglenophyceae is so far mainly based on light microscopically recognizable morphological characteristics. In the genus *Euglena* more than 200 species have been described to date. Many species, however, are similarly shaped and exhibit strong metabolic movement, making observation difficult. Therefore, transmission and scanning electron microscopes have been applied to clarify details of organelles.

Several groupings or subdivisions have been attempted before by Pringsheim [1], Gojdics [2], taking into consideration cell shape or type of chloroplasts. These were later taken over or changed again [3].

Within the genus *Euglena* the subgroup" Serpentes " [1] is characterized by species with long, slim cell bodies, which move without flagellum by snake-like locomotion in the detritus or in the mud, or swim freely in the water with a flagellum. They have characteristic chloroplasts with so-called naked pyrenoids, meaning without adjacent paramylon. The reserve carbohydrate is found outside the chloroplasts as paramylon grains of different shapes within the cytoplasm.

Our work aims to better characterize this whole group, to better distinguish between the diverse species, and to provide a basis for further molecular-genetic analysis of the phylogeny and relationship of the *Euglena* species.

Within the subgroup Serpentes two major groups can be distinguished, one characterized by a straight-ending front piece without protruding flagellum, the other with a lateral canal opening at the front piece with one flagellum protruding sideways. See Figures 1-4.

One important distinguishing feature among the species is the form of the pellicula. It can differ in thickness or cross-sectional shape (e.g. **A-**, **M-**, or plateau-like) and can have various arrangements of microtubules and ER mucus-vesicles [4].

Some *Euglena* species were first identified by morphological studies, later confirmed by molecular genetic analysis [5].

1. E.G. Pringsheim, Nova Acta Leopoldina 18 Nr. 125 (1956), p.168.
2. M. Gojdics, The Genus Euglena. Madison, The University of Wisconsin Press (1953), p. 268.
3. A. Batko, B. Zakrys, Algolog. Studies 79 (1995), p.1.
4. R.E. Triemer et al., J. Phycology 42 (2006), p. 731.
5. B. Marin et al., Protist 154 (2003), p 99.

A. Aretz, B. Hermanns-Sachweh, J. Mayer (Eds.): EMC 2008, Vol. 3: Life Science, pp. 151–152,
DOI: 10.1007/978-3-540-85228-5_76, © Springer-Verlag Berlin Heidelberg 2008

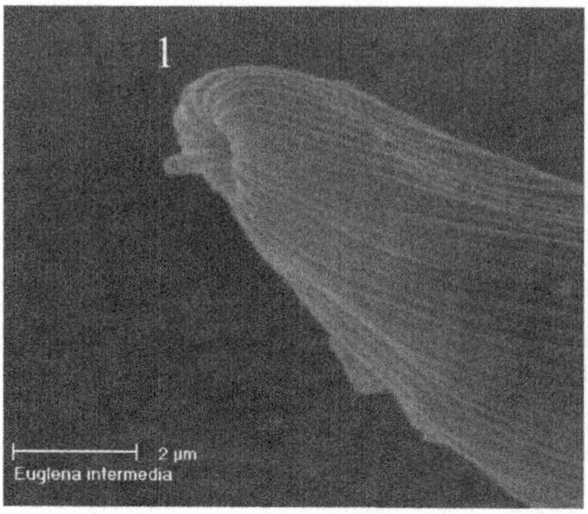

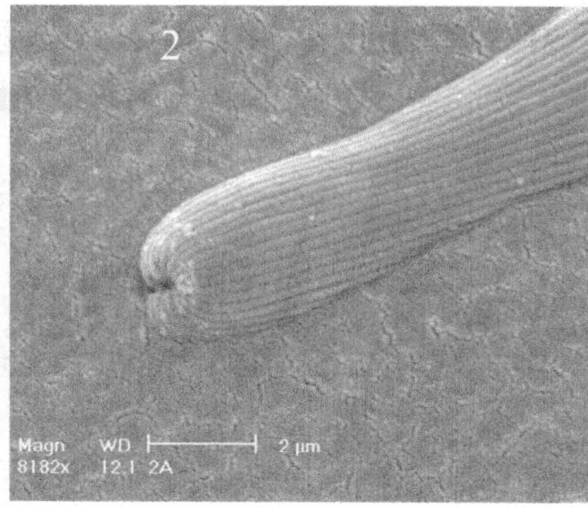

Figure 1. *Euglena deses* var. *intermedia*, SEM micrograph of anterior end of the cell, lateral canal opening with flagellum.protruding sideways.

Figure 2. *Euglena satelles*, SEM micrograph of anterior end of the cell, apical front opening of the canal without protruding flagellum.

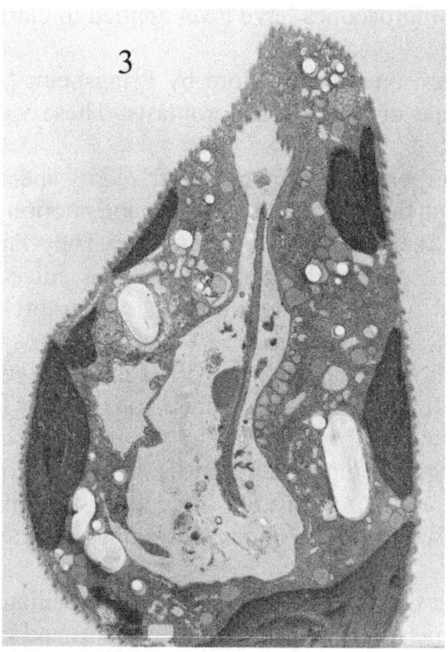

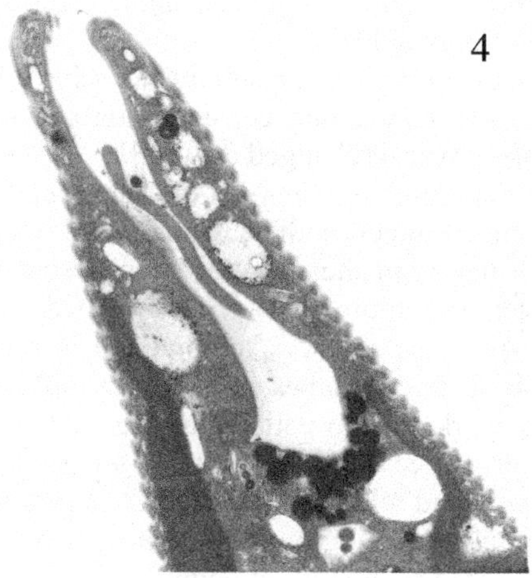

Figure 3. *Euglena deses*, TEM micrograph of anterior end of the cell, section near the lateral canal opening (left side) showing flagellum, stigma and paraflagellar body, chloroplasts reaching the end of the cell.

Figure 4. *Euglena carterae*, TEM micrograph of anterior end of the cell, section showing front canal opening with flagellum ending within the canal, chloroplasts not reaching the end of the cell.

Confocal Microscopy Reveals Molecular Interactions Between DNA-binding Drugs and Chromatin in Live Cells

J.W. Dobrucki and K. Wojcik

Division of Cell Biophysics, Faculty of Biochemistry, Biophysics and Biotechnology, Jagiellonian University, ul. Gronostajowa 7, 30-387 Krakow, Poland

jurek.dobrucki@mol.uj.edu.pl
Keywords: chromatin, DNA, histone, intercalator, minor groove binder, confocal microscopy

Background: A number of drugs are capable of penetrating plasma membrane of live cells and binding to DNA. Some anticancer drugs, like daunomycin, are known to bind to DNA. Daunomycin is thought to cause cytotoxicity by inhibiting topoisomerase II and DNA polymerase [1]. Chromatin of live cells is commonly visualized using specific fluorescent probes that exhibit affinity to DNA. Most of these probes are intercalators or minor groove binders. A fluorescent probe bound to chromatin reveals the presence, spatial distribution and a local concentration of DNA in nuclei of live cells. However, binding of a probe to DNA may cause changes in chromatin or nuclear structure, which are dependent on chemical nature and concentration of the probe.

Goal: This study aims at understanding the influence of DNA intercalators, like DRAQ5 or daunomycin, as well as minor groove binders, like Syto17, on chromatin higher order structures and histone-DNA interactions, and replication, transcription and gene expression.

Methods: We used fluorescence confocal microscopy to study HeLa cells expressing eGFP-tagged histones (linker (H1) and its mutants [2] as well as core H2B, H3, and H4), challenged by intercalators like acridine orange, DRAQ5, daunomycin or a minor groove binder Syto17.

Results: 3D imaging, time lapse and FRAP studies demonstrated changes of nuclear and chromatin structure and revealed molecular events. DRAQ5 (anthracycline derivative), at a concentration of 1 μM, caused microscopically detectable changes of nuclear architecture [3]. Following DRAQ5 intercalation into DNA, chromatin aggregated into distinct areas and foci. The loss of 3D chromatin distribution was exerted via interference with a dynamic exchange of a linker histone (H1), which is a known chromatin stabilizing factor. At higher concentrations (3- and 7.5 μM) DRAQ5 interfered with binding of H2B core histones to DNA. Similar effects were detected following exposure to daunomycin (at clinically relevant concentrations) (Fig. 1, 2, 3). Interestingly, replication was not abolished even at high daunomycin concentrations (2,5 uM). Dissociation of histone H1 was efficient in the presence of DNA intercalators, but not minor groove binders.

Discussion: We postulate that when a drug, like daunomycin or DRAQ5, intercalates into DNA, unwinds the helix and pushes bases apart, the affinity of H1 for DNA decreases. Detachment of H1 from DNA causes a collapse of higher order chromatin structures. Since daunomycin has been shown to bind to histone H1 and

A. Aretz, B. Hermanns-Sachweh, J. Mayer (Eds.): EMC 2008, Vol. 3: Life Science, pp. 153–154, DOI: 10.1007/978-3-540-85228-5_77,

change its structure and properties [4], it is possible that the altered, mobile histone is recognized as damaged and degraded [5]. Since higher order chromatin structures are known to be involved in control of gene expression, we postulate that cytotoxic action of daunomycin is exerted via adverse influence on control of gene expression.

1. D. Gewirtz, Biochem. Pharmacol. 57 (1999) 727.
2. D.T. Brown, T. Izard, T. Misteli, Nat. Struct. Mo.l Biol. 13(3), (2006) 250
3. K. Wojcik , J. Dobrucki. Cytometry, (2008) accepted.
4. S.J. Zargar, A. Rabbani, Int. J. Biol. Macromol. 28 (2000) 75
5. O. Ullrich, T. Grune, Free Radic. Biol. Med. 31 (2001) 887

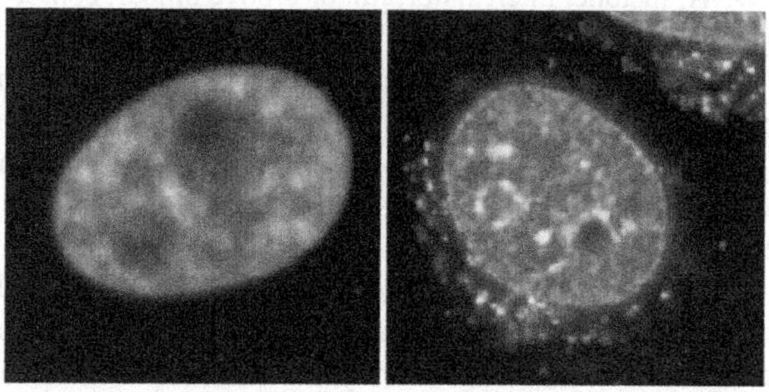

Fig. 1. Chromatin in untreated nucleus (left) and in daunomycin-treated nucleus (right); green -H2B-GFP, red - daunomycin

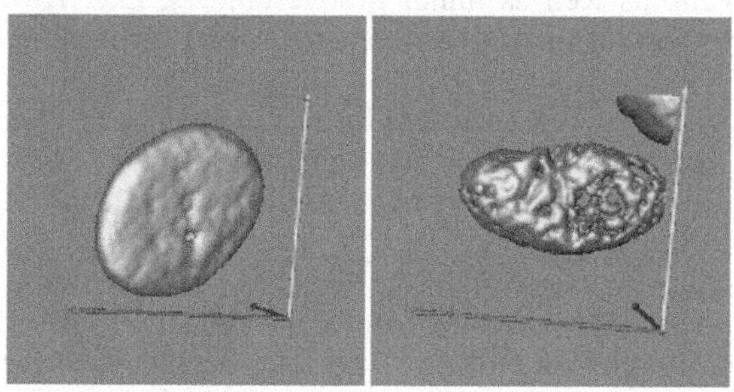

Fig. 2. 3-dimensional image of untreated nucleus (left) and daunomycin-treated nucleus (right)

Fig. 3. Postulated structure of chromatin in untreated (left) and daunomycin-treated cells (right), histone H1 dissociates from DNA following drug intercalation

Fluorescence microscopy methods for measuring the mobility and stability of molecules in 3-D samples

K. Braeckmans[1], D. Vercauteren[1], H. Deschout[1], D. Mazza[2], A. Diaspro[2], S.C. De Smedt[1], and J. Demeester[1]

1. Lab. General Biochemistry & Physical Pharmacy, Ghent University, Harelbekestraat 72, 9000 Gent, Belgium
2. LAMBS-Microscobio, Physics Department, University of Genoa, Via Dodecaneso 33, 16146 Genova, Italy

kevin.braeckmans@ugent.be
Keywords: fluorescence microscopy, FRAP, SPT

In the last decade, confocal laser scanning microscopy (CLSM) has become a very popular tool for biological, medical, pharmaceutical and material research. Because of its optical sectioning ability, a confocal microscope can give insight in the three-dimensional structure of fluorescently labelled samples. Many CLSMs of the latest generation have the additional ability of scanning user-defined regions. This very feature makes them a convenient tool for fluorescence recovery after photobleaching (FRAP) experiments. FRAP is a fluorescence microscopy technique for examining the mobility of fluorescently labelled molecules. In a FRAP experiment, the fluorescent molecules are at first photobleached in a user-defined region of the sample by a powerful laser beam. Immediately after the bleaching process, the fluorescence inside the bleached area will recover due to the inward diffusion of fluorescent molecules from the surrounding unbleached areas. As the rate of fluorescence recovery reflects the local mobility of the fluorescent molecules, analysis of the recovery curve allows to calculate their diffusion coefficient at that particular location in the system. While the principle of FRAP is essentially quite simple, it is not so evident to come to correct quantitative diffusion measurements. Here we present three methods for performing quantitative FRAP measurements on a confocal or multi-photon laser scanning microscope. The first method is based on the bleaching of a large disk bleached by a CLSM with an objective lens of low numerical aperture (NA) [1]. This method is especially suited for performing diffusion measurements in 3-D extended samples, such as solutions, gels or extracellular matrices. The second FRAP method is based on the photobleaching by a CLSM of a line segment and can be used for diffusion measurements in small samples, such as the cytoplasm or nucleus of living cells [2]. The third FRAP method is for use with a multi-photon laser scanning microscope and can be applied to both small and large samples [3]. These three complementary FRAP methods form a convenient but powerful set of tools for performing local quantitative diffusion measurements in virtually any kind of sample compatible with fluorescence microscopy.

A complementary method to FRAP is single particle tracking (SPT), which is a relatively new fluorescence microscopy method capable of visualising the movement of individual fluorescently labelled molecules/particles with a high tempo-spatial

A. Aretz, B. Hermanns-Sachweh, J. Mayer (Eds.): EMC 2008, Vol. 3: Life Science, pp. 155–156,
DOI: 10.1007/978-3-540-85228-5_78,

resolution. Although the molecules themselves are below the spatial resolution limit and have a very low intensity, their movement can be visualised with sub-resolution accuracy using efficient laser illumination in combination with a fast and sensitive electron-multiplying CCD (EMCCD) cameras. In a pharmaceutical context, SPT can play an important role by characterising the dynamic behaviour of individual drug molecules in e.g. living cells [4]. By being capable of analyzing the movement of individual molecules, SPT can reveal details of the dynamics that cannot be resolved by ensemble averaging techniques such as FRAP or fluorescence correlation spectroscopy). In this presentation, the different aspects of designing an SPT set-up and analyzing the data will be discussed. Apart from specific hardware requirements, software is needed to determine the exact location of the individual particles in each frame of an SPT movie. This set of coordinates can subsequently be used to calculate the trajectories of individual particles using a fast nearest neighbour algorithm. Finally, the individual trajectories can be analyzed in terms of mode of motion (diffusion, directed transport, restricted diffusion) and the relevant quantitative parameters (diffusion coefficient, transport velocity, compartment size).

1. K. Braeckmans, L. Peeters, N.N. Sanders, S.C. De Smedt, and J. Demeester. Biophys. J. **85** (2003), p. 2240-2252.
2. K. Braeckmans, K. Remaut, J. Demeester, and S.C. De Smedt. Biophys. J. **92** (2007), p. 2172-2183.
3. D. Mazza, K. Braeckmans, F. Cella, I. Testa, D. Vercauteren, J. Demeester, S.C. De Smedt, and A. Diaspro. *submitted* (2008).
4. R. Bausinger, K. von Gersdorff, K. Braeckmans, M. Ogris, E. Wagner, C. Bräuchle, and A. Zumbusch. Angew. Chem. Int. Edit. **45** (2006), p. 1568-1572.
5. Kevin Braeckmans is a Postdoctoral Fellow of the Fund of Scientific Research, Flanders (Belgium).

Phasor-FLIM analysis of FRET for homotypic and heterotypic non-covalent interactions: stimulated membrane receptors in live cells

V.R. Caiolfa[1,2], E. Gratton[3] and M. Zamai[2]

1. Centro Nacional de Investigaciones Cardiovasculares, Madrid, Spain
2. Scientific Institute San Raffaele, Milano, Italy
3. Laboratory for Fluorescence Dynamics, UCI, Irvine, CA, USA

vcaiolfa@cnic.es; valeria.caiolfa@hsr.it
Keywords: FLIM, FRET, non-covalent interactions

It is commonly believed that the best approach to measure FRET in cells is through the measurement of the fluorescence lifetime quenching of the donor molecules as done in Fluorescence Lifetime Imaging (FLIM). FLIM-based FRET has the benefit that the results are obtained from a single-lifetime measurement, it does not need calibration in different cells, and it is non-destructive. Moreover, it can resolve the interacting and non-interacting donor fractions.

However, FLIM is still restricted to few laboratories because of the cost of the instrumentation and the difficulty of data analysis. Several assumptions are always introduced in the analysis, such as the use of average lifetime. Since many fluorescent proteins and cellular autofluorescence show multi-exponential decays, at every pixel in a FLIM image there might be contributions from several fluorescent species, and each one can be multi-exponential. Furthermore, the light can be collected only for a limited amount of time, 100-200 ms/pixels, which results in ≈500-1000 photons/pixels. Such a number of photons is barely enough to distinguish a double exponential from a single exponential decay! Therefore, resolving the decay at each pixels in multiple components involves fitting to a function, and it is a complex computational task.

We need to go to a new "space": the Phasor space [1], using a different representation of the fluorescence decay where each molecular species has its own unique representation, and where each process (e.g., FRET) is easily identified. We conceived a method for analysis and visualization of the decay components in FLIM, which is based on the representation of moments of the decay distribution [2, 3].

Our method applies to any technique for acquiring lifetime images such as time correlated single photon counting (TCSPC) and frequency-domain methods. Each decay component or molecular species can be represented by a single vector in a phasor plot, irrespective of the number of exponentials needed for its description (Figure 1.A). The "Phasor" is a new coordinate system so that the single-lifetime components add linearly to produce the total measured signal. Changing the data representation from the classical time delay histogram to the phasor representation provides a global view of the fluorescence decay at each pixel of an image.

In the phasor representation we easily recognize the presence of different molecular species in a pixel or the occurrence of FRET also when the cellular background is

A. Aretz, B. Hermanns-Sachweh, J. Mayer (Eds.): EMC 2008, Vol. 3: Life Science, pp. 157–158, DOI: 10.1007/978-3-540-85228-5_79,

relevant (Figure 1.B). We show the application of phasor-FLIM to determine the homo-dimerization of the urokinase plasminogen GPI-anchored receptor, uPAR, in stable clones of HEK293 cells [4]. We also followed the effects of stimulatory and inhibitory agents on the assembly of the endothelial adhesive platforms members, VCAM and ICAM receptors with their tetraspanins partners in human primary endothelial cells, having variable and transient co-expression of donor and acceptor.

1. M.A. Digman, V.R. Caiolfa, M. Zamai, and E. Gratton, Biophys J. **94**(2008), p. L14-6.
2. A.H. Clayton, Q.S. Hanley, and P.J. Verveer, J Microsc. **213**(2004), p. 1-5.
3. G.I. Redford and R.M. Clegg, J Fluoresc. **15**(2005), p. 805-15.
4. V.R. Caiolfa, et al., J Cell Biol. **179**(2007), p. 1067-82.

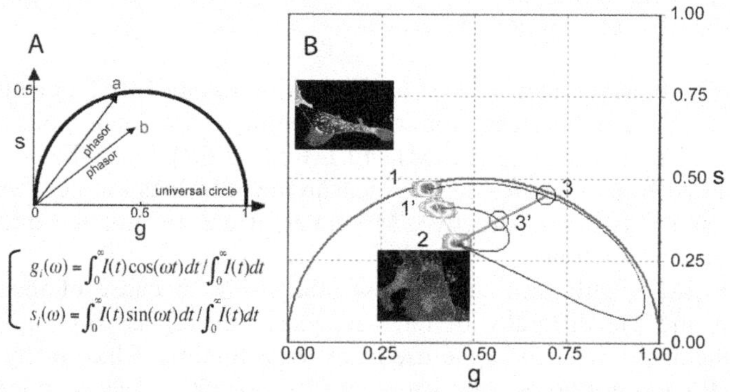

Figure 1. (A) A phasor is a quantity like a vector. Fourier-transformed time-domain decays can be described by phasor coordinates in a polar plot representation, where ω is the pulse frequency and I(t) is what it is measured. As the modulation frequency increases, the end of the phasor describes a semicircle of radius 0.5 and centered at (0.5, 0). Only single-exponential decays have phasors that fall on the universal circle. Decay components made of multi-exponentials can fall any where inside the circle. (B) Phasor ensembles in contour plots from two basal membranes in uPAR-mEGFP HEK293 cells with 1% (1) and 37% (1') autofluorescence respectively. The contour plot in position 2 describes the phasor ensemble of the autofluorescence acquired on a untrasfected HEK293 cell (2). The green line represents the experimental mean phasor distribution for unquenched donors with different contributions of cell auto-fluorescence. The curved trajectories (black lines) describe all possible positions of the donor phasors of cell 1 and 1' in the presence of quenching due to FRET. The red line marks the expected positions of the donor phasors when FRET is 50%.

In vitro immunofluorescence and ultrastructural analysis of the expression of noncollagenous matrix proteins in human pulpal cells induced by mineralizing factors

V.E. Arana-Chavez, E.F. Martinez and T.A.G. Donato

Laboratory of Mineralized Tissue Biology, Institute of Biomedical Sciences, University of São Paulo, 05508-900 São Paulo, SP, Brazil

vearana@usp.br
Keywords: pulp cells, mineralizing factors, gingival fibroblasts

The dental pulp is a typical loose connective tissue in which fibroblasts are the most abundant cells. Pulpal fibroblasts exhibit some particularities in relation to fibroblasts present in other connective tissues such as the expression of tenascin and osteonectin, two mineralized tissue-related proteins (1). *In vitro* studies have shown that dental pulp cells may form mineralized matrix after the addition of some mineralizing factors such as ascorbic acid and β-glycerophosphate to the culture medium (2). The purpose of this *in vitro* study was to verify by immunofluorescence the expression of some mineralized tissue-related extracellular matrix proteins [tenascin (TN), osteonectin (ONEC) and osteopontin (OPN)] and fibronectin (FN) in pulp cells induced to form mineralized matrix in comparison to gingival fibroblasts, as well as to osteogenic cells from rat calvaria. The ultrastructural aspects of the induced cells were also examined.

The human dental pulp and gingival cells were obtained from *explant* technique, each one from three different donors. The osteogenic cells were isolated by calvarial bone from 2 to 4-day-old Wistar rats. The obtained cells were cultured in DMEM and for mineralizing induction, the medium was supplemented with 10mM β-glicerophosphate and 50μg/ml ascorbic acid. For immunofluorescence, cells were incubated with primary monoclonal or polyclonal FN, TN, ONEC and OPN antibodies. For ultrastructural analysis, cells were grown on Thermanox®, embedded in Spurr resin and examined in a JEOL 1010 transmission electron microscope.

The results revealed that TN, ONEC and OPN were only present in pulpal, odontoblast-like, and osteoblast-like cells as shown in Figure 1; gingival fibroblasts did not express those proteins, even when induced by mineralizing factors. FN were also present in these cells (Figure 1), but in gingival fibroblasts was the only protein expressed (Figure 2). Ultrastructural aspects of the cells are shown in Figures 3 and 4.

In conclusion, the present results showed that mineralized tissue-related matrix proteins (TN, ONEC and OPN) were only present in cells committed to form mineralized tissue as pulpal, odontoblast-like and osteoblast-like cells.

1. Martinez EF, Araujo VC, Int Endod J **37** (2004), 749-55.
2. Couble ML, Farges JC, Bleicher F et al., Calcif Tissue Int **66** (2000), 129-38.
3. This work was supported by Fapesp (grant # 06/50153-4) and CNPq (Brazil).

A. Aretz, B. Hermanns-Sachweh, J. Mayer (Eds.): EMC 2008, Vol. 3: Life Science, pp. 159–160,
DOI: 10.1007/978-3-540-85228-5_80,

Figure 1

FN TN ONEC OPN

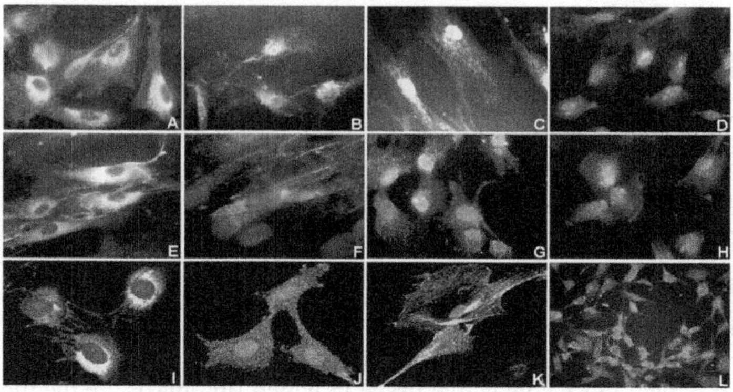

Figure 2

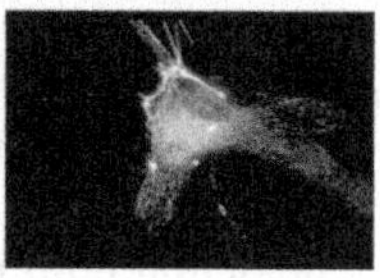

Figure 1. Immunofluorescence. Expression of FN, TN, ONEC and OPN in pulp cells (A-D), odontoblast-like (E-H) and osteoblast-like cells (I-L). 630X.

Figure 2. Immunofluorescence. Expression of FN in gingival fibroblasts. 630X.

Figure 3

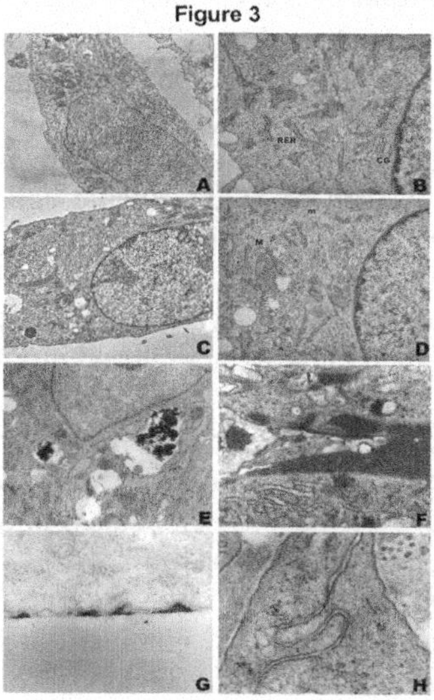

Figure 4

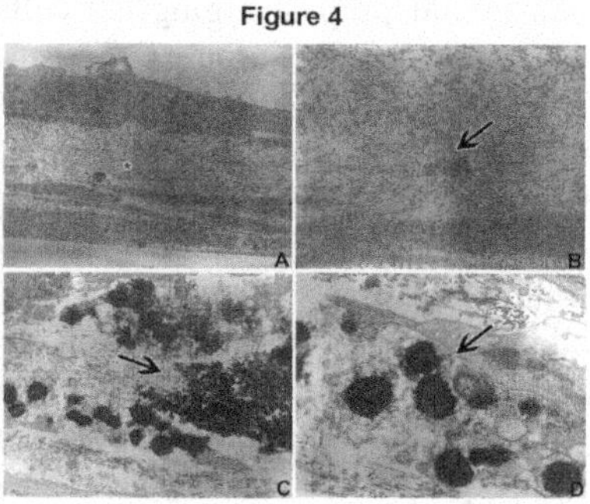

Figure 3. Transmission electron microscopy. Pulpal cells (A, B), gingival fibroblasts (C, D), and odontoblast-like cells (E-H). In A-D, well developed rough endoplasmic reticulum (RER), Golgi apparatus (GC), mitochondria (M), as well as microtubules (m). In E-F, odontoblast-like cells exhibit intra and extracellular mineralized material, as well in the cell surface in contact with Thermanox® (G). In H, a gap junction between odontoblast-like cells. A: 10000X; B and E: 20000X; C: 8000X; D: 15000X; F and H: 25000X; G: 50000X.

Figure 4. Transmission electron microscopy. Osteoblast-like cells in 10 (A and B) and 14 days (C and D) of culture. Note the presence of collagen fibrils in A (*), as well as mineralizing globules in B, C and D (arrows). A and C: 7500X; B: 25000; D: 20000X.

Differentiation-dependent Golgi fragmentation in the bladder urothelial cells in vitro

Kristijan Jezernik[1], Mateja Erdani Kreft[1] and Alexander A. Mironov[2]

1. Institute of Cell Biology, Faculty of Medicine, University of Ljubljana, Lipičeva 2, SI-1000 Ljubljana, Slovenia
2. Department of Cell Biology and Oncology, Consorzio Mario Negri Sud, Via Nazionale, 66030 Santa Maria Imbaro (Chieti), Italy

kristijan.jezernik@mf.uni-lj.si
Keywords: urothelial cell, Golgi, differentiation

Four integral membrane proteins uroplakins, i.e., UPIa, UPIb, UPII, and UPIIIa construct almost the entire apical surface of the differentiated superficial cell layer in the urinary bladder urothelium. They form 16-nm particles packed into urothelial plaques [1]. Since the assembly of uroplakins to 16-nm particles takes place in the Golgi complex [2], we hypothesize that the subcellular position of the Golgi in urothelial cells is highly regulated. The aim of this study was therefore to examine the organization and subcellular localization of Golgi during urothelial cell differentiation.

Normal pig urothelial cell cultures from the 3rd to the 10th passage were analysed. Immunofluorescence experiments and fluorescence recovery after photobleaching (FRAP) experiments were performed by laser scanning confocal microscopy. The following antibodies were used: rabbit polyclonal antibody against UPIIIa, UPIa, UPIb and UPII (a kind gift of Dr T.-T. Sun, Department of Dermatology, New York University Medical School); mouse monoclonal antibodies against GS15 and GS28, (both BD Transduction Laboratories) and against giantin (a kind gift of Dr H. P. Hauri, Department of Pharmacology, University of Basel, Basel, Schweiz). Anti-rabbit and anti-mouse antibodies conjugated with Alexa Flouor 488 or Alexa Fluor 546 were from Molecular Probes. Purified plasmid coding for fusion protein between recombinant yellow fluorescent protein (YFP) and β1,4-galactosyltransferase (GalT) was microinjected into cell nuclei using an Eppendorf microinjecting system.

By immunofluorescence analyzing of Golgi-related markers (GS 15, GS 28, giantin) and urothelial differentiation-related markers (uroplakins), we observe that in non-differentiated, uroplakin-negative urothelial cells the Golgi is in ribbon structure, while in differentiated, uroplakin-positive urothelial cells, the Golgi is fragmented and dislocated to the cell periphery "Figure 1". The FRAP experiments on the urothelial cells transfected with GalT-YFT show that Golgi-resident enzyme cycles between Golgi elements, when the Golgi is in ribbon. When the Golgi is fragmented, the fluorescence recovery after photobleaching is diminished, suggesting that Golgi fragments are at least to some degree independent functional subunits of cellular Golgi.

These findings provide a new view of Golgi fragmentation associated with the process of urothelial cell differentiation. We suppose that Golgi fragmentation is one of

A. Aretz, B. Hermanns-Sachweh, J. Mayer (Eds.): EMC 2008, Vol. 3: Life Science, pp. 161–162,
DOI: 10.1007/978-3-540-85228-5_81,

the key events, which promote the uniform delivery of uroplakins over the entire apical plasma membrane of superficial urothelial cells.

1. X-R. Wu, M. Manabe, J. Yu and T-T. Sun, J Biol Chem **265** (1990), 19170-19179.
2. C-C. Hu, F-X. Liang, G. Zhou, L. Tu, C-H. Tang, J. Zhou, G. Kreibich and T-T. Sun, Mol Biol Cell **16** (2005), 3937-3950.
3. We kindly acknowledge the technical assistance of Dr. Daniele Di Giandomenico, Department of Cell Biology and Oncology, Consorzio Mario Negri Sud, Italy.

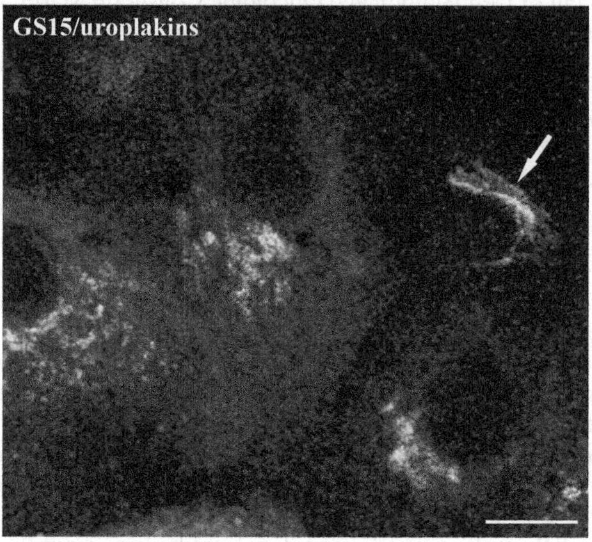

Figure 1. Double immunofluorescence of GS 15 (green) and uroplakins (red) in normal urothelial cell culture. Anti-GS15 labelling shows that in non-differentiated, uroplakin-negative urothelial cell the Golgi is in ribbon structure (arrow), while in differentiated, uroplakin-positive urothelial cells, the Golgi is fragmented. Scale bar, 10 μm.

Monitoring Mitochordria Dynamics in Living Cell System by Confocal Microscopy

C.-Y. Lin[1], L.-T. Li[1]

1. Core Facilities, Institute of Cellular and Organismic Biology, Academia Sinica, Taipei, Taiwan, R.O.C.

chuangyu@gate.sinica.edu.tw
Keywords: mitochondria, live cell image, Confocal microscopy

Paramecium is one of the most abundant unicellular ciliate Protozoa in fresh water. Fresh water environment is a hypotonic condition to cells. The plasma membrane of Paramecium, similar to other organisms, allows for continuous entrance of water to cells, however, slows the permeability. The contractile vacuole (CV) of Paramecium is an evolutionary adaptation for osmoregulation. There are generally two contractile vacuoles in a cell, one in anterior and the other in posterior. The CV bails water out to keep constant osmosis in the cells [1][2]. This transporter-associated osmoregulatory system is energy consuming [3]. As we know, the mitochondrion is the major power supplier in cellular physiological process. The purpose of this experiment is to monitor the mitochondrial dynamics of Paramecium during the energy-consumption osmosis regulation.

In this experiment *Paramecium multimicronucleatum* is used in a living cell system for monitoring the mitochondrial dynamics. The vital dye used for tracking mitochondria was MitoTracker® Red CMXRos. For vital sign monitoring, we also use Hoechst 33342 and Wheat Germ Agglutinin 647 for the nucleus and the plasma membrane staining, respectively. One percent of methylcellulose was added after incubating *Paramecium multimicronucleatum* with vital dyes at 25°C for 30 minutes, which is to block movement of Paramecium and also function as an osmosis imbalance reagent. The images were acquired by a confocal microscope (Leica TCS-SP5) with proper excitation laser wavelengths. The X-Y-Z-T mode was performed during image acquisition. To avoid cross-talk effect, the images were taken by sequential scan. The images were later processed and analyzed by LAS AF and MetaMorph softwares.

The results show the time series pictures of cell swelling, plasma membrane protruding, and collapse as osmosis regulation losing control (Figure 1). We found the mitochondrial signal intensity increased with the cellular demand of osmoregulation. This result indicates the mitochondria might migrate close to the CV from neighborhood to provide energy for cellular osmosis regulation (Figure3). The preliminary data show that the correlation of osmoregulation and mitochondrial dynamics can be monitored by microscopy analysis.

1. N. A. Campbell and J. B. Reece, "Biology", 7 th ed. (2005). P. 132-33.
2. E. Vivier in "Paramecium - A Current Survey", ed. W. J. Van Wagtendonk, (1974), p. 1-35.
3. C. Stock, H. K. Grønlien, R. D. Allen and Y Naitoh, J. Cell. Sci. 115 (2002), p. 2339-48.

A. Aretz, B. Hermanns-Sachweh, J. Mayer (Eds.): EMC 2008, Vol. 3: Life Science, pp. 163–164, DOI: 10.1007/978-3-540-85228-5_82,

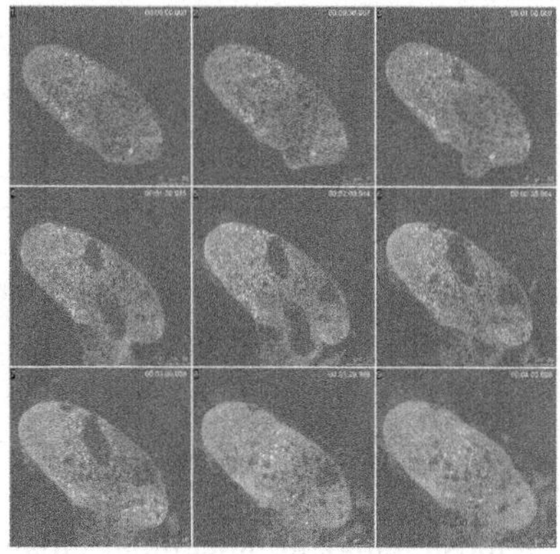

Figure 1. Time series overlay images of the mitochondria (green), nucleus (blue) and plasma membrane (red) in *Paramecium multimicronucleatum* in medium containing 1% methylcellulose.

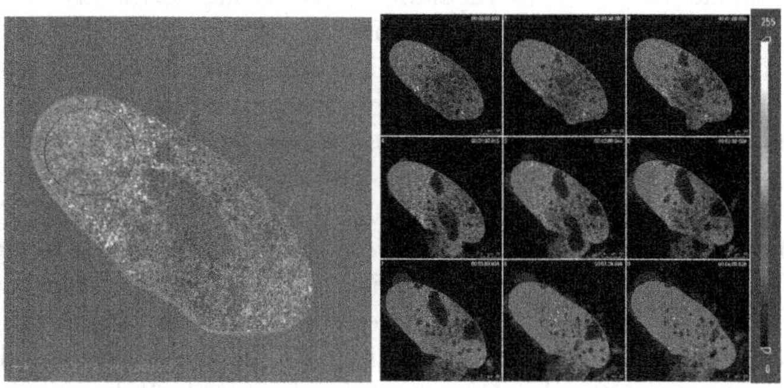

Figure 2. The left panel shows the ROIs for image analysis. Red arrows indicate the location of contractile vacuole (CV). Blue circle is ROI1 and purple circle is ROI2.The right panel shows time series images of mitochondrial signal increased in ROI1 and ROI2 as represented by color-coded mode (White stands for the highest signal intensity and black the lowest)in *Paramecium multimicronucleatum*.

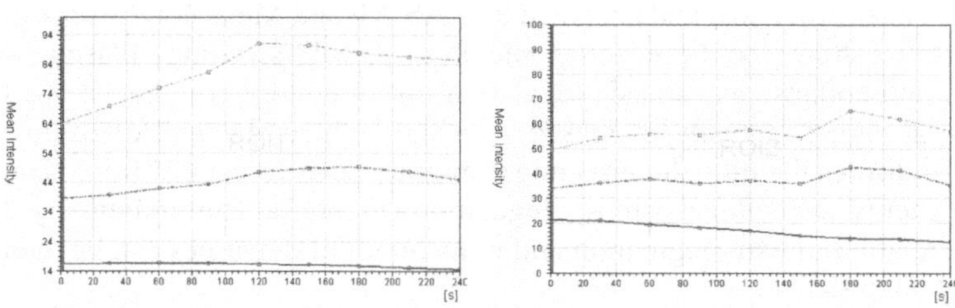

Figure 3. The charts show the mitochondrial intensity increased progressively (green line) during cellular osmosis regulation. The nucleus (blue) and the plasma membrane (red) represent vital signs of *Paramecium multimicronucleatum* .

The freeze-fracture replica immunolabeling technique for the study of lipid droplet biogenesis and function

H. Robenek[1], I. Buers[1], O. Hofnagel[1], M. J. Robenek[2], D. Troyer[1] and N. J. Severs[3]

1. Leibniz Institute for Arteriosclerosis Research, Domagkstr. 3, 48149 Münster, Germany
2. Department of Medicine, Division of Cardiology, Pulmonary Diseases and Angiology, Heinrich-Heine-University, Moorenstr. 5, 40225 Düsseldorf, Germany
3. National Heart and Lung Institute, Imperial College, London, Dovehouse Street, London SW3 6LY, UK

robenek@uni-muenster.de
Keywords: freeze-fracture immunocytochemistry, lipid droplets, PAT proteins

Of all the techniques in freeze-fracture cytochemistry attempted and tested over the last three decades, the freeze-fracture replica immunolabeling technique (FRIL) [1], has proven to be the most successful, and is currently having a substantial impact in solving questions in cell biology that have hitherto been impossible to address with other ultrastructural, cell biological or molecular approaches. For FRIL the specimen is rapidly frozen, freeze-fractured and platinum/carbon replicas made following the standard protocol used for freeze-fracture electron microscopy. However, instead of removing the biological material from the replicas with bleach as in the standard technique, the replica is treated with sodium dodecylsulphate (SDS). The aim of SDS treatment is to dissociate most of the biological material from the replica so that structure is visible by electron microscopy, leaving just sufficient material adherent to the replica (ideally a single molecular layer) to retain epitopes for labeling. By applying immunogold techniques, the spatial distribution of the targeted epitopes can be viewed superimposed upon standard planar freeze-fracture views of the membrane interior.

To illustrate the value of FRIL in addressing topical questions in cell biology, we will here summarize some of its current contributions to our understanding of the biogenesis and function of the lipid droplet [2-6].

It is widely held that the formation of lipid droplets involves accumulation of neutral lipids within the lipid bilayer of the ER membrane followed by budding-off into the cytoplasm. It is through this budding off process that the lipid droplet is presumed to acquire its phospholipid monolayer and associated proteins from the P half of the ER membrane. Results from FRIL refute this model on several counts. First, freeze-fracture demonstrates that at sites of close association, the lipid droplet is not situated within the ER membrane, but adjacent to it. Both ER membranes clearly lie external to and follow the contour of the lipid droplet, enclosing it in a manner akin to an egg-cup (the ER) holding an egg (the lipid droplet). Application of FRIL demonstrates that the PAT (perilipin, adipophilin, TIP47) family protein adipophilin is concentrated in prominent clusters in the P half of the ER membrane at the site of the closely apposed lipid droplet, as well as in the lipid droplet surface apposed to the ER. FRIL demonstrates

A. Aretz, B. Hermanns-Sachweh, J. Mayer (Eds.): EMC 2008, Vol. 3: Life Science, pp. 165–166,
DOI: 10.1007/978-3-540-85228-5_83, © Springer-Verlag Berlin Heidelberg 2008

unequivocally that lipid droplets originate and develop adjacent and external to specialized domains of the ER membrane enriched in adipophilin, not within the bilayer of the ER as previously supposed (Figure1).

A related area to which FRIL has shed new light is the mechanism of milk fat globule secretion. The milk fat globule has a lipid bilayer derived from the plasma membrane of the mammary epithelial cell lying external to the phospholipid monolayer that encloses the lipid core. The molecular mechanism of the secretory process is proposed to involve formation of complexes between butyrophilin in the plasma membrane with cytosolic xanthine oxidoreductase; the resulting complexes are then believed to interact with adipophilin on the outer surface of the lipid droplet to enclose the secretory granule in plasma membrane. FRIL however, demonstrates that the topological distribution of the relevant proteins would make the proposed mechanism impossible. Adipophilin is found in the plasma membrane domains to which secretory granules are apposed in the mammary epithelial cell, in the bilayer surrounding the secreted milk fat globule and in the monolayer enclosing the lipid droplet. Xanthine oxidoreductase is diffusely distributed in the lipid droplet monolayer. Intensive labeling of butyrophilin occurs on both the P face of the monolayer and the E face of the bilayer. The findings from FRIL suggest that while adopophilin-rich domains in the plasma membrane may be linked to secretory granule positioning at the cell surface, butyrophilin-butyrophilin interactions between monolayer and bilayer are responsible for envelopment of the granule by the plasma membrane and its release from the cell.

1. K. Fujimoto, J. Cell Sci. **108** (1995), p. 3443-3449.
2. H. Robenek and N. J. Severs, Biol. Proc. Online **10** (2008), p. 8-20.
3. H. Robenek, O. Hofnagel, I. Buers, M. J. Robenek, D. Troyer and N. J. Severs, J. Cell Sci. **119** (2006) p. 4215-4224.
4. H. Robenek, S. Lorkowski, M. Schnoor, and D. Troyer, J. Biol. Chem. **280** (2005), p. 5789-5794.
5. H. Robenek, M. J. Robenek, I. Buers, S. Lorkowski, O. Hofnagel, d. Troyer and N. J. Severs, J. Biol. Chem. **280** (2005), p. 26330-26338.
6. H. Robenek, O. Hofnagel, I. Buers, S. Lorkowski, M. Schnoor, M. J. Robenek, H. Heid, D. Troyer and N. J. Severs, Proc. Natl. Acad. Sci. USA **103** (2006), p. 10385-10390.

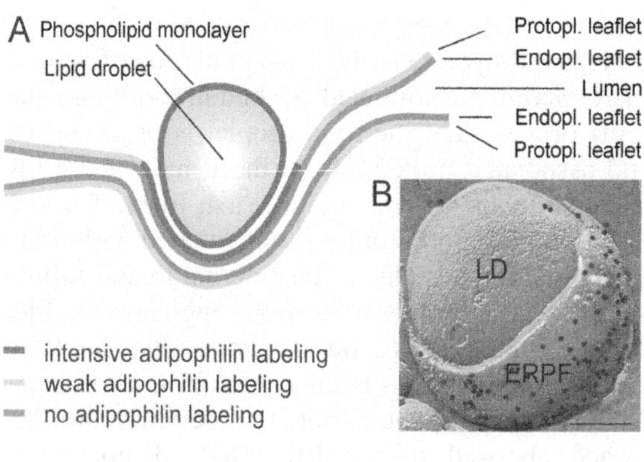

Figure 1. Lipid droplet-ER membrane associations after immunogold labeling for adipophilin in lipid-laden macrophages. A. Summary of principal features of putative sites of lipid droplet biogenesis in ER. B. In convex fractures of lipid droplets and associated ER membranes, prominent labeling for adi-pophilin is visable on the P face of the ER membrane (ERPF) adjacent to the lipid droplet (LD). Bar: 0.2 μm

Dynamics of epigenome duplication, accessibility and translation

M. C. Cardoso

Max Delbrueck Center for Molecular Medicine, Robert Roessle Str. 10, 13125 Berlin, Germany

cardoso@mdc-berlin.de
Keywords: DNA replication, DNA methylation, live-cell microscopy

Most cells of multicellular organisms contain identical genetic information but differ in their epigenetic information. We are interested on how genetic and epigenetic information is replicated every time a cell divides and how the epigenetic information is accessed and "translated" to define specific cell types with specific sets of active and inactive genes, collectively called the epigenome. We are using a combination of in vivo approaches including photodynamic assays (fluorescence photobleaching and activation) in living mammalian cells and, more recently, single molecule microscopy.

Replication of the mammalian genome starts at tens of thousands of origins that are activated at specific times during S phase. The spatio-temporal progression of DNA replication is inherited through consecutive cell division cycles, raising the question how this replication program is coordinated. We are studying the coordination of the multiple enzymatic activities involved in the replication of the genome preceding every mitotic division. With fluorescent fusion proteins and high-resolution multidimensional time-lapse microscopy, we showed that replication patterns within the nucleus change in a characteristic manner throughout S phase [1]. We are further investigating the temporal and spatial dynamics of the replication machinery components in living mammalian cells by a combination of biochemical in situ fractionations and fluorescence photobleaching/activation techniques. Our results suggest that processivity and fidelity of this complex enzymatic machinery is not achieved by stable interactions between its components. Our data is rather consistent with the existence of a stable core in vivo composed of the PCNA clamp (proliferating cell nuclear antigen) bound to the replicating DNA while other factors transiently associate with this core [2, 3]. We are currently extending this hypothesis by measuring the kinetics of the various replication factors and their response to challenges during genome replication.

Nuclear DNA is organized together with structural proteins into dynamic higher order chromatin structures. Chromatin can be subdivided into eu- and heterochromatin, depending on its condensation state, transcriptional activity and the modification of associated chromatin organizing proteins. Whereas euchromatin is generally assumed to be actively transcribed and less condensed, heterochromatin condensation is thought to be similar to mitotic chromosomes in which DNA metabolism (e.g. transcription and replication) has stopped. It is unclear whether and how changes in the chromatin compaction state affect the mobility of chromatin organizing proteins and the access of proteins to chromatin. To address these questions, we are using a combination of live-

A. Aretz, B. Hermanns-Sachweh, J. Mayer (Eds.): EMC 2008, Vol. 3: Life Science, pp. 167–168,
DOI: 10.1007/978-3-540-85228-5_84,

cell chromatin labels and high-speed single molecule tracing microscopy in living mammalian cells [4]. Our results indicated that all nuclear subcompartments were easily and similarly accessible for an average-sized probe protein (streptavidin) and even condensed heterochromatin did neither exclude single molecules nor impede their passage. The only significant difference was a higher frequency of transient trappings in heterochromatin lasting though, only tens of milliseconds. The streptavidin molecules, however, did not accumulate in heterochromatin suggesting comparatively less free volume. Our results suggest that genome metabolism may not be regulated at the level of physical accessibility but rather by local concentration of reactants and availability of binding sites.

Epigenetic information in the form of histone/DNA modifications is "translated" into higher-order chromatin organization by proteins such as the family of methyl-cytosine binding proteins (MeCP2, MBD1-4) that recognize and bind to sites of DNA methylation. MeCP2, the founding member of the MBD family, is mutated in most Rett syndrome patients, which is the second most common neurological disorder after Down syndrome. We have found that MBDs induce large-scale heterochromatin reorganization during cellular differentiation [5, 6]. Based on this finding, we are dissecting the mechanisms responsible for this chromatin reorganization and are testing the molecular composition and role of these heterochromatin compartments in genome expression/silencing during differentiation and in disease. This should help to elucidate the role of genome topology in cellular differentiation, and provide new ways to manipulate the phenotypic plasticity of cells for application in cell replacement therapies in regenerative medicine.

1. Leonhardt, H., Rahn, H.-P., Weinzierl, P., Sporbert, A., Cremer, T., Zink, D. and Cardoso, M. C. (2000). Dynamics of DNA replication factories in living cells. J. Cell Biol. 149, 271-280.
2. Sporbert, A., Gahl, A., Ankerhold, R., Leonhardt, H. and Cardoso, M. C. (2002). DNA polymerase clamp shows little turnover at established replication sites but sequential de novo assembly at adjacent origin clusters. Mol. Cell, 10: 1355-1365.
3. Sporbert, A., Domaing, P., Leonhardt, H. and Cardoso, M. C. (2005). PCNA acts as a stationary loading platform for transiently interacting Okazaki fragment proteins. NAR, 33: 3521-3528.
4. Grünwald, D., Martin, R. M., Buschmann, V., Bazett-Jones, D. P., Leonhardt, H., Kubitscheck, U. and Cardoso, M. C. (2008). Probing Intranuclear Environments at the Single Molecule Level. Biophys. J. 94: 2847-2858.
5. Brero, A., Easwaran, H. P., Nowak, D., Grunewald, I., Cremer, T., Leonhardt, H. and Cardoso, M. C. (2005). Methyl CpG binding proteins induce large-scale chromatin reorganization during terminal differentiation. J. Cell Biol., 169: 733-743.
6. Agarwal, N., Hardt, T., Brero, A., Nowak, D., Rothbauer, U., Becker, A., Leonhardt, H., Cardoso, M. C. (2007). MeCP2 interacts with HP1 and modulates its heterochromatin association during myogenic differentiation. NAR 35: 5402-5408.
7. We kindly acknowledge the financial support of the German Research Council and the Volkswagen Foundation.

Controlled light exposure microscopy reveals telomeric microterritories throughout the cell cycle

W. H. De Vos[1], R.A. Hoebe[2], G. Joss[3], E. Manders[4] and P. Van Oostveldt[2]

1. Department of Molecular Biotechnology, University of Ghent, Coupure links 653, 9000 Ghent, Belgium
2. Centre for Advanced Microscopy, Swammerdam Institute for Life Sciences, University of Amsterdam, Kruislaan 316, 1098 SM Amsterdam, The Netherlands.
3. Department of Biological Sciences, Macquarie University, North Ryde, NSW 2109, Sydney, Australia
4. Department of Cell Biology and Histology, Academic Medical Center, University of Amsterdam, Meibergdreef 15, 1105 AZ Amsterdam, The Netherlands

winnok.devos@ugent.be
Keywords: telomeres, TRF2, stable transfectants, CLEM, cell cycle

Telomeres are the natural ends of linear chromosomes. In mammalian cells they consist of a double stranded array of simple TTAGGG repeats ending in a single stranded overhang that folds back to form a T-loop structure [1]. In combination with sufficient telomere repeats, a telosome complex of indirect and direct telomere binding proteins, dubbed shelterin, assures proper telomere function [2]. Telomeres confer structural integrity and positional stability to human chromosomes [3].

Despite the pivotal role of these structures in ageing and cancer [4] there is no clear view on telomere organization in cycling human cells and their dynamic behavior throughout the cell cycle. Recently, efforts have been made to shed more light on this topic, however, most studies were limited to immunostainings and FISH, both methods inherently limited to fixed material [5,6].

Using stable transfectants, expressing a fusion protein of citrin and telomere binding protein TRF2, and confocal microscopy we investigated the spatiotemporal organization of telomeres in living human fibroblasts and endothelial carcinoma cells (ECV). New CLEM technology (Controlled Light Exposure Microscopy [7]) strongly reduced phototoxicity which enabled us to observe their dynamic behavior throughout the cell cycle. Figure 1 illustrates the beneficial effect of CLEM on image quality preservation.

Based on our results, a view emerges in which two to three telomeres share a small territory at the interface of eu- and heterochromatin, where they show a dynamic interplay with each other (Figure 2). The territories themselves show an overall constrained diffusive motion, although temporally increased and directed mobility were observed as well. Our study supports the current view on nuclear organization based on compartmentalization of chromosome territories and clustering of nuclear proteins in nuclear bodies.

1. C. Greider, Cell **97** (1999), p. 419.
2. T. De Lange, Genes & Development **19** (2005), p. 2100.

A. Aretz, B. Hermanns-Sachweh, J. Mayer (Eds.): EMC 2008, Vol. 3: Life Science, pp. 169–170,
DOI: 10.1007/978-3-540-85228-5_85, © Springer-Verlag Berlin Heidelberg 2008

3. J. Hackett, D; Feldser and C. Greider, Cell **106** (2001), p275.
4. M. Blasco, Nature Reviews Genetics **6** (2005), p. 611.
5. S. Ermler, D. Krunic, T. Knoch, S. Moshir, S. Mai, K. Greulich-Bode and P. Boukamp, European Journal of Cell Biology **83** (2004), p. 681.
6. R. Nagele, A. Velasco, W. Anderson, D. McMahon, Z. Thomson, J. Fazekas, K. Wind and H. Lee, Journal of Cell Science **114** (2001), p. 377.
7. R. Hoebe, C. Van Oven, T Gadella, P Dhonukshe, C Van Noorden and E. Manders, Nature Biotechnology **25** (2007), p. 249.

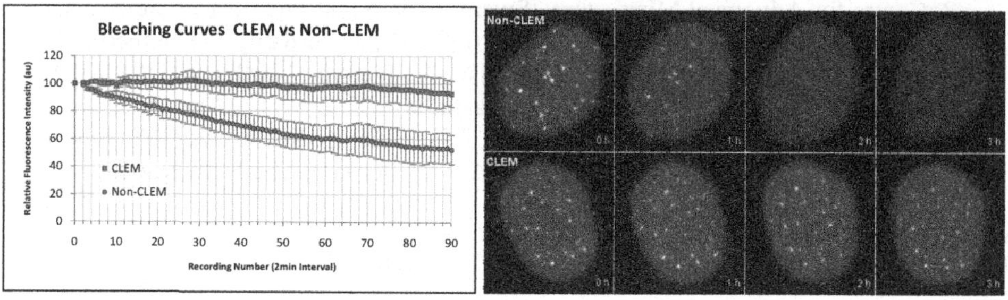

Figure 1. Comparison of CLEM and non-CLEM. Left: bleaching curves illustrate the superior conditions of CLEM. Right: image quality and telomere-resolvability under controlled light exposure are maintained throughout a 3h imaging period, whereas non-CLEM illumination causes rapid bleaching and image quality reduction.

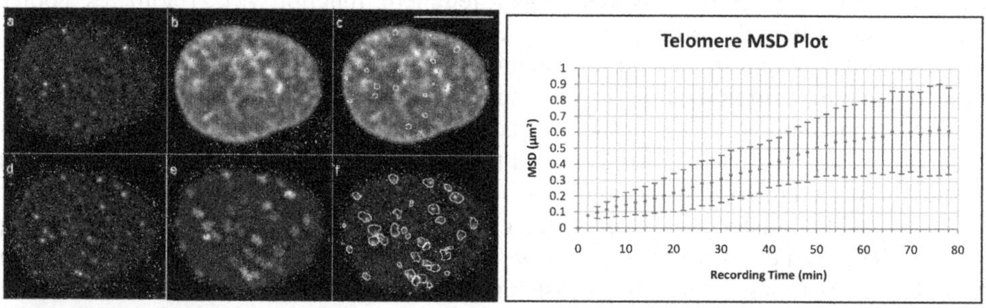

Figure 2. Telomere organisation and dynamics. Left, top panel: predominant localization of telomeres at the interface of hetero- and euchromatin shown in a single confocal section of an ECV-TRF2 nucleus. a. TRF2-citrin, b. H2A-mCherry, c. Image b overlaid with (telomere) outlines from image a to show local chromatin environment. Left, lower panel: confined mobility domain of telomere territories. d. Maximum projection of the first time point, e. Maximum projection of 40 time points (80 min), f. Image d. overlaid with the (territory) outlines from image e to compare telomere with domain size. Right: telomeres show a constrained motion, characterized by a typical MSD plot. Scale bar: 10μm.

Calibration in fluorescence correlation spectroscopy for measurements of stem cell differentiation kinetic

D. Dumas[2], B. Riquelme[1], H. Castellini[3], L. Basciano[1,2], N. de Isla[2] and JF Stoltz[2]

1. Physics Area at the Faculty of Biochemical and Pharmacology Sciences and Biology Applied Optics Groups at the Rosario Physical Institute, CONICET-UNR, Argentina
2. Lab. Mecanique et Ingenierie Cellulaire et Tissulaire – UMR CNRS 7563 LEMTA et IFR 111 CNRS – Université Henry Poincaré, Nancy, France
3. Dpto. Física, Fac. Cs Exactas Ing. y Agr, UNRosario, Argentina

bibiana_riquelme@fbioyf.unr.edu.ar
Keywords: mesenchymal stem cells, fluorescence correlation spectroscopy, cell differentiation

Several works have recently shown that Mesenchymal Stem Cells (MSC) collected from bone marrow can differentiate *in vitro* into cartilage cells under the effects of transforming growth factors or critical transcription factors [1]. The transforming growth factor TGF-β is a family of multifunctional cytokines controlling cell growth, differentiation, and extra cellular matrix deposition. The biological effects of TGF-β are mediated by type I (TBR-I) and II (TBR-II) receptors. In this work we presented a preliminary study to characterize the localization and mobility of TBR-II using mesenchymal stem cells derived from bone marrow by Fluorescence Correlation Spectroscopy (FCS). This sensitive technique is mainly used for biological applications for measuring dynamic processes (number density, interaction fractions and molecular dynamics) in a fluorescent signal on the nanosecond to second time scales [2]. To apply this analytical method to living cells, some steps of standardization have to be optimized in the case of indirect immunolabeling.

FCS parameter analysis was performed by using two different primary antibodies revealed with a secondary antibody and the influence of power laser, objective and molecule concentration in solution were compared. The autocorrelation functions $G(\tau)$ from the fluorescence intensity fluctuations of diluted suspensions of Fab2 fragments conjugated with Alexa Fluor 488, were determined to obtain the average diffusion time taken by the molecule to cross the observation volume, along with the average number of molecules present. The characteristics of the system utilized for this analysis is very important because a non-Gaussian observation volume will result in an altered form for the diffusion autocorrelation function measured by FCS and this altered functional form can easily mislead FCS users into mistakenly attributing fitting residuals to a chemical kinetic process, a second diffusing species, or some type of non-standard diffusion [3]. For calibration on TCS SP2 AOBS- FCS2 Leica Microsystems, 40x/NA 0.8 and 60x/NA 1.2 water immersion objectives, a diffusion coefficient for Fab2 fragments were obtained with an effective volume. Double labelling was used for TBR-II detection (primary antibody: mouse anti-TBR-II human, secondary antibody: IgG anti-mouse Alexa 488), and DiI was used for labeling of membranes and lipid vesicles. The results show that TBRII is localized on cellular surface and in the cytoplasm internalized in

A. Aretz, B. Hermanns-Sachweh, J. Mayer (Eds.): EMC 2008, Vol. 3: Life Science, pp. 171–172,
DOI: 10.1007/978-3-540-85228-5_86,

cylindrical non-lipid vesicles of 3 to 5 μm diameter and 8 to 10 μm longitude (figure 1). These vesicles can be acting as reservoir or recycle function.

Despite the presence of some artefacts, it is possible to get correct FCS parameters under a calibration process. The observation volumes have been calculated with free antibodies in solution for two objectives dedicated to monophoton or multiphoton excitation mode (respectively 11.8 ± 1.2 fL for 40x/0.8 and 1.73 ± 0.12 fL for 60x/1.2). For 40x/0.8 objective, the results indicate a non Gaussian observation volume which will result in an altered form for the diffusion autocorrelation function. According to the laser power delivered (15% or 31%), mean effective volume for 60x/1.2 objective were respectively 1.10 ± 0.03 fL and 2.29 ± 0.07 fL. The washing step has been optimized to avoid all aggregates of varying sizes which are a problem in FCS measurements. The real concentration of antibodies (Qdot 655 goat F(ab')2 anti-mouse IgG conjugate (H+L) Q11021MP Lot 44749A) was recalculated by knowing the number of molecules and the observation volume with 60x/1.2 objective, laser power at 31% and effective volume of 2.29 ± 0.07 fL. For anti-TBR-II human primary and secondary antibodies in solution, time diffusion calculated increased according to time incubation (from 0.93 ± 0.18 to 1.6 ± 0.29 ms) related to a decreased number of molecules in observation volume (from 50 ± 5 to 25 ± 4 molecules). These results, as well as the knowledge of the dimension of the confinement regions will be included for the modelization of signal transduction during the cellular differentiation. Applied to living MSC, this calibration step will be very useful to elucidate the critical role of the TGF-β in chondrogenesis.

1. Pittenger, M.F., Mackay, A .M., Beck, S.C., Jaiswal, R.K., Douglas, R., Mosca, J.D., Moorman, M.A., Simonetti, D.W., Craig, S. and Marshak, D.R. (1999) Multilineage potential of adult human mesenchymal stem cells. Science 2, 143-147
2. D Dumas, B Riquelme, E. Werkmeister, N. de Isla, JF Stoltz. Multimodality of Microscopy Imaging Applied to Cartilage Tissue Engineering. In. Ostearthitis, Inflamation and Degradation: A Continum , Capitulo XVI, J. Buckwalter et al. (Eds.), IOS Press, 2007, pp. 254-260
3. S. Hess and W. Webb. "Focal Volume Optics and Experimental Artifacts in Confocal Fluorescence Correlation Spectroscopy". Biophysical Journal Volume 83 October 2002 2300–2317.

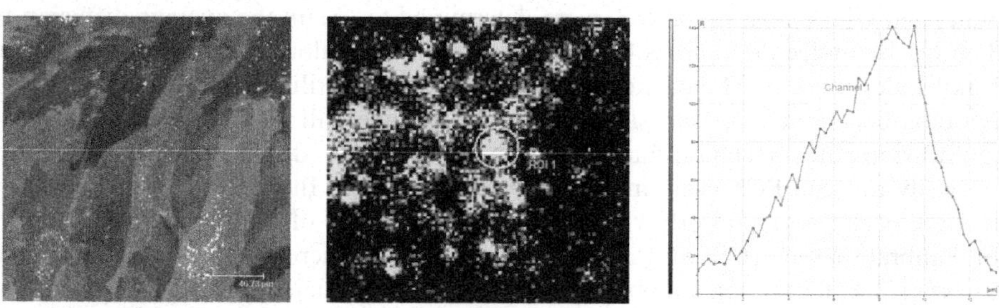

(a) **(b)** **(c)**

Figure 1. Determination of vesicle dimensions in MSC: (a) lipid in red labelled with DiI; (b) Receptor II in green labelled with Alexa 488; (c) intensity profile.

Spectral methods for cells and particles identifications

N. Koltovoy

Labmetod, Moscow, Russia

koltovoi@mail.ru
keywords: cell, single cell absorption spectroscopy, cell identification, microspectrophotometer, multispectral imaging, image processing.

For single cell identifications was created UV-visible-NIR microscope-spectrometer. It was used to measure spectra of microscopic samples (individual cells). This system consists of microscope connected with spectrometers. It is possible to registrations spectrum of transmission (absorption) or reflectance for microscopic sample (individual cells). Systems can be used for cytology, histology and for living cells. It is possible to study static or dynamics.

Microscope connected with spectrometers by means of fiber optics. Fiber optics cables are connected to the optical axis of microscope. The absorption spectra are recorded from ultraviolet (UV) to near infrared (IR) (from 350nm to 1100 nm). It is possible observation of cellular absorption by ultraviolet, optical and infrared light. Range of field for spectral measurements – is from 0.004 mm to 1 mm. It is possible to measure spectra for cell or some components of cell.

System applications – individual cells (stem cells) identification and sorting.

The first step of the procedure is the calibration. This is performed by inserting dark shutter in the optical path, and by setting the dark-carnet amplifier to zero.

Next step – light source calibrations.

Next step – educations – spectrum registrations for a priory known type of cells. It is possible for identifications one or more type of cells. For each type of cells it is need to registration big amount of spectrum examples. After educations it was created single spectrum for each type of cells.

Next step – cell identifications. We produced spectrum registrations for unknown cell. This spectrum is compared with spectrum of educational cells. It was used pattern recognition methods for spectrum identifications. To further assist in the spectral classification of individual cells, multivariate analysis was also applied.

Next step – sorting. It is used handle sorting. Microscope have left and right double holder microinjection. It is possible to sort cell by four class. Then individual cell was identified it was safe in suit microinjectors.

It is possible to use systems as microfluorimeter for registration full fluorescence spectra of cell. In this case it is need to remove emission (barrier) filter. For fluorescence registration it is need to increase time of accumulations.

A. Aretz, B. Hermanns-Sachweh, J. Mayer (Eds.): EMC 2008, Vol. 3: Life Science, pp. 173–174,
DOI: 10.1007/978-3-540-85228-5_87,

UV and IR spectrum registration is a modification on UV and IR microscopy. The near IR range of an absorption spectrum contains features relating to the vibration absorption of the molecular bounds within the cell being examined.

If we use XY stage motors – it is possible to create full 2-dimentional spectral imaging. In this case we will get 3-dimentional imaging – X, Y and Lambda. For image processing in this case it was used special algorithms for cell identifications. It is need to use special methods for multispectral image processing. It is need specials methods for segmentations and image processing.

3D realistic visualization of supramolecular assemblies

D. Larivière[1] and E. Fourmentin[1]

1. Scientific Foundation Fourmentin-Guilbert, 2 avenue du Pavé Neuf, 93160 Noisy-Le-Grand, France

fondation@fourmentinguilbert.org
Keywords: 3D visualization, supramolecular structures, cryotomogram segmentation.

Thanks to NMR, X-ray crystallography and electron microscopy, it is possible to resolve 3D structure of most cell components. Even though many softwares help biologists to visualize the 3D shape of individual molecules, several authors have been calling for a visual discovery tool which can link the individual component view with the system view [1, 2] (Fig.1).

As a first step in this direction, we have developed a prototype which enables to map cell components (reconstructed from their PDB structures) and navigate in real-time within the cell at various scales from a complete cell down to a single macromolecule. The tool has 4 main features : 1/ A user interface to navigate inside a cell at various scales in different ways 2/ A graphical tool to build 3D scenes and manipulate objects (translation, rotation) 3/ An annotation tool to organize and share information about a 3D scene or its component 4/ A data repository of biological data linked to existing databases.

Real-time 3D rendering is a software challenge. A typical cell has millions of molecules and thousands of macromolecules. Such a crowded environment cannot be rendered in a fully realistic way. We have thus implemented a specific visualization mode (orthographic display) when a crowded biological scene is to be displayed. We have also designed a 3D molecular representation such that each molecule can be rendered as quickly as possible while being distinguishable and recognizable. The viewer is entailed to look at the overall ultra structure rather than to the specific atomic details of a molecule [3].

An unlimited number of biological models can be created (Fig.2). In the field of structural biology specifically, 3D reconstructions resulting from segmentation of cryo-electron tomograms do not reveal the entire proteome of a cell. This is due to low signal-to-noise ratios and the dense packing of the cytoplasm which hamper protein recognition. The software technology described in this abstract can be used to complete such 3D reconstructions with macromolecular structures initially unrevealed but recognized or supposed as being involved.

The software is licensed under an open source license schema. It runs on Intel platforms with a 3D graphical card running either Microsoft Windows or Linux. The project page is http://www.lifeexplorer.eu.

1. H. Breithaupt, EMBO Reports **volume 7** (2006), p.467–470.
2. S. Nickell, C. Kofler, A.P. Leis, W. Baumeister, Molecular Cell Biology **volume 7** (2006), p. 225–230.
3. D.S. Goodsell, Structure volume 13 (March 2005), p. 347–354.

A. Aretz, B. Hermanns-Sachweh, J. Mayer (Eds.): EMC 2008, Vol. 3: Life Science, pp. 175–176,
DOI: 10.1007/978-3-540-85228-5_88,

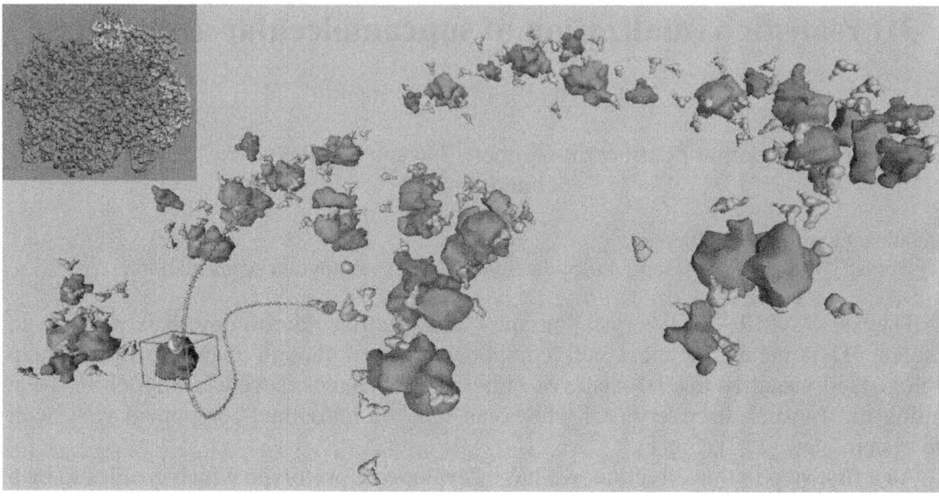

Figure 1. The close-up view shows individual RNA polymerase visualized with the rendering software Qutemol from PDB 1L9U. This RNA polymerase has been replaced with LifeExplorer in an environment containing about 200 proteins (mainly transfer RNA, tRNA synthetases and ribosomes).

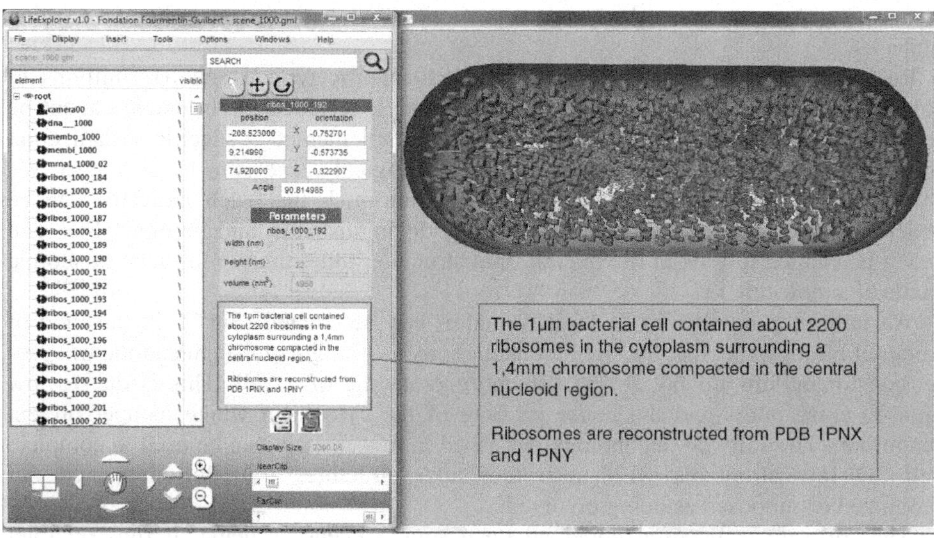

Figure 2. The LifeExplorer software user interface with annotation and data organization tools.

Heterosis in Arabidopsis thaliana: Structural aspects in mature and germinating seeds from hybrid and parental lines

E. Maximova, A. Blacha, Th. Altmann, M. Udvardi

Max-Planck Institute of Molecular Plant Physiology, am Mühlenberg 1, 14476 Golm, Germany

Maximova@mpimp-golm.mpg.de
Keywords: Arabidopsis, heterosis, hybrid, reserve mobilization

Heterosis is a biological phenomenon of great significance to agriculture and breeding [1,2,3]. The widespread occurrences of heterosis in the model plant *Arabidopsis thaliana (L.)* [4,5] open the possibility to investigate the genetic basis of this phenomenon using the tools of genetical genomics and metabolomics. In order to define the time in development at which point the F1 and parents diverge in structure, composition and rates of growth, it is important to supplement qualitative observations with quantitative structural information.

In our study the metabolic analysis were complemented by microscopic analysis of *Arabidopsis thaliana (L.)* mature and germinating seeds from hybrid and parental lines. Mature seeds of *Arabidopsis thaliana (L.)* from parents and hybrids obtained by manual pollination: Col-0xCol-0 (self parent), C24xC24 (self parent), Col-0xC24 (F1 hybrid), C24xCol-0 (F1 hybrid) were studied. We have analysed the mature seeds and germinating patterns in *Arabidopsis thaliana (L.)* during reserve mobilization qualitatively and quantitatively and related development to the time scale “hours after sowing” (HAS) “Figure 1”. Protein bodies vary structurally and chemically, and oil bodies mainly chemically. Protein bodies vary between genotypes in size, number, structure, inclusions. Probably, proteins may be used as means of identifying parents of putative F1 hybrids, as well as cultivars and species.

Here, using the light microscopy and histochemistry, we showed the cellular basis of *Arabidopsis thaliana (L.)* seeds from different genotypes during dormancy and post embryonic development. In initial growth under long-day conditions the *Arabidopsis thaliana* Col-0xCol-0 and C24xCol-0 seeds displayed the faster germination by contrast to others. Histochemical study of pre-and post-germinated *Arabidopsis thaliana (L.)* seeds showed the difference in the mobilization /utilization rates of storage proteins. More rapid mobilization and utilization of seed reserves were in Col-0xCol-0 at the initial growth and the latest was hybrid C24xCol-0.

It was established, that the metabolically more active seeds of Col-0xCol-0 demonstrate the greater degree of the mobilization of cell and tissue reserves what cause their preservation in active state at the initial growth until 48 HAS. On contrast, the less metabolically active seeds of *Arabidopsis thaliana (L.),* C24xCol-0, demonstrate the greater degree of protein reserves utilization at the latest stages of post-embryonic development, what is displayed later in bigger sizes and mass of the seedlings (stronger heterosis).

A. Aretz, B. Hermanns-Sachweh, J. Mayer (Eds.): EMC 2008, Vol. 3: Life Science, pp. 177–178,
DOI: 10.1007/978-3-540-85228-5_89,

1. Srivartana H.K. (1983). Heterosis. Reappraisal of theory and practice. Monographs on Theoretical and Applied Genetics 6: 260-277.
2. Berlyn GP. (1972). Seed germination and morphogenesis. In: TT Kozlowskii ed. Seed biology. v.2, Academic Press, New York.
3. Werker E. (1997). Seed anatomy. In: Encyclopedia of Plant Anatomy, v. X, part 3, ed. S.Carlquist et al., Berlin, Stuttgart.
4. Pyke KA, Marrison JL, Leech RM. (1991). J. of Experimental Botany, v.42 : 1407-1416.
5. Mansfield SG, Briarty LG. (1996). J Plant Sci., v.157 (3) : 280-295.

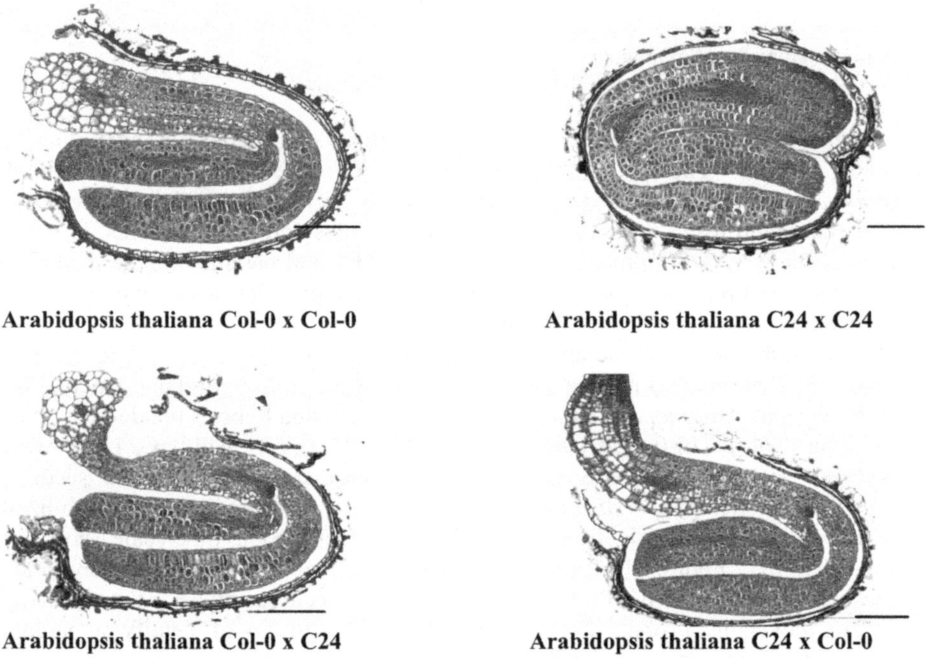

Figure 1. Light micrographs of seeds 48 hrs after sowing (HAS). Semi-thin longitudinal sections stained with Toluidine blue 0, acid Fuchsin (proteins) and Sudan III (lipids) for histochemical localisation of storage components in hybrids and parental lines.
Scale bar = 100μm

Colour visualization of red blood cells in native smears by the new method reflected light microscopy

A. Paiziev and V. Krakhmalev

Lab of Applied Physics, Institute of Electronics Uzbek Academy of Science, Tashkent 100125 F.Khodjaeva str.33, Uzbekistan

adxam_payziev@rambler.ru
Keywords: light microscopy, red blood cells, interference

As a rule to get color contrast image of biological samples under optical microscope need to use a variety of dyes and fluorescent substances but it leads to artificial staining of sample, destructive modification and loss very important structural information its native structure. Usually in medical practice conventional bright-field microscopy let us see black and white image of separate morphological elements of blood smears only (Figure 1a). Proposed in this paper the new nondestructive method of optical microscopy allows to examine the structures of living cells in their natural colors without its staining by using a specially designed substrate for deposition of biological sample and observing a native blood smears in reflected light. This method based on physical phenomena of white light interference reflected from sample surface and special supporter on which this sample is deposited. It allows to occur at the image plane converting previously invisible gradients of refractive index within the specimen in to intensity gradients in the image. Color interference contrast image is achieved due to special condition of experiment is connected with chose of angle of incidental light, wave length of light of reflected ray, chemical composition of sample, thickness of sample, refractive index of sample, refractive index of substrate, chemical composition of substrate [1,2]. If any conditions of the experiment are fixed beside of chemical content of sample we can say that color contrast is caused by chemical compounds. On the Figue1(b) red color of erythrocytes correspond hemoglobin. Therefore from this picture we can see visually content of hemoglobin in blood smear. Similar we can determine other chemical compounds after calibration color scale by alternative methods. To demonstrate the potential usefulness of this method, we provide qualitative data describing color image of healthy and pathological damaged cells for alive and dry blood smears (Figure 1a-1e).

Comparison Figure1a and Figure1b for same samples but obtained by conventional bright-field microscopy and by using new method correspondently showed distinguishing in color not only separate red blood cells but distinguishing difference parts in area separate erythrocytes too.

Usually for healthily individuals a albuminous aureole around the erythrocytes are mainly white-yellow (Figure 1c), but for cancer cells (core rectal cancer) the aureole color is quite different and reflects significant changes in chemical composition of both internal, and external contents of erythrocytes (Figure 1d, 1e).

A. Aretz, B. Hermanns-Sachweh, J. Mayer (Eds.): EMC 2008, Vol. 3: Life Science, pp. 179–180,
DOI: 10.1007/978-3-540-85228-5_90,

Easy detection of organic shells around blood cells in our case is evident. Operations by fixing, smear coloring, prolonged processing, the availability for phase-contrast or interference microscope, special illuminators, radiating the exciting short-wave light beams are not required. Interferometric coloring of blood elements occurs on a surface of specially selected substrate. Corresponding colored images of blood elements are formed due to interference phenomena occurring under interaction of light beams reflected from front and back surfaces of blood elements, smeared on a substrate.

1. V. Krakhmalev, A. Paiziev, Noninvasive Color Visualization of Blood Cells. J. Nondestructive Testing, **11**, (2006), p.6
2. M.S. Abdullakhodjaeva, V.A. Kakhmalev and A.A. Paiziev,. Investigation of a medical-biological samples patten by the method of light reflected microscopy. Medical journal of the Uzbekistan **4**, (2007), p.34-39.

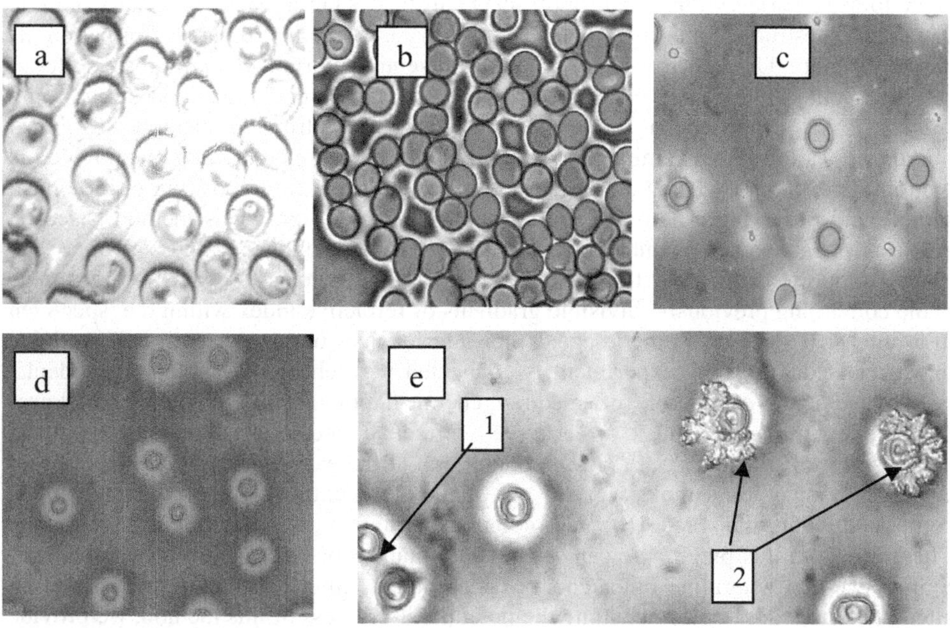

Figure 1. a- bright-field image of erythrocytes of healthy individuals; b-color image same blood sample; c-erythrocytes of healthy individuals. Different colors can be seen in certain areas of an individual red blood cell. d- and e- living erythrocytes of a patient with diagnosed core rectal cancer. Images show a multi-layered pattern of interference picture. Differences between the coloration in left (d) and right image (e) caused by different experimental conditions. We can see double cells (1) and damaged cells (2), both are pathologic.

In vitro culture of *Trigonella foenum-graecum* plantules and their anatomic characterization

C. Pribac[1], A. Ardelean[1]

1. Vasile Goldis West University, Arad, Romania

gpribac@uvvg.ro
Keywords: Trigonella, in vitro culture

Trigonella foenum-graecum, also known as fenugreek is an annual herb found in southern Europe and eastern Asia having off-white flowers and aromatic seeds used medicinally and in condiments. Fenugreek seeds are a rich source of the polysaccharide galactomannan. They are also a source of saponins such as diosgenin, yamogenin, gitogenin, tigogenin, and neotigogens. Other bioactive constituents of fenugreek include mucilage, volatile oils, and alkaloids such as choline and trigonelline.

Fenugreek is frequently used in the production of flavoring for artificial syrups. The taste of toasted fenugreek is additionally based on substituted pyrazines, as is cumin. By itself, it has a somewhat bitter taste. Dried fenugreek seed is mainly used as digestive aid. Fenugreek seed is widely used as a galactagogue, by nursing mothers to increase inadequate breast milk supply. Supplements of fenugreek seeds were shown to lower serum cholesterol, triglyceride, and low-density lipoprotein in human patients and experimental models of hypercholesterolemia and hypertriglyceridemia [1].

Some studies were conducted on the Trigonella seeds but there is no structure studies on plantules so far [2]. The fruit contains 10-20 comprised, prismatic seeds, with a length of 4-5 mm and width of 3-4 mm, of a yellow colour, smooth and with a cumin smell (Figure 1). Plantules were cultivated *in vitro* under aseptic and non-sterile conditions with 12% hummidity (Figure 2 and 4). These plantlets were produced in vitro on Murashige-Skoog (1962) mineral basal medium [3]. To this medium we added various organic compounds, usually used in similar experiments, as FeEDTA, growth regulators, myo-inositol, vitamins, saccharose, agar-agar etc. On this kind of medium, the cultivated Trigonella had to be sub-cultivate after a longer cultivation period (about 3-4 months).

To induce the cultivation we have used *Trigonella* seeds, that were sterilized using immersion for 1 minute in 70% ethanol and then in a solution of 9% calcium hypochlorite for 25 minutes. After that, the seeds were washed for 3 times with sterile distilled water. Seeds were inoculated on a nutrition medium using the Murashige and Skoog recipe (Ms, 1962), without hormons, in a ½ concentration [3] (Figure 3). The value of pH of all media were ajusted to 5.5 before the sterilization of the medium recipients at 121°C for 60 minutes. From the obtained plantules we have prepared the samples and examined them using electron microscopy. For more conclusive data we have made a correlation between the results obtained on in vitro plantules and those cultured on non-sterile media. There were no significant differencies at the anatomic

A. Aretz, B. Hermanns-Sachweh, J. Mayer (Eds.): EMC 2008, Vol. 3: Life Science, pp. 181–182, DOI: 10.1007/978-3-540-85228-5_91,

level on laminar leaf and petiole but on the radices there were small modifications for the in vitro plantules.

1. E. Basch et al. (2003) in "Therapeutic applications of fenugreek". Altern Med Rev **8** (1): 20-27.
2. X. Gong, G.W. Bassel, A. Wang, J.S. Greenwood and J.D. Bewley, The Emergence of Embryos from Hard Seeds is Related to the Structure of the Cell Walls of the Micropylar Endosperm, and not to Endo-b-mannanase Activity, Annals of Botany **96**: 1165–1173, 2005.
3. T. Murashige and F. Skoog, A revised medium for rapid growth and bioassays with tabaco tissue culture, Phisiol. Plant **15** (1962), p. 473-497.

Figure 1. Seeds of *Trigonella foenum-graecum*

Figure 2. *In vitro* germination of Trigonella seeds cultivated under aseptic conditions on culture medium Ms

Figure 3. *Trigonella* plantules obtained through *in vitro* culture

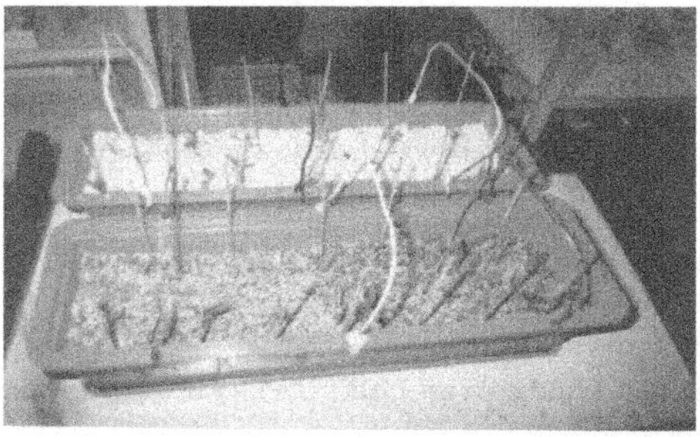

Figure 4. Germination of *Trigonella* seeds under non-sterile conditions on a substrate composed of perlite, sand and dirt

Live cell imaging by SEM-hosted X-ray microscope in "water window" energy range

A. Sasov

SkyScan, Kartuizersweg 3B, 2550 Kontich, Belgium

sasov@skyscan.be
Keywords: cell biology, X-ray microscopy, soft X-ray imaging, SEM

Using soft X-ray radiation in the so called "water window" energy range between absorption edges of oxygen (water) and carbon (proteins, etc.) (Fig.1) allows visualizing of cell structure and provides information which is complimentary to electron and optical microscopy. Till now such type of X-ray imaging was possible only by synchrotron based set-ups [1,2] or by expensive laser-pumped plasma sources [3,4].

An SEM-hosted soft X-ray microscope has been designed using a titanium target to produce 450eV radiation, which is capable for "water window" imaging. Using a special specimen container to keep the biological sample under normal environmental conditions allows imaging of living cells without any fixation or preparation. The suggested set-up was tested in two types of SEM (JSM840 and JSM7000F, JEOL, Japan) with different electron beam density, showing the possibility of live cell imaging with submicron resolution.

The SEM hosted soft X-ray microscope uses an SEM focused electron beam to generate X-rays from titanium target with characteristic energies 450eV/460eV and 4.51/4.93keV. The calculated X-ray spectrum from a Ti target under a 5.5keV electron beam is shown in Fig.2. The practical possibility to generate an X-ray line inside the "water window" range was verified by EDX analysis using different accelerating voltages in the SEM. The measured emission spectra for 4, 5, 6, 8 and 10keV electron beams accelerating voltages are shown in Fig.3.

An experimental setup is shown in Fig.4. Inside the SEM specimen chamber a focused electron beam (1) hits the Ti target (2) to produce an x-ray beam (3). The X-ray beam passes through the object container (4) with the object (5) installed in the 0.2mm air gap and separated from the vacuum by two 50nm Si_3N_4 membranes (6). Behind the object container a special magnetic trap (7) deflects scattered and secondary electrons from the path to the cooled back-illuminated X-ray camera (8), which is working as energy-selective photon counting two-dimensional detector.

First testing has been done by imaging onion skin cells. Using the JSM7000F SEM with distance from emission point to the CCD around 330mm, the average counting rate (only "water window" photons) was around 120 photons per second. One of the images acquired using "water window" X-ray energies is shown in Fig.5. Pixel size is 800nm, exposure - 60min, 5.2kV accelerating voltage, $6x10^{-8}$A beam current on the target.

Described technique allows "water window" imaging of living cells to be obtained using an inexpensive setup for any conventional SEM.

A. Aretz, B. Hermanns-Sachweh, J. Mayer (Eds.): EMC 2008, Vol. 3: Life Science, pp. 183–184,
DOI: 10.1007/978-3-540-85228-5_92,

1. W. Chao et al., Proc. 8th Int. Conf. X-ray Microscopy (2005), IPAP Conf. Series 7, pp.4-6A.
2. M.A. Le Gros, G. McDermott, C.A. Larabell, Current Opinion in Structural Biology, **15**, 5 (2005), p.593-600.
3. M. Berglung, et al., J. of Micr., **197**, Pt.3 (2000), pp 268-273.
4. Z. Chang, et al. IEEE J. of Selected Topics in Quantum Electronics, **4**, 2 (1998), pp.266-270.

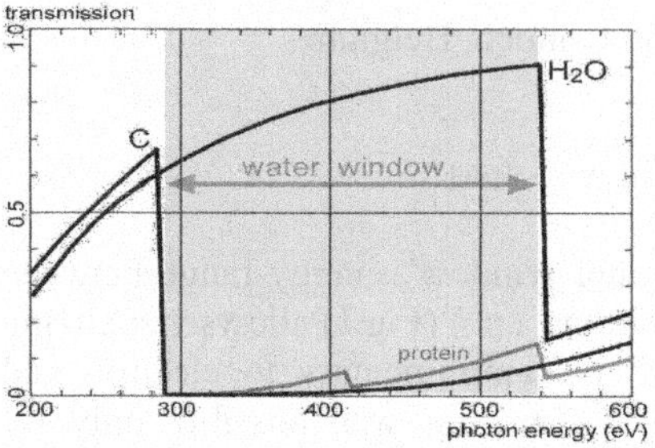

Figure 1. "Water window" energy range.

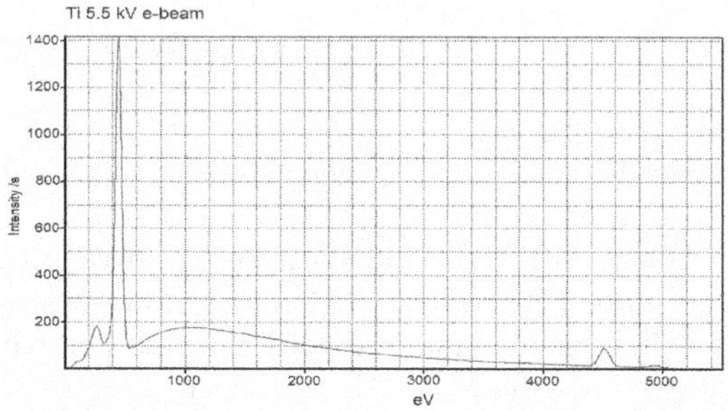

Figure 2. Calculated spectrum from Ti target.

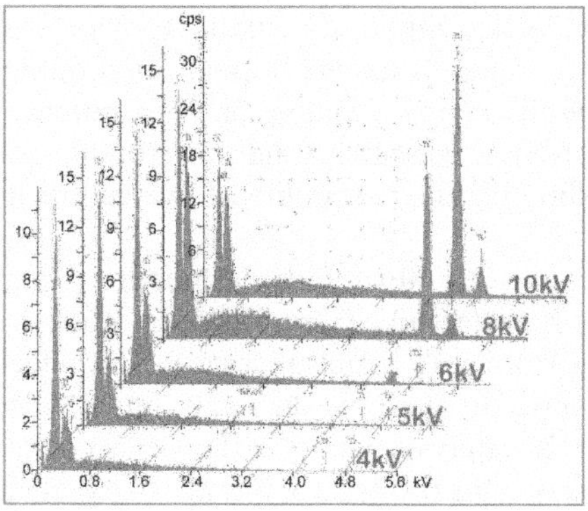

Figure 3. Measured X-ray spectra for different accelerating voltages.

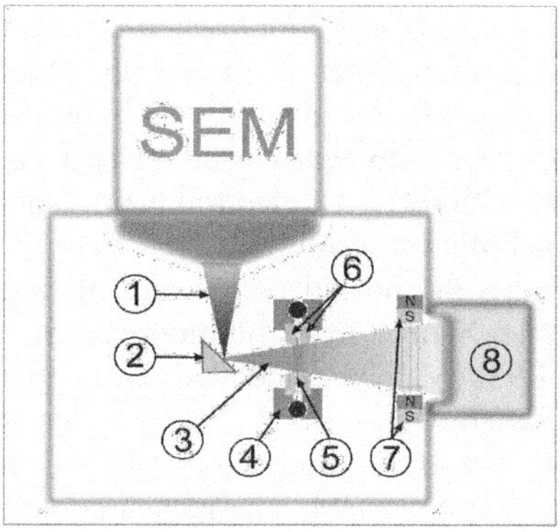

Figure 4. Experimental setup.

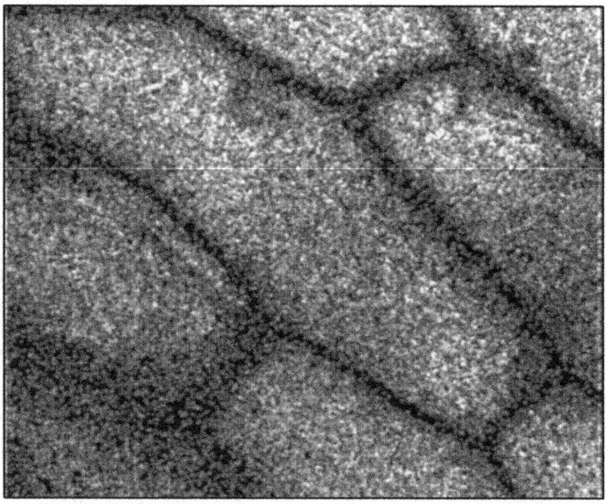

Figure 5. X-ray image of onion skin cells obtained in SEM using 450eV photons energy.

Microscopic anatomy and mineral composition of cuticle in amphibious isopods *Ligia italica* and *Titanethes albus* (Crustacea:Isopoda)

J. Štrus[1], N. Žnidaršič[1], S. Hild[2], A. Ziegler[2]

1. Department of Biology, Biotechnical Faculty, University of Ljubljana, Večna pot 111, SI-1000 Ljubljana, Slovenia
2. Central Facility for Electron Microscopy, University of Ulm, Albert-Einstein-Allee 11, D89069 Ulm, Germany

jasna.strus@bf.uni-lj.si
Keywords: calcium, magnesium, cuticle

Terrestrial isopods are well adapted to terrestrial life and inhabit different areas with abundant organic matter. Structural and biochemical characteristics of their cuticle enable them to survive also in the driest environments, including deserts. Amphibious species are mostly found in the humid areas of litoral and cave habitats. The structure and composition of the cuticle might reflect special ecophysiological and/or phylogenetic relationships within the group of oniscidean isopods.

Tergal cuticles of intermoult specimens of *Ligia italica* from the litoral areas of the Adriatic sea and *Titanethes albus* from Planina cave in Slovenian Karst, were prepared for microscopic and EPMA analyses. In *Ligia italica* pieces of cuticle were fixed in 3.5% glutaraldehyde in 0.2M cacodylate buffer, incubated in 5% lead acetate and postfixed in 1% osmium tetroxide. Samples of cuticle in *Titanethes albus* were fixed in 2.5% glutaradehyde in 0.1M phosphate buffer and postfixed in 1% osmium tetroxide. Ultrathin sections were stained with uranyl acetate and lead citrate. High pressure frozen and freeze dried or liophilized samples for EPMA analysis were examined with a SEM equipped with EDAX detector system.

Cuticle of *Ligia italica* is often restructured due to frequent moulting [1,2] while cuticle of moulting *Titanethes albus* was not described yet. Ultrastructurally both cuticles exhibit a multilayered structure composed of epicuticle, mineralized exocuticle and lamellar endocuticle with unmineralized membranous layer (Fig.1).

A thin epicuticle is composed of several electron dense layers which are more prominent in *Titanethes albus*. Ultracytochemical detection of unsaturated protein-bound lipids in the inner epicuticle and free unsaturated lipids in the surface waxy layers was reported in the terrestrial isopod *Oniscus asellus* [3]. The precise location of free and bound lipids in crustacean epicuticles remains mostly unknown. The mineralized exocuticle consists of helicoidally aranged chitin fibers associated with proteins. Distal layers of the exocuticle in *Ligia italica* exhibit dark electron dense areas of parallel chitin/protein fibres, mineralized with calcium carbonate while distal exocuticular layers in *Titanethes albus* contain electron dark and light areas. The lamellar endocuticle of *Ligia italica* and *Titanethes albus* is electron lucent with an electron dense membraneous layer in *Titanethes albus*.

A. Aretz, B. Hermanns-Sachweh, J. Mayer (Eds.): EMC 2008, Vol. 3: Life Science, pp. 185–186,
DOI: 10.1007/978-3-540-85228-5_93, © Springer-Verlag Berlin Heidelberg 2008

EPMA analysis of cuticle samples in *Ligia italica* and *Titanethes albus* revealed the presence of calcium, magnesium and phosphorus in addition to carbon and oxygen. Low amounts of magnesium were present in *Titanetes albus*, whereas in *Ligia italica* magnesium content is higher and comparable to more terrestrial isopods. Raman spectroscopy and confocal micro-Raman spectroscopic imaging was employed to allocate calcite, amorphous calcium carbonate and organic matrix within the tergite cuticle of *Titanethes albus* and *Ligia italica.* The results show that the cuticle of both amphibious species are thinner and contain less calcite than the cuticles of most isopods that are adapted to more terrestrial habitats. In the cuticle of *L. italica* the layer of calcite is about twice as thick as that of *T. albus*.

1. J. Štrus and A. Blejec, in Isopod systematics and evolution, Crustacena Issues 13, eds. B. Kensley and R.C.Brusca (A.A.Balkema, Brookfield) (2001), p.343-352.
2. Ziegler, M.Hagedorn, G.A.Ahearn and T.H.Carefoot, J.Comp.Physiol. B **177** (2007), p.99-108.
3. P. Compere, S.Defize and G.Goffinet, ICEM 13-Paris, 17-22 July (1994), p.417, 418.
4. The work was supported by Slovenian Research agency within the programe P1-0184 and by the Deutsche Forschungsgemeinschaft within the programe Principles of Biomineralisation (SPP 1117, EP 22/16-3 and Zi 368/4-3).

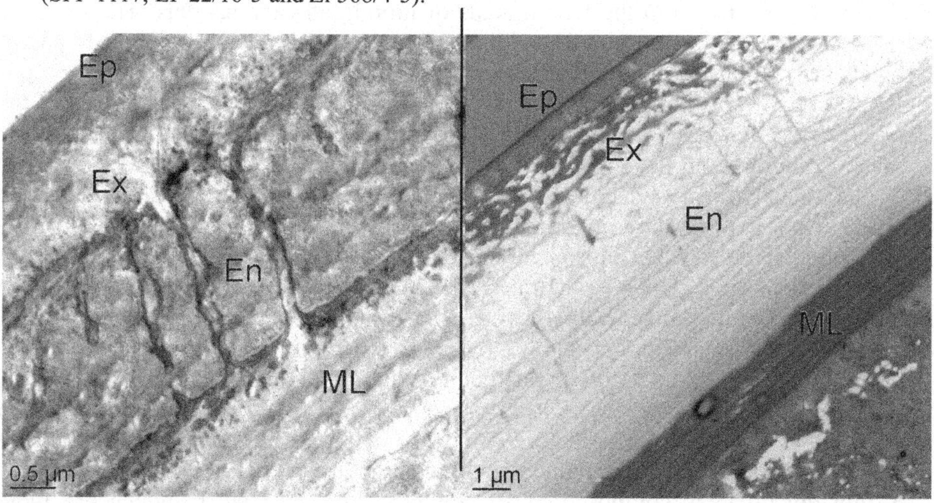

Figure 1. TEM micrographs of cuticle structure in intermoult *Ligia italica* (left) and *Titanethes albus* (right). Ep-epicuticle, Ex-exocuticle, En-endocuticle, ML-membraneous layer

In vivo imaging and quantification of the continuous keratin filament network turnover

R. Windoffer[1,3], A. Kölsch[3], S. Wöll[1], T. Würflinger[2], T. Aach[2] and R.E. Leube[3]

1. Department of Anatomy and Cell Biology, Johannes Gutenberg-University, Becherweg 13, 55128 Mainz, Germany
2. Institute of Imaging & Computer Vision, RWTH Aachen University, 52056 Aachen, Germany
3. Institute of Molecular and Cellular Anatomy, RWTH Aachen University, Wendlingweg 2, 52074 Aachen, Germany

windoff@uni.mainz.de
Keywords: cytoskeleton, life cell imaging, image processing

Keratin polypeptides are major components of the epithelial cytoskeleton forming a filamentous 3D-network. Like intermediate filament polypeptides of other cell types, keratins make up a stable, but elastic network that is responsible for mechanical stress resilience. At the same time the keratin network is able to change its shape during development, cell division, metastasis and cell migration.

Our long term goal is to elucidate and characterize mechanisms that control the formation, shaping and remodelling of the keratin network to establish a general model of dynamic intermediate filament network organisation. Therefore we established a workflow to monitor fluorescently labelled keratins in living cultured cells under standard and experimental conditions. We isolated several epithelial and non-epithelial cell lines stably expressing fluorescent protein-tagged keratins alone or in combination with other fluorescently-tagged proteins for simultaneous monitoring. Time lapse series of fluorescence microscopical images were generated with the help of widefield or confocal fluorescence microscopy.

Unexpectedly and in contrast to the existing text book model, a topologically-defined constant turnover of the keratin cytoskeleton was observed [1]. Detailed high resolution analyses of z-stacks in combination with fluorescence recovery after photobleaching (FRAP) experiments and excitation of photoactivable keratin-GFP fusions led us to propose a dynamic model of the keratin network turnover cycle that is independent of protein biosynthesis. Central to this model is the continuous and regulatable polymerization of keratin filaments which is primarily restricted to the cell periphery. After nucleation filament network precursors further elongate and move from the cell cortex to the cell interior by continuous, actin-dependent transport thereby reaching the peripheral pre-existing network into which the precursors are integrated. Filaments continue to migrate toward the perinuclear region where they dissolve giving rise to soluble subunits that can be re-utilized in the cell periphery for another round of the keratin turnover cycle.

To substantiate the model and to work out aspects of individual steps of the proposed keratin network treadmill, we developed and applied image processing tools.

A. Aretz, B. Hermanns-Sachweh, J. Mayer (Eds.): EMC 2008, Vol. 3: Life Science, pp. 187–188,
DOI: 10.1007/978-3-540-85228-5_94,

In the first set of analyses, the conspicuous network flow was examined. To this end, quantification was performed on a time-lapse series of 69 z-stacks (22 focal planes at 1024x256 pixels) recorded at 60 s intervals. Global cellular movement was compensated for by an image intensity-based method [2] using the normalised sum of squared intensity differences (SSD). Robustness was increased by embedding the registration in a multiscale Gaussian pyramid framework. Visual assessment of the results revealed no misalignments. Based on the aligned image series, analyses could be restricted to intracellular keratin motion. A z-axis projection of each volume was superimposed by a grid of regions of interest (ROIs) at a 20 pixel spacing. The overlapping ROIs were of 40 pixel width. Each ROI was rigidly registered to the previous frame by the SSD criterion using exhaustive search for translations of ±20 pixels. SSD was modified by a weighted window function on the ROIs which dropped linearly from the centre to zero. Thereby the non-rigid character of the data was accounted for while enough structural information was captured by the ROIs, similarly to [3]. Assuming long-term local stability of the keratin flow, the medians of the translation vectors were computed for each grid point over time revealing coordinated centripetal keratin flow all the way to the nucleus (Fig. 1). Different zones characterized by different transport rates could be defined with incremental decrease from the periphery to the cell centre. Taken together, these data strongly support our proposed centripetal keratin treadmill model.

1. R. Windoffer, S. Wöll, P. Strnad, R.E. Leube, Identification of novel principles of keratin filament network turnover in living cells. Mol. Biol. Cell (2004) **15**, 2436-2448.
2. T. Würflinger, M. Bergmeister, A. Böcking, D. Meyer-Ebrecht, T. Aach, Image registration for multimodal cell analysis. Cell. Oncol. (2008) **30**:166. ISCO, Amsterdam, NL.
3. K.P. Wilkie and E.R. Vrscay, Mutual information-based methods to improve local region-of-interest image registration. Lecture Notes in Computer Science, Proc. ICIAR (2005) 3656.

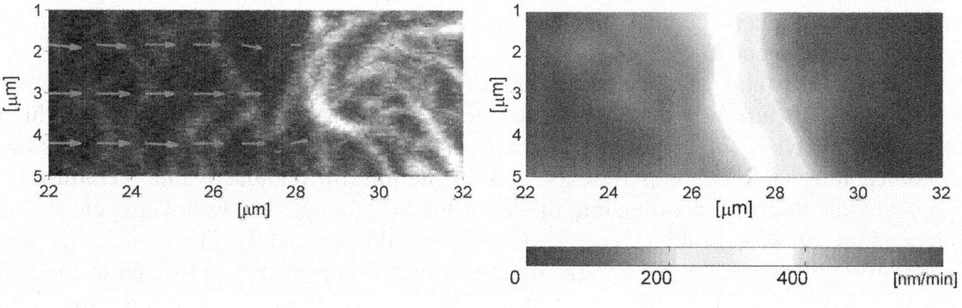

Figure 1. Determination of direction and speed of keratin flow from a time-lapse fluorescence image series. (Left) Vector visualization superimposed on projection of initial fluorescence images at time point 0. (Right) False colour representation of velocities determined in different regions (cell periphery at left, perinuclear region at right).

Necrotic cell death, a controlled way of cellular explosion

Tom Vanden Berghe[1,2], Nele Vanlangenakker[1,2], Eef Parthoens[3], Wies Deckers[3], Michael Devos[1,2], Nele Festjens[1,2] and Peter Vandenabeele[1,2]

1. Molecular Signaling and Cell Death Unit, Department for Molecular Biomedical Research, Flanders Institute for Biotechnology (VIB), Technologiepark 927, B-9052 Zwijnaarde, Gent, Belgium.
2. Department of Molecular Biology, Ghent University (UGent)
3. Microscopy Core Facility, Department for Molecular Biomedical Research, (VIB).

Peter.Vandenabeele@dmbr.ugent.be
Keywords: necrosis, apoptosis, time laps microscopy

Three major morphological types of cell death have been described. Type I or apoptotic cell death is mediated by caspases (a family of cysteine-dependent aspartate-specific proteases) and characterized by cellular shrinkage, membrane blebbing, chromatin condensation and DNA degradation. Type II cell death is associated with the formation of autophagic vacuoles inside the dying cell. Type III or necrotic cell death is characterized by cellular swelling, plasma membrane rupture and the subsequent loss of the intracellular content.

Necrosis has long time been described as accidental and uncontrolled cell death as a consequence of physico-chemical stress. Recently, it became obvious that TNF-induced necrotic cell death is a well-controlled process consisting of defined cellular events and that it is a cell death mode with important pathological and physiological relevance. Use of a wide variety of inhibitors (ROS scavengers, complex I inhibitors, Ca^{2+} chelators, calpain inhibitors, cathepsin inhibitors and lipoxygenase inhibitors) has revealed that necrotic cell death involves multiple subcellular compartments such as mitochondria, endoplasmic reticulum and lysosomes (for review see [1]).

The RIP serine/threonine kinases have emerged as essential sensors of cellular stress. The different members integrate both extracellular and intracellular stress signals, such as pathogen infections, inflammation, T cell receptor stimulation and DNA damage [2]. RIP1 bears a C-terminal death domain (DD) belonging to the structural related death domain superfamily. Seminal work of the group of Jürg Tschopp demonstrated that RIP1 kinase activity is crucial for the induction of necrotic cell death of FADD deficient Jurkat cells stimulated with TNF [3]. The group of Junying Yuan has recently identified inhibitors of necrotic cell death [4], which apparently block RIP1 kinase activity [5].

Hydrogen peroxide induced oxidative stress and death receptor-induced necrosis are often considered similar and to rely on overlapping signaling pathways. We show here that although both forms of cell death are caspase-independent and exhibit similar necrotic morphology, different molecular mechanisms are responsible for the propagation of both forms of cell death. Hydrogen peroxide-induced necrosis in L929 fibrosarcoma cells occurs independently of typical mediators of tumor necrosis factor induced necrosis such as RIP1, phospolipase A2 and mitochondria derived reactive

A. Aretz, B. Hermanns-Sachweh, J. Mayer (Eds.): EMC 2008, Vol. 3: Life Science, pp. 189–190, DOI: 10.1007/978-3-540-85228-5_95,

oxygen species. Instead, Fenton reaction-mediated lipid peroxidation of lysosomal membranes seems the crucial initiating event in hydrogen peroxide, but not tumor necrosis factor induced necrosis.

High-resolution live cell images were made using the Application Solution Multi-Dimensional Workstation (AS MDW) (Leica Microsystems, Mannheim, Germany). This system includes a DM IRE2 microscope equipped with an HCX PL APO 63x/1.3 glycerin corrected 37°C objective and a 12 bit Coolsnap HQ Camera. The objective is equipped with a PIFOC P- Piezo element allowing z-direction images in a precise and reproducible manner. The time laps analysis reveals major differences in reactive oxygen species generation, changes of the mitochondrial membrane potential and rupture of lysosomal membranes between tumor necrosis factor and hydrogen peroxide induced necrosis.

1. Festjens, N, Vanden Berghe, T, Vandenabeele, P. Necrosis, a well-orchestrated form of cell demise: Signalling cascades, important mediators and concomitant immune response. *Biochim Biophys Acta - Bioenergetics* 1757, 1371-1387, 2006.
2. Festjens, N., Vanden Berghe, T. and Vandenabeele, P. RIP1 at the crossroads of cell's decisions to live or die. *Cell Death and Differentiation* 14, 400-410, 2007.
3. Holler, N., *et al.* Fas triggers an alternative, caspase-8-independent cell death pathway using the kinase RIP as effector molecule. *Nat Immunol* 1, 489-495, 2000.
4. Degterev, A., Hitomi, J., Germscheid, M., Ch'en, I.L., Korkina, O., Teng, X., Abbott, D., Cuny, G.D., Yuan, C., Wagner, G., *et al.* Identification of RIP1 kinase as a specific cellular target of necrostatins. *Nature chemical biology* 1, 112-119, 2005.
5. Degterev, A., Huang, Z., Boyce, M., Li, Y., Jagtap, P., Mizushima, N., Cuny, G.D., Mitchison, T.J., Moskowitz, M.A., and Yuan, J. Chemical inhibitor of nonapoptotic cell death with therapeutic potential for ischemic brain injury. *Nature chemical biology* 4, 313-321, 2008.

Dissecting mitochondrial protein distributions using sub-diffraction resolution fluorescence microscopy

S. Jakobs

Mitochondrial Structure and Dynamics, Dept. of NanoBiophotonics, Max Planck Institute for Biophysical Chemistry, Am Fassberg 11, 37077 Goettingen, Germany

sjakobs@gwdg.de
Keywords: Optical nanoscopy, STED, mitochondria

In most animal cells mitochondria are tubular organelles that feature a smooth outer membrane and a highly folded inner membrane. Typically, mitochondria have a diameter of 200 to 400 nm. Because the resolution of conventional fluorescence microscopy is limited by diffraction to about 180 nm in the focal plane and to about 500 nm along the optic axis, the small size of mitochondria has previously largely prevented the light microscopic visualization of protein distributions in these organelles. To overcome this problem we have employed stimulated emission depletion (STED) microscopy to enable diffraction-unlimited far field fluorescence microscopy [1-3]. We have analyzed the sub-mitochondrial distribution of several mitochondrial proteins. A number of proteins display an inhomogeneous distribution within mitochondria which strongly supports the view of a complex compartmentalization of these small organelles.

1. G. Donnert et al., Two-color far-field fluorescence nanoscopy. Biophys J. **92**, 2007, p. L67-L69.
2. S. W. Hell, Far-field optical nanoscopy. Science **316**, 2007, p. 1153-1158.
3. S. W. Hell, M. Dyba, and S. Jakobs, Concepts for nanoscale resolution in fluorescence microscopy. Curr. Opin. Neurobiol. **14**, 2004, p. 599-609.

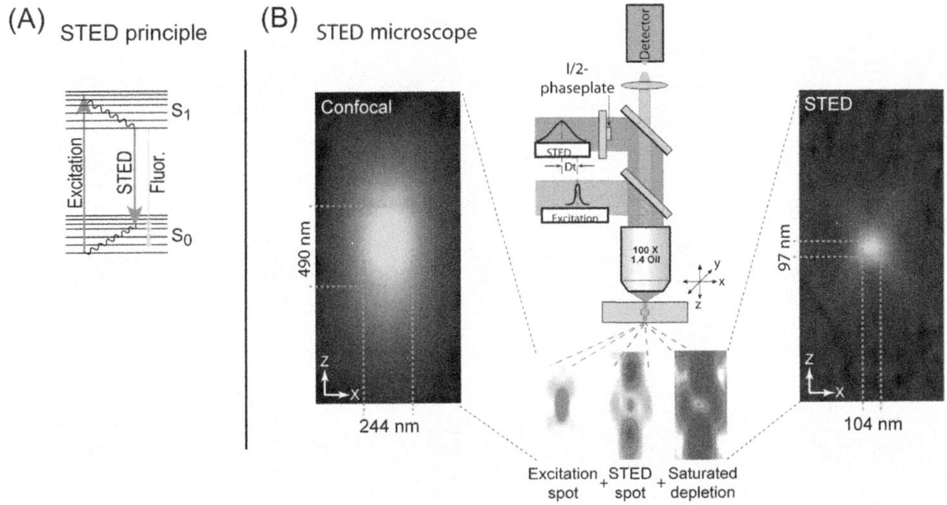

A. Aretz, B. Hermanns-Sachweh, J. Mayer (Eds.): EMC 2008, Vol. 3: Life Science, pp. 191–192,
DOI: 10.1007/978-3-540-85228-5_96, © Springer-Verlag Berlin Heidelberg 2008

Figure 1. Principle of STED-microscopy. (A) Molecules in the fluorescent state S_1 return to the ground state S_0 by spontaneous fluorescence emission. Return to S_0 may also be optically enforced through stimulated emission. (B) Schematic of a point scanning STED-microscope. Excitation and STED are accomplished with synchronized laser pulses focused by a lens into the sample, sketched as green and red beams, respectively. Intensity distributions in the focus are shown at the bottom: The diffraction limited excitation spot (left) is overlapped with the doughnut shaped STED spot featuring a central intensity zero (center). Saturated depletion by the STED beam reduces the region of excited molecules (right) to the very zero-point, leaving a fluorescent spot of subdiffraction dimensions. The (measured) images on the left and right depict the confocal and the sub-diffraction-sized spot left by STED, respectively. In spite of using diffraction limited beams, the concept of STED-fluorescence microscopy eliminates the resolution-limiting effect of diffraction. Adapted from (3).

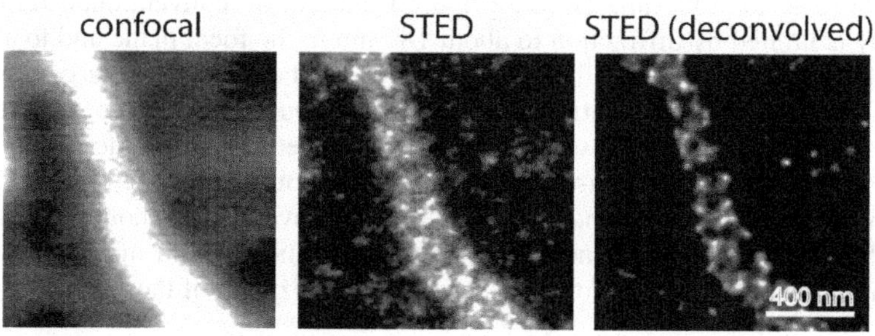

Figure 2. Mitochondria labeled with antibodies specific for Tom20, a subunit of the translocase of the outer membrane of mitochondria (TOM complex). Unlike the confocal image, the STED image reveals individual clusters of the TOM complex.

High Thoughput, High Content Tissue Cyotometry

Peter T. C. So[1]

1. Department of Mechanical and Biological Engineering, Massachusetts Institute of Technology, NE47-279, 77 Mass Ave, Cambridge, MA 02139, USA

ptso@mit.edu
Keywords: microscopy, cytometry, tissue

The study of pathogenesis of many diseases requires quantification of cell-cell and cell-extracellular matrix interactions on the tissue level. A particularly promising assay technology is high throughput multiphoton tissue cytometry. This technique is based on mulitphoton microscopy (MPM) allowing minimally invasive 3D resolved imaging deep inside tissues. MPM not only can provide tissue and cellular morphological information but can also provide genetic (via fluorescence in situ hybridization), proteomic (via antibody labeling) and biochemical (via fluorescence spectroscopy with appropriate biochemical probes) information. A major limitation of microscopy is its relatively low speed and only a small tissue volume can be assayed greatly limiting statistical power of the assay. We have developed 3D tissue cytometry based on high speed MPM imaging by producing and simultaneously scanning multiple two-photon excitation foci simultaneously. This approach parallelizes the raster scanning process by acquiring data over multiple regions in the specimen. The advantage of this method is that the imaging speed is enhanced by the degree of parallelization but the pixel dwell time can in principle remain the same as in the conventional multiphoton microscope preserving image signal to noise ratio. An implementation is shown in Figure 1. A lenslet array (MLA) is used to produce an array of foci from the laser (Ti-Sap) and the beam expander (T). The excitation beamlet is send to the objective (OL) via a dichoric mirror (DCM1) and galvanometric scanner (GSM). The GSM provides lateral raster scanning while the axial translation is achieved with a piezo-positioner (ZP). Specimen is placed on a 3D translation stage (TS). The signal is detected by a multianode PMT (MA-PMT) via dichroic (DCM2), lens, and barrier filter (TPB). To overcome the depth penetration limitation of TPM in studying thicker specimens, we integrated an automated, motorized (M), microtome (MT) into a high-speed TPM system. By alternating and overlapping optical sectioning with mechanical sectioning, it is possible to rapidly image samples with arbitrary thickness. The application of this system has been demonstrated in two studies. In the first study in collaboration with Drs. H. Huang and R. Lee in Brigham and Women's Hospital, we seek to understand how genetic mutation of certain proteins, such as desmin, may lead to cardiac hypertrophy by studying the structure of the heart hierarchically from cellular level up to organ level. Whole organ imaging of a wide type mouse heart is presented in Figure 2. In a second study in collaboration with Dr. R. Gilbert of MIT, lingual muscle mechanics can be studied by diffusion tensor MRI imaging (DTI). DTI produces images with fibers that appear to characterize the organ musculature while the cellular origin of these fibers remains unclear (Figure 3, Left). By correlating DTI images with microscopic images

A. Aretz, B. Hermanns-Sachweh, J. Mayer (Eds.): EMC 2008, Vol. 3: Life Science, pp. 193–194, DOI: 10.1007/978-3-540-85228-5_97,

obtained with 3D image cytometer (Figure 3, Right), we can accertain that the DTI fibers represents the mean orientation direction of the myocytes and, in the future, may allow a better understanding of muscular pathology detected with MRI to be correlated with cellular morphological and biochemical changes. These data represents one of the pioneering efforts in 3D whole organ imaging with subcellular resolution.

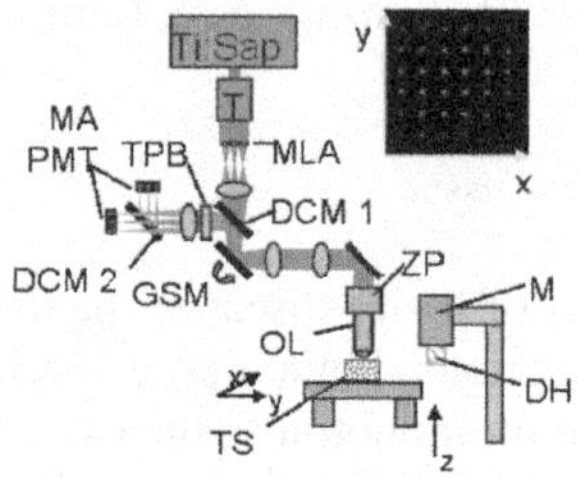

Figure 1. Schematic of a 3D image cytometer. For the explanation of symbols used, see main text. The inset image shows a fluorescent image of a fluorescein sample excited by the multifoci array.

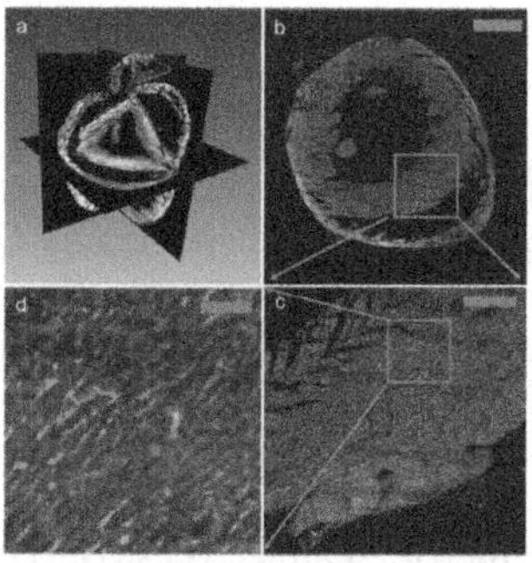

Figure 2. (a) 3D slices through a mouse heart. A horizontal plane is shown in (b) and successive magnifications of the inner ventricular wall are shown in (c) and (d). We can see that the microvasculature (green), cardiomyocyte (red) orientation follow the known spiralling architecture of the ventricular wall, and the nuclei of the cardiomyocytes (blue). The scale bars in the figure are 1 mm, 200 μm, and 20 μm for (b), (c), and (d).

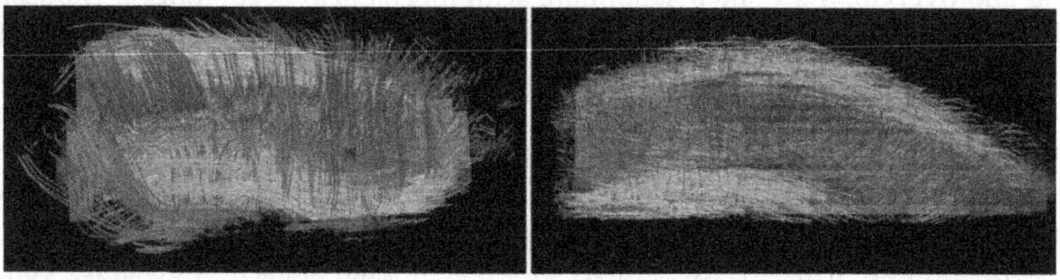

Figure 3. (Left) a DTI image and (Right) a 3D tissue cytometry image of two mouse tongues.

Focus on the vascular wall: Imaging of large arteries using two-photon microscopy

M.A.M.J. van Zandvoort[1], R.T.A. Megens[1], D.W. Slaaf[1], M.G.A. oudeEgbrink[1]

1. Cardiovascular Research Institute Maastricht, Maastricht University, Universiteitssingel 50, 6229 ER, Maastricht, the Netherlands.

Mamj.vanzandvoort@bf.unimaas.nl
Keywords: two-photon microscopy, atherosclerosis, collagen, in vivo

Blood vessels are part of a subtle regulatory system with differential properties along the vascular tree. Alterations in the vessel wall may lead to various diseases such as atherosclerosis. Current knowledge of vascular alterations is mostly based on histological studies of isolated samples that have lost viability. Functional properties of various compounds in the atherosclerotic vessel wall are still largely unknown. Better understanding of the functionality of the atherosclerotic arterial wall, and, thus, increased insight in (development of) atherosclerotic lesions requires studying of these properties *in vivo* or in viable arteries *ex vivo*. This talk focuses on imaging of the vessel wall of large murine arteries using two-photon laser scanning microscopy (TPLSM).

First, the basic principles of conventional (single-photon) microscopy and TPLSM are discussed and compared. Then, we evaluate applicability of TPLSM in large elastic and small muscular arteries under physiological conditions. To this purpose we developed a perfusion chamber model in which the arteries were mounted, their viability and functionality was assessed and structural components such as elastin, collagen, nuclei, and endothelial glycocalyx were visualized. We conclude that TPLSM enables visualization and quantification of subcellular structures in vital and functional elastic and muscular murine arteries, allowing unraveling of structure-function relationships in healthy and diseased arteries.

In advanced atherosclerosis, collagen content is considered to be a major determinant for plaque stability. We therefore applied the novel fluorescent collagen marker CNA35/OG488 to assess the distribution of collagen in mounted viable large arteries of control (C57BL6/J) and atherosclerotic (ApoE$^{-/-}$) mice. To that end, the mounting method was adapted to visualize atherosclerotic lesions in the carotid artery bifurcation. The luminal uptake of CNA35/OG488 in mounted healthy arteries appeared to be limited by an intact endothelial layer and elastic laminae. However, in mounted atherosclerotic arteries, CNA35/OG488 did penetrate into the vessel wall and labeled the collagen structure in atherosclerotic lesions. Next, we visualized early and advanced atherosclerotic lesions in the carotid artery bifurcation as well as the direct relationship between collagen and inflammatory cells in intact and viable mounted carotid arteries of control and atherosclerotic mice, at a subcellular level. Isolated carotid arteries of ApoE$^{-/-}$ mice, aged 15 or 21 weeks, after 7 and 13 weeks on a western diet, and of C57BL6/J control mice fed a chow diet were mounted in a perfusion chamber, pressurized, and

A. Aretz, B. Hermanns-Sachweh, J. Mayer (Eds.): EMC 2008, Vol. 3: Life Science, pp. 195–196, DOI: 10.1007/978-3-540-85228-5_98,

labeled with specific fluorescent markers for cell nuclei, inflammatory cells, collagen, and lipids. Special attention was paid to the fibrous cap regions and the various vessel wall layers. Control vessels had intact endothelium without adhering blood cells or significant intimal collagen labeling. In contrast, in ApoE$^{-/-}$ mice already at 15 weeks inflammatory cells adhered to the endothelium, while at the same time increased labeling of collagen was observed in the tunica intima both at lesion-prone and non-lesion-prone sites, both indicating endothelial activation. In plaques, located exclusively in carotid bifurcations of ApoE$^{-/-}$ mice, internalized inflammatory cell density increased with age and plaque progression. Interestingly, this was observed both in the tunica adventitia and in the tunica intima, while in the tunica media no inflammatory cells were detected at both ages. In the whole plaque, aging or plaque progression did not alter the direct relation between inflammatory cells and collagen. However, within the fibrous caps specifically, direct contact between inflammatory cells and collagen increased with age. Altogether, this study indicated that visualization of viable and structurally intact arteries provides new and detailed insight in the complex relationship between inflammatory cells and collagen in fibrous caps and whole plaques during atherogenesis and suggests involvement of the tunica adventitia in atherogenesis.

In vivo imaging of especially large arteries using TPLSM results in images (optical sections) that severely suffer from motional artifacts within each image and between subsequent images. We applied TPLSM image acquisition *in vivo*, triggered on cardiac and respiratory cycle; in addition, image acquisition was accelerated in order to reduce the impact of motion between subsequent images and within each image. Carotid arteries of anesthetized C57BL6/J mice and renal arteries of anesthetized WKY-rats were surgically exposed, labeled with specific fluorescent markers for cell nuclei and cytoplasm, and imaged using TPLSM. Blood pressure and respiratory signals were recorded as triggering signals for (accelerated) image acquisition. In both mice and rats, timing of the trigger moment at the diastolic phase in the cardiac cycle combined with acquisition rates of 4 Hz or higher, resulted in stable images with reduced motional artifacts. Moreover, subsequently acquired images contained a similar and stable image of the artery. Application of the Fourier transformation further improved overall image quality. The presented method creates new opportunities for *in vivo* studying of various structural and functional properties of the (diseased or damaged) arterial wall at subcellular level.

We conclude that the described techniques offer a new and different view on healthy and diseased arteries and provide new insight in various structural and functional properties of atherogenesis. Further development of these techniques holds potential for future applications in both a scientific and clinical environment.

High resolution confocal Ca^{2+} imaging of the pulmonary neuroepithelial body microenvironment in lung slices

I. Brouns, I. De Proost, I. Pintelon, R. Lembrechts, J.-P. Timmermans, D. Adriaensen

Laboratory of Cell Biology and Histology, University of Antwerp, Groenenborgerlaan 171, 2020 Antwerp, Belgium

Inge.brouns@ua.ac.be
Keywords: calcium imaging, confocal microscopy, in vitro, slices

The microenvironment of airway neuroepithelial bodies (NEBs) contains densely innervated groups of neuroendocrine cells that are shielded from the airway lumen by so-called Clara-like cells (for review: [1]), and a variety of other epithelial cells (Clara cells, ciliated cells and small polyhedric basal cells). The NEB microenvironment has recently been assigned a potential pulmonary stem cell niche [2], but conclusive data on the nature of physiological stimuli for NEBs are unavailable because of the lack of a reliable *in vitro* model. We previously reported that in vibratome slices of live mouse lungs, NEBs can selectively be visualised by 4-Di-2-ASP, a fluorescent styryl pyridinium dye [3]. The aim of the present study was to validate this *ex vivo* lung slice model for Ca^{2+} imaging as a valuable tool for the visualisation of physiological reactions in the airway epithelium in general, and in NEB cells in particular.

Immunohistochemistry of fixed slices demonstrated that the NEBs are almost completely covered by tight junction-linked Clara-like cells. Changes in the intracellular free calcium concentration ($[Ca^{2+}]_i$) in airway epithelial cells, identified by their typical 4-Di-2-ASP staining pattern [4], were monitored using the calcium indicator Fluo-4 on an Ultra*VIEW* ERS dual spinning disk confocal microscope (PerkinElmer; Seer Green, UK), which allows for fast capturing of high-resolution images with minimal photobleaching and low phototoxicity. Experimental short-term application of ATP (10µM; 10s) was seen to evoke an increase in Fluo-4 fluorescence, indicative of a rise in $[Ca^{2+}]_i$, in Clara-like cells, Clara cells and to a lesser extent in ciliated epithelial cells. Substance P (10µM; 10s) selectively activated all ciliated epithelial cells. Short-term addition of an elevated extracellular potassium concentration ($[K^+]_o$; 20mM; 5s), which evokes plasma membrane depolarisation, invariably showed a cytoplasmic $[Ca^{2+}]_i$ increase in all NEB cells, while the surrounding Clara-like cells typically displayed a delayed (±4 s) $[Ca^{2+}]_i$ increase, suggestive of an indirect, NEB-mediated activation (Figure 1).

These data provide the first evidence that reactions to applied stimuli of cells within the pulmonary NEB microenvironment can selectively and reproducibly be visualised in real-time in live lung slices using calcium indicators. Moreover, our calcium imaging set-up, based on confocal microscopy, allows for the simultaneous measurement of physiological parameters in large numbers of nearby Clara cells and ciliated cells present in the lung slice, therefore enabling not only the functional study of NEBs, but also of the potential interactions between NEBs and their environment. The proposed

A. Aretz, B. Hermanns-Sachweh, J. Mayer (Eds.): EMC 2008, Vol. 3: Life Science, pp. 197–198,
DOI: 10.1007/978-3-540-85228-5_99,

model clearly opens new perspectives for further unravelling the functions of the NEB microenvironment in health and disease.

1. D. Adriaensen, I. Brouns, I. Pintelon, I. De Proost, J-P. Timmermans, J Appl Physiol 101 (2006), 960-970.
2. R.I. Linnoila, Lab Invest 86 (2006), 425-444.
3. Pintelon, I. De Proost, I. Brouns, H. Van Herck, J. Van Genechten, F. Van Meir, J-P. Timmermans, D. Adriaensen, Cell Tissue Res 321 (2005), 21-33.
4. De Proost, I. Pintelon, I. Brouns, A.K.A. Kroese, D. Riccardi, P.J. Kemp, J-P. Timmermans, D. Adriaensen, Am J Respir Cell Mol Biol (2008) in press

5. This work was supported by the following research grants: Fund for Scientific Research-Flanders (G.0085.04 and G.0081.08 to D.A.); NOI BOF 2003 and GOA BOF 2007 (to D.A.), and KP BOF 2006 (to I.B.) from the University of Antwerp.

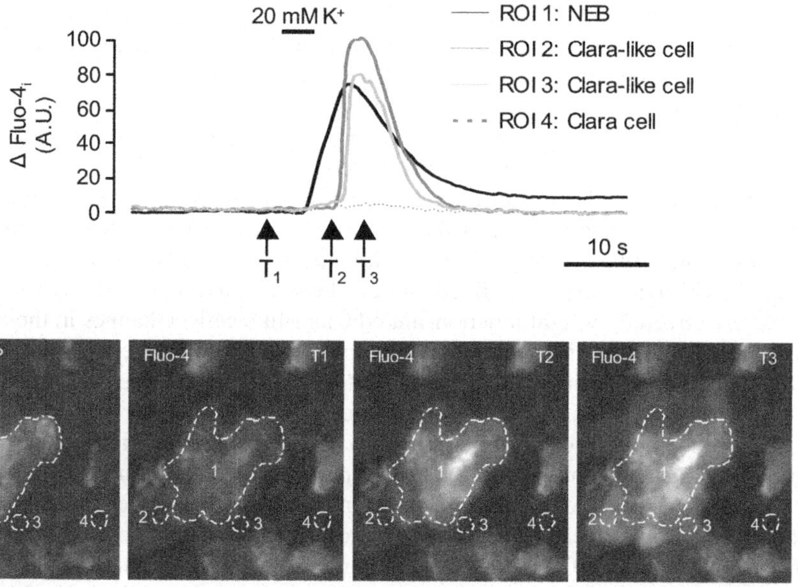

Figure 1. Lung slices are stained with 4-Di-2-ASP to locate NEBs, and subsequently loaded with the Ca^{2+} probe Fluo-4 to visualise Ca^{2+}-dependent reactions. Short-term stimulation with high $[K^+]_o$ is used as a positive control. Stimulation of 4-Di-2-ASP stained, Fluo-4 loaded lung slices evokes an increase in Fluo-4 fluorescence intensity [ΔFluo-4$_i$; expressed as arbitrary units (A.U.)] in NEBs (ROI 1; T2), indicative of a rise in $[Ca^{2+}]_i$. Remarkably, the surrounding Clara-like cells (ROI 2,3) display a delayed activation (T3). These observations suggest that high $[K^+]_o$ evokes release of neurotransmitters from NEBs that subsequently activate Clara-like cells.

Two photon excitation microscopy and SHG imaging as a tool for visualisation of type I and type II collagen, and their use in tissue engineering

Z. Burdíková[1,5], E. Filová[2,3], M. Rampichová[2,3], P. Bianchini[4], M. Čapek[1,6], E. Amler[2,3], L. Kubínová[1]

1. Department of Biomathematics, Institute of Physiology, Academy of Sciences of the Czech Republic v.v.i., Vídeňská 1083, 142 20 Prague, Czech Republic
2. Department of Tissue Engineering, Institute of Experimental Medicine, Academy of Sciences of the Czech Republic v.v.i., Vídeňská 1083, 142 20 Prague, Czech Republic
3. Institute of Biophysics, 2nd Faculty of Medicine, Charles University, Prague, Czech Republic
4. LAMBS, Dept. of Physics, University of Genoa, Genoa, Italy
5. Institute of Geology and Palaeontology, Faculty of Science, Charles University, Prague, Czech Republic
6. Faculty for biomedical engeneering, CTU in Prague, Sitna sq. 3102, Kladno, Czech Republic

burdikova@biomed.cas.cz
Keywords: SHG imaging, type I collagen, type II collagen, autofluorescence

Tissue engineering is a technology which provides tissues for replacing damaged tissues and organs for functional restoration. Collagen, as a basic substance in this field, can be divided into several types, namely type I and type II collagen are studied in this work. Type I collagen is present in fibrous tissues (e.g. tendon) and fibrocartilage, whereas type II collagen proves hyaline character of cartilage [1]. Their relative abundance in cartilage tissues is related to age, degeneration rate, and other qualities of a sample [2]. The goal of our experimental work is to explore the capabilities of Second Harmonic Generation (SHG) Imaging and Two Photon Excitation Microscopy (TPEM) to distinguish type I and type II collagen.

In this study we examined cartilage and tendon tissues from pig, rabbit, and rat, as well as repaired articular cartilage four weeks after making the defect. The samples were fixed with frozen methanol, immunocytochemically labelled with monoclonal antibody against type I, and type II collagen, and subsequently with Alexa Flour 633 and Cy 3 conjugated secondary antibody, respectively. Moreover, autofluorescence of both native and fixed samples was measured.

We studied co-localization between 1- photon collagen fluorescence and SHG imaging and TPEM signal using Leica TCS SP2 AOBS laser scanning microscope coupled with Chameleon Ultra femtosecond laser system.

Using 1-photon excitation, the cartilage samples emitted strong autofluorescence in the range of 530 - 600 nm when excited at 488 and 514 nm. No autofluorescence was detected using excitation wavelengths of 543 and 633 nm. The tendon excited at 488 nm

A. Aretz, B. Hermanns-Sachweh, J. Mayer (Eds.): EMC 2008, Vol. 3: Life Science, pp. 199–200,
DOI: 10.1007/978-3-540-85228-5_100,

showed a strong autoflurescence signal between 500 and 530 nm. Autofluorescence at other excitation wavelengths was not detected.

Using 2-photon excitation, the cartilage samples showed autofluorescence in the range of 500-620 nm and 680-730 nm when excited at 860 nm. A very strong SHG signal appeared at the wavelengths 430±5 nm. With the same excitation the tendon samples showed autofluorescence between 680 and 730 nm only. Again, intensive SHG signal was present.

In addition, the potential use of SHG imaging for the detection of type I and II collagen in cell-seeded scaffolds was examined. In comparison to 1-photon excitation and TPEM, SHG imaging was able to provide the signal from the deep parts of the samples. The same observation was found both for type I and type II collagens.

We conclude that detection of type I and type II collagens using SHG and TPEM is promising for determination of the hyaline character of cartilage in tissue engineering applications.

1. J.A. Buckwalter, H.J. Mankin: Articular cartilage: tissue design and chondrocyte-matrix interactions. *Instr Course Lect* **47** (1998a), pp. 477-486our first reference enters here in the style given in [2,3].
2. J.A. Buckwalter, H.J. Mankin: Articular cartilage: degeneration and osteoarthritis, repair, regeneration, and transplantation. *Instr Course Lect* **47** (1998b), pp. 487-504
3. The presented study was supported by the Academy of Sciences of the Czech Republic (Research projects No. AV0Z50110509, AV0Z 50390512, and AV0Z 50390703), Ministry of Education, Youth and Sports of the Czech Republic (research program LC06063, NPV II 2B06130, MSM6840770012), Grant Agency of the Czech Republic (No. 102/08/0691), and Grant Agency of the Academy of Sciences No. IAA500390702.

Testing calibration standards for confocal and two-photon microscopy

O. Chernyavskiy[1] and L. Kubínová[1]

1. Department of Biomathematics, Institute of Physiology ASCR, v.v.i., Videnska 1083, 14220 Prague, Czech Republic

cernavsky@biomed.cas.cz

Keywords: confocal microscopy, fluorescence microscopy, SIPcharts, fluorescence photobleaching, quantitative microscopy, calibration, two-photon microscopy

Confocal fluorescence microscopy with one-photon excitation (1PE) or two-photon excitation (2PE) has become a powerful tool, which is extensively used in biomedical research. Its optical sectioning capabilities can be exploited for measurements of geometrical characteristics of biological structures and their 3D reconstructions, enabling spatial characterization of tissue organization and morphology. Further, in the last two decades, confocal microscopy finds increasing utilization in various quantitative microscopic techniques. By means of FRAP, FLIP, FRET techniques based on fluorescence intensity measurements, biological studies of cell membrane diffusion and protein binding, spatial and temporal distribution of protein associations inside living cells [1, 2] are possible.

In order to measure spatial proportions as well as changes in fluorescence intensity precisely, a reliable calibration method should be applied. It should take into account not only the signal detection efficiency, but also inhomogeneities caused by irregularities of illumination in the field of view as well as spherical and chromatic aberrations.

In the present study we tested different standards, such as calibration procedures based on uniform thin fluorescent layer [3] or fluorescent microspheres applied to calibration of 1PE as well as 2PE microscope systems. We compared the different types of calibration measurements with the aim to find a reliable and easy-to-use tool for routine checking of the system. We conclude that some very simple samples, such as usual coverslip can be useful for basic checking of system setting and performance, however, for proper calibration of the confocal systems usually a combination of several more complex calibration measurements is necessary.

1. B.L. Sprague, R.L. Pego, D.A. Stavreva and J.G. McNally, Biophys.J. **86** (2004), p. 3473.
2. M. Elangovan, H. Wallrabe, Y. Chen, R.N. Day, M. Barroso, and A. Periasamy, Methods **29** (2003) p. 58–73.
3. G.J. Brakenhoff, G.W.H. Wurpel, K. Jalink, L. Oomen, L. Brocks and J.M. Zwier, (2005) Journal of Microscopy **219** (2005), p. 122.
4. The presented study was supported by the Academy of Sciences of the Czech Republic (grant IAA500200510 and AV0Z50110509) and Ministry of Education, Youth and Sports of the Czech Republic (research program LC06063).

A. Aretz, B. Hermanns-Sachweh, J. Mayer (Eds.): EMC 2008, Vol. 3: Life Science, pp. 201–202,
DOI: 10.1007/978-3-540-85228-5_101,

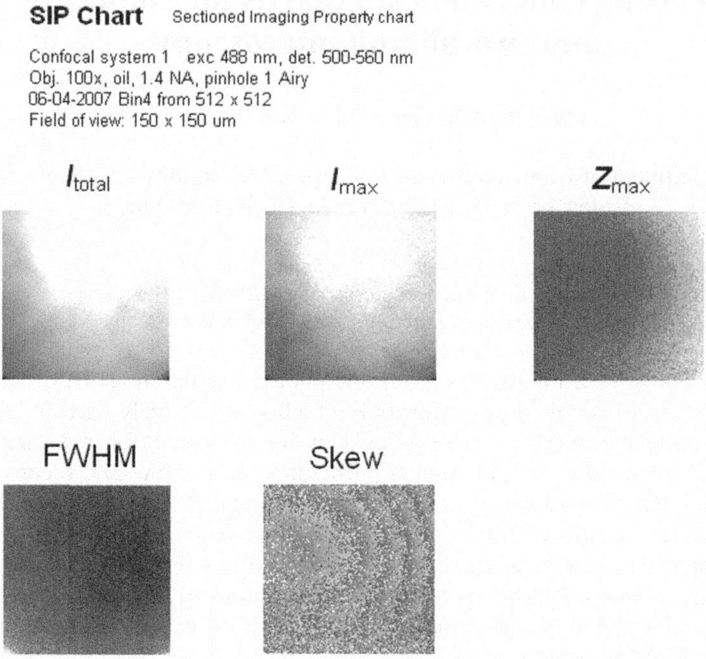

Figure 1. Sectioned Imaging Property chart [3] for a confocal system. This fast and simple yet reliable and universal approach contains important information about LCSM system and can be useful for characterization of different confocal microscopes.

Distribution of the voltage-gated delayed-rectifier K^+ subunits, $K_v1.1$ and $K_v1.2$, in the adult murine enteric nervous system

A. Costagliola[1], L. Van Nassauw[2] and J.-P. Timmermans[2]

1.Dept of Biological Structures, Functions & Technology, University of Naples Federico II, Via F. Delpino, 80137 Naples, Italy
2. Research Group of Cell biology and Histology, University of Antwerp, Groenenborgerlaan 171, 2020 Antwerpen, Belgium

costagli@unina.it
Keywords: voltage-gated delayed-rectifier K^+ channel, enteric glia, mouse

The voltage-gated K^+ (K_v) channels, forming the largest family of K^+ channels, are tetrameric transmembrane channels specific for potassium and sensitive to voltage changes in the cell's membrane potential. They regulate cell membrane potential and excitability in neurons and other cell types [1]. Electrophysiological studies of enteric neurons and glial cells in isolated myenteric ganglia of the guinea-pig small intestine, have shown that these cells display several types of voltage-sensitive ion channels, including voltage-gated delayed-rectifier K^+ channels [2]. A pharmacological study in the guinea-pig has revealed that $K_v1.1$ subunits are expressed in the enteric nervous system (ENS) [3], while a morphological and functional study reported on the presence of $K_v1.1$ in enteric neurons and interstitial cells of Cajal (ICC) in gastrointestinal tissues of the dog, guinea-pig and mouse [4]. In dog, using Northern blot analysis, $K_v1.2$ subunit expression has been found to be restricted to gastrointestinal smooth muscle [5], while in rabbit, using immunoprecipitation, $K_v1.2$ subunits were found to be expressed in colonic epithelium [6]. Due to the lack of detailed data concerning K_v channels in enteric glia and the entangled results concerning K_v subunit distribution in the gastrointestinal wall, we aimed to elucidate the expression of voltage-gated delayed rectifier K^+ channels in the murine gastrointestinal wall with immunocytochemical techniques and confocal microscopy.

In the present study, the ileum was collected of aged-matched mice of different strains (Swiss, BALB/c and C57/Black6). A small distal part of each ileum was processed for cryosectioning, while the remaining part was processed for whole-mount preparations containing either the myenteric plexus or the submucous plexus. Single- and double-immunolabelling experiments were conducted on the cryosections and the whole-mount preparations, using an indirect immunofluorescence method and antibodies directed against $K_v1.1$ and $K_v1.2$ subunits (Alomone Labs Ltd.), and against neuronal markers (PGP, peripherin), glial markers (S100, GFAP), and an ICC marker (c-Kit protein). The immunostainings were evaluated and analysed by conventional fluorescence and confocal microscopy.

In all murine strains examined, a similar result was obtained for $K_v1.1$ and $K_v1.2$ expression in the ileal wall. No $K_v1.1$ or $K_v1.2$ immunoreactivity (IR) was observed in

A. Aretz, B. Hermanns-Sachweh, J. Mayer (Eds.): EMC 2008, Vol. 3: Life Science, pp. 203–204,
DOI: 10.1007/978-3-540-85228-5_102,

ICC. The combined immunostaining for S100 and $K_v1.1$ (Figure 1) demonstrated $K_v1.1$ IR in some S100-expressing glial cells located at the periphery of myenteric ganglia, in almost all S100-positive glial cells along the interganglionic, intramuscular and vascular nerve fibers, and in the S100-positive glial cells of the submucosal plexus. The combined immunostaining for GFAP and $K_v1.1$ showed $K_v1.1$ IR in some GFAP-expressing glial cells at the periphery of myenteric ganglia, and in GFAP-positive glial cells of the submucosal internodal strands. Simultaneously detection of S100 or GFAP and $K_v1.2$, demonstrated $K_v1.2$ IR in a part of the intramuscular S-100-expressing glial cells, and in almost all submucosal S-100-positive and a few GFAP-expressing glial cells. Moreover, $K_v1.2$ IR was expressed in some myenteric and submucosal nerve cell bodies.

In conclusion, $K_v1.1$ channel was not observed in enteric neurons nor in ICC as observed in previous studies. Both voltage-gated channels appear to be predominantly expressed in enteric glial cells, while only $K_v1.2$ subunits were expressed on few neuronal somata. The distribution of $K_v1.1$ and $K_v1.2$ in combination with the glial markers S100 and GFAP, indicates that multiple glial phenotypes may exist in the enteric glial network. The present result support the active role of glial cells in intestinal motility.

1. Y. Li, S.Y. Um and T.V. McDonald, The Neuroscientist **12** (2006), 199-210.
2. M. Hanani, M. Francke, W. Härtig, J. Grosche and A. Reichenbach, Am. J. Physiol. Gastrointest. Liver Physiol. **278** (2000), 644-651.
3. R. Vianna-Jorge, C.F. Oliveira, M.L. Garcia, G.J. Kaczorowski and G. Suarez-Kurtz, Br. J. Pharmacol. **138** (2003), 57-62.
4. W.J. Hatton, H.S. Mason, A. Carl, P. Doherty, M.J. Latten, J.L. Kenyon, K.M. Sanders and B. Horowitz, J. Physiol. **533** (2001), 315-327.
5. P.J. Hart, K.E. Overturf, S.N. Russell, A. Carl, J.R. Hume, K.M. Sanders and B. Horowitz, Proc. Natl. Acad. Sci. USA **90** (1993), 9659-9663.
6. M. Grunnet, H.B. Rasmussen, A. Hay-Schmidt and D.A. Klaerke, Biochim. Biophys. Acta **1616** (2003), 85-94.

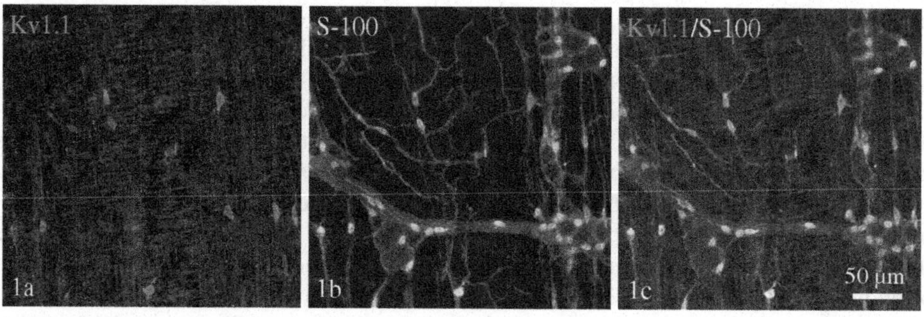

Figure 1. Double immunolabelling with antibodies directed against $K_v1.1$ (**a**) and S100 (**b**) revealed in the ileum of a Swiss mouse that $K_v1.1$ is expressed in some glial cells of the myenteric plexus. Merged picture (**c**).

Observation of Ventricular Myocyte Morphology in Long Term Culture using High Resolution Confocal Imaging

K. Hammer, S. Ruppenthal, Martin Oberhofer, A. Vecerdea, L. Kaestner and P. Lipp

1. Institute for Molecular Cell Biology, Saarland University, 66421 Homburg/Saar, Germany

k.hammer@uks.eu

Keywords: myocyte subcellular structure, confocal microscopy, long term culture

Previously we have introduced a single cell system that allows long term culturing (1 week) of adult rat ventricular myocytes while maintaining their morphology and cross striation as well as contractile behavior and calcium-signaling.

The aim of the present study was to further characterize the subcellular morphology of the myocytes, including the Golgi apparatus, endoplasmic reticulum, plasma membrane and mitochondria during long term culturing. To analyze these structures, two different methods were used: confocal fluorescence imaging and the patch-clamp technique. Cells were isolated from adult rats following a standard procedure using a collagenase-based enzyme mix and plated on coated glass cover slips.

For confocal microscopy organelles and structures were labelled using targeted expression of fluorescent proteins such as dsRed1 fused to subunit VIII of human cytochrome C oxidase for mitochondria and YFP fused to a GPI-anchor for plasma membranes accessible throughout the whole cell or staining with fluorescent dyes such as di-8-ANEPPS for the plasma membrane (Figure 1) and MitoTracker Green for mitochondria (Figure 2), a YFP fused to ts045G for Golgi apparatus and dsRed2 fused to calreticulin for the endoplasmic reticulum. 3-dimensional stacks of single cells were acquired using a nipkow-disc based confocal microscope. Naïve cells were used for the electrophysiological determination of the total cell surface by capacitance measurements. One population of the cells was treated with the cytoskeleton disruptor cytochalasin D to further improve the morphological conservation.

The analysis of the plasma membrane using the fluorescent dye di-8-ANEPPS and the fluorescent protein YFP fused to a GPI-anchor revealed a loss of t-tubules during culturing. This loss was tremendously reduced by cytochalasin D treatment. This finding was strongly supported by cell capacitance measurements.

Mitochondria in cardiac myocytes are very promintent and densly packed. They showed no major changes during the culturing time.

The endoplasmic reticulum remained unaltered (not shown here) whereas the golgi apparatus showed a significant redistribution until day 6 in culture (not shown here).

Only a multidisciplinary approach as presented here facilitates the characterization of ventricular myocyte morphology. Here we provide strong evidence that in cultured rat ventricular myocytes major cellular structures are conserved. Therefore such a model appears suitable for screening experiments.

A. Aretz, B. Hermanns-Sachweh, J. Mayer (Eds.): EMC 2008, Vol. 3: Life Science, pp. 205–206,
DOI: 10.1007/978-3-540-85228-5_103,

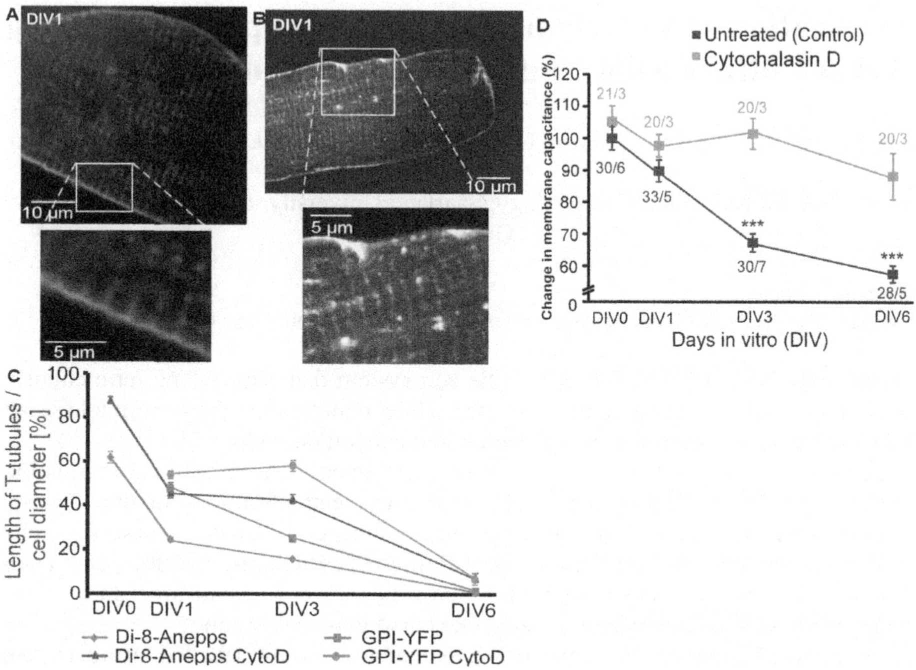

Figure 1. Plasmamembrane with T-tubules visualized with di-8-ANEPPS (A) and GPI-YFP (B) during the time of culture (DIV = days in vitro). T-tubules were shortened but not directly cut off from the outer membrane; they were visible for a longer period of time when labelled with GPI-YFP (C). Electrophysiological studies substantiated such results (D).

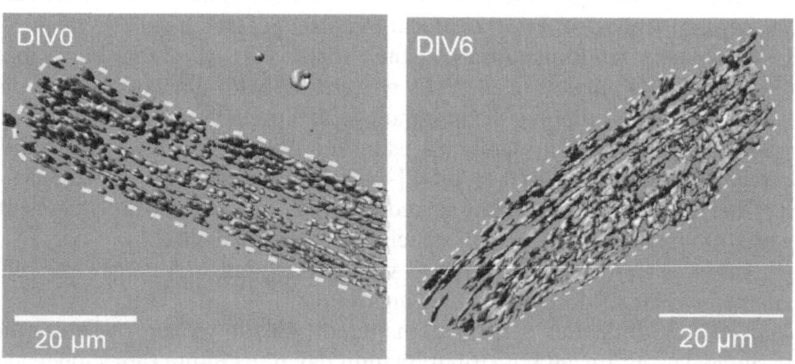

Figure 2. Mitochondria were labelled with MitoTracker Green or Mito-dsRed (not shown) and 3D-rendered after recording confocal image stacks. There were no morphological differences visible by using two different labelling methods. The longitudinal and cross striation is clearly visible. On day 6 the cross striation pattern is lost, most probably due to the loss of T-tubules.

A magnetic field enhances the inmunofluorescence signal in a lymphocytic cell

N.V. Joshi [1], Reyes [1, 2], L. Barboza [3] & L. Berrueta [3]

1. Department of Physiology, University of Los Andes, Mérida Venezuela
2. Department of Morphological Sciences, University of Los Andes, Mérida Venezuela
3. Institute of Clinical Immunology, University of Los Andes, Merida, Venezuela

Joshi @ula.ve
Keywords: magnetic field, lymphocyte, inmunofluorescence

The effects of a magnetic field on the intensity of fluorescence has been discussed earlier and both enhancement and quenching effects are expected depending upon the fluorophore and its environment, particularly a protein which is attached to it [1,2] . Using this asset, several details of neutrophil cells have been examined earlier by fluorescence confocal microscopy and useful information has been reported [3]. In the present investigation, we have extended the application of this technique to the visualization and detection of signaling proteins involved in the activation process of lymphocytic cells.

For this purpose, peripheral mononuclear cells were purified using ficoll hypaque density gradients, fixed with paraformaldehyde, permeabilized with triton X-100 and subsequently stained with specific primary and rhodamine B conjugated secondary antibodies to detect signaling proteins. Double staining was performed by incubating the cells with cholera toxin B (CTB)-FITC to locate the lipid domains. The cells were examined with the help of an Olympus confocal microscope (fluoview 200 laser confocal system) with and without a magnetic field. The confocal system was controlled through supplied software and images were collected using an oil-immersion lens (60×10 numerical aperture), and then they were digitalized. We observed that in this system, the use of a magnetic field enhances the fluorescent signal considerably and facilitates the detection of proteins (in the present case, CD3ζ protein), a fact that may be determinant when an important component of the signaling cascade is not abundant and it is difficult to document the presence of this protein by conventional confocal microscopy. The images obtained in the present case are shown in figures 1 and 2.

The obtained enhancement in the intensity of fluorescence is discussed in the light of the alteration of energy levels of a fluorophore in the magnetic field. Because of the Zeeman Effect, singlet and triplet states are split giving rise to a more complex energy level scheme. At room temperature, vibrational states also play a substantial role in the distribution of electrons in the excited states. The magnitude of the splitting and the shift in the positions of the energy levels in the magnetic field are not important in the present investigation as we do not examine the spectral variation. However, these changes affect the radiative and non-radiative path ways [4]. There will be an increase or decrease in the internal conversion and the rate of the intersystem crossing. The

A. Aretz, B. Hermanns-Sachweh, J. Mayer (Eds.): EMC 2008, Vol. 3: Life Science, pp. 207–208,
DOI: 10.1007/978-3-540-85228-5_104,

process is complex and the variation in the intensity (quenching and enhancement) depends upon a particular protein and the fluorophore.

Thus in a magnetic field, the contrast in fluorescence image will increase and this makes visible certain details which otherwise are not possible. Obviously, the application of magnetic field in confocal fluorescence microscopy will have far reaching impact in morphological sciences.

1. A. Matsusaki and S. Nagaksura , Chem. Phys. Letters, **37** (1976),p.204
2. N. Geacintov, M. Pope, and F. Vogel , Phys. Rev. Letters, **22** (1969),p 593
3. E. Reyes,W. Molina, S.Salmen, L. Berrueta and N.V. Joshi Microscopy and Analysis 2008 (In Press)
4. A. Catherine Biophysical Journal. **68** (1995), p.1191

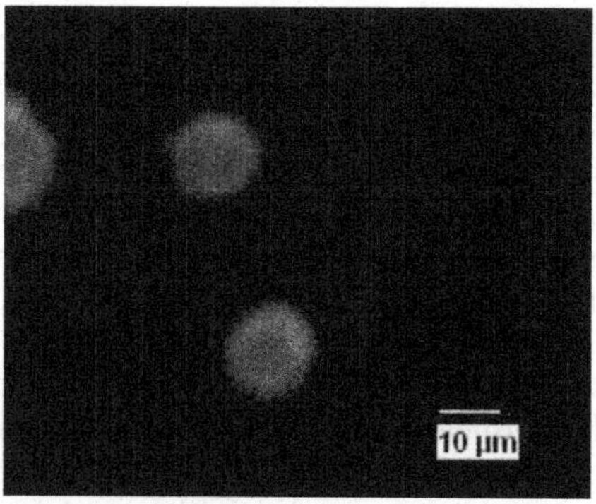

Figure 1. Confocal image of lymphocytes Cell without magnetic field

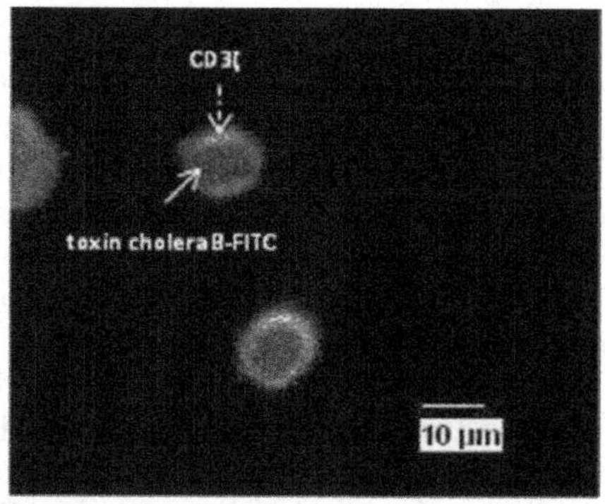

Figure 2. Confocal image of lymphocytes cell with magnetic field

Novel 515 nm DPSS-laser brings excitement in life cell imaging

L. Kaestner[1], P. Lipp[1], D. von Gegerfelt[2], H. Karlsson[3], E. Illy[3] and J. Hellström[3]

1. Institute for Molecular Cell Biology, Saarland University, Building 61, 66421 Homburg/Saar, Germany
2. von Gegerfelt Photonics, Fröbelstrasse 54, 64625 Bensheim, Germany
3. Cobolt AB, Vretenvägen 13, 17154 Stockholm, Sweden

anlkae@uks.eu
Keywords: life cell imaging, laser technology, mitochondria, protein translocation

The use of solid-state lasers as excitation sources in bioanalytical instrumentation technology, as alternatives to gas lasers, has grown increasingly popular over the last few years.The reasons are a multitude of advantages in comparison to gas lasers, such as compact size, less heat generation and power consumption, no vibrations or noise, high beam stability and longer lifetime.

A continuous-wave diode-pumped solid-state (DPSS) laser operating at a fixed wavelength of 515 nm and with an output power of up to 50 mW was developed. The design includes an Yb-doped laser crystal emitting at 1030 nm and proprietary intra-cavity frequency conversion. Spectral filters enable the laser to operate in single longitudinal mode (<30 MHz bandwidth), which results in very low noise and stable output power over a wide temperature range. The use of a separate curved cavity mirror ensures a very high transversal mode quality in the output beam at all output power levels (cp. Fig. 1).

Many fluorescent dyes have their absorption maximum around the 514.5 nm argon-ion laser line. These include e.g. anti-body markers like Alexa Fluor 514 and Oregon Green 514, the yellow fluorescent protein eYFP as well as fluorescent protein-based fusion proteins such as the calcium indicator Premo Cameleon. Further markers with absorption maximum close to 515 nm are e.g. SYBR Safe, a substance, which can replace the toxic ethidium bromide for DNA gel staining, 6-JOE, used in automated DNA sequencing, or FIAsH, a minimally invasive tetracystein-based protein tag.

We tested the laser by coupling it into a kilobeam array scanner for confocal life cell imaging [1]. The two example applications based on the yellow fluorescent protein eYFP are 3D-imaging of mitochondria as seen in Fig. 2 and proteinkinase C translocation [2]. The novel 515 nm laser will enable to perform CFP-YFP Förster resonance energy transfer (FRET) based quantitative imaging.

1. L. Kaestner, and P. Lipp, In: S. L. Shorte and F. Frischknecht (eds.) Imaging Cellular and Molecular Biological Function, Towards Imaging the Dynamics of Protein Signalling (Springer, Berlin, Heidelberg, New York, 2007).
2. G. Reither, M. Schaefer, and P. Lipp, P., "PKC$_\alpha$: a versatile key for decoding the cellular calcium toolkit", Journal of Cell Biology, **174**, 521-533 (2006).

A. Aretz, B. Hermanns-Sachweh, J. Mayer (Eds.): EMC 2008, Vol. 3: Life Science, pp. 209–210,
DOI: 10.1007/978-3-540-85228-5_105,

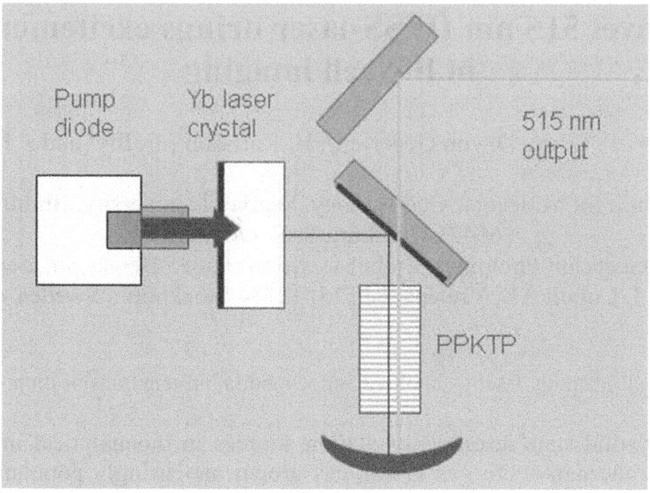

Figure 1. Schematic design of novel 515 nm DPSS laser (details see text)

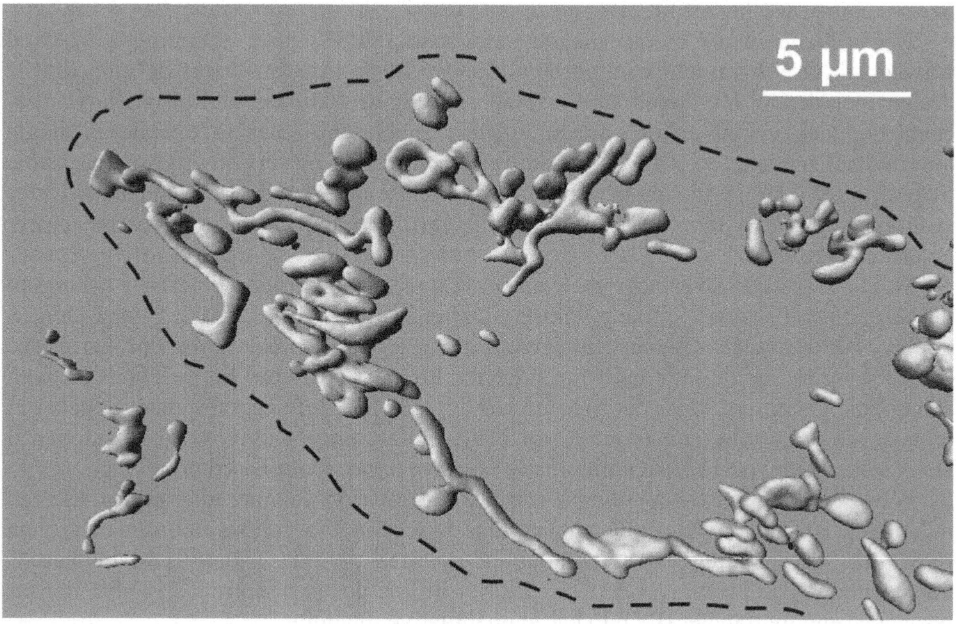

Figure 2. 3D-organell morphology in living COS-1 cells. Mitochondria are labelled with eYFP. The dotted line outlines the cell shape. Mitochondria show highly complex shapes.

Confocal laser scanning microscopy study on human fissure caries stained with alizarin red

S.M. Masudi[1], N.I.M. Salleh[1]

1. School of Dental Sciences, University of Science Malaysia, 16150 Kubang Kerian, Malaysia

sam@kb.usm.my
Keywords: CLSM, dental fissure, human caries, alizarin red

Dental caries results from imbalance between demineralization and remineralization process in which a net loss of mineral occur, leading to dissolution of dental hard tissue [1]. Under light microscope, fissure caries have been describe as cone-shaped lesion with 4 different zones: surface zone, body of lesion, dark zone and translucent zone [2]. Alizarin red is used in a biochemical assay to determine the presence of calcific deposition by cells of an osteogenic lineage, quantitatively [3].

The aims of this study is to describe the microscopic findings of human fissure caries stained with Alizarin red by using CLSM as well as to improve our understanding of the enamel early carious lesion.

This is a descriptive laboratory study involving sixteen extracted premolar/molar teeth with early fissure caries and stained fissure. All the teeth were fixed in neutral formalin solution for 1-7 days, rinse under water and stored in calcium free phosphate buffer saline (PBS) at 4^0C for 24 hours. All the specimens were processed as undecalcified ground section. They were dehydrated with 60, 80, 96, 100% alcohol and infiltrated by Alcohol/Technovit 7200 with ratios of 70/30; 50/50; 30/70 and 0/100 for 24 hours in each steps. Each specimen were embedded and polymerized in resin. Specimens were fixed on glass slides by using Spurr's-glue (England) and cut into thick sections (100-350 μm) by using Exakt (Germany) hard tissue cutter and surfaces was smoothened by grinding with Exakt polisher. They were stained with Alizarin red S (100 μg/ml) in 0.5 M HCl for 24 hours at 37^0C. The specimens were rinse under water, dried and the surface is polished again by using liquid polisher before viewed under CLSM (Leica TCS SP II, Germany). All specimens were examined under CLSM with Ar/He-Ne (argon, helium and neon) ion radiation source [4]. The excitation wavelengths were 488 nm and 514 nm.

Results: Under CLSM, demineralized enamel appeared dark as alizarin red cannot detect the presence of calcific deposition. Eight specimens with a deep and narrow fissure had demonstrated both spheroid and conical shaped lesion with base towards the outer surface of enamel. An area of different contrast was identified with the outer layer darker compared to inner layer which determined degree of mineral lost of that area. The other six specimens with a shallow and wide fissure did not show area of demineralization.

A. Aretz, B. Hermanns-Sachweh, J. Mayer (Eds.): EMC 2008, Vol. 3: Life Science, pp. 211–212,
DOI: 10.1007/978-3-540-85228-5_106,

Conclusion: Tissue destruction can be seen in teeth with early fissure caries or stained fissure. In conclusion, pit and fissure of posterior teeth particularly the one with deep and narrow fissure is susceptible to the development of caries lesion.

1. EAM Kidd and O Fejerskov O. What constitute dental caries? Histopathological of cariousenamel and dentin related to the action of cariogenic biofilm. *J Den Res.* 83 (2004): 35-38.
2. KA Galil and AJ Gwinnet. Three dimensional replicas of pits and fissures in human teeth: scanning electron microscope study. *Arch Oral Biol,* **20** (1988)**:** 493-499.
3. T Nishikawa, S Yoshida, A Tanaka, H Zoellner and DM Walker. Histological Aspect of Human Enamel Fissure Caries studied by CLSM. *Asia Pac Micros and Anal*, 31 (2003): 9-11.
4. Al-Salihi KA, Samsudin AR and Al-Raouf A (2003). Histopathological and confocal laser scanning microscopy observation of structural changes in dental caries. *Annals of Microsc* **3:** 26-30.

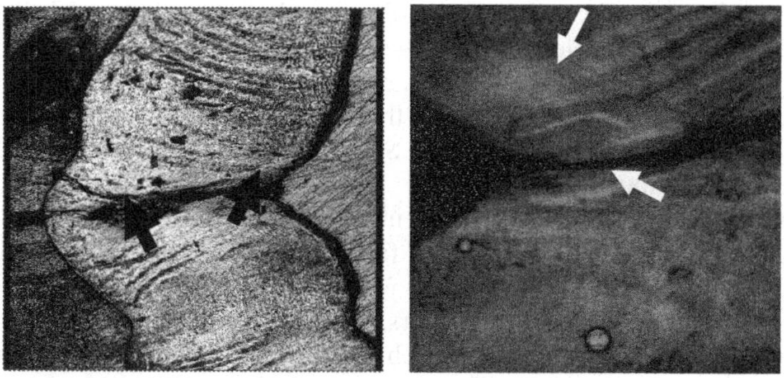

Figure 1. A. Specimen with narrow and deep fissure. Both show a cone shape lesion with the base towards the outer surface of enamel. Lesion is seen at the deepest part of fissure (black arrow). B. Area of different contrast was identified with the outer layer darker compared to inner layer (white arrow).

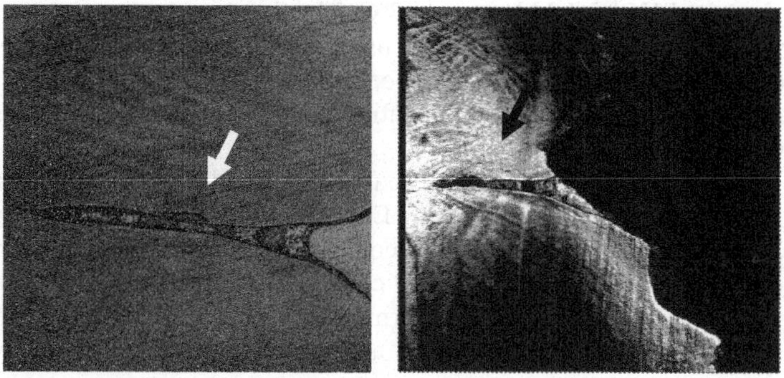

Figure 2. Specimen with narrow and deep fissure showing spheroid shaped lesion.

Non-invasive skin tissue characterization using non-linear spectral imaging microscopy

J.A. Palero[1], H.S. de Bruijn[2], A. van der Ploeg van den Heuvel[2], H.J.C.M Sterenborg[2] and H.C. Gerritsen[1]

1. Molecular Biophysics, Utrecht University, P.O. Box 80000, 3508 TA, Utrecht, The Netherlands
2. Center for Optical Diagnostics and Therapy, Erasmus MC, University Medical Center Rotterdam, P.O. Box 2040, 3000 CA, Rotterdam, The Netherlands

j.palero@phys.uu.nl
Keywords: live tissue imaging; nonlinear microscopy, spectral imaging; autofluorescence imaging; second harmonic generation; nonlinear spectral imaging; multiphoton microscopy; optical biopsy.

Live tissue nonlinear microscopy based on multiphoton-excited autofluorescence and second harmonic emission originating from endogenous fluorophores and noncentrosymmetric-structured proteins is rapidly gaining interest in biomedical applications. The advantage of this technique includes high imaging penetration depth and minimal phototoxic effects on tissues.

Here, we have developed nonlinear spectral imaging microscopy (NSIM) based on a homebuilt multiphoton microscope, a prism spectrograph, and a high-sensitivity CCD camera for detection (Figure 1) [1]. The sensitivity of the microscope was optimized for autofluorescence and second harmonic imaging over a broad wavelength range. Spatially resolved tissue emission spectrum based on the autofluorescence from both cellular and extracellular structures, second-harmonic signal from collagen, and a narrowband emission related to Raman scattering of collagen were recorded. Visualization of the spectral images by wavelength-to-RGB color image conversion allowed us to identify and discriminate tissue structures such as epidermal keratinocytes, lipid-rich corneocytes, intercellular structures, hair follicles, collagen, elastin, and dermal cells (Figure 2). Application to the imaging of *ex vivo* and *in vivo* mouse skin tissues showed clear differences in spectral emission between skin tissue layers as well as biochemically different tissue components [2].

Nonlinear spectral imaging microscopy (NSIM) offers a simple yet versatile method of non-invasive characterization of unstained skin tissues. It provides a wealth of information that surpasses what conventional multiphoton microscopy or autofluorescence spectroscopy offers.

1. J.A. Palero, H.S. de Bruijn, A. van der Ploeg-van den Heuvel, H.J.C.M. Sterenborg, and H.C. Gerritsen, Opt Express 14 (2006). pp 4395-4402.
2. J.A. Palero, H.S. de Bruijn, A. van der Ploeg van den Heuvel, H.J.C.M. Sterenborg, and H.C. Gerritsen, Biophys J 93 (2007). pp 992-1007.

A. Aretz, B. Hermanns-Sachweh, J. Mayer (Eds.): EMC 2008, Vol. 3: Life Science, pp. 213–214,
DOI: 10.1007/978-3-540-85228-5_107,

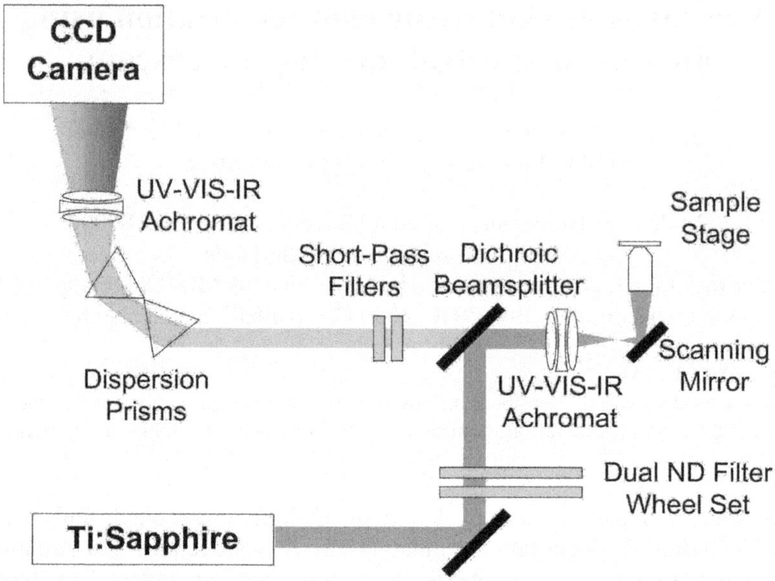

Figure 1. Schematic diagram of the spectral imaging setup. The intrinsic emission of a sample is epi-detected without an aperture. The signal is spectrally dispersed by two prisms, focused with an achromat, and detected with a back-illuminated CCD array.

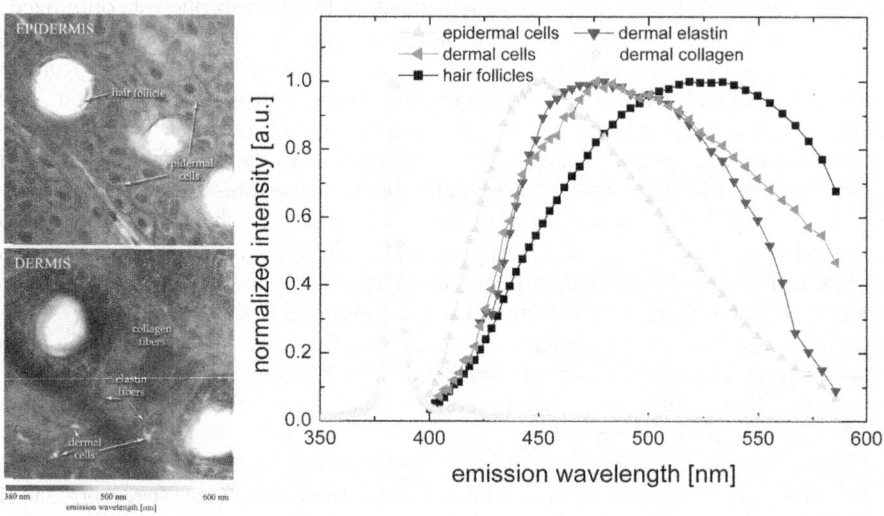

Figure 2. RGB real color spectral images of mouse skin epidermis (*A*) and dermis (*B*) and corresponding spectra (*C*). The images are 120 μm × 120 μm and the excitation wavelength is 764 nm.

Force Spectroscopy reveals mechanical structure of capsids

G.J.L. Wuite[1], W.H. Roos[1], C. Uetrecht[2], N. Watts[3], P. Wingfield[3], A.C. Steven[3], A.J.R. Heck[2], J.L. Carrascosa[4], C.F. Schmidt[5]

1. Natuur- en Sterrenkunde, Vrije Universiteit, Amsterdam, the Netherlands
2. Biomoleculaire Massa Spectrometrie, Bijvoet Instituut, Universiteit Utrecht, Utrecht, the Netherlands
3. Laboratory of Structural Biology Research, NIH Bethesda, USA
4. Department of Structure of Macromolecules, Centro Nacional de Biotecnología, Consejo Superior de Investigaciones Científicas, Spain
5. Physikalisches Institut, Fakultät für Physik, Göttingen, Germany

gwuite@nat.vu.nl
Keywords: Force Spectroscopy, Biomaterials, Virus

Force spectroscopy measurements are effectively used in the research of single-molecules. The tools and methods of this technique can also be used for the investigation of complex protein assemblies such as viral capsids. Viral capsids need to encapsidate their genome to protect it from external influences. Therefore they need to withstand considerable forces from the outside, but also from the inside, as the genome is often packaged at near crystalline densities. To get a better understanding of the mechanics of these nanocontainers, force spectroscopy is an ideal tool to obtain insight into their mechanical structure. Force spectroscopy on viruses can be preformed by indenting viral capsids with an atomic force microscope (AFM). Unexpectedly, most experimental studies on the deformation of empty viral capsids demonstrate a linear relation between indentation depth and force. For thin shells this might be expected for deformation in the order of the thickness of the shell but experimentally a similar behavior has been shown to exist for thick shells. Numerical simulations guided by the Föppl- von Kármán (FvK) number (a dimensionless number relating the "in-plane" elasticity of the shell to its "out-of-plane" bending rigidity) have been able to explain these results for shells with either small (>150) or large (<~500) FvK numbers. For shells with an FvK number between those values a non-linear response is expected. Here we report nanoindentation experiments on different viral capsids to investigate this elastic deformation. Moreover, we explore when elastic deformation fails resulting in capsids deformation and collapse.

Elastic deformation was investigated using Hepatitis B Virus (HBV). HBV was chosen because its capsids can exist in two different morphologies, a smaller T=3 and a bigger T=4 configuration that have FvK values in the linear and the non-linear regime. Both morphologies can be directly compared since they self-assemble from many copies of the same single protein. We demonstrate that the HBV T=3 capsid shows mainly linear behavior while the T=4 capsid react strongly non-linear to nanoindentation. Both responses can be very well understood in relation to their FvK values, illustrating the strength of continuum elastic theory to describe nanometer size objects.

A. Aretz, B. Hermanns-Sachweh, J. Mayer (Eds.): EMC 2008, Vol. 3: Life Science, pp. 215–216,
DOI: 10.1007/978-3-540-85228-5_108,

Linear elastic response parameters are relevant for small applied forces and deformations. Of equal importance are structural failure thresholds which correspond to non-linear and often irreversible response. Empirical correlations exist with linear response parameters and failure thresholds (tensile strength 5-10% of the Young's modulus). In macroscopic bulk materials, failure depends on defects such as cracks. In shells, in contrast, localized non-linear instabilities (buckling) are often the dominant mode of failure. We explored how such self-assembled biomaterials of nanometer size behave under extreme mechanical conditions and have tested in how far classical materials science applies at the molecular scale.

For viral capsids with large FvK values buckling (and capsid failure) is expect at indentations above 30% of it radius. Interestingly, for HBV buckling is not expected and indeed we do not report breaking events. Instead HBV undergoes permanent plastic deformation at these large indentations indicating that a rearrangement of capsid proteins takes place. The FvK number of Φ29 bacteriophage is > 1000, thus buckling is believed to be the major cause of failure. We imaged the rigidity and the patterns of fracture of the relatively brittle shells of the Φ29 bacteriophage by probing with the tip of an atomic force microscope. While gently scanning we observed patterns in the local elasticity with pentameric symmetry. When we irreversibly destroyed the shells with higher applied forces, we found that the capsids fractured along well-defined lines with respect to their surface lattice. The mechanically stable building block of the protein shell was a trimer of monomers, organized according to the triangulation net of the icosahedron. Moreover, we determined that the interactions between proteins in the equatorial belt were only about half as resilient against fracture as the interactions within the icosahedral end-caps. Combining the tools to analyze the mechanics of biomaterials on molecular scales with well established concepts of macroscopic materials science is an important step towards nanoscale bioengineering.

Manipulating and imaging molecular motors with optical traps, single-molecule fluorescence and atomic force microscopy

C.F. Schmidt [1], L.C. Kapitein [2], B.H. Kwok [3], J.S. Weinger [3], T. M. Kapoor [3], E.J.G. Peterman [2], I.A.T. Schaap [4], C. Carrasco [5], Pedro J. de Pablo [5], Stefan Lakämper [1]

1. III. Physikalisches Institut, Fakultät für Physik, Georg-August-Universität, Friedrich-Hund-Platz 1, 37077 Göttingen, Germany
2. Department of Physics and Astronomy and Laser Centre, Vrije Universiteit, De Boelelaan 1081, 1081 HV Amsterdam, The Netherlands
3. Laboratory of Chemistry and Cell Biology, The Rockefeller University, New York, NY 10021, USA
4. National Institute for Medical Research, The Ridgeway, Mill Hill, London, NW7 1AA, UK
5. Departamento de Física de la Materia Condensada C-III, Universidad Autónoma de Madrid, 28049 Madrid, Spain

cfs@physik3.gwdg.de
Keywords: motor proteins, optical traps, atomic force microscopy

Motor proteins move cargos in cells and are involved in coordinated large-scale force generation e.g. in cell locomotion, cell division or muscle contraction. Optical traps in conjunction with high-resolution light microscopy has been a tool of choice to analyze dynamic properties of motors, such as power strokes, step lengths or unitary forces. Much detail about microscopic function has also been learned by direct imaging.

I will report here on recent results with the mitotic Eg5 kinesin which slides microtubules apart in the mitotic spindle [1,2] and displays an intriguing feature of regulation by cargo binding. The motor is activated to move processively only when bound between two microtubules (Figure 1). It also appears that a single Eg5 molecule can support considerable less force than a Kinesin 1 motor (Figure 2) [3]. Furthermore I will report on imaging and mechanical characterization of microtubules, the tracks of kinesins, by AFM [4,5], and on the use of the AFM to image moving kinesin motors with nanometer resolution (Figure 3).

1. L. C. Kapitein, , E. J. G. Peterman, B. H. Kwok, J. H. Kim, T. M. Kapoor, C. F. Schmidt, Nature, **435** (2005), p. 114-118.
2. B. H. Kwok, L. C. Kapitein, J. H. Kim, E. J. G. Peterman, C. F. Schmidt, T. M Kapoor, Nat. Chem. Biol., **2** (2006), p. 480-485.
3. M. J .Korneev, S. Lakamper, C. F. Schmidt, European Biophys. J. /Biophys. Lett., **36** (2007), p. 675-681.
4. P. J. de Pablo, I. A. T. Schaap, F. C. MacKintosh, C. F. Schmidt, Phys. Rev. Lett., **91** (2003), 098101.

A. Aretz, B. Hermanns-Sachweh, J. Mayer (Eds.): EMC 2008, Vol. 3: Life Science, pp. 217–218,
DOI: 10.1007/978-3-540-85228-5_109,

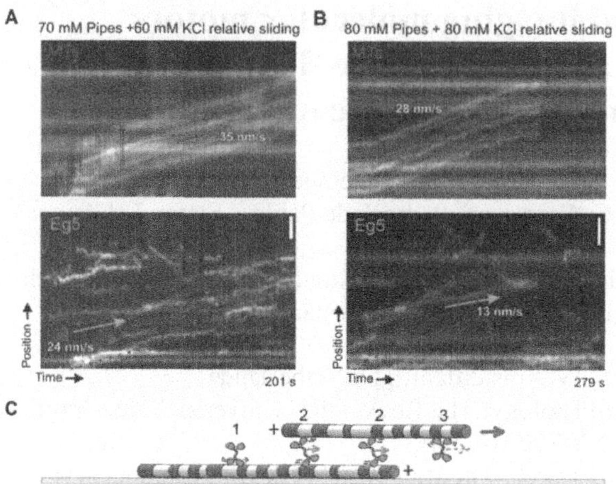

Figure 1. Eg5-GFP switches from diffusive to directional motility upon binding to a second microtubule. **A-B** Upper kymographs: sliding of a microtubule, lower: sliding of Eg5-GFP (C) Cartoon illustrating three different microtubule interaction geometries of the Eg5 motors: 1. Eg5 diffuses on a single surface-attached microtubule, 2. Eg5 moves directionally between two microtubules, 3. Eg5 diffuses on the sliding microtubules

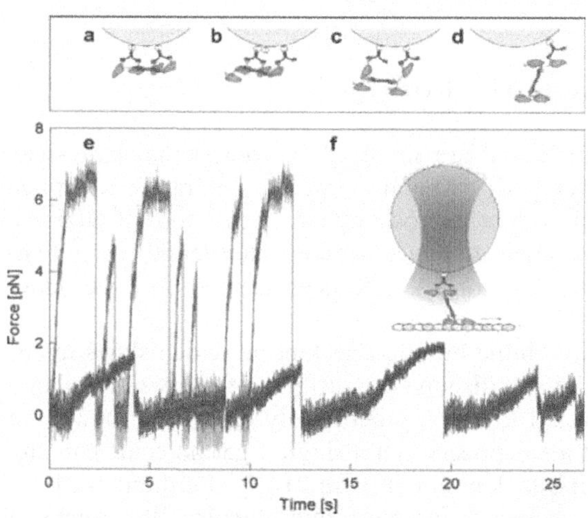

Figure 2. Eg5 bead optical trap assays: **a-d, f** motor attachments to beads via N-terminal His-tag. **e** Bead motility generated by individual Eg5 (green) and Kinesin-1 (grey) motors. The averaged and median-filtered signal is overlaid in red over both traces. The trap stiffnesses were 0.03 pN/nm (Kinesin-1) and 0.013 pN/nm (Eg5). **f** One dimer is bound, one dimer is free; sketch of a silica-sphere with motor held in the laser-trap, such that it interacts with a surface-attached microtubule track. (From ref.)

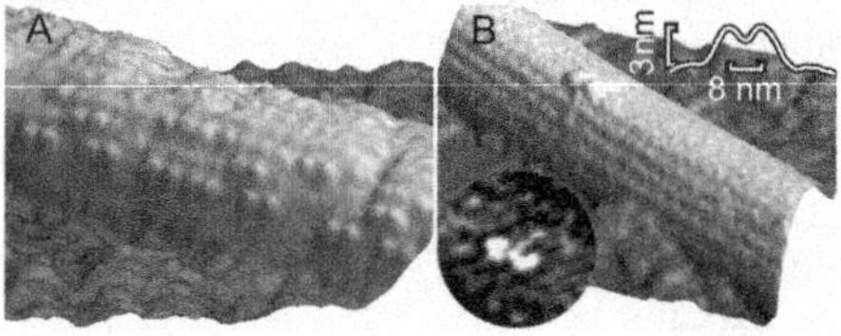

Figure 3. AFM image of individual kinesin motors immobilized on MTs in the presence of 15 μM AMP-PNP. **A**, intermediate kinesin concentrations (< 1 motor/tubulin dimer), **B**, low kinesin concentrations (<1 motor/ 10 tubulin dimers). Upper inset: averaged axial profile of 17 kinesin molecules, Lower inset: Laplacian filtered image of kinesin.

Mechanical stability of ethanol-induced A-DNA

J.R. Arias-Gonzalez[1,2], S. Hormeño[1], B. Ibarra[1], J.M. Valpuesta[1] and J.L. Carrascosa[1]

1. Centro Nacional de Biotecnología, Consejo Superior de Investigaciones Científicas.
2. Instituto Madrileño de Estudios Avanzados en Nanociencia (IMDEA-Nanociencia). Campus de Cantoblanco, 28049 Madrid. Spain.

ricardo.arias@imdea.org
Keywords: A-DNA, structural transition, single molecule

The DNA double helix is known to adopt the *B* conformation in physiological conditions, but this is not an immutable structure. In reality, double-stranded DNA (dsDNA) has the significant ability of changing its form. This structural variability has been observed to take place through changes in its environmental conditions, such as the ionic strength and the relative humidity, the base sequence or its interaction with some proteins, all of which inducing transitions between the *B* and the *A* forms. In contrast, RNA double helix adopts the *A* form and it is structurally conservative. RNA-DNA hybrids, which are formed during transcription, also adopt the A form.

Low-relative humidity environments induce A-form dsDNA. For the purpose of investigating this structure, dried DNA fibres and DNA in ethanol-water mixtures above 70% ethanol concentration were used in bulk experiments [1]. The binding of polycationic molecules, such as spermine and $Co(NH_3)_6^{3+}$, probed to stabilize A-DNA al lower ethanol concentrations [2]. We have used single-molecule optical tweezers [3], which allows the study of individual dsDNA molecules under the simultaneous effects of tension and environmental changes. The differences in the *A* and *B* structures, which stem from the sugar puckering in the nucleotides (C2'-endo for *B* form and C3'-endo for *A* form), provides a different length and elastic response on each double helix type.

We have stretched individual 8-kbp dsDNA molecules above the overstretching transition [4], in both Tris buffer and ethanol-water mixtures. Our results show a shortening of the molecule's contour length when we replace the Tris buffer by an ethanol solution above 80% concentration. This shortening level is higher in a 60% ethanol-concentration solution with spermine. In both cases, the compaction of the molecule is expected since DNA transitions from the *B* to the *A* form, as proven by circular dichroism experiments. However, our measured contour length reduction with single-molecule force-spectroscopy is considerably smaller than that predicted by bulky assays [5]. In the latter experiments, DNA molecules of a few hundred base pairs or less are used in order to prevent both strong condensation and aggregation.

Our results show, firstly, that an individual dsDNA molecule does not transition to the *A* form in all its segments under stress. As shown in [6, 7], B-to-A transition is markedly dependent on base sequence: GC-rich regions are more stable in the A-form than AT-rich regions. In low relative-humidity environments, therefore, GC-rich contents change to the *A* form whereas AT-rich contents remain in the *B* form when a force is applied. In addition, other secondary structures, such as *C* and allomorphs, may play a more significant role under the influence of lengthwise tension on the DNA polymer.

A. Aretz, B. Hermanns-Sachweh, J. Mayer (Eds.): EMC 2008, Vol. 3: Life Science, pp. 219–220, DOI: 10.1007/978-3-540-85228-5_110, © Springer-Verlag Berlin Heidelberg 2008

DNA aggregation, and also condensation –which is the aggregation of one molecule around itself due to the polymer's bending elasticity–, favour a certain conformation by intermolecular and intramolecular contacts, respectively, in the presence of a binding agent. A binding agent generally works either by decreasing repulsions between DNA segments or by making DNA-solvent interactions less favourable. Due to the nature of our experiments, aggregation is not possible since we only grab one molecule at a time, and condensation can be prevented by controlling the extension of the molecule. Stretching assays in the presence of spermine, as a binding agent, reveal a low-force plateau due to condensation and an increase in the force at which the molecule overstretches in buffer, yielding values characteristic of high monovalent salt [8].

Secondly, our results further show that in the presence of spermine, on the one hand, the level of the DNA condensation plateau increases when a buffer solution is replaced by an ethanol-water mixture, thus demonstrating that both DNA condensation and aggregation are stronger in ethanol environments. On the other hand, they show that the overstretching force-increase tendency, which is observed in buffer, is maintained in the ethanol-water mixture. Both single-molecule facts are in accord with spermine acting as a structure stabilizer. The former fact additionally implies that DNA condensation in ethanol and spermine cannot be prevented in bulk –since DNA remains at zero force–, unless molecules shorter than the persistence length are used.

We conclude that the B-to-A transition is strongly dependent on a sequence under tension, and that both *B* and *A* forms –whose cooperative transition is initiated by local (base-pair step) sugar-puckering processes–, are stabilized by both condensation and aggregation, which are global and collective processes, respectively. Stabilization of *A* form by the binding of some molecules to DNA support a mechanism for some regulatory and motor proteins, such as RNA polymerases, in which they help in locally inducing *A* structure, both by removing water molecules around dsDNA and by making intermolecular contacts [2, 9].

1. V.I. Ivanov, L.E. Michenkova, E.E. Minyat, M.D. Frank-Kamenetskii and A.K. Schyolkina, J. Mol. Biol. **87** (1974), p. 817.
2. E.E. Minyat, V.I. Ivanov, A.M. Kritzyn, L.E. Michenkova and K. Schyolkina, J. Mol. Biol. **128** (1978), p. 397.
3. S. Hormeño and J.R. Arias-Gonzalez, Biol. Cell **98** (2006), p. 679.
4. S.B. Smith, Y. Cui and C. Bustamante, Science **271** (1996), p.795.
5. W. Saenger, "Principles of nucleic acid structure", 2nd ed. (Springer-Verlag) (1984), p. 229.
6. A.F. Usatyi and L.S. Shlyakhtenko, Biopolymers **13** (1974), p. 2435.
7. A.G.W. Leslie and S. Arnott. J. Mol. Biol. **143** (1980), p. 49.
8. C.G. Baumann, S.B. Smith, V.A. Bloomfield, C. Bustamante, Proc. Natl. Acad. Sci. USA **94** (1997), p. 6185.
9. X.J. Lu, Z. Shakked and W.K. Olson, J. Mol. Biol. **300** (2000), p. 819.
10. This work has been supported by the International Human Frontier Science Program (grant RGP63/2004 and J.R.A.-G.'s fellowship LT00085/2004-C) and the Consejo Superior de Investigaciones Científicas. S.H. acknowledges a scholarship from the Consejería de Educación de la Comunidad de Madrid and the European Social Fund.

Manipulation of the mechanical properties of a virus by protein engineering

P. J. de Pablo[1], C. Carasco[1], M. Castellanos[2], M. G. Mateu[2]

1. Departamento de Física de la Materia Condensada Universidad Autonoma de Madrid, Madrid, 28049 Madrid, Spain
2. Centro Nacional de Biologia Molecular "Severo Ochoa" (CSIC-UAM). Universidad Autónoma de Madrid, 28049 Madrid, Spain.

p.j.depablo@uam.es
Keywords: virus mechanics, atomic force microscopy, nano-indentation

In a previous study [1] we showed that the DNA molecule within a spherical virus, the minute virus of mice (figure 1), exerts an architectural role by increasing the stiffness of the virus. A finite element model predicted that this mechanical reinforcement is a consequence of the interaction between crystallographically visible, short DNA patches and the inner capsid wall. We have now tested this model by using protein engineering. Selected amino acid side chains have been truncated to specifically remove major interactions between the capsid and the visible DNA patches, and the effect of the mutations on the stiffness of virus particles has been measured using atomic force microscopy (figure 2). The mutations do not affect the stiffness of the empty capsid, while they significantly reduce the difference in stiffness between the DNA-filled virion and the empty capsid. The results reveal that intermolecular interactions between individual chemical groups contribute to the mechanical properties of a supramolecular assembly, and identify specific protein-DNA interactions as the origin of the anisotropic increase in the rigidity of a virus. This study [2] demonstrates that it is possible to control the mechanical properties of a protein nanoparticle by the rational application of protein engineering based on a simple mechanical model.

1. C. Carrasco, A. Carreira, I. A. T. Schaap, P. A. Serena, J. Gómez-Herrero, M. G. Mateu. PNAS **103** (2006) 13706.
2. C. Carrasco, M. Castellanos, P. J. de Pablo, M. G. Mateu. PNAS **105** (2008) 4150.

A. Aretz, B. Hermanns-Sachweh, J. Mayer (Eds.): EMC 2008, Vol. 3: Life Science, pp. 221–222,
DOI: 10.1007/978-3-540-85228-5_111,

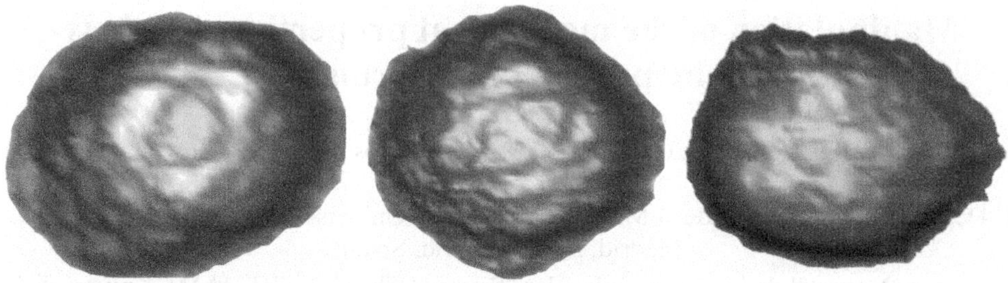

Figure 1. High-resolution Atomic Force Microscopy images of MVM virus particles along three different axes of symmetry.

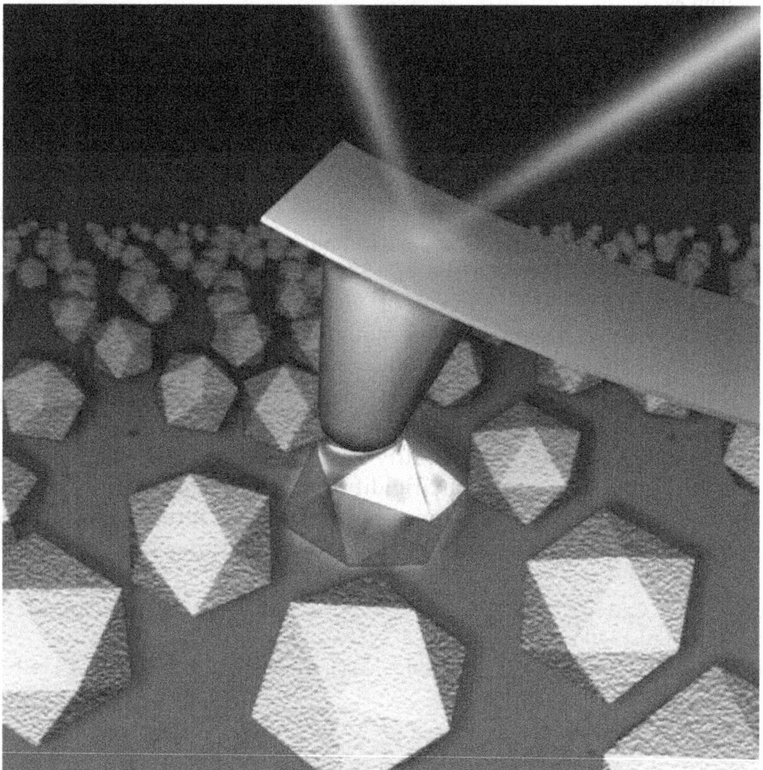

Figure 2. Artist´s rendition of an AFM cantilever squeezing an icosahedrical virus. The colour map represents the stress distribution on the virus capsid as calculated by a finite elements model.

A single-molecule mechanical assay to study DNA replication coupled to DNA unwinding

B. Ibarra[1], J. A. Morin[1], R. Arias[1], S. Hormeño[1], M. Salas[2], J.M. Valpuesta[1] and J.L. Carrascosa[1]

1. Departamento de Estructura de Macromoléculas, Centro Nacional de Biotecnología (C.S.I.C.), Canto Blanco, Madrid 28049, Spain
2. Centro de Biología Molecular "Severo Ochoa" (C.S.I.C.-U.A.M), Canto Blanco, Madrid 28049, Spain

borja.ibarra@gmail.com
Keywords: Optical Force Microscopy, single molecules, DNA replication

Mechanical force at the molecular level is involved in the action of many enzymes. For example, during DNA replication the mechanical unwinding of the DNA helix is required to separate the two complementary strands which are used by the DNA polymerases to generate two identical copies of DNA [1]. A tight coupling between these two activities (unwinding and replication) is required for the proper progress of the reaction. Some organisms like the bacteriophage Phi29 have coupled these two processes within the same protein. The Phi29 DNA polymerase unwinds the double-stranded DNA (dsDNA) helix as it replicates in a processive manner one of the DNA strands [2]. This polymerase works as a molecular motor converting the energy provided by the incoming nucleotides (dNTPs) into mechanical work as it moves unidirectionally along the DNA unwinding its helical structure.

Using Optical Force Microscopy or Optical Tweezers [3] we have developed a single molecule mechanical assay to elucidate the molecular mechanism by which this particular motor works. A single DNA molecule is attached between two polystyrene beads; one strand of the molecule is bound to a bead on the optical trap and the other strand to a bead placed on a mobile micropipette, "Figure 1". By moving the pipette relative to the optical trap it is possible to apply force to the DNA strands and unwind the polymer. Mechanical unwinding occurs between 12-16 pN, is a reversible process and resembles accurately the sequence heterogeneity of the DNA "Figure 1". Using this experimental set-up when the Phi29 DNA polymerase is flowed into the reaction chamber, it loads onto the 3' end of the primer and in the presence of dNTPs replicates one strand while displacing the other. As the polymerase works its activity can be measured in real time, since the distance between the beads increases due to the generation of dsDNA and the simultaneous release of the complementary strand "Figure 2". The response of the replication rate to increasing tensions on the DNA and to the DNA sequence will allow us to determine the molecular mechanism of this molecular motor.

1. Benkovic, S.J., Valentine, A.M., and Salinas, F., Annu. Rev. Biochem. **70** (2001), p.181.
2. Blanco. L., Bernad, A., Lazaro, J.M., Martin, G., Garmendia, C., and Salas, M., J. Biol. Chem. **264** (1981), p. 8935.
3. Smith, S.B., Cui, Y., and Bustamante, C., Methods Enzymol. **361** (2003), p.134

A. Aretz, B. Hermanns-Sachweh, J. Mayer (Eds.): EMC 2008, Vol. 3: Life Science, pp. 223–224,
DOI: 10.1007/978-3-540-85228-5_112,

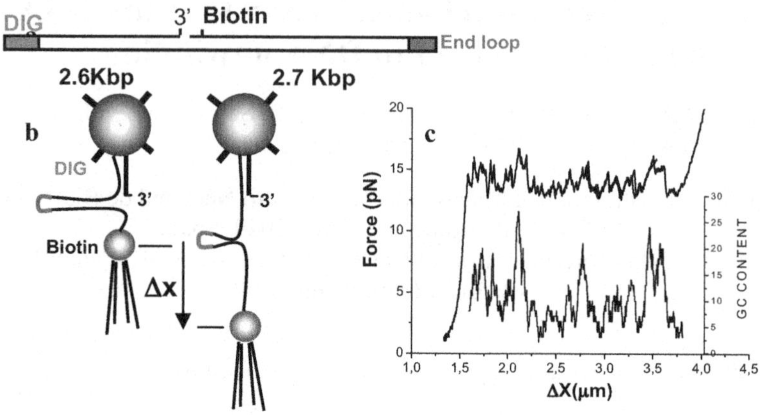

Figure 1. a) Diagram illustrating the DNA construct containing digoxigenin (DIG) and biotin for binding between the beads. On the same strand near the biotin label is a nick in the dsDNA. **b)** Schematic representation of the experimental set-up (not to scale). The nick allows the DNA to be mechanically unwound when the beads are forced apart increasing the distance between then (Δx). The End loop allows rezipping. **c)** Experimental curve showing that increasing forces promote the DNA unwinding between 12-16pN. Lower curve shows the GC density plot of the unwound DNA.

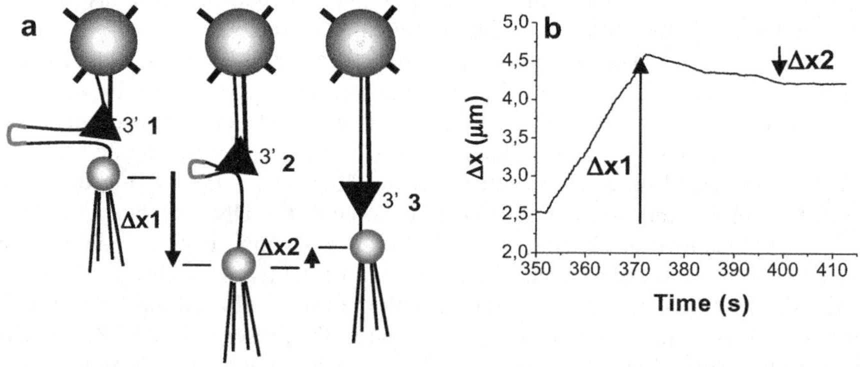

Figure 2. a) Schematic drawing representing the activity of the Phi29 DNA polymerase when holding the force constant between the beads. The polymerase (triangle) loads onto the 3′end of the DNA (1) and starts replication through the DNA hairpin increasing the extension between the beads, Δx1 (2). Upon DNA unwinding, conversion of the ssDNA portion to dsDNA (3) shortens the polymer extension, Δx2. **b)** Representative experiment at constant force (8pN) showing the changes in extension between the beads due to the activity of a single polymerase molecule.

Generation of multiple DNA lesions at subnuclear resolution by multi-photon irradiation

D. Träutlein[1], M. Deibler[2], E. Ferrando-May[2] and A. Leitenstorfer[1]

1. Department of Physics and Center for Applied Photonics, University of Konstanz, Universitätsstr. 10, D-78457 Konstanz, Germany
2. Depertmanet of Biology, Center for Applied Photonics and Bioimaging Center, University of Konstanz, Universitätsstr. 10, D-78457 Konstanz, Germany

daniel.traeutlein@uni-konstanz.de
Keywords: DNA damage, DNA repair, multi-photon

In the field of DNA repair, multi-photon absorption is becoming increasingly popular as a tool to induce DNA damage with high spatial resolution. Typically, cell nuclei of live cells are subjected to a defined pattern of irradiation at a wavelength around 800 nm, which is the characteristic output of femtosecond Ti:Sapphire lasers. This allows analyzing the spatial and temporal aspects of the recruitment of nuclear proteins, in particular repair factors, at the site of damage. Various types of DNA lesions, like UV-photoproducts [1], strand breaks [2] and reactive oxygen species [3, 4] were reported by different studies using femtosecond pulses with comparable parameters. This method thus appears to lack specificity with respect to the type of DNA damage inflicted [5] hampering the analysis of individual DNA repair pathways. In this study we take advantage of a tunable Er:fiber laser source [6] to investigate DNA damage induction by multiphoton absorption at $\lambda > 800$ nm and assess its specificity. To this end, we have increased the output power of our system by integrating a homebuilt Yb:fiber amplifier and a subsequent grating compressor. The output of the Yb-amplifier has a center wavelength of 1050 nm with a full width half maximum (FWHM) of 50 nm, the pulse duration in the focal plane of the objective-lens is sub-100 fs and the average power after the grating compressor is 250 mW. We present a comparative analysis of DNA-damage induced by multi-photon absorption at 774 nm (Figure 1.) and 1050 nm based on damage-specific immunostainings and the recruitment of various DNA repair factors. Potential mechanisms underlying the difference in the observed spectrum of lesions will be discussed.

1. R. A. Meldrum, S. W. Botchway, C. W. Warton and G. J. Hirst, EMBO reports **4** (2003), p. 1144.
2. P.-O. Mari, B. I. Florea, S. P. Persengiev, N. S. Verkaik, H. T. Brüggenwirth, M. Modesti, G. Giglia-Mari, K. Bezstarosti, J. A. A. Demmers, T. M. Luiders, A. B. Houtsmuller and D. C. van Gent, PNAS **103** (2006), p.18597-18602.
3. B. A. King and D. H. Oh Photochemistry and Photobiology, **80** (2004), p1-6.
4. P. E. Hockberger, T. A. Skimina, V. E. Centoze, C. Lavin, S. Chu, S. Dadras, J. K. Reddy and J. G. White, PNAS **96** (1999), p. 6255-6260.

A. Aretz, B. Hermanns-Sachweh, J. Mayer (Eds.): EMC 2008, Vol. 3: Life Science, pp. 225–226, DOI: 10.1007/978-3-540-85228-5_113,

5. C. Dinant, M. de Jager, J. Essers, W. A. van Cappellen, R. Kanaar, A. B. Houtsmuller and W. Vermeulen, Journal of Cell Science **120** (2006), p.2731-2740.
6. D. Träutlein, F. Adler, K. Moutzouris, A. Jeromin, A. Leitenstorfer and E. Ferrando-May, Journal of Biophotonics, **1** (2008), p. 53-61.

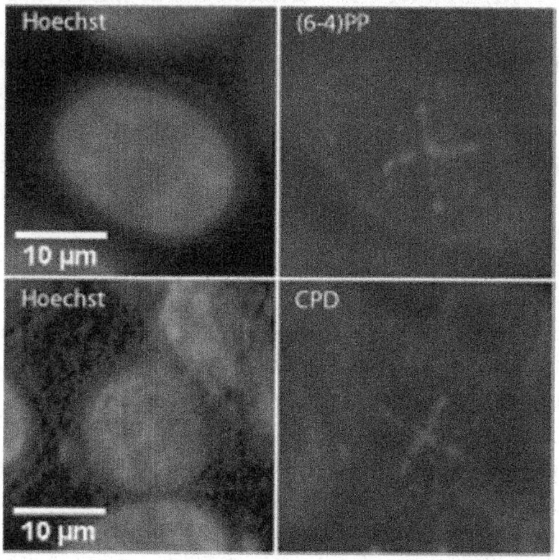

Figure 1. HeLa 229 cells were irradiated with a wavelength of 774 nm for a dwell time of 44 ms per pixel. The pulse duration was 230 fs and the average power at the sample was 14.3 mW. Cells were fixed directly following the irradiation.

Light and electron microscopy of phagosomes

Gareth Griffiths, David Liebl , Christopher Bleck, Andreas Merz, Paul Walther, Jacques Dubochet and Benoit Zuber

EMBL, Meyerhofstrasse 1, 69117 Heidelberg, Germany.

griffiths@embl.de
Keywords: phagocytosis, macrophages, Mycobacteria, live cell imaging, EM thin sections

Phagocytosis is the process whereby cells such as macrophages take up particles larger than 0.3 uM. The ingested particle can range from inert material such as carbon or asbestos to live pathogens such as *Mycobacterium tuberculosis*. After engulfment by the plasma membrane of the phagocytic cell the engulfed particle becomes surrounded by a membrane in a specialized intracellular compartment - the phagosome. This compartment fuses sequentially with early endosomes, late endosomes and lysosomes, a process referred to as phagosome maturation, in which the phagosome acquires acid hydrolases and a low pH that together facilitate pathogen clearance. Actin assembly by phagosomes plays an important role in phagosome maturation. Once internalized into the phagocytic cell *M.tuberculosis* has evolved the ability to block actin assembly and maturation of phagosome, that facilitates the ability of this pathogen to survive and grow within macrophages. This presentation will focus on two aspects of phagosome biology, in order to emphasize the importance of light microscopy and EM in analyzing the cell biology of phagosomes. In the first part the use of live cell video microscopy will be described in order to reveal striking dynamic process of transient assembly of actin by latex bead phagosomes in GFP-actin expressing mouse macrophages. The second part will address the use of different EM sectioning approaches to visualize different mycobacteria in macrophage phagosomes and will conclude that the best sectioning method to monitor the native structure of free mycobacteria and mycobacteria within phaagosomes is the Cryo EM of Vitrified Sections (CEMOVIS).

A. Aretz, B. Hermanns-Sachweh, J. Mayer (Eds.): EMC 2008, Vol. 3: Life Science, pp. 227–227,
DOI: 10.1007/978-3-540-85228-5_114, © Springer-Verlag Berlin Heidelberg 2008

The Use of Electron Microscopy in the Diagnosis of Human Microsporidial Infection – the Manchester, UK, Experience

Alan Curry

Health Protection Agency, Clinical Sciences Building, Manchester Royal Infirmary, Manchester, M13 9WL, UK

alan.curry@cmmc.nhs.uk

Keywords: microsporidia, emerging pathogens, electron microscopy

Microsporidia are a group of tiny obligate intracellular parasites with a unique mode of infecting cells via an eversible polar tube within a highly resistant spore. Infection within the animal kingdom is common, but pre-1985 human infection by microsporidia was considered to be rare, as only a handful of cases had been published in the scientific literature. In 1985 a new species of microsporidian, *Enterocytozoon bieneusi*, was described from a French AIDS patient with symptoms of chronic diarrhoea [1]. Following this publication, more microsporidial infections have been recognised in humans and electron microscopy of biopsy samples has proven to be an important diagnostic method in such investigations [2]. Microsporidia are emerging pathogens and at least sixteen species are now recognised from humans.

The commonest infections found in AIDS patients are enteric, but these pathogens have been identified in many other organs, including the kidney and urinary tract, eyes, skeletal and cardiac muscle, respiratory system, brain etc. A number of non-enteric cases associated with AIDS have been investigated over several years. Species such as *Trachipleistophora hominis* (Figure 1), *Encephalitozoon hellem* (Figure 2) and *Encephalitozoon cuniculi* have been identified from various sites.

With the advent of antiretroviral therapy few cases now come from HIV-infected individuals. Currently most samples requiring investigation for microsporidia are from individuals who are not HIV-positive and the most common site of infection is ophthalmic [3]. *Encephalitozoon spp.* (probably *E. hellem*), *Vittaforma corneae*, *Trachipleistophora hominis* and several species of *Nosema* have been identified in ophthalmic samples. Sources of microsporidial infection remain obscure, but some ophthalmic infections may be from insect sources, such as the aquatic larval stages of mosquitoes [4].

1. I. Desportes, Y. Le Charpentier, A. Galan. et. al. J. Protozool. 32 (1985), p.250-3.
2. A. Curry in Topley and Wilson's "Microbiology and Microbial Infections", eds F. E. G. Cox, D. Wakelin, S. H. Gillespie and D. D. Despommier, (Hodder Arnold, London, Sydney, Aukland), chapter 25.
3. S. Rauz, S. Tuft, J. K. G. Dart. et. al. J. Med. Microbiol 53 (2004), p. 775-81.
4. A. Curry, H.S. Mudhar, S. Dewan. et. al. J. Med. Microbiol. **56** (2007), p. 1250-2.

A. Aretz, B. Hermanns-Sachweh, J. Mayer (Eds.): EMC 2008, Vol. 3: Life Science, pp. 229–230, DOI: 10.1007/978-3-540-85228-5_115,

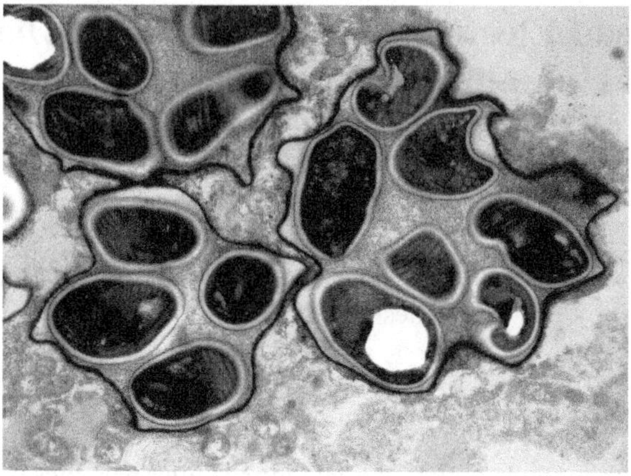

Figure 1. Sporophorous vesicles of *Trachipleistophora hominis* in muscle of an AIDS patient.

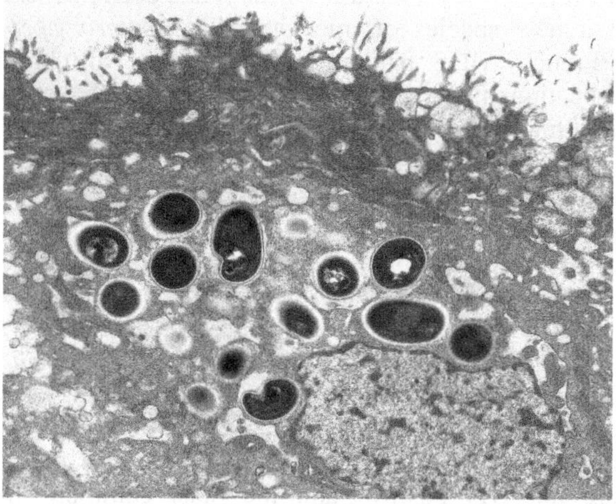

Figure 2. An epithelial cell in the sinus of an AIDS patient showing the microsporidian, *Encephalitozoon hellem*.

Overview of the impact of biological microanalysis in health and disease

Ann LeFurgey and Peter Ingram

Veterans Affairs and Duke University Medical Centers, Durham, NC 27710 USA

a.lefurgey@cellbio.duke.edu
Keywords: Microanalysis, x-rays, biology

The general trend of microscopical investigation in biology from the 1950's to the early 1970's was towards obtaining structural information. This goal initially was met using heavy metal and/or aldehyde fixatives, room temperature dehydration with polar organic liquids, embedding with epoxy and acrylate resins, and thin sectioning at room temperature. By the mid 1970's, a perceptible change occurred in the direction of both light and electron microscopy towards investigation of the chemical reactivity and composition of structures made visible with increasingly better spatial resolution for light and electron microscopes. Between the mid 1980's and the present, innovations in microanalytical techniques, including x-ray microscopy and microanalysis, secondary ion mass spectrometry, laser microprobe mass analysis, the scanning probe microscopies, and confocal/multi-photon microscopy, and development of compounds for visualization of molecular and ionic sites within individual living cells as well as membranes, redefined the goal of microscopical preservation: to stabilize cell *structure and composition* as they exist in the living state.

Analytical microscopy is used here mainly in the context of acquisition and analysis of compositional data at cellular and subcellular spatial resolutions using an electron microscope. Clearly this is only one aspect of a very broad field of scientific inquiry which generally falls under the rubric of Compositional Imaging. In biological microscopy subtle distinctions exist between "structural microanatomy" and "chemical microanatomy". These microanatomies can be significantly different. For example during the early stages of cell injury changes in the elemental distribution within a cell can become readily apparent before any observable changes occur in morphology via "conventional" transmission electron microscopy.

In addition, it is important to recognize that in general there are only 2 *possible* ways to form an image: either each picture point is formed simultaneously or each is formed sequentially. In either scenario quantitative information is retrievable; however from a practical point of view it is much simpler to perform this digitally i.e. on a pixel by pixel basis. Both electron energy-loss spectroscopy (EELS) and x-ray spectroscopy in the form of either energy or wavelength dispersive spectroscopies (EDX/WDX) are routinely carried out on biological specimens in the (scanning) electron microscope [(S)TEM]. It is now convenient to collect and store entire EELS and/or x-ray spectra at each pixel for either on-line or *post facto* analysis.

A variety of frequently encountered problems in health and disease lend themselves readily to investigation by analytical electron microscopy (AEM). i.e. STEM/EDX. The most common application is identification of xenobiotics or exogenous substances, such

A. Aretz, B. Hermanns-Sachweh, J. Mayer (Eds.): EMC 2008, Vol. 3: Life Science, pp. 231–232,
DOI: 10.1007/978-3-540-85228-5_116, © Springer-Verlag Berlin Heidelberg 2008

as localization and quantitation of inorganic particulates in lung tissues in patients with pneumoconiosis; identification of foreign materials within granulomas; and analysis of foreign bodies. EDX is also useful in the study of tissue reactions to various surgical implants of foreign materials. A variety of metals and other elements may be detected with EDX, including copper in tissues of patients with Wilson's disease, thorium, technetium and gadolinium in patients injected with radiographic contrast agents, or gold in chrysotherapy treated patients, and trace metals as essential cofactors in proteins.

Endogenous particulates such as urinary calculi, gallstones, intraarticular and periarticular crystalline deposits in patients with rheumatic disease, dystrophic or metastatic calcifications, and hemosiderin may be analyzed rapidly and efficiently by means of EDX. Certain organometallic drugs such as amiodarone (iodine) or sodium stibogluconate (antimony) may also be detected in human tissues. AEM has been important to forensic pathology for many years in diverse areas such as identification of trace evidence constituents or detection of arsenic or lead in victims with heavy metal poisoning. The detailed elucidation of anatomic, physiologic, and pathologic conditions provided by AEM is a useful investigative tool in clinical medicine; the analytical results often have diagnostic, therapeutic, and/or medicolegal implications. This technology growing rapidly in its sophistication and use as it is complemented by other techniques such as mass spectrometry, and laser Raman and infrared microspectroscopy.

Experimental approaches to quantitate and localize free and bound ions within subcellular compartments require the application of cryopreparative techniques. By definition, cryofixation or instantaneous lowering of cell temperature stops all metabolic and diffusional processes, and significantly reduces molecular transformational and vibrational movements. Chemical fixatives, solvents or embedding media employed at any stage of preparation do alter content and location of elements which are not tightly bound. Only cryopreparative techniques provide unambiguous retention of all diffusible substances. . Attention should be given to (a) the rate of cryopreservation relative to the rate of the metabolic, structural or physiological process under investigation; (b) the physiological and biochemical requirements for *in situ* maintenance of cellular constituents prior to, during, and subsequent to, cryofixation; (c) for a given cell/tissue configuration and geometry, the optimal methods of cryofixation [cold metal block freezing, plunge freezing, jet freezing, spray freezing, high pressure freezing], cryosectioning and freeze drying; and (d) the criteria for assessment of cryopreservation and cryosection quality. Exceptions exist: insoluble crystalline deposits and covalently bound elements, such as those found in many diagnostic pathology specimens, may be qualitatively preserved by routine chemical methods.

In summary, different specific tactical approaches, tailored for the cell system and the hypothesis in question, have been adopted over the past 40 years to successfully convert the molecular, ionic and elemental constituents and aqueous compartments of cells and tissues into a state suitable for microscopy and analytical imaging.

1. P. Ingram et al., *Biomedical Applications of Microprobe Analysis,* Academic Press (1999)
2. J Feng, AV Somlyo & AP Somlyo J. Microscopy 215, 92-99, (2004)
3. A Warley, *X-Ray Microanalysis for Biologists*, Portland Press, London (1997)

Immunolocalization of FOXP3 in HCV-infected liver biopsies. Preliminary observations.

F. D'Amico[1], A. Amoroso[1], M. Consolo[2], E. Skarmoutsou[1], B. Mauceri[2], G. Malaponte[1], F. Stivala[1], C.M. Bruno[2], S. Neri[2] and M.C. Mazzarino[1]

1. Department of Biomedical Sciences, University of Catania, via Androne 83, 95124 Catania, Italy
2. Department of Internal Medicine and Systemic Diseases, University of Catania, Via S. Sofia 86, Catania, Italy

f.damico@unict.it
Keywords: FOXP3, HCV, liver, immunohistochemistry

Regulatory T cells (Tregs) represent a subset of T lymphocytes. This cell subpopulation plays an important role in controlling immune response as well as in inducing and maintaining immunologic tolerance [1]. The identification of this cell population in pathological condition, such as in hepatitis C virus (HCV) infected patients, may be useful for a better understanding of pathology.

In HCV-infected patients, peripheral Tregs are increased [2] and are able to suppress specific cell immune response to the virus [3].

A specific phenotypic marker for the identification of Tregs is the transcription factor FOXP3 [4]. Forkhead proteins represent a family of transcription factors that have been implicated in a variety of cellular processes. FOX transcription factors, such as FOXP3, are involved in the regulation of the adaptative immune response [4].

We investigated the expression of FOXP3 in HCV-infected liver biopsies.

Immunohistochemical staining on formalin-fixed and paraffin-embedded biopsies from liver of HCV-infected patients, was performed with the use of a mouse monoclonal antibody anti FOXP3 (Santa Cruz Biotechnology) and peroxidase-streptavidin complex technique.

Our preliminary studies suggest that the expression of FOXP3 in HCV infected liver is localized exclusively on periportal lymphocyte infiltrates, which characterize the biopsy of HCV liver infection. Only a very small fraction of the entire population is immunolabelled. The labelled cells are scattered throughout the infiltrate and do not appear to be clustered (fig. 1).

As suggested by [5], HCV-infected liver Tregs seem to have a direct physical access in the liver for the suppression of T cell response. Our results would support the fact that the occurrence of Tregs in HCV-infected liver may influence the immune response against the infection.

A better understanding of mechanisms involved in this pathology would contribute to the development of therapeutic strategies to overcome Tregs activity in the treatment of viral infections.

Finally, our immunocytochemical reaction would contribute to the standardization process of the protocol for the immunolocalization of this protein in paraffin section. A

A. Aretz, B. Hermanns-Sachweh, J. Mayer (Eds.): EMC 2008, Vol. 3: Life Science, pp. 233–234,
DOI: 10.1007/978-3-540-85228-5_117,

microwaved retrieval approach with 10mM citric buffer, pH 6, for 15 min seems to be the optimal choice for the antigen unmasking.

1. N. Cools N, P. Ponsaerts, V.F. Van Tendeloo and Z.N. Berneman, Clinical and Developmental Immunology **2007** (2007), p. 89195.
2. R. Cabrera, Z. Tu, Y. Xu, R.J. Firpi, H.R. Rosen, C. Liu and D.R. Nelson, Hepatology **40** (2004), p. 1062.
3. T. Boettler, H.C. Spangenberg, C. Neumann-Haefelin, E. Panther, S. Urbani, C. Ferrari, H.E. Blum, F. von Weizsäcker and R. Thimme, Journal of Virology **79** (2005), p. 7860.
4. S. Hori, T. Nomura and S. Sakaguchi, Science **299** (2003), p. 1057.
5. S.M. Ward, B.C. Fox, P.J. Brown, J. Worthington, S.B. Fox, R.W. Chapman, K.A. Fleming, A.H. Banham and P. Klenerman, Journal of Hepatology **47** (2007), p. 316.

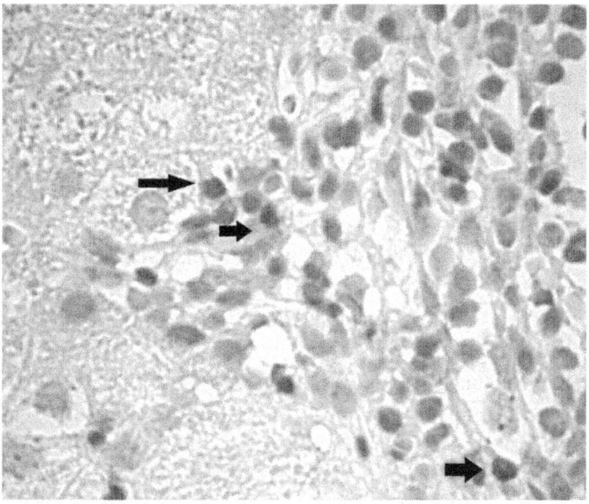

Figure 1. FOXP3 is localized on Tregs (arrows). Original magnification: 400x

Ultrastructural and Intracellular Elemental Composition Analysis of Human Hematopoietic Cells During Cold Storage in Preservation Solutions

F. Arrebola, F.J. Cañizares, M.A. Cubero, M.M. Serrano, M.A. Robles, and E. Fernández-Segura

Department of Histology, Faculty of Medicine, University of Granada, E-18071 Granada, Spain.

fav@ugr.es
Keywords: cell preservation, hypothermia, ultrastructural changes, X-ray microanalysis

Hematopoietic cell transplantation is a medical therapeutic procedure which aims to reconstitute the hematopoietic activity of bone marrow. Stem cells from bone marrow, mobilized peripheral blood or umbilical cord blood cells are used for these transplants. These transplant modalities require hypothermic storage and cryopreservation. However, cellular injury still occurs after extended cold storage. Despite large number of studies, however, optimal non-frozen clinical storage conditions for hematopoietic cells have not been established. Optimization of cell preservation protocols to maintain the viability and quality of hematopoietic progenitor cells has been an important task for tissue banks, and several attempts have been made to create alternative solutions for preservation of cells for transplantation. Evaluation of the storage solutions has been carried out using morphological techniques, studies of cell viability, functional studies as cellular ATP levels as well as by clinical assessment. However, few studies determining changes in electrolyte composition of cells during cold storage have been carried out [1]. In this study, we propose electron probe X-ray microanalysis (EPXMA) as a gold standard method for evaluation of the effect of preservation solutions on the intracellular elemental composition of cells for transplantation. For this, we evaluated the time-related elemental changes during cold storage in tissue culture medium in comparison with EuroCollins and lactated Ringer's solutions. In addition, we examined the time-related ultrastructural changes of hematopoietic cells in preservation solutions.

As experimental hematopoietic model, the human myelomonocyte cell line U937 was used. For preservation experiments, the following storage solutions were tested: EuroCollins and lactated Ringer's solutions. RPMI 1640 culture medium was used as control. After removing the culture medium and washing the cells with Hank's balanced salt solution, cold preservation solutions or culture medium (4° C) was added and cells were stored to 4°C for 2, 4, 8 and 24 h. At different times, cells were removed from preservation solution and processed for light microscopy, transmission electron microscopy (TEM), and X-ray microanalysis as described in detail elsewhere [2].

Figure 1 shows the structural alterations of U937 cells (mitochondria, endoplasmic reticulum and nuclei) after cold storage in RPMI 1640 culture medium, EuroCollins, and lactate Ringer's solutions for 4 h. Figure 2 shows the time-dependent changes in intracellular concentrations (mmol/kg dry weight) of Na, Cl, K from human U937 cells

A. Aretz, B. Hermanns-Sachweh, J. Mayer (Eds.): EMC 2008, Vol. 3: Life Science, pp. 235–236,
DOI: 10.1007/978-3-540-85228-5_118,

hypotermically preserved in RPMI 1640 culture medium, EuroCollins, and lactated Ringer's solutions.

1. I. Kozlova, and G.M. Roomans, Am J Transpl 3 (2003), p. 697.
2. E. Fernández-Segura, F.J. Cañizares, M.A. Cubero, A. Warley and A. Campos, J Micros 196 (1999), p. 19.

3. This work was supported by Instituto de Salud Carlos III (Contract grant number FIS PI061836).

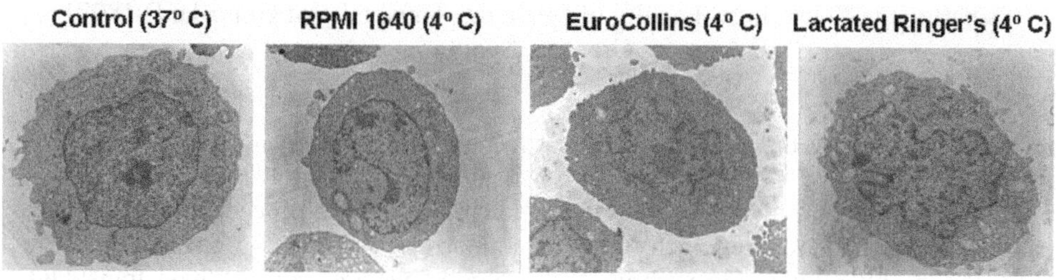

Figure 1. Transmission electron microscopy. (A) U937 cells of the control group with regular morphology. (B) Ultrastructural changes in U937 cells hypothermically preserved in RPMI 1640 culture medium for 4 h. (C) Cells after storage in EuroCollins solution for 4 h. (D) Cells after storage in lactated Ringer's solution for 4 h.

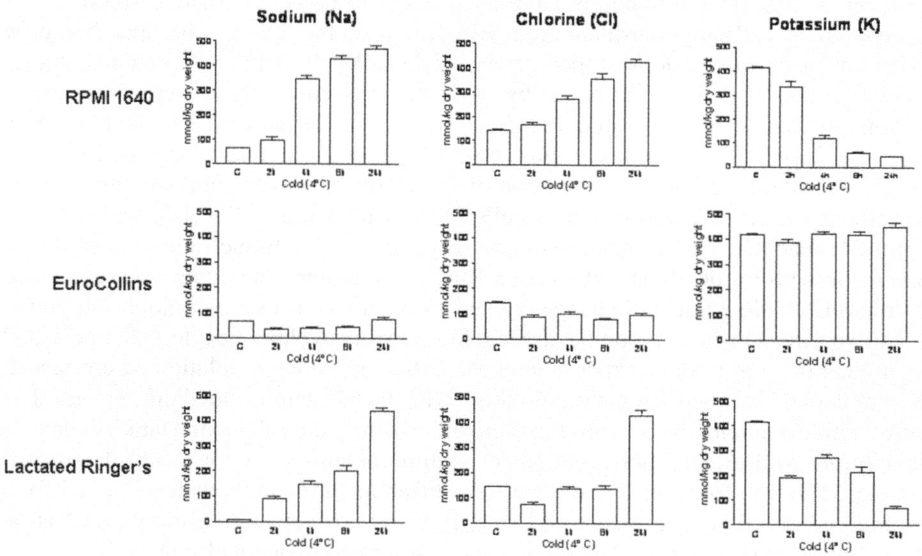

Figure 2. Effect of preservation time on sodium (Na), chlorine (Cl), and potassium (K) content in U937 cells incubated at 4° C for 2, 4, 8 and 24 hours.

Reconstitution and morphological characterisation of an original Human Endogenous Retrovirus

N. Bannert[1], N. Beimforde[1,2], L. Möller[1], G. Holland[1], M. Laue[1] and R. Kurth[1]

1. Robert Koch Institute, Centre for Biological Safety 4, Nordufer 20, 3353 Berlin, Germany
2. Charité - Universitätsmedizin Berlin, Hindenburgdamm 30, 12203 Berlin, Germany

BannertN@rki.de
Keywords: HERV, Betaretrovirus, Reverse Transcriptase, Cancer

Human endogenous retroviruses (HERVs) are relicts of infectious exogenous retroviruses that invaded the germ line of our ancestors and have since been transmitted vertically from generation to generation [1]. In contrast to several other species, no infectious endogenous retroviruses have so far been identified in the human genome, although there is support for reinsertion of active HERV-K(HML-2) elements in our recent evolutionary history as well as a record of evidence for its tumorigenic and other pathogenic properties.

One of the youngest and most complete HERV-K(HML-2) proviruses with preserved open reading frames is HERV-K113 [2]. To facilitate molecular and functional studies of these elements and the human factors restricting their proliferation, we have cloned the entire HERV-K113 provirus into a small plasmid vector and analyzed its ability to form infectious particles [3]. In addition, by aligning HERV-K113 with other human specific HERV-K elements we identified non-synonymous postinsertional mutations in the sequence. These mutations were reverted by site-directed mutagenesis to reconstitute the original provirus. Human cell lines transfected with a plasmid expressing the reconstituted HERV-K113 proteins produced and released complete retroviruses with B-type morphology and reverse transcriptase activity. The efficiency of this protein and particle production was significantly enhanced by codon-optimization. This allowed us, for the first time, to study the assembly and the release of these particles from human cell lines by light- and electron microscopy "Figure 1"as well as other properties and morphological criteria important for its identification in healthy tissues and tumors.

1. N. Bannert and R. Kurth Proc Natl Acad Sci U S A. 101, (2004), p. 14572
2. G. Turner , M. Barbulescu, M. Su et al. Curr. Biol. 11, (2001), p. 1531
3. Beimforde N, Hanke K, Ammar I, Kurth R, Bannert N., Virology, 371, p. 216

A. Aretz, B. Hermanns-Sachweh, J. Mayer (Eds.): EMC 2008, Vol. 3: Life Science, pp. 237–238, DOI: 10.1007/978-3-540-85228-5_119,

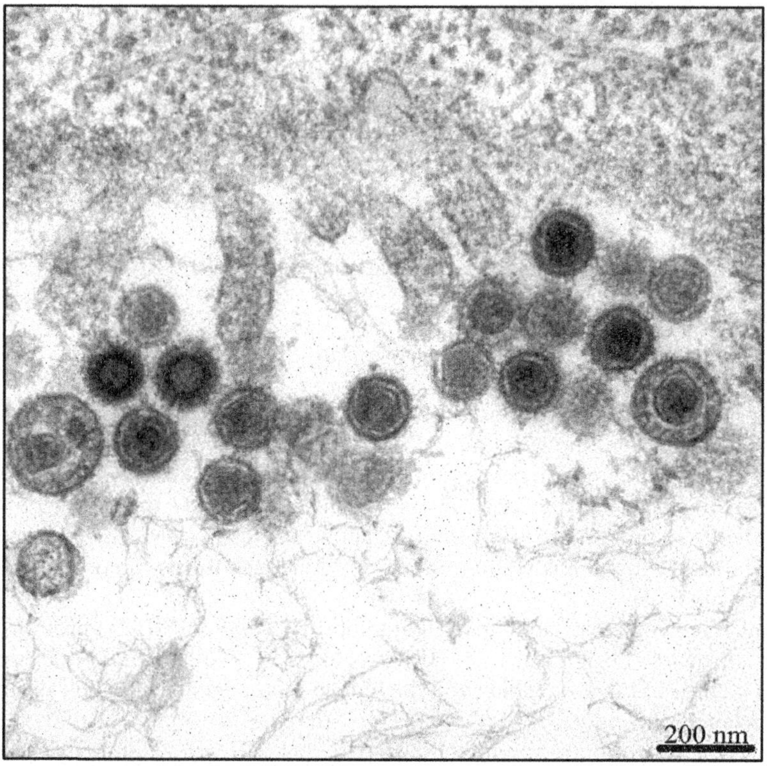

Figure 1. Viral particles produced from HEK 293T cells transfected with a plasmid encoding the reconstituted HERV-K113 provirus.

Cell death in human articular chondrocyte: an ultrastructural study in micromass model.

M. Battistelli[1], A. D'Emilio[1], E. Olivotto[2], S. Pagani[2], R. Borzì[2], A. Facchini[2], E. Falcieri[1,3]

1. Istituto di Scienze Morfologiche, Università degli Studi di Urbino "Carlo Bo", Urbino,
2. Laboratorio di Immunologia e Genetica e 3. Istituto di Genetica Molecolare, CNR, Istituti Ortopedici Rizzoli, Bologna, Italy.

m.battistelli@uniurb.it
Keywords: micromass, chondrocyte, apoptosis, osteoarthritis

Introduction: Chondrocyte apoptosis is known to contribute to articular cartilage damage in osteoarthritis (OA) (1, 2, 3) and is correlated to a number of cartilage disorders. Human chondrocyte primary monolayer culture does not recover a proper microenvironment comparable with "in vivo" extracellular matrix (ECM), and so it cannot be considered a reliable experimental model (4). Differently, micromass cultures represent a convenient means for studying chondrocyte biology, and, in particular, their death (5). In this study, we present the first ultrastructural analysis on chondrocyte death experimentally induced by different agents, all known to be powerful apoptotic triggers.

Materials and Methods: Chondrocytes were obtained from subjects undergoing joint arthroplasty (6). At the end of the maturation, they were treated as follows: 10 or 30μM etoposide for 24h, UV-B for 30 min followed by 4h recovery, 200 or 500nM staurosporine for 24h, hyperthermia for 1h at 43°C followed by 4h recovery. They were processed for TEM and immunofluorescence as previously described (7).

Results: Chondrocyte morphology in control condition appears very similar to that of human articular cartilage (A, B). Cells are rounding or slightly elongated with plurilobated nucleus and diffusely dispersed chromatin. It, frequently, appears eccentrically located due to a relevant amount of lipid vacuoles. Proteoglycans and collagen fibers are present in the intercellular space, indicating a good ECM production (C, D). Etoposide, when effective, induces necrosis (E). UV-B treated cells show chromatin condensation and pore clustering, typical of apoptotic nuclei (F). Hyperthermia seems to induce apoptosis in the presence of abundant ECM (G), giving, differently, a general necrosis when ECM is scarse (H). Staurosporine has no effect at 200nM concentration (I), but frequently, at 500nM, chromatin condensation appears (L). In all conditions in which we observed chromatin condensation, it appears different from that of classical apoptotic models. Roach et al. (8, 9) proposed the term "chondroptosis" to indicate this type of cell death, which was observed in the majority of articular hypertrophic chondrocytes in vivo. In our system chondrocytes seem to undergo apoptosis in a non-classical manner, but in a way that seems typical for them (chondroptosis). As in commonly reported apoptosis, cells appear shrunk and the nucleus contains clumped chromatin. However, the chromatin is not marginated into large solid masses, mostly cup-shaped, but in small patches, mainly at nuclear periphery.

A. Aretz, B. Hermanns-Sachweh, J. Mayer (Eds.): EMC 2008, Vol. 3: Life Science, pp. 239–240,
DOI: 10.1007/978-3-540-85228-5_120, © Springer-Verlag Berlin Heidelberg 2008

Conclusion: Our observations demonstrate that:
-human chondrocytes appear able to undergo apoptosis after UV-B treatment.
-staurosporine and hyperthermia induce condroptosis-like cell death.
-hyperthermia is effective dependently on ECM quantity.
-etoposide is a necrotic trigger.
Therefore, apoptotic/necrotic/chondroptotic response shown by micromass chondrocytes represents an important tool to understand articular cartilage disorders.

1. Scher et al., Curr. Rheumatol. Rep., 9,9,2007;
2. John et al., Histol. Histopatol., 22, 469,2007;
3. Olivotto et al., J. Cell Physiol., 210,417,2007;
4. Kavalkovich et al., Vitro Cell Biol. Anim., 38,457,2002;
5. Kafienah et al., Arthritis. Rheum., 56,177,2007;
6. Olivotto et al., Arthritis. Rheum., 24,99,2008;
7. Battistelli et al., Microsc. Res. Tech., 67,286,2005;
8. Roach et al., Apoptosis, 9,265,2004;
9. Perez et al., Apoptosis, 10,1105,2005.

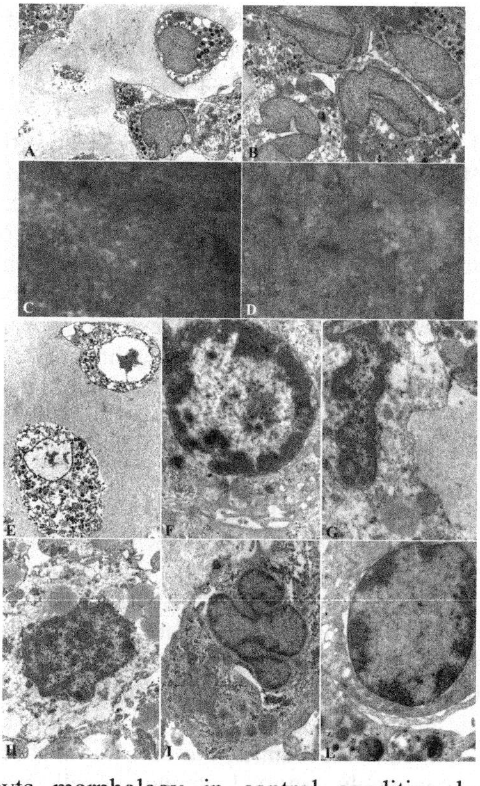

Figure 1. Chondrocyte morphology in control condition by TEM (A, B) and immunofluorescence of proteoglycans (C) and collagen (D). Micromass treatment with etoposide (E), UV-B (F), hyperthermia (G, H) and staurosporine (I, L).

Pathomorphological diagnostic of paraffin embedded versus epon embedded cardiac tissues with transmission electron microscope analysis

J. Bornemann[1], B. Hermanns-Sachweh[1] and N. Gaßler[1]

1. Electron Microscopic Facility, University Hospital, Rheinisch-Westfälische Technische Hochschule Aachen, Pauwelsstraße 30, 52074 Aachen, Germany

jbornemann@ukaachen.de
Keywords: amyloid, heart biopsy, cardiac tissue

Aggregation and deposition of proteins are found in several diseases, e.g. atherosclerosis and amyloidosis. Deposition of amyloid proteins, characterized by fibers with abnormal structural folding in various organs, is thought to be responsible for diseases like Alzheimer disease, Huntington disease, Parkinson disease and systemic amyloidosis, which frequently affects cardiac and renal tissues. Limitations in the diagnosis of early forms of amyloid deposits in formalin-fixed and paraffin embedded tissues with light microscopy and Congo red staining can be overcome by transmission electron microscopy using glutaraldehyd-fixed tissues. Here we describe a powerful method to perform transmission electron microscopy on primary paraffin-embedded cardiac tissue. Primary glutaraldehyd-fixed cardiac tissues from the same patient were used as control.

The paraffin embedded tissue samples were released from paraffin by xylene and subsequently prepared for epon embedding. Afterwards tissue samples were fixed in 3 % glutaraldehyde (in 0.1 M B2 buffer [pH 7.4; 13 mM NaH_2PO_4 x H_2O; 87 mM Na_2HPO_4 x $2H_2O$]) for 14 hours, washed with 0.1 M B2 buffer overnight, followed by 1 h in 1 % OsO_4 (in 17 % sucrose buffer [pH 7.4; 88 ml 0.1 M B2 buffer; 12 ml aqua dest., 17 g sucrose]), rinsed with water and dehydrated with ethanol (30 % - 100 %) and propylenoxide (100 %). Finally tissues were processed for embedding in epon, polymerized 48 h at 60 °C, cut into 80-100 nm thick slices and contrasted with uranyl acetate and lead citrate. The samples were analyzed with a PHILIPS EM 400 T at 60 kV and micrographs were taken by an OLYMPUS CCD-Camera MORADA.

The analyses of re-embedded cardiac tissues showed clusters of fibril deposits with diffuse formation in longitudinal and cross sections, depicted in "Figure 1". These results were very similar to primary glutaraldehyd-fixed epon embedded cardiac tissues "Figure 2". Morphological features were identical and no tissue damage after re-embedding was found. Both tissues were highly useful in morphological diagnosis of cardiac amyloidosis.

In conclusion, this demonstrated method is a useful tool in the diagnostic synergism of light microscopy and transmission electron microscope analysis using paraffin embedded tissues.

1. We thank Hiltrud Königs und Heide Kühnle for their technical assistance.

A. Aretz, B. Hermanns-Sachweh, J. Mayer (Eds.): EMC 2008, Vol. 3: Life Science, pp. 241–242,
DOI: 10.1007/978-3-540-85228-5_121,

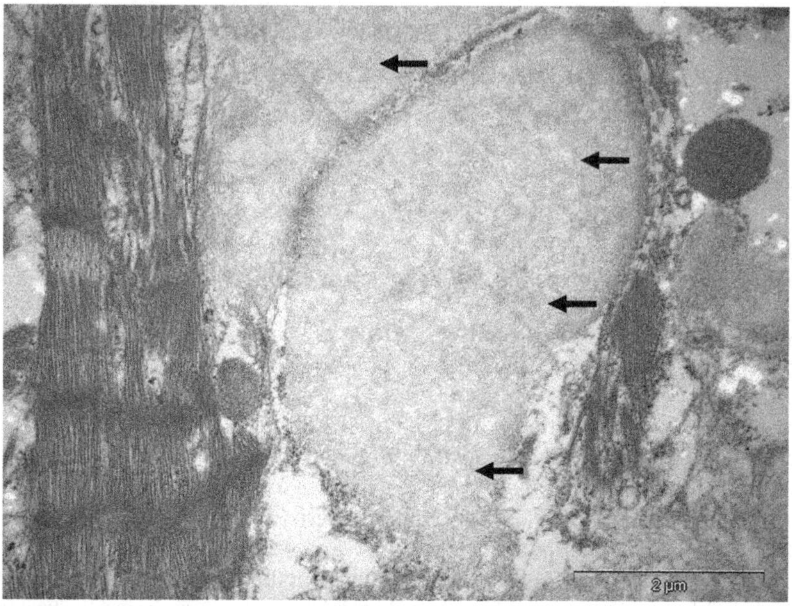

Figure 1. Part of an epon re-embedded myocardial area with high dense amyloid-like deposits (arrows, mag. 17000x).

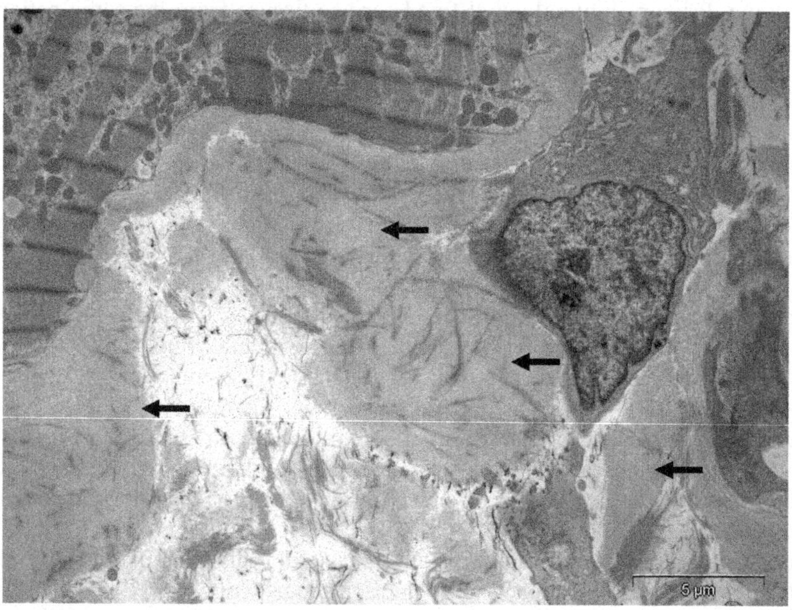

Figure 2. Part of glutaraldehyd-fixed epon embedded cardiac tissue with dark spots of high dense amyloid-like deposits (arrows, mag. 4600x).

Interaction of carbon nanotubes with macrophages: STEM and EELS approach

Cyrill Bussy[1,2], Nathalie Brun[1], Julien Cambedouzou[1], Sophie Lanone[2], Alain Grodet[3], Vasile Heresanu[1], Jorge Boczkowski[2], Pascale Launois[1] and Claudie Mory[1]

1. Laboratoire de Physique des Solides, UMR CNRS 8502, université Paris-Sud, Orsay, France.
2. Institut Mondor de Recherche Biomédicale, INSERM U841, faculté de médecine H. Mondor, 51 Av. M. De Lattre de Tassigny, Créteil, FR.
3. Centre de Recherche Biomédicale Bichat-Beaujon, INSERM U773, Atelier de Microscopie Electronique à Transmission IFR02, faculté de médecine X. Bichat, 16 rue H. Huchard, Paris, FR.

bussy@lps.u-psud.fr
Keywords: macrophages, carbon nanotubes, TEM and STEM, EELS analysis.

Carbon nanotubes (CNT) are cylinders with diameters in the nanoscale range, possessing unique properties that make them candidates to promising applications. The forecast increase of CNT use in manufacture, which will lead to an increase of human exposure, highlights the urgent need for a better understanding of the potential human health impact of CNT. Since the respiratory tract could be the main route of exposure to these volatile materials, studies of the effects of CNT on cells involved in the immune respiratory response – murin macrophages in the present study- are of special relevance.

Characterization of CNT interaction with cells is a difficult issue since CNT, and single-walled CNT in particular, are difficult to visualize in biological environments. Indeed, optical microscopy observations of single-walled CNT (SWCNT) are intrinsically limited by the spatial resolution, while low contrast makes electron microscopy observations very difficult. SWCNT have only been successfully imaged in cells, by making use of sophisticated electron microscopy-based methods [1], or by using the intrinsic near-infrared fluorescence of SWCNT, but only with treated, individualized CNT [2,3]. An alternative method, recently used [4], is synchrotron-based X-ray fluorescence microscopy. The analysis of the fluorescence signal of metal catalyst nanoparticles bound to the CNT allowed detecting the incorporation of SWCNT inside cells, even for very small amount of catalyst.

In the present study, we follow a strategy relying on the use of scanning transmission electron microscopy (STEM) and electron energy loss spectroscopy (EELS) approaches to unambiguously visualize CNT within phagocytic cells. The first step consists on checking the image contrasts of raw and purified grade SWCNT (i.e. with the accompanying metallic catalyst particles or with a reduced number of them after acidic treatment) (figure 1). The second step is to compare HAADF (high annular angular dark field) STEM imaging in stained or unstained biological cells. Figure 2 show SWCNT mostly present in vacuoles with contrast enhanced by remaining iron nanoparticles used as catalyst. In the final step, we will map the iron distribution together with other biologically significant elements (P, Ca, N, O). The STEM

A. Aretz, B. Hermanns-Sachweh, J. Mayer (Eds.): EMC 2008, Vol. 3: Life Science, pp. 243–244,
DOI: 10.1007/978-3-540-85228-5_122,

techniques offer unambiguously localization capabilities for monitoring the exact positions and the modifications of the CNT and their catalyst particles within their environment inside cells.

1. Porter, A. E., M. Gass, *et al.*, Nature Nanotechnology, **2** (2007), p. 713-717.
2. Cherukuri, P., S. M. Bachilo, *et al.*, J Am Chem Soc, **126**-48 (2004), p.15638-9.
3. Cherukuri, P., C. J. Gannon, *et al.*, Proc Natl Acad Sci U S A, **103**-50 (2006), p.18882-6.
4. Bussy, C., J. Cambedouzou, *et al.*, submittted
5. C. Bussy is granting by the nanoscience competence center of Paris Region, C'Nano-IdF. The authors thank C. Colliex for his fruitful discussion.

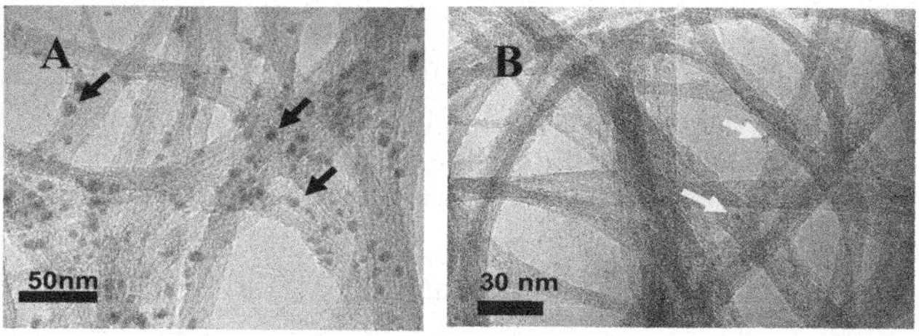

Figure 1. Conventionnal TEM bright field images of SWCNT. Remaining iron catalyst particles are visible (arrows). A) Raw SWCNT. B) Purified SWCNT.

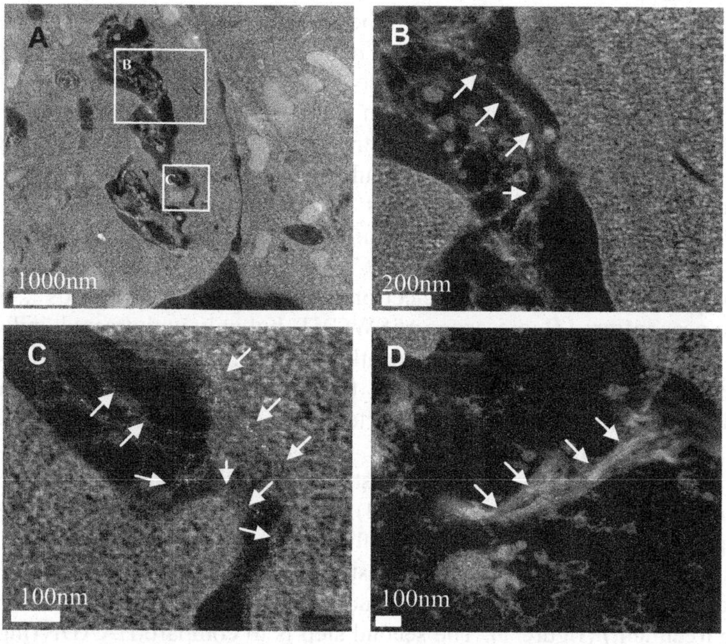

Figure 2. STEM HAADF images of stained sections of cells containing SWCNT. The catalyst particles (gloss white tiny points) have a high contrast which allows visualizing CNT in cell vacuoles. A), B), C) Raw SWCNT, different magnifications. C) Purified SWCNT, higher magnification.

The ultastructural effect of nitrogen mustard gas on rat brain and therapeutic value of proanthocyanidin

Can B[2], Tekiner A[1], Köse Sargın A[2], Yücel O[3], Bayar M. A[1]

1. Ministry of Health, Training and Research Hospital, Department of Neurosurgery, Ankara, Turkey
2. Ankara University School of Medicine, Department of Histology and Embryology, Ankara, Turkey.
3. Gülhane Military Medical Academy, Department of Thoracic Surgery, Ankara, Turkey

belgincan@yahoo.com
Keywords: Nitrogen mustard, Proanthocyanidin, Rat brain.

Mustard gas functions as a highly cytotoxic blistrogen for both humans and animals [1]. Proanthocyanidins, have been reported to possess therapeutic activities against free radicals and oxidative stress [2]. In this study, ultastructural effect of nitrogen mustard gas on rat brain and therapeutic value of proanthocyanidin was investigated.

Adult male 32 rats weighing 140-160 g were used. The rats were assigned to four groups. Group I: Control, Group II: Proanthocyanidin induced, Group III: Mustard gas, Group IV: Mustard gas + Proanthocyanidin induced group. All rats were sacrificed and brain cortex tissue were removed and fixed in 2.5% glutaraldehyde and post-fixed in 1% osmium tetroxide. Semi-thin sections were stained with toluidin blue-azur II and examined under Zeiss Axioscope photomicroscope. Ultrathin sections stained with uranyl acetate and lead citrate were examined on LEO 906 E transmission electron microscope.

Light and electron microscopic investigation showed that there was no significant difference between control and proanthocyanidin induced group (Figures 1C, D). Degenerative changes indicating disintegration of internal structure were observed in the neurons of the mustard gas group. Capillary dilatation and erythrocyte plugging were prominent. The network of axons and dendrites surrounding pericaryons were severely damaged (Figure 1E). In mustard gas + proanthocyanidin induced group, the mesh of neuronal and glial processes was normal in appearance and similar to the control group findings (Figure 1J).

These findings may imply that structural changes induced by mustard gas can be prevented and restored by proanthocyanidin treatment although in normal rats proanthocyanidin does not effect the structure of brain significantly.

1. Papirmeister B, Gross CL, Meier HL, Petrali JP, Johnson JB, Fundam Appl Toxicol, **5** (1985), p.134.
2. Debasis B, Manashi B, Sidney J, Dipak K , Toxicology **148** (2000), p.187.

A. Aretz, B. Hermanns-Sachweh, J. Mayer (Eds.): EMC 2008, Vol. 3: Life Science, pp. 245–246, DOI: 10.1007/978-3-540-85228-5_123,

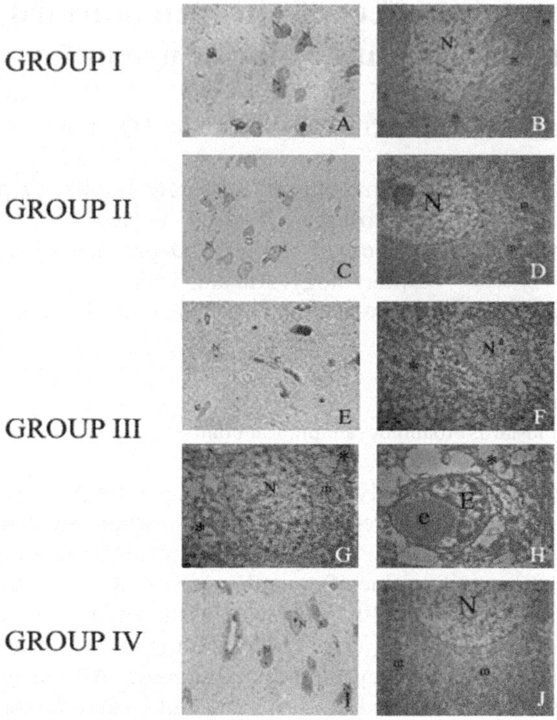

Figure 1, Photomicrographs and electronmicrographs of the section of brain cortex. Figure 1A, B: Group I; C, D: Group II; E, F, G, H: Group III; I, J: Group IV N: nucleus, c: capillary, *: mesh of neuronal and glial processes, m: mitochondria
Figure 1A, C, E, I, Toluidin blue-Azur II x250
Figure 1B, D, F, G, H, J, Uranyl Acetate- Lead citrate
Figure 1B, D, G, J x2784; Figure 1F x1293; Figure H x3597

Assembly and maturation of pestiviruses

S. Deike, M. König and H.-J. Thiel

Institut für Virologie, FB Veterinärmedizin, Justus-Liebig-Universität Giessen, Germany

stefanie.deike@vetmed.uni-giessen.de
Keywords: pestiviruses, virion morphogenesis, subcellular localization of viral proteins

The genus *Pestivirus* belongs to the family *Flaviviridae* together with the genera *Hepacivirus* and *Flavivirus*. Members of the genus *Pestivirus* are bovine viral diarrhea viruses (BVDV-1 and BVDV-2), classical swine fever virus (CSFV) and border disease virus (BDV), all of which have a significant economic impact on farming worldwide. Because of their close relationship to Hepatitis C virus (genus *Hepacivirus*), pestiviruses are studied as a model for this important human pathogen.

Pestiviruses are enveloped viruses with a single stranded RNA genome of positive polarity. Virions have a size of 40-50 nm (Figure 2) and have been proposed to show an icosahedral symmetry. In ultrathin sections of infected cells virus particles exhibit a hexagonal outline of the envelope [1]. The viral capsid is composed of a single core protein and the viral RNA. The symmetry of the viral capsid has not yet been determined. Glycoproteins termed E1, E2 and E^{rns} are associated with the host cell derived lipid membrane, together they form the viral envelope (Figure 1). For members of the genus *Flavivirus* an icosahedral symmetry of the envelope could be demonstrated by image reconstructions from cryo-electron micrographs [2].

Morphogenesis of pestiviruses is poorly understood. Assembly of virions is assumed to take place by budding presumably at ER-derived intracellular membranes. It is assumed that virions are transported via the membrane systems of the host cell secretory pathway to the cell surface and released by exocytosis. Cellular enzymes located within the ER lumen are required for processing and folding and hence for synthesis of virions [3, 4]. Macovei et al. could show that treatment of pestivirus infected cells with Brefeldin A inhibits release of virions into the supernatant while infectious virus particles are still present in cell lysates [5]. In comparison with many other virus systems studies on morphogenesis of pestiviruses at the ultrastructural level are impeded by the low rate of virion synthesis. Only once pestivirus particles have been demonstrated inside ER structures in infected cells by transmission electron microscopy (TEM) [6].

In the focus of our study was the localisation of pestivirus assembly and the subsequent maturation at the ultrastructural level in infected culture cells. First an infection model based on virus growth kinetics and quantitative real-time RT-PCR was established. Conventional embedding techniques (epoxid resin) were applied as well as the Tokuyasu technique [7, 8]. One aspect of our project concerned alterations of intracellular membranes after infection with pestiviruses as it has been shown for cells infected with Kunjin virus of the genus *Flavivirus* [9] and in cells expressing parts of the Hepatitis C Virus polyprotein [10]. Interestingly throughout the intensive analysis of

A. Aretz, B. Hermanns-Sachweh, J. Mayer (Eds.): EMC 2008, Vol. 3: Life Science, pp. 247–248,
DOI: 10.1007/978-3-540-85228-5_124,

infected cells no significant alterations of intracellular membranes were observed. Anyhow the ultrastructural examination revealed new aspects of virion localisation as well as virus release from infected cells and substantiates the current model of pestiviral morphogenesis. Furthermore immunogold labeling of ultrathin cryosections was performed. Using this technique we could demonstrate intracellular distribution of pestivirus proteins on the ultrastructural level for the first time.

1. F. Weiland, E. Weiland, G. Unger et al., J. Gen. Virol. **80** (1999): p. 1157-1165.
2. Y. Zhang, J. Corver, P. R. Chipman et al., EMBO J. **22** (2003): p. 2604-2613.
3. N. Branza-Nichita, D. Durantel, S. Carrouée-Durantel et al., J. Virol. **75** (2001): p. 3527-
4. D. Durantel, S. Carrouée-Durantel, N. Branza-Nichita et al., Antimicrob. Agents Chemother. **48** (2004): p. 497-504.
5. Macovei, N. Zitzmann, C. Lazar et al., Biochem Biophys Res Commun **346** (2006): p. 1083-90.
6. E. W. Gray and P. F. Nettleton, J. Gen. Virol. **68** (1987): p. 2339-2346.
7. K. T. Tokuyasu, J Cell Biol **57** (1973): p. 551-65.
8. W. Liou, H. J. Geuze and J. W. Slot, Histochem Cell Biol **106** (1996): p. 41-58.
9. E. G. Westaway, J. M. Mackenzie, M. T. Kenney et al., J. Virol. **71** (1997): p. 6650-6661.
10. D. Egger, B. Wölk, R. Gosert et al., J. Virol. **76** (2002): p. 5974-5984
11. This project was supported by the "Stiftung der Eheleute Dr. med. vet. Hans-Joachim und Gertrud Engemann" of the Justus-Liebig-Universität Giessen. We thank Dr. M. Hardt (EM Unit of the ZBB, Justus-Liebig-Universität Giessen) for providing the technical facilities for cryoultramicrotomy.

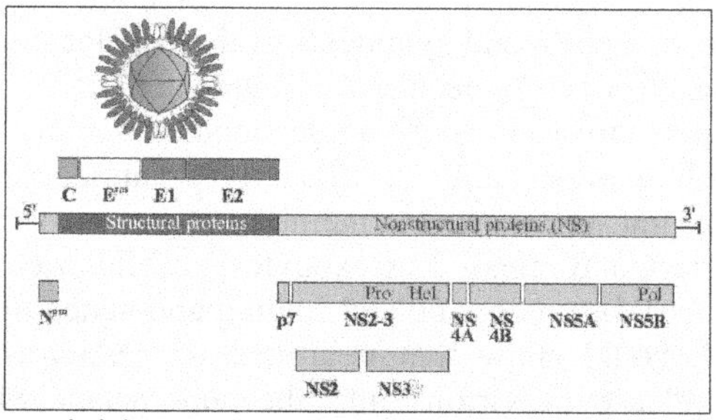

Figure 1. Model of a pestivirion and genome organisation of pestiviruses

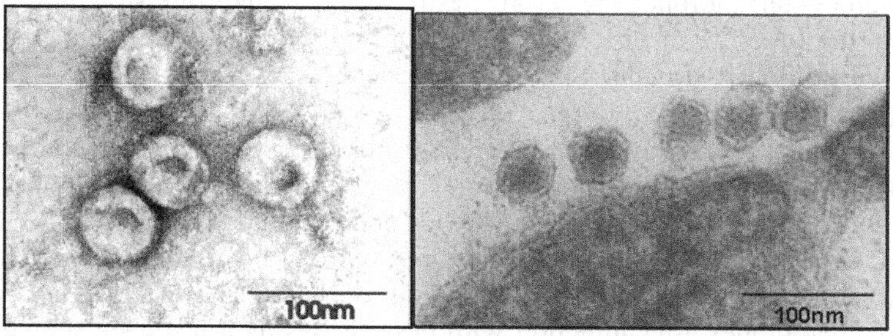

Figure 2. Electron micrographs of pestiviruses. **Left:** Negative staining of purified pestivirions. **Right:** Thin section of infected MDBK cells with extracellular virions.

Analysis of radiation-induced bystander effects using high content screening

B. Dieriks[1], W. De Vos[1] and P. Van Oostveldt[1]

1. Molecular Biotechnology, Cell biology and Molecular Cytology, Faculty of Bio-engineering University Ghent, Coupure Links 653, 9000, Belgium

birger.dieriks@ugent.be
Keywords: γH_2AX, bystander effect, high content screening

When cells are exposed to (ionising) radiation there is a rapid phosphorylation of a minor nucleosomal histone protein, H_2AX, at the sites where double stranded breaks (DSB) occur. This phosphorylation is one of the earliest events in the repair cascade and extends over several mega base pairs surrounding the break. Nowadays it is generally accepted that the formation of γH_2AX functions as a signal enhancer. Using immuno histochemistry we can visualise this phosphorylation as foci in the nucleus, where each foci represents a DSB [1].

For our research we use normal human primary fibroblasts (NHDF's) to study the so called radiation-induced bystander effects which refer to the responses induced in non-irradiated cells, when neighbouring cells are irradiated. Although the exact pathways of transmission are yet to be determined, studies have shown that gap junction-mediated transport and secretion of soluble extracellular factors play an important role [2].

To exclude variation we first tried synchronisation of the fibroblasts using nocodazole or aphidicolin. Our attempts did not produce the desired synchronisation level. In addition, recent reports doubt the effectiveness of these products in cell synchronisation [3]. To resolve this problem, we used high content screening of cells together with specific cell cycle markers. One of these markers is 5-bromo-2-deoxyuridine (BrdU). BrdU, a synthetic nucleoside, is an analogue of thymidine that can be incorporated in replicating cells and specifically label S-phases [4].

Cells are cultured on membrane inserts, with a pore size of 0,4µm allowing soluble factors to pass but preventing the cells to interchange. These cells are irradiated with different doses and subsequently placed together with NHDF that are grown on cover glasses (see figure 1). Depending on the objectives BrdU is added 20-40 minutes before fixation.

We found a differential pattern for γH_2AX that we could specifically link to the cell cycle. During the S phase γH_2AX is significantly more induced than during other phases of the cell cycle (see figure 2). This is probably due to the increased vulnerability caused by the unwinding of DNA during replication.

1. S.H.Macphail, J.P.Banath, T.Y.Yu, E.H.Chu, H.Lambur, P.L.Olive, Int.J.Radiat.Biol. **79** (2003) P. 351-358.
2. H.Yang, N.Asaad, K.D.Held, Oncogene **24** (2005) p. 2096-2103.
3. S.Cooper, G.Iyer, M.Tarquini, P.Bissett, Cell Tissue Res. **324** (2006) p.237-242.
4. R.T.O'Keefe, S.C.Henderson, D.L.Spector, J.Cell Biol. **116** (1992) p.1095-1110.

A. Aretz, B. Hermanns-Sachweh, J. Mayer (Eds.): EMC 2008, Vol. 3: Life Science, pp. 249–250,
DOI: 10.1007/978-3-540-85228-5_125,

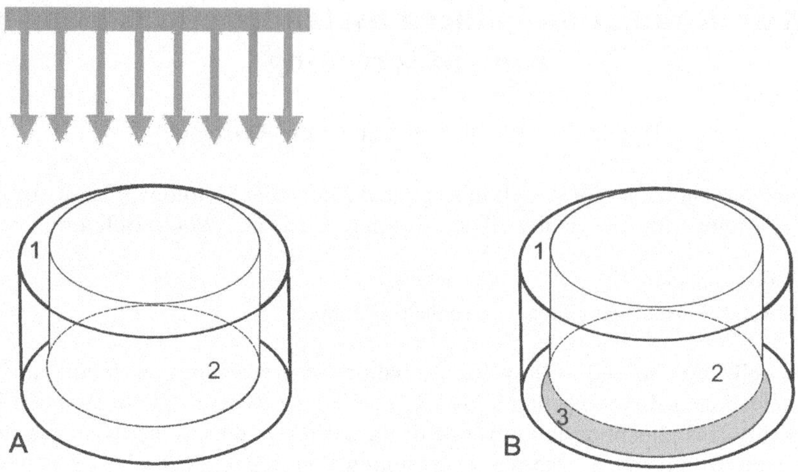

Figure 1. experimental setup A: NHDF's are cultured in a 12well plate (1) on membrane inserts (2) and subsequently irradiated at ambient temperature with 250kV X-rays, 13mA and 1mm Cu filter. B: The membrane inserts were transferred to a new well so that they would be in close proximity with non-irradiated NHDF's on cover slips (3)

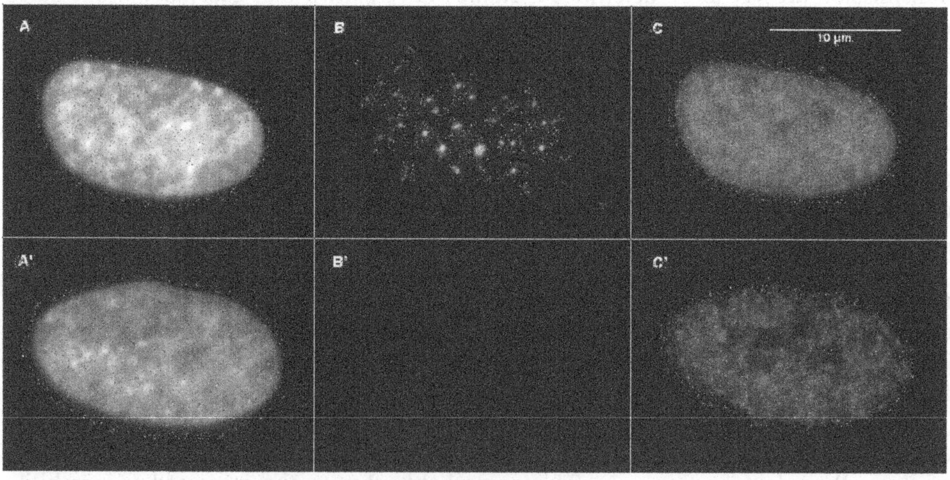

Figure 2. Immunostaining of NHDF pulsed labelled with 20μM BrdU for 20min followed by 15min UV irradiation. DAPI, BrdU, γH_2AX in an S phase (respectively A, B and C) and a non S phase cell (A', B' and C').

Electron microscopic investigations on normal and dexamethasone applied rat placentas

H. Er[1], E.T. Korgun[2], Ç. Çelik-Özenci[2], M. Saçan[2], M. Asar[2]

1. Akdeniz University Medical Faculty Electron Microscopy Unit (TEMGA)
2. Akdeniz University Medical Faculty Histology and Embryology Department

hakaner@akdeniz.edu.tr
Keywords: IUGR, placenta, rat, TEM

During the growth of foetus in prenatal life glucocorticoids are very important, especially in cell maturation and differentiation events. Today it is known that glucocorticoids lead to intrauterine growth restriction (IUGR) by antiproliferative effects [1].

In this project, we aimed to define the effects of dexamethasone -a glucocorticoid- on placental development by transmission electron microscopy (TEM). We composed two groups of pregnant rats as control (8 female) and experiment (8 female).

In this study in experiment group we have applied same doses of dexamethasone acetate as reported before [2]. In order to compose IUGR on rats via dexamethasone, every pregnant rat in experiment group was injected a bolus dose of 100μg dexamethasone acetate subcutaneous on the 13th day of pregnancy and 200μg/kg between 14th-19th days of gestation. Between the 13th-19th days of gestation vehicle was applied to control rats. On the 20th day of gestation blood samples were taken from rats under ether anaesthesia in order to measure plasma glucocorticoid levels. Then dissected placentas and foetuses were weighed. Electron microscopic samples were fixed in 4% glutaraldehyde and following in 1% O_sO_4.

It was determined that dexamethasone leads to statistically significant weight loss on rat placenta and foetus ($p<0,005$) “Figure 1”. Both in labyrinthine and junctional zones apoptotic cells were observed by TEM “Figure 2”.

In conclusion, we can say that dexamethasone may have an antiproliferative effect on rat placenta by stimulating apoptosis.

1. Hahn, T., et al., “Placental glucose transporter expression is regulated by glucocorticoids.”, J Clin Endocrinol Metab, 1999. 84(4): p. 1445-52
2. Sugden, M.C. and M.L. Langdown, “Possible involvement of PKC isoforms in signalling placental apoptosis in intrauterine growth retardation.”, Mol Cell Endocrinol, 2001. 185(1-2): p. 119-26
3. We thank to Sibel Ozer and Arife Demirtop for technical assistance, also thank to Akdeniz University Medical Faculty Electron Microscopy Unit (TEMGA) and Akdeniz University Medical Faculty Histology and Embryology Department for enabling us to held this experiment no their laboratories. This study was supported by Akdeniz University Research Fund (Grant Number: 2005.02.0122.003)

A. Aretz, B. Hermanns-Sachweh, J. Mayer (Eds.): EMC 2008, Vol. 3: Life Science, pp. 251–252,
DOI: 10.1007/978-3-540-85228-5_126,

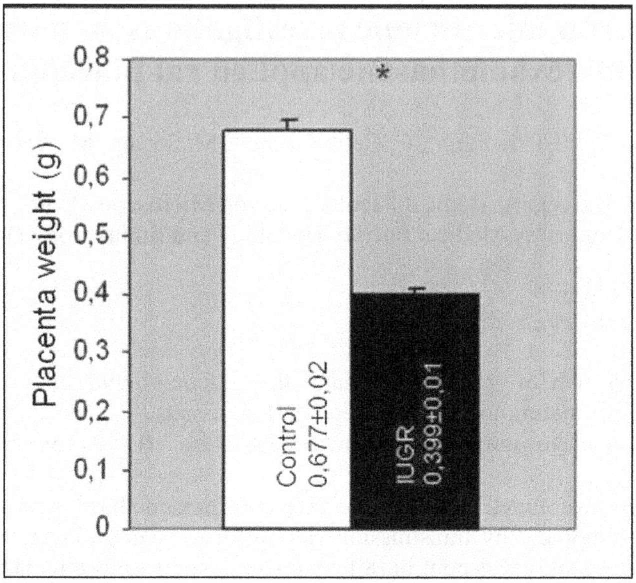

Figure 1. Weight histogram of control and experiment groups' rat placentas (* p<0,001).

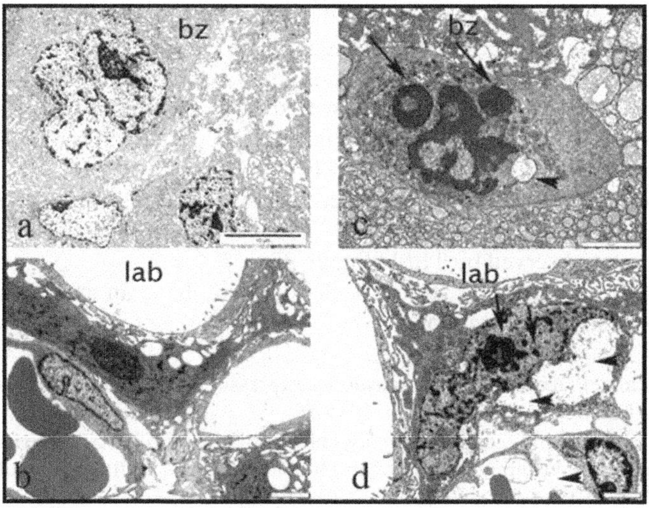

Figure 2. Electron micrographs of some cells of labyrinth (lab) and junctional (bz: basal zone) zones of control **(a,b)** and experiment **(c,d)** groups' placentas. In **a** and **b** images normal cell structure was seen but in **c** nuclear fragmentation (arrow) and deposition (arrow head) in **d** chromatin condensation (arrow) and cytoplasm with vacuoles (arrow head) were seen.

Changes in dental enamel birefringence after CO_2 laser irradiation through fluoride gel –a pilot study

M. Esteves-Oliveira[1,2], C. P. Eduardo[2], F. Lampert[1], C. Apel[1]

1. Department for Conservative Dentistry, Periodontology and Preventive Dentistry (ZPP), RWTH Aachen University, Aachen, Germany.
2. Restorative Dentistry Department, School of Dentistry, University of São Paulo (USP), São Paulo, Brazil.

marcella@usp.br
Keywords: CO_2 laser, dental enamel, birefringence, light polarized microscopy

The benefits of the CO_2 laser irradiation for enamel caries prevention are well established[1, 2].Also the enhancement of this effect may be expected when the laser irradiation is combined with topic fluoride application.[3]. Although it is well known that laser irradiation with or without fluoride may render enamel more acid-resistant , the modifications caused in enamel structure after CO_2 laser irradiation are not sufficiently understood. Therefore the aim of the present study was to evaluate the structural changes occurring in enamel surface and subsurface after CO_2 laser irradiation through fluoride gel. Ten blocks of 4x4 mm obtained from bovine enamel were used. The surfaces were polished using Al_2 O_3 papers and divided in 2 groups: control (GC), not treated and laser group (GL) treated with fluoride and laser. The surfaces in group GL were covered with fluoride gel (NaF 1,23%) for 4 minutes and after that they were irradiated with a CO_2 laser (10.6 μm) with 14 mJ, 0.3 J/cm^2, 154 Hz and 2308 overlapped pulses of 10 μs duration. All the samples were than submitted to an 8-day pH-cycling regime in order induce caries lesion formation. Thin slices of 100 μm were prepared and morphological changes and caries lesion formation were assessed with a light polarized microscope. The micrographs of the laser-treated group (Fig 1, 2 and 3) showed birefringence changes below the irradiated surface and no carious lesion formation. In the control, carious lesions were observed and there was no alteration in the enamel structural aspect (Fig.4). Based on the results of the present in vitro study it may be concluded that the CO_2 laser irradiation with these settings used causes changes in enamel structure and birefringence and can indeed inhibit caries lesion formation.

1. 1. J. D. B. Featherstone, N. A. Barrett-Vespone, D. Fried, Z. Kantorowitz and W. Seka, J. Dent. Res **77** (1998), 1397-1403.
2. 2. C. Y. Hsu, T. H. Jordan, D. N. Dederich and J. S. Wefel, J.Dent. Res **79** (2000), 1725-30.
3. 3. L. K. Rodrigues, M. Nobre Dos Santos and J. D. Featherstone, J Dent Res **85** (2006), 617-21.

A. Aretz, B. Hermanns-Sachweh, J. Mayer (Eds.): EMC 2008, Vol. 3: Life Science, pp. 253–254,
DOI: 10.1007/978-3-540-85228-5_127,

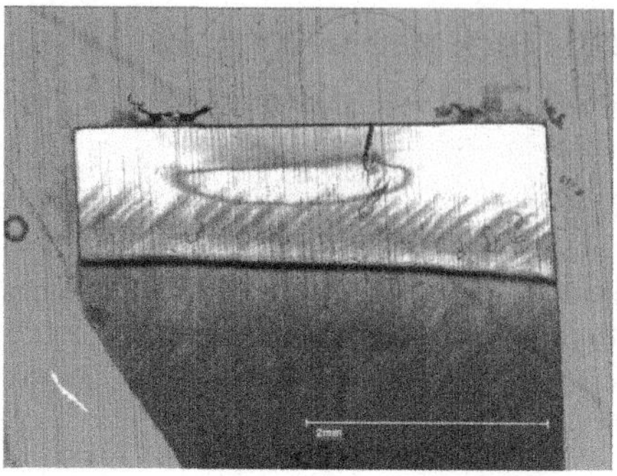

Figure 1. Slice of a bovine tooth showing enamel and dentin (upper and lower part). A CO_2 laser was used to irradiate the center of the enamel surface through a fluoride gel. The changes in the birefringence at the enamel subsurface can be seen only under the irradiated area, suggesting structural modifications in the tissue.

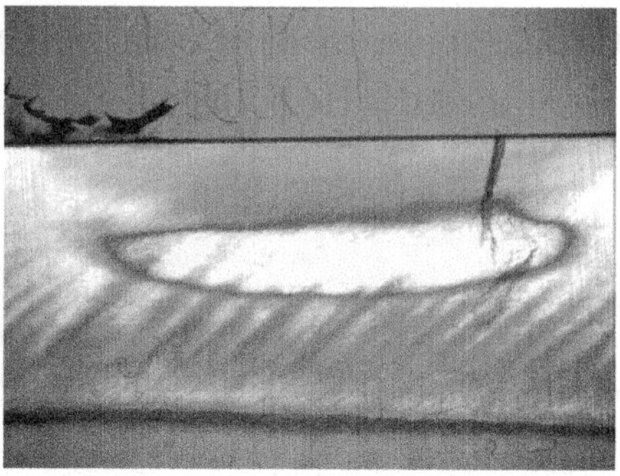

Figure 2. The affected area in the subsurface of enamel observed in higher magnification, where it is possible to see that the gradients of alterations formed in the tissue are similar to the profile of energy distribution inside the laser beam. Therefore the intensity of the tissue modification is higher in the center then in the margins of the irradiated area.

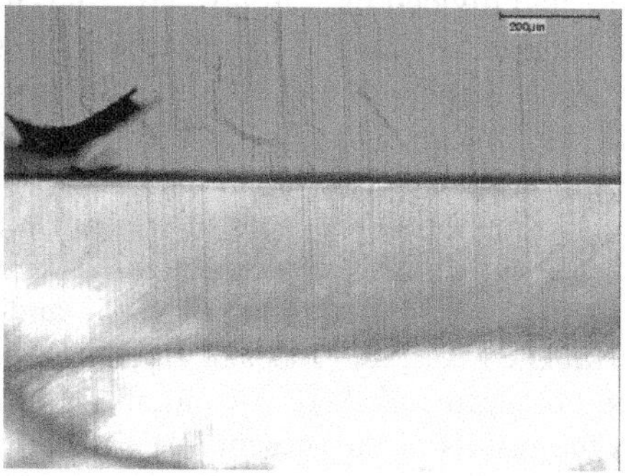

Figure 3. The same sample observed in higher magnification, showing that these changes in enamel birefringence might have turned the enamel structure more resistant to acid dissolution, since no signs of caries lesion formation are observable near the surface.

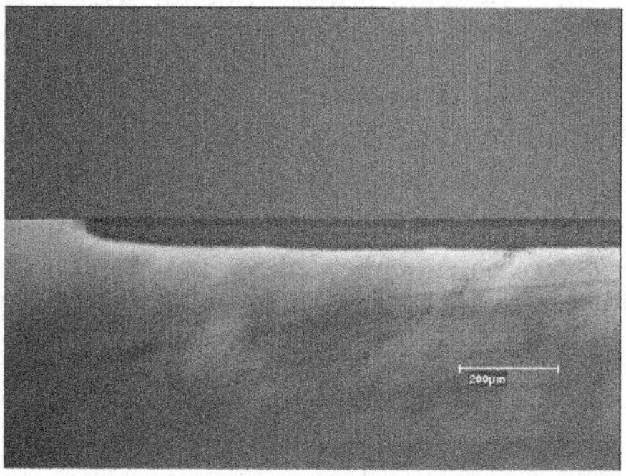

Figure 4. Control sample, which was not irradiated, showing subsurface caries lesion formation and no alteration of enamel morphology

SEM and light microscopic examination and morphometric analysis of temporal changes in intimal and medial connective tissue component of human carotid arteries

Evirgen O.[1], Zeyrek T.[2], Apaydin N.[3], Ilkay H.[1], Semiz O.[4]

1. Ankara University, School of Medicine, Department of Histology and Embryology, Ankara, Turkey
2. Dr. Zekai Tahir Burak Maternity Hospital, Ankara, Turkey
3. Ankara University, School of Medicine Department of Anatomy, Ankara, Turkey
4. Sakarya University, School of Health, Sakarya, Turkey

oya.evirgen@gmail.com
Keywords: conduit arteries, ageing, SEM, elastic lamella

INTRODUCTION

Major changes occur in elastic artery wall structure, with age [1, 2]. Barrier function fails due to endothelial dysfunction; thus, LDL cholesterol collects under subendothelium and atherosclerotic lesions develop. Medial changes include mechanical and enzymatic break down of elastic lamellae, increase in collagen content, decrease in vascular smooth muscle tonus, and degeneration of smooth muscle cells. These changes result in intimal-medial thickening and atherosclerosis, which lead to cardiovascular complications [3]. This study aims to examine intimal and subintimal changes, arrangement of elastic lamellae in human carotid arteries obtained from cadavers, and define changes with age.

MATERIALS and METHODS

Seventeen cadavers which were grouped into 3 decades; G1: 30-39 years, G2:40-49 years, G3: >50 years. Common carotid artery segments were prepared for light microscopy, stained with Haematoxylin-Eosin and orcein-indigocarmin (Fraenkel's method) for differentiation of elastic lamellae and connective tissue. Digital photomicrographs were taken and morphometric measurements were made with Adobe® Photoshop CS2® image analyzing software. For scanning electron microscopic study, two consecutive segments from all samples were macerated, postfixed and coated with gold-palladium, observed under Leo438VP (England) SEM and digital images were taken.

RESULTS and DISCUSSION

G1 showed intimal thickening and foamy cell infiltration and was thinner than G2 and G3. IEL was continuous. In G2, cholesterol crystals were embedded in atherom plaques. Atherosclerotic lesions were seen as crescents, sometimes filling more than half of the lumen, in G3 (fig. 1). IEL showed occasional fragmentations and duplications at sites of atherom plaques in G2 and G3. IEL surface was smooth and contained few oval fenestrations in G1 and G2 whereas G3 had many polymorphous and larger fenestrae (fig. 2) [4]. This indicates the failure in remodelling of elasic

A. Aretz, B. Hermanns-Sachweh, J. Mayer (Eds.): EMC 2008, Vol. 3: Life Science, pp. 255–256,
DOI: 10.1007/978-3-540-85228-5_128,

lamellae. Intima, media and the whole vessel wall elastic fibre content decreased with age but there was no statistically significant difference between groups, which suggests that elasticity is preserved during ageing, due to the stable nature of elastin [5]. While intimal area was increasing with age, medial area measurements revealed no difference between groups. This indicates that intimal thickening contributes more to intima-medial measurement of elastic arteries by USG in order to assess cardiovascular/cerebrovascular risk factors [6, 7].

1. S.E. Greenwald, J Pathol **211** (2007), p.157-172
2. E.G. Lakatta, Heart,Lung and Circulation **11** (2002), p.76-91
3. T. Jogestrand, O. Eiken, J. Nowak, Clin Physiol & Func Im **23** (2003), p.134-137
4. M. Raspanti, M. Protasoni et al, Micron **37** (2006), p.81-86
5. F.H. Sims, Pathology **32** (2000), p.171-175
6. H.I. Manninen, H.Räsänen, et al , Radiology **206** (1998), p.65-74
7. J. Persson, J. Formgren, et al, Arteriosclerosis and Thrombosis **14** (1994), p.261-264

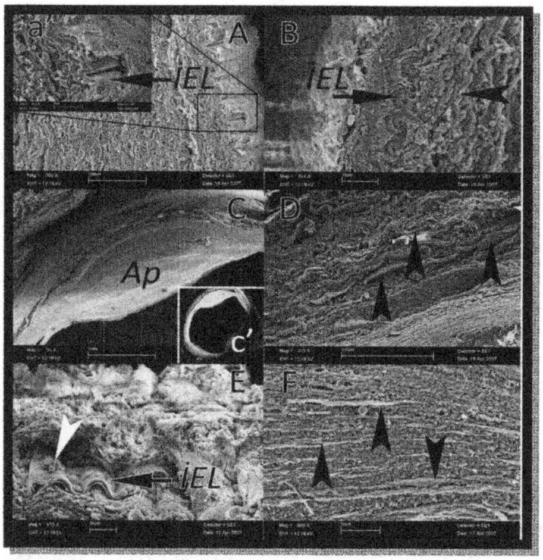

Figure 1. Scanning electron microscopic appearance of 3 groups. Group 1: A, a', B. Group 2: C, c', D. Group 3: E, F. **IEL**: Internal elastic lamina **Ap**: Atherom plaque, *: Foamy cell (macrophage) White arrowhead: Cell infiltration, Black arrowheads: Elastic lamellae,

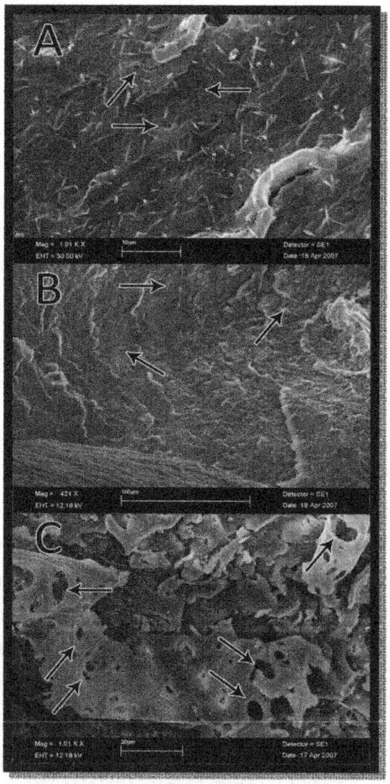

Figure 2. Scanning electron microscopic appearance of fenestrae in Group 1: A, Group 2: B and Group 3: C. Thin arrows indicate fenestrae

Ultrastructure of human amniotic membrane covering villous chorion and smooth chorion in preeclampsia

Evirgen O.[1], İlkay H.[1], Akkuş S.[1], Semiz O.[2], Heper Okçu A.[3], Aytaç R.[4]

1. Ankara University, School of Medicine, Department of Histology and Embryology, Ankara, Turkey
2. Sakarya University, School of Health, Sakarya, Turkey
3. Ankara University, School of Medicine Department of Pathology, Ankara, Turkey
4. Ankara University, School of Medicine Department of Obstetric and Gynecology, Ankara, Turkey

oya.evirgen@gmail.com
Keywords: Placenta, amniochorionic membrane, basement membrane

INTRODUCTION

Amniotic membrane consists of amniotic epithelium, avascular compact layer and fibroblastic layer. The fusion of amniotic membrane with chorion occurs in 12th week of pregnancy. Amniotic fluid and uterin wall deep layers nourishes amnion epithelium. Basal face infoldings extend to basal lamina which is rich in proteoglicans and heparan sulphate that acts as a permeability barrier for macromolecules. Preeclampsia is a maternal disorder characterised by hypertension and proteinuria after 20th week of gestation.Decreased placental perfusion causes hypoxia and release of cytotoxic factors. [1, 2] The aim of this study was to evaluate the possible structural changes of amniochorionic membrane taken from the area of placental disc (central) and from the free reflected membranes (peripheric) of preeclamptic placentas by light and electron microscopy.

MATERIAL and METHOD

Placentas were obtained upon the preeclemptic women consent after normal vaginal deliveries. The amniochorionic tissue randomly collected from the the placental disc and from the free reflected membranes were fixed in Bouin's solution.. 5 μm paraffine sections were stained with standard hematoxylene–eosin and PAS (Periodic Acid Schiff). For morphometric analysis basement membrane thickness,amnion epithelial cell height and amniochorionic mesoderm thickness were measured from the placental disc and from the free reflected membrane in each placenta using an eyepiece ocular micrometer adapted light microscopy under oil immersion .For electron microscopy tissue samples fixed in 2.5% gluteraldehyde were processed and embedded in araldite.. Ultra thin sections were observed under LEO 906 E Transmission electron microscope.

RESULTS

By transmission electron microscopy, basal lamina of the placental disc amnion epithelium was found to be thick and condensed. Within the basal lamina a meshwork of branched podocyte like cell projections extends deeply (fig. 1). Controversely in reflectedperipheral amnion basal lamina was thin and loosely organized that lamina lucida and lamina densa could be distinguished clearly. Short and numereous basal cell

A. Aretz, B. Hermanns-Sachweh, J. Mayer (Eds.): EMC 2008, Vol. 3: Life Science, pp. 257–258, DOI: 10.1007/978-3-540-85228-5_129,

surface invaginations containing hemidesmosomes were observed (fig. 2) . Epithelial heights were 8,26μm in placental and 7,99μm in reflected amnion, amniochorionic mesoderm thickness was increased in placental region (51.88μm) compared to reflected amnion (41,6 μm)

1. E.C. C. Castro, M.A. Reis, V.P.A. Teixeira European J Obstetrics & Gynecology and Reproductive Biology **114** (2004), p.171-176
2. G. D. Bryant-Greenwood, Placenta **19** (1998), p.1-11
3. S. Matsubara, R. Iwasaki Acta Histochem. Cytochem **37** (2004), p.407-409

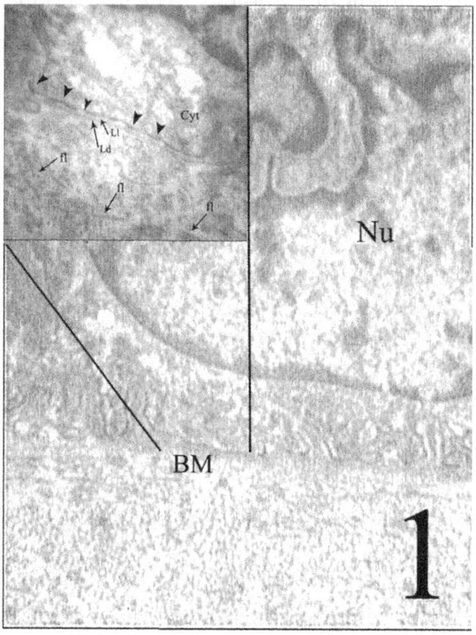

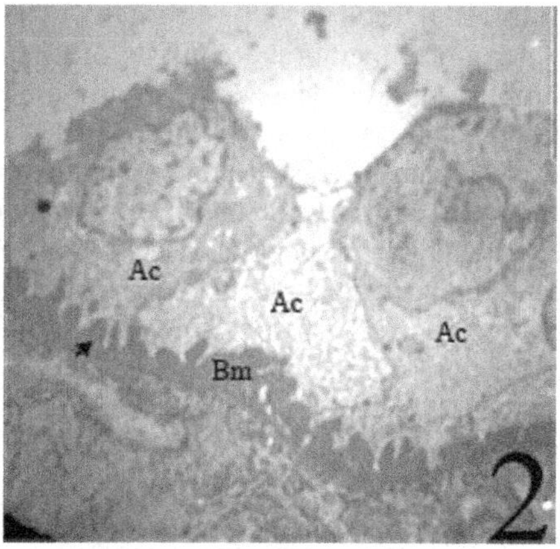

Figure 1. TEM of reflected peripheric amnion membrane **Nu**:amniotic epithelium nucleus,**BM:** basal lamina and underlying extracellularmatrix. X4.646 **İNLET:** higher magnification X10.000 arrowheads: hemidesmosomes, cyt: cytoplasm, **Lı:** lamina lucida, **Ld:** lamina densa of basal lamina, **fl:** cross striations are visible on fibrils in underliying extracellular matrix

Figure 2. TEM of placental disc central amnion **Ac:** threeamniotic epithelial cells two with darker cytoplasm and one with lighter cytoplasm, **Bm:** basal lamina, **arrow:** podocyte like cell projections extending deeply in a thick and condensed basal lamina. X 1000

Changes in Intracellular Sodium, Chlorine, and Potassium Content in Hematopoietic Cells after Hypotermic Storage

E. Fernández-Segura[1], F. Arrebola[1], M.A. Cubero[1], F.J. Cañizares[1], M.A. Robles[1], P. Navarrete[2]

1. Department of Histology, Faculty of Medicine, University of Granada, E-18071 Granada, Spain.
2. Critical Care and Emergency Department, "Virgen de las Nieves" University Hospital, Granada, Spain.

efsegura@ugr.es
Keywords: cell death, hypothermic preservation, electron probe X-ray microanalysis

Hematopoietic precursor cells (HPCs) hold tremendous potential in the emerging field of cell-based therapies. The purpose of this therapy is to replace, repair or enhance the biological function of damaged tissue or organs. This can be achieved by the transplantation of cells in sufficient number and quality to restore function. Transplantation of HPCs requires the ability to preserve cells. Cell preservation technology involves both hypothermic methods for short-term storage and cryopreservation for long-term storage. However, both methods they are not exempt to induce cell damage and therefore to diminish the engraftment kinetics. Oxidative stress, mechanical injury due to ice crystal formation, altered physical properties of cellular structures, osmotic injury, and disturbed ion homeostasis are responsible for cell damage. However, in spite of the not clear correlation with capacity of engraftment, quality control assays currently used are based on methods revealing the loss of integrity of plasma membrane or ex vivo expansion capacity through clonogenic assays. Based on these premises, we propose to utilize the elemental composition as an indicator of cell viability and quality previous to cell transplantation.

To address the above issue, we used electron probe X-ray microanalysis (EPXMA) which is an electron microscopy technique that allows multiple elements to be measured simultaneously in single, individual cells. In addition, we evaluated cell death during hypothermic-thaw process by lactated dehydrogenase release, expression of cleaved caspase-3 and poly(ADP)-ribose polymerase and DNA electrophoresis.

As experimental model, U937 cells were cultured in RPMI 1640 medium supplemented with 10% heat-inactivated bovine fetal serum and 2 mM L-glutamine. Cells were maintained at 37 ºC in a humidified atmosphere. The cells were washed with RPMI 1640 and were then incubated at 4º C. After periods of cold incubation, the cells were rewarmed to 37º C in an incubator containing a humidified atmosphere of 95% air/5% CO_2. X-ray microanalysis was performed on whole freeze-dried U937 cells in a Philips XL30 scanning electron microscope. The concentration of elements in the specimen was obtained with the peak-to-local-background (P/B) ratio method. Cell element concentrations were obtained with reference to 20% dextran standards containing known amounts of inorganic salts.

A. Aretz, B. Hermanns-Sachweh, J. Mayer (Eds.): EMC 2008, Vol. 3: Life Science, pp. 259–260, DOI: 10.1007/978-3-540-85228-5_130,

Our data suggested that EPXMA is a sensitive and reliable quantitative method that can be used to evaluate cell viability and quality in cell-based therapies

1. This work was supported by Fondo de Investigaciones Sanitarias (FIS), Instituto de Salud Carlos III (Contract grant number FIS PI061836).

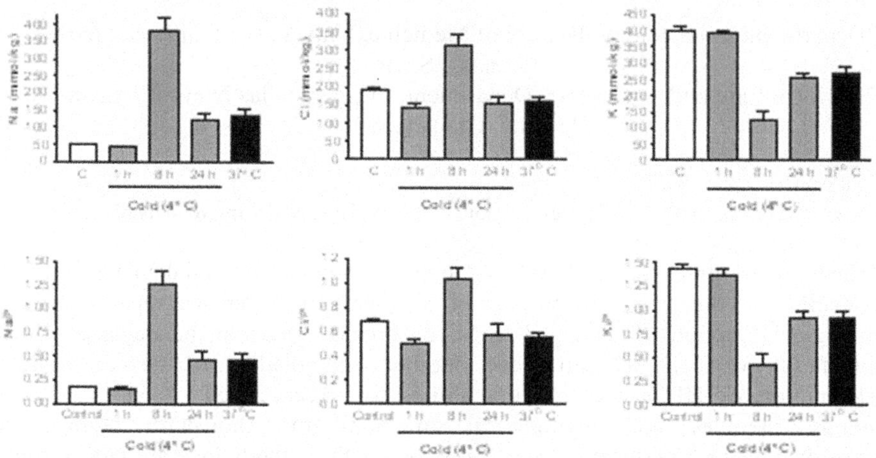

Figure 1. Effect of hypothermic preservation (4° C) on sodium (Na), chlorine (Cl) and potassium content in U937 cells. Cells were maintained at 37 °C in a humidified atmosphere. The cells were washed with RPMI 1640 and were then incubated at 4° CO_2.

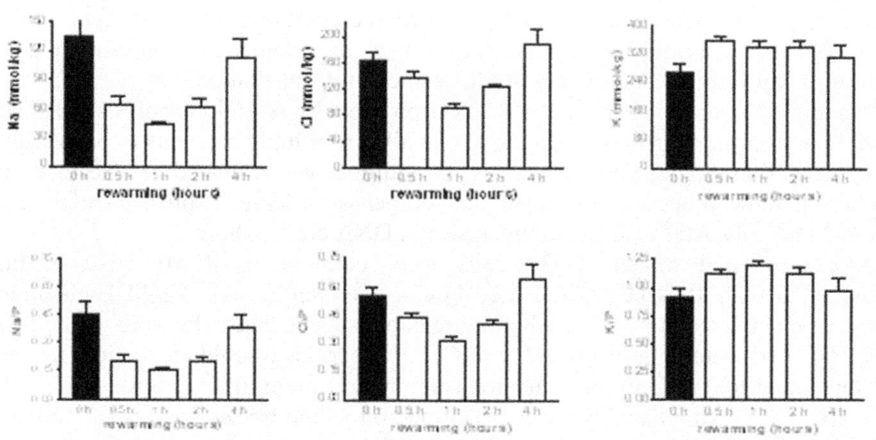

Figure 2. Effect of rewarming (37° C) on sodium (Na), chlorine (Cl) and potassium content in U937 cells. After periods of cold incubation, the cells were rewarmed to 37° C in an incubator containing a humidified atmosphere of 95% air/5% CO_2

Requirements for a bright future of electron microscopy in the rapid laboratory diagnosis of infectious diseases

H. R. Gelderblom

Robert Koch-Institut, Nordufer 20, D-13353 Berlin, Germany

gelderblomh@rki.de
Keywords: infectious diseases, rapid diagnosis, diagnostic EM

Pan-microbial nucleotide assays, multiplex-PCRs, spectroscopic techniques etc. are changing the conventional laboratory diagnosis. Such "modern" techniques can be run with high sample through-put and sensitivity, in an automated, quality controlled way. In view of these advantages, the question arises for the future of electron microscopy (EM) in the lab diagnosis of infectious diseases.

With the direct visualization of viral pathogens, 70 years ago, also diagnostic EM started [1]. Visualization does not require particularly demanding techniques like cryo-EM. Mainly transmission EM of ultrathin sections is performed after resin embedding of, e.g., biopsies and cell cultures, and much simpler and more often, negative-contrast staining of particle suspensions to provide for the required thin samples [2, 3].

Different from alternative lab methods, EM depends on high degrees of motivation, skills and expertise. When performed properly, EM excels by speed, high resolution (in praxi down to 2 nm) and "open view": i.e. EM as a „catch all"- method detects all agents present in a sample, down to the smallest viruses - also double and unexpected infections. Diagnostic EM does not depend on agent-specific reagents. Particle size and fine structure are the criteria to assign the agent observed directly to one of the 31 vertebrate-specific virus families [2, 3]. It is not confined to virology, diagnostic bacteriology and parasitology will benefit, too [2-5]. EM is applied to scrutinize cell cultures, blood, biopsies, skin lesions (Figure 1) and body secretions like urine, stool, liquor, tears, throat-washings. From the arrival of the sample, a negative stain-based diagnosis requires seldom more than 20 min [2, 3]. This speed is particularly required in emergencies, when clinical symptoms do not point to a particular etiology or when bioterrorism is suspected. EM excludes or confirms rapidly the presence of a specific pathogen. As a differential diagnosis, EM often suffices to start antiviral therapy and/or isolation measures. If a type-specific diagnosis is required, an EM diagnosis focuses the fine diagnosis rapidly on the relevant group/family of agents resulting in a clinically significant gain of time. E.g., a rapid diagnosis is required for a timely antiviral therapy or to defuse anxieties and panic potential if a bioterrorist attack is suspected. The "open view" helps also in quality control, pathology and dermatology [6]. Diagnostic EM should be used front-line to make full use of its advantages [7, 8].

Limiting can be the need for particle concentrations >10^6 particles/ml, though many infections contain much higher concentrations. Moreover, enrichment techniques often help for readily detectable particle numbers on the grid [2, 3, 7]. If negative staining does not lead to satisfying results, thin section EM should be applied using rapid

A. Aretz, B. Hermanns-Sachweh, J. Mayer (Eds.): EMC 2008, Vol. 3: Life Science, pp. 261–262, DOI: 10.1007/978-3-540-85228-5_131, © Springer-Verlag Berlin Heidelberg 2008

embedding routines [9]. Diagnostic EM is supported by telemicroscopy [10] and immuno-EM techniques. Provided specific antisera are available, an isolate can be typed by EM – in less than one hour [2].

Internal and external means can guarantee the future of diagnostic EM. Conditions are an appropriate staff and lab organisation, a decent sample through-put and "continuous education" by lab courses, External Quality Assurance scheme and workshops [11, 12; detailed info at http://www.rki]. The EM lab should develop links to academia, hospitals and Public Health in order to utilize and propagate the unique advantages of and raise interest in EM. Thus, based on expertise and motivation, diagnostic EM also in the future will exert competitive performance within the conventional and molecular-genetic diagnostic armament [2, 11, 13].

1. DH. Krüger, P. Schneck and HR. Gelderblom, LANCET **355** (2000), 1713-1717
2. SS. Biel and HR. Gelderblom, J Clin Virol **13** (1999), 105-119
3. PR. Hazelton and HR. Gelderblom, Emerg Inf Dis **9** (2003), 294-303
4. Curry, H. Appleton and B. Dowsett, Micron **37** (2006), 91-10
5. HR. Gelderblom and N. Bannert in "Europ. Micr. Soc. 2005", ed. P. Hawkes, (2005) 47-54
6. HR. Gelderblom, N. Bannert,W. Muss and CR. Madeley, Brit J Derm **154** (2006), 158-560
7. HR. Gelderblom, Bundesgesundheitsbl Gesundheitsforsch **46** (2003), 984-988
8. CR. Madeley, Lancet **361** (2003), 97-98
9. M. Laue, B. Niederwöhrmeier and N. Bannert, J Microb Meth **70** (2007), 45-54
10. JA. Schröder, E. Voelkl and F. Hofstaedter, Ultrastruct Pathol **25** (2001), 307-310
11. SS. Biel and D. Madeley, J Clin Virol **22** (2001), 1-9
12. SE. Miller, Ultrastr Pathol **27** (2003), 133-140
13. Kurth and A. Nitsche, Future Med **2** (2007), 467-479
14. http://www.rki.de/EN/Content/Institute/DepartmentsUnits/NRC/CONSULAB/consulab__node.html

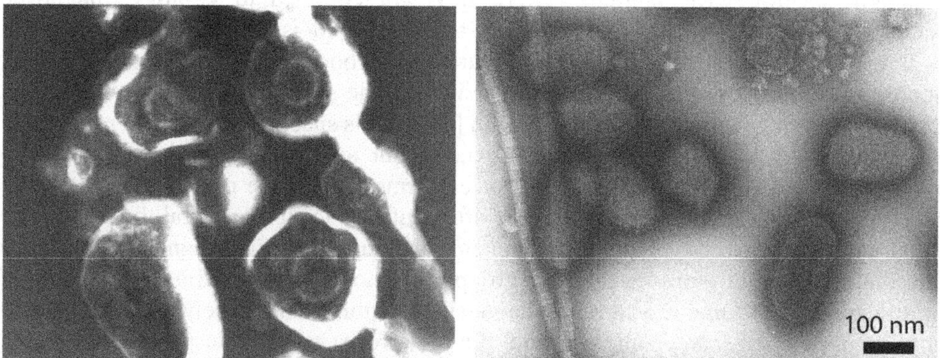

Figure 1. TEM differential diagnosis of febrile vesicular rash diseases. Negative staining of skin lesions using 1% uranyl-acetate. Herpes- (left), parapoxviruses (right).

Changed thickness of zona pellucida as a result of stimulated ovulation

B. Jodłowska-Jędrych[1], M. Jędrych[2], W. Matysiak[1], K. Czerny[1]

1. Departament of Histology and Embryology, Medical University of Lublin, 20-080 Lublin, Poland
2. Department of Mathematics and Biostatistics, Medical University of Lublin, Poland

barbara.jjedrych@am.lublin.pl
Keywords: zona pellucida, stimulation of ovulation

The zona pellucida (ZP) is the extracellular matrix that surrounds oocytes (mammalian eggs). ZP is composed of three sulphated glycoproteins [1,2]. It protects the oocyte and embryo in its early stage of development [3]. It is an essential element of fertilization [4].

In this study we present measurements of ZP in stimulated ovulation in model females of Wistar rats.

The test used two-month-old females of 250 – 300 grams in weight. The study involved only those females that had two regular 4- day estrus cycles, determined by means of vaginal smears. On the third day of the cycle the animals were administered chorionic gonadothropin hCG (Choragon 1500) in a single intramuscular injection of 0,1 ml. The rats from group C1 were given 7,5 IU, C2 received 15 IU, and C3 received 22,5 IU. The animals from the control group were administered physiological saline (0,1 ml). 14 hours after the injection the animals were destroyed by decapitation, then pairs of ovaries were obtained and prepared for inspection. Patches of 0,5 μm in thickness, were morphometrically analysed using MultiScan software. The results were statistically processed by means of ANOVA I [5]. Differences were found to be significant at $p<0,05$.

Analysis of mean values demonstrated that zona pellucida was the thinnest in group C2 - 3,274 μm, while the highest mean value was found in group C1 – 4,816 μm, see Figure 1. The lowest value of 2,70 μm was recorded also in group C2, namely the one that received 15 units of hCG. The highest value (ZP thickness) was in group C1 – 7,50 μm, see Figure 2 and Figure 3. All means were significantly lower in comparison with those of the control group.

Changes in the thickness of zona pellucida depended on the dose of hCG. The zona pellucida was the thickest in the animal group that received human – therapeutic doses (7,5 IU, the lowest), see Figure 2. The lowest values were recorded with hCG at 15 IU. It is then to be supposed that zona pellucida is a dynamic structure that reacts to human chorionic gonadothropine.

1. S. Conner, L. Lefievre, D. Hughes, C. Barratt, Hum Reprod 20 (2005), 1148.
2. S. Prasad, S. Skinner, C. Carino, N. Wang, J. Cartwright, B. Dunbar, Cell Tissues Organs 166 (2000), 148.
3. B. Dunbar, T. Timmons, S. Skinner, S. Prasad, Biol Reprod 65 (2001), 951.

A. Aretz, B. Hermanns-Sachweh, J. Mayer (Eds.): EMC 2008, Vol. 3: Life Science, pp. 263–264,
DOI: 10.1007/978-3-540-85228-5_132,

4. T. Rankin, S. Soyal, J. Dean, Molecular and Cellular Endocrinology 163 (2000), 21.
5. S. Brandt, „Analiza danych. Metody statystyczne i obliczeniowe." Ed. Wydawnictwa Naukowe PWN, (Warszawa), (1998), 31

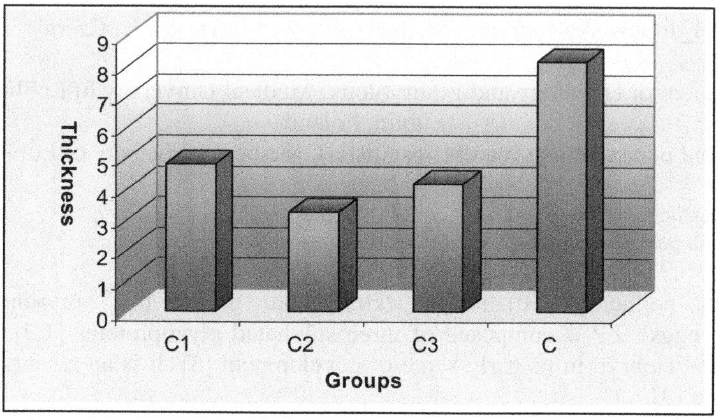

Figure 1. Mean values of thickness of zona pellucida

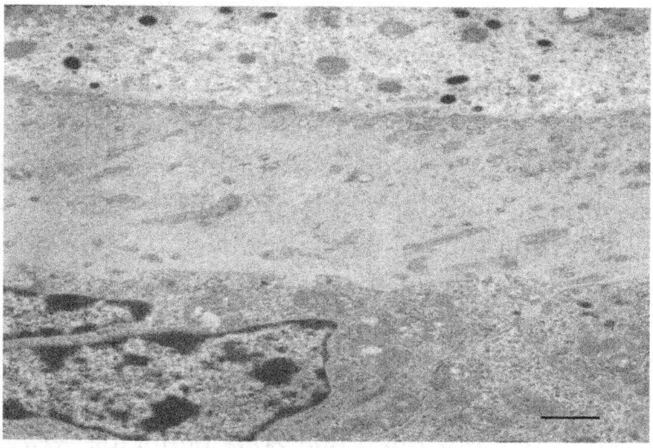

Figure 2. Zona pellucida – control group. Bar - 0,6 μm

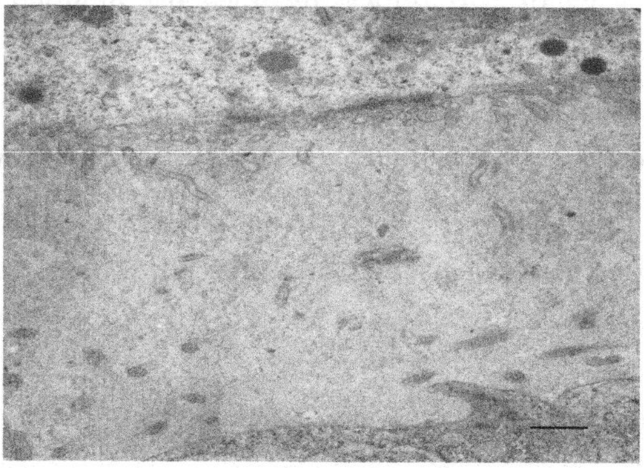

Figure 3. Zona pellucida - experimental group C1. Bar – 0,6 μm

Mozaic microscopy of pancreatic islets

D. Kalicharan, K.A. Sjollema, F. Dijk, J.J. L. van der Want and B. N. G. Giepmans

Molecular Imaging and Electron Microscopy, Dept. of Cell Biology,
University Medical Center Groningen, 9713 AV-1 Groningen, The Netherlands
Medical Center Groningen, The Netherlands

d.kalicharan@med.umcg.nl
Keywords: pancreas, islets, beta cells, mosaic microscopy, correlated microscopy

The interplay between molecules that regulate (sub)cellular processes and between cells and tissues that regulate organisms is highly fascinating. Precise knowledge of the contribution of these molecules and cells helps us to understand the basis of biology under (patho)-physiological conditions. Microscopic techniques have aided in visualization of molecules, cells and tissues for many years. However, choosing a certain microscopic technique often limits our ability to look into detail at different scales. This is especially important when studying complex, heterogeneous tissue where cells of interest are sparsely distributed. We study beta-cells in health and disease (diabetes). Beta-cells are the insulin-producing cells located in the Islets of Langerhans in the pancreas. Other cell-types are present in these islets, including hormone-secreting alpha-cells and delta cells and Islets are highly vascularized and innervated. Islets are heterogeneous, and can consist of a few up to several thousands of cells. Islets are scattered through the exocrine (acinar) pancreas, and contribute only a few percent to the total pancreas, making quantitative imaging challenging.

We apply mosaic microscopy to study cell-types and proteins within the islet to help to understand the regulation of beta-cell turnover and insulin secretion in a quantitative and high-throughput manner. Automated light or fluorescent mosaic microscopy [1];(Fig. 1) enables high-resolution image-analysis of tissue sections up to ~one square cm. Using confocal microscopy, depth of 10s of microns can be achieved, while serial sectioning might further increase the volume of tissue to be imaged.

To further study the molecules *in situ*, tissue can be studied at the electron microscope (EM) level. High-throughput imaging has been made feasible by the digital revolution in acquisition of images both at light and electron microscopes. Images of tissue are stitched together post-processing (Fig. 2), a process which is aided by microscopy software and freeware software such as *ImageJ.*

Ultimately, we will combine light and electron microscopy aided by fluorescent probes that stain certain organelles feasibly visible with EM (i.e. Hoechst and nuclei) and probes that can be visualized on both fluorescent microscopes and electron microscopes (like immuno-Quantum dots [2], or genetically targeted probes (reviewed in [3] to visualize multiple molecules with high-throughput at multiple scales.

1. Chow SK, Hakozaki H, Price DL, MacLean NA, Deerinck TJ, Bouwer JC, M.E.Martone, S.T.Peltier, and M.H.Ellisman. 2006. Automated microscopy system for mosaic acquisition and processing. *J. Microsc.* 222:76-84.

A. Aretz, B. Hermanns-Sachweh, J. Mayer (Eds.): EMC 2008, Vol. 3: Life Science, pp. 265–266,
DOI: 10.1007/978-3-540-85228-5_133,

2. Giepmans, B.N., T.J.Deerinck, B.L.Smarr, Y.Z.Jones, and M.H.Ellisman. 2005. Correlated light and electron microscopic imaging of multiple endogenous proteins using Quantum dots. *Nat. Methods* 2:743-749.
3. Sosinsky, G.E., B.N.Giepmans, T.J.Deerinck, G.M.Gaietta, and M.H.Ellisman. 2007. Markers for correlated light and electron microscopy. *Methods Cell Biol.* 79:575-591

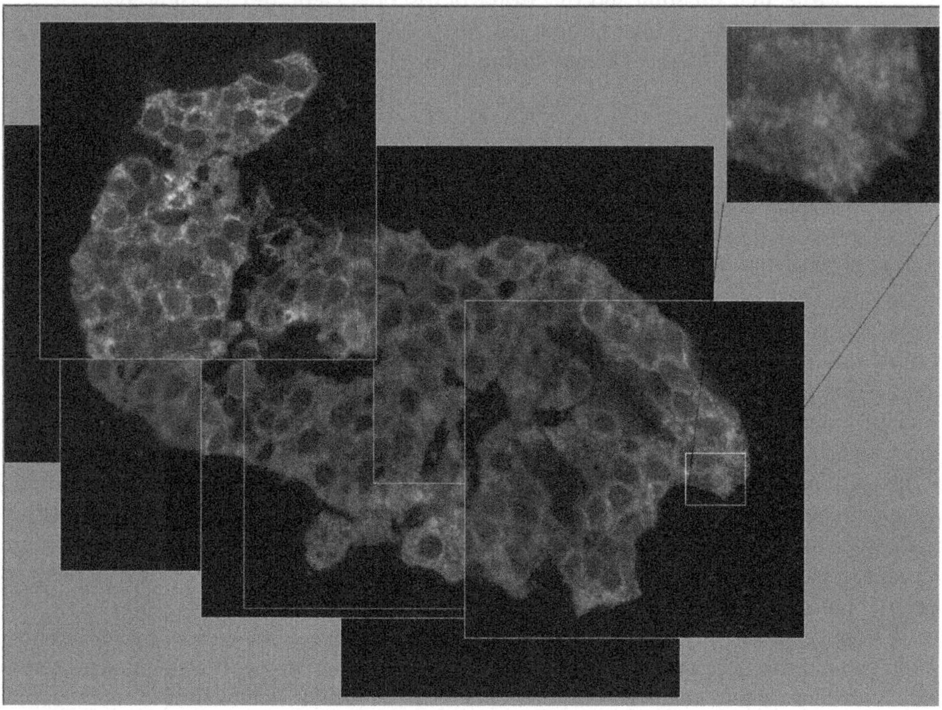

Figure 1. Mosaic microscopy allows high-resolution inspection of areas of interest (i.e. islets), because they were all taken at high resolution. Glucagon immuno-staining of rat pancreas, tissue kindly donated by Jeroen T. J. Visser.

Microscope history Database

N. Koltovoy

Labmetod, Moscow, Russia

koltovoi@mail.ru
keywords: cell, absorption spectrum

It was created "Microscope History Data Base". It was collected images of antique and new microscopes from difference firms, from difference country.

Time interval is– from 1619 to 2007.

Country – Australia, Austria, Belarus, Canada, China, Czech, Denmark, England, France, Germany, Holland, Hungary, India, Italy, Japan, Poland, Russia, Spain, Sweden, Switzerland, Ukraine, USA.

It is 22.000 images for 12.000 microscopes in database.

№	Country	Firms (Makers)	Microscopes
1	Germany	75	3.179
2	USA	71	2.136
3	England	72	1.937
4	Japan	56	1.320
5	Russia	24	811
6	France	28	741
7	China	21	381
8	Austria	10	337
9	Italy	22	179
10	Holland	29	151
11	India	10	141
12	Czech	7	75
13	Poland	4	53
	Another		559
Total			12.000

Firms:

1. Leitz (Germany)– 912 microscopes,
2. Zeiss (Germany)– 839 microscopes,
3. Baush&Lomb (USA)- 460 microscopes,
4. LOMO (Russia) – 402 microscopes,
5. AO/Spencer (USA)- 348
6. Nikon (Japan)– 339 microscopes,
7. Olympus (Japan)– 304 microscopes.
8. Reichert (Austria) – 282

A. Aretz, B. Hermanns-Sachweh, J. Mayer (Eds.): EMC 2008, Vol. 3: Life Science, pp. 267–268,
DOI: 10.1007/978-3-540-85228-5_134,

9. Watson (UK) – 183
10. Nachet (France) – 174
11. Swift&Sons (UK) - 128

Database include next fields:
1-Country, 2-Town, 3-Firms (makers), 4-Model of microscope, 5- Serial Numbers of microscope, 6- Yeas of productions, 7-Some images of microscopes.
It is very interesting to see evolution for construction different type of microscope:
Polarization type - 450 microscopes,
Inverted type –345 microscopes.

Interaction of H-Ras Transformed Cell Line with Folic Acid Modified Magnetic Nanoparticles and Detection by Transmission Electron Microscope

M. Kutlu[1], N. Saltan[1,3], R. Say[2,3] and A. İşcan[3]

1. Anadolu University, Faculty of Science, Department of Biology, 26470, Eskişehir, TURKEY
2. Anadolu University, Faculty of Science, Department of Chemistry, 26470, Eskişehir, TURKEY
3. Anadolu University, Plant, Drug and Scientific Research Center 26470, Eskişehir, TURKEY

hmkutlu@anadolu.edu.tr
Keywords: Cancer, Magnetic nanoparticle, TEM (Transmission Electron Microscope)

Cancer is usually classified according to the tissue from which the cancerous cells originate. Most cancers can be treated and some cured, depending on the specific type, location, and stage. Cancer can be treated by surgery, chemotherapy, radiation therapy, immunotherapy, monoclonal antibody therapy or other methods. Cancer cells also produce lots of receptors and biomarkers. Targeted therapy is a type of medication which blocks the growth of cancer cells by interfering with specific targeted molecules needed for carcinogenesis and tumor growth, rather than by simply interfering with rapidly dividing cells. Targeted cancer therapies may be more effective than current treatments and less harmful to normal cells.

Interaction of H-*Ras* transformed cell line (Rat embryo fibroblast cells) with folic acid modified magnetic nanoparticles were determined using TEM, FEI Tecnai™ G^2 Spirit TWIN / BioTWIN. They were deposit on Formvar-coated 200-300 mesh copper grids and dried. Cells were fixed with %2,5 gluteraldehyde in 0,1M phosphate buffer and left in phosphate buffered saline (PBS) overnight at +4°C. After being embedded in Agar and stained in 2 % osmium tetroxide. Cells were dehydrated in graded ethanol. Then cells were embedded in EPON 812 epoxy. They were thin sectioned using a diamond knife to a maximum thickness of 100 nm. The sections were stained with lead citrate and uranyl acetate.

In this study, cancer cells with magnetic nanoparticles which include folic acid of interaction was estimated. Folate targeting to tumor-specific drug delivery systems are one of the most succesful anticancer strategies in cancer. We aimed to investigate in cellular and ultrastructural levels if these particles were taken by cancer cells by endocytosis or not. At first stage, nanoparticles were conjugated with folic acid and that complex can be transported into tumour tissues. Cancer cells have more vitamin receptors than normal cells. Folic acid was connect to folat receptor existing on the surface of tumoral cells. Thus, nanoparticles to target spesific cells was facilitated. Cancer cells with folic acid modified magnetic nanoparticles of interaction was displayed on TEM. Furthermore, Transmission Electron Microscope was used to determine ultrastructure of the cells.

A. Aretz, B. Hermanns-Sachweh, J. Mayer (Eds.): EMC 2008, Vol. 3: Life Science, pp. 269–269,
DOI: 10.1007/978-3-540-85228-5_135, © Springer-Verlag Berlin Heidelberg 2008

Sedimentation of suspensions for diagnostic thin section electron microscopy

M. Laue and N. Bannert

ZBS 4, Robert Koch Institute, Nordufer 20, 13353 Berlin, Germany

lauem@rki.de
keywords: rapid embedding, ultracentrifugation, particle enrichment

Diagnostic thin section electron microscopy plays a significant role in pathology and the diagnosis of emerging infectious diseases [1]. In this context preparation time is critical and has to be minimized because of obvious reasons. Several strategies have been developed to reduce the preparation time for a diagnosis. Recently, we have presented a rapid thin sectioning protocol, which allows a diagnosis within about 2 h [2]. However, time is not the sole factor affecting the diagnosis. Sensitivity in terms of detection likelihood is also of significant importance.

Suspensions are extremely critical in this respect. Usually they are diluted by at least the fixation and occasionally by other preparation procedures. To omit this dilution effect and thus to improve the detection likelihood, we use sedimentation of suspensions by ultracentrifugation with a Beckman table desk Airfuge at approx. 100.000 g for 10 min. However, the tiny pellet is usually invisible and not very stable over time. We therefore firstly added colloidal gold particles to the suspension to colour the pellet with a red stain and secondly, we either stabilized the pellet by adding low-melting point agarose or we introduced the wet pellet into a cellulose micro capillary.

In order to find out which stabilization method is most effective, we have compared the following protocols using mixtures of either a vaccinia virus or a bacterial endospore suspension (115 µl) with gold colloid (5 µl of 6 nm colloids) as a model: (1) resuspension of the pellet with 5 µl of low-melting point agarose and casting of a gel of 0.3 mm thickness [2]; (2) centrifugation in a agarose gel cushion (0.5% in distilled water) (Fig. 1); (3) dissolving the pellet with the tip of pipette and aspiration into an Alcian blue coated cellulose capillary using the method described in [3]. Dehydration and embedding in LR White were conducted as described in [2].

The red stain of the pellet facilitated the detection of the pellet, which was retained even in the final plastic block using the gel cushion method. A qualitative comparison of the particle density revealed that the gel cushion method gave the highest particle density (Fig. 2), followed by the cellulose capillary and finally by the resuspension method.

1. P.R. Hazelton and H.R. Gelderblom, Emerg. Infect. Diseases 9 (2003), p.294.
2. M. Laue, B. Niederwöhrmeier, and N. Bannert, J. Microbiol. Meth. 70 (2007), p.45.
3. T. Müller-Reichert, M. Srayko, A. Hyman, E. O´Toole and K. McDonald, Meth. Cell Biol. 79 (2007), p.101.

A. Aretz, B. Hermanns-Sachweh, J. Mayer (Eds.): EMC 2008, Vol. 3: Life Science, pp. 271–272,
DOI: 10.1007/978-3-540-85228-5_136,

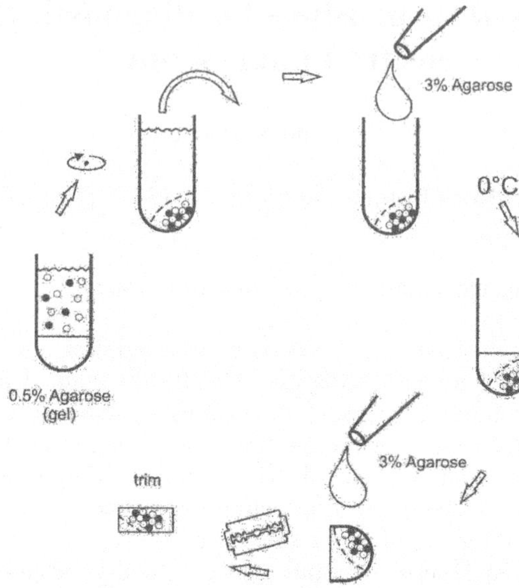

Figure 1. Flow chart of the gel cushion method for the stabilization of tiny particle pellets following Airfuge ultracentrifugation.

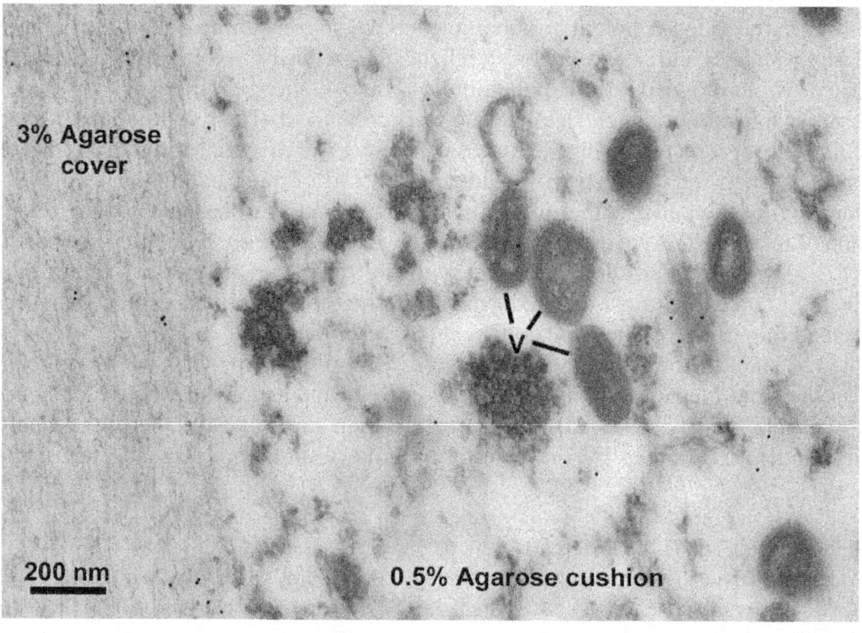

Figure 2. Ultrathin section through vaccinia virus particles (v) after Airfuge ultracentrifugation in an agarose gel cushion and rapid processing in LR White.

Evaluation of rofecoxib administration influence on ultrastructural image of the kidney

W. Matysiak, B. Jodłowska-Jędrych, E. Kifer-Wysocka, J. Romanowska-Sarlej, K. Czerny

Department of Histology and Embryology, Medical University of Lublin, Radziwiłłowska Street 11, 20-080 Lublin, Poland

wlodzimierz.matysiak@am.lublin.pl
Keywords: kidney, ultrastructure, rofecoxib

Long term administration of non-steroidal anti-inflammatory drugs in the treatment of inflammatory diseases, despite its effectiveness, poses a considerable risk of adverse effects. Launching of new class of anti-inflammatory drugs onto the market – coxibs (selective inhibitors of cyclooxygenase-2 (COX-2, cyclooxygenase-2) aroused hopes for increasing safety of anti-inflammatory therapies. Initial enthusiasm generated by the appearance of coxibs, considerably waned after the analysis and publication of the data from large-scale randomized clinical trials which were conducted using double-blind test method [1,2,3].

The experiment was conducted on male rats of Wistar strain, with body mass of 350g. Experimental animals received rofecoxib (Vioxx®, MSD, N.Y. USA) intragastrically administered through a tube, in the form of suspension in physiological saline (1ml) in the period of 4 weeks– experimental group I, and in the period of 8 weeks – experimental group II. The drug was administered in the maximum 24-hour dose (0,125mg/rat). Specimens of the kidney were prepared for observation in a transmission electron microscope. Ultra-thin strips were observed in TEM Tesla BS500.

The observed changes confirm the observations and suggestions which have been made and stored in bibliographical data bases hitherto concerning the effects of high doses of coxibs on the function and morphology of the kidney [4,5,6], and especially on the ultrastructural image of podocytes described in this paper.

In the analyzed images renal corpuscles after four-week administration of rofecoxib to the animals, demonstrated altered structure in comparison to the images observed in the kidneys of rats from the control group. The lack of secondary processes of the podocytes was conspicuous, and podocytes leaned against the basement membrane with wide flat processes, which sometimes created an impression that the cell body stuck directly to the capillary vessel. Epithelial cells of proximal tubules frequently had more abundant apical vacuoles. Lysosomes were more plentiful, irregular in shape and contained material of varied electron density (Figure 1). There were also observed cells which had a loosened cytoplasmic structure – rarefactions of cytoplasm were visible in the apical and basal parts of cells. Basement membranes of tubule epithelium were irregularly thickened. Mitochondria often demonstrated features of slight swelling.

Especially after eight-week administration of rofecoxib in therapeutic doses there appeared changes in the structure of nephron proximal segment (atrophy of secondary

A. Aretz, B. Hermanns-Sachweh, J. Mayer (Eds.): EMC 2008, Vol. 3: Life Science, pp. 273–274, DOI: 10.1007/978-3-540-85228-5_137,

processes of the podocytes, more abundant occurrence of lysosomes, rarefaction of cytoplasm of major tubule cells and irregular thickening of basement membranes).

Kitahara [7] conducted an experiment aimed at checking the effects of the action of selective inhibitors COX-2 in the inflammatory process of the renal corpuscles. The authors observed that rofecoxib led to mesangium disintegration and intensification of albuminuria. Non-steroidal anti-inflammatory drugs are most frequently used medicines in the world. It must be remembered, however, that their long term administration can be limited because of the occurrence of adverse effects, especially in the alimentary tract and kidneys [5,6,7,8,9,10]. The history of coxibs shows that selective inhibition of the pathway of arachidonic acid metabolism – apart from an undoubtedly favourable effect on safety in relation to the alimentary tract – also poses a risk of complications in the cardiovascular system, which eventually led to the withdrawal of the drug Vioxx® from the pharmaceutical market.

1. C. Bombardier, N. Engl. J. Med. 343 (2000), 1520.
2. D. C Brater, C. Harris, J. S. Redfern and B. J. Gertz, Am. J. Nephrol. 21 (2001), 1.
3. F. Catella-Lawson, B. McAdam and B. W. Morrison, J. Pharm. Exp. Ther. 289 (1999), 735.
4. A. Dey and Ch. Maric, Clin. Sci. 107 (2004), 561.
5. M. Hermann, Hypertension 45 (2005), 193.
6. P. Juni, Br. Med. J. 324 (2001), 1287.
7. M. Kitahara, J. Am. Soc. Nephrol. 13 (2002), 1261.
8. J. L. Rocha and J. Fernandez-Alonso, Lancet 357 (2001), 1946.
9. F. E. Silverstein, JAMA 284 (2000), 1247.
10. M. M. Skelly and C. J. Hawkey, Int. J. Clin. Pract. 57 (2003), 301.

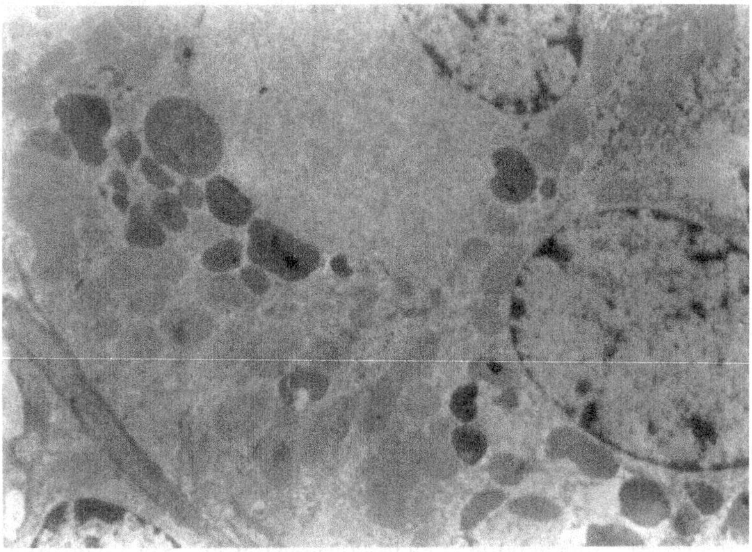

Figure 1. Experimental group. Rat kidney. Rarefication of cytoplasm in proximal tubule cells, irregularly thickened basement membrane. Magn. ca 4000x.

Microstructural changes produced by Pulsed Electric Fields in liquid whole egg studied by Transmission Electron Microscopy (TEM)

I. Pérez-Munuera, R. Marco-Molés, I. Hernando, A. Quiles and A. Puig

Departamento de Tecnología de Alimentos, Universidad Politécnica de Valencia, Camino de Vera s/n 46022 Valencia, Spain.

iperez@tal.upv.es
Keywords: Pulsed Electric Fields, whole liquid egg, Transmission Electron Microscopy

The application of new technologies for food processing and preservation aims to meet to the increasing number of catering companies due to the growing demand for fresh foods, high quality and ready-to-eat meals. Among the non-thermal technologies for food preservation, the application of Pulsed Electric Fields (PEF) is receiving big attention over the past few years, as it reduces the detrimental changes on physical and sensorial properties of foods and offers fresh fluid foods of high quality, taste, nutritional value and improved shelf-life [1, 2].

The aim of this work is to study the microstructural changes that occur in liquid whole egg treated by PEF (Table I). The effect of the electric field strength and the homogenization, through a perforated surface (∅ ≅0.4 mm), are studied.

Samples are stabilized with agarose [3], fixed with glutaraldehyde and osmium tetroxide, dehydrated with ethanol, contrasted with uranyl acetate and embedded in epoxy resin to be observed in a Philips EM 400 (Philips, Eindhoven, Holland transmission electronic microscope at 80 kV.

Non-treated samples (Figure 1A) shows a continuous matrix mainly composed of a protein network, from the egg white, and lipids, from the yolk, incorporated into the protein network during beating, with small yolk protein granules (∅ ≅1.5 μm) inserted in the continuous protein matrix. LowPEF (19kV) produces a protein aggregation effect (Figure 1B) that is increased when high field strength is applied (32kV). The lipoporotein matrix is specially affected in HighPEF treated samples, the continuity of the protein network is broken and some protein granules from the yolk are degradated (Figure 1C).

When non-treated samples are homogenized (Figure 2A) the protein granules size increase. The samples homogenized and treated with HighPEF (Figure 2B) shows the effect of both treatments with bigger protein granules (Figure 2A), although the lipoprotein matrix seems not to be so degradated than in the non-homogenized ones (Figure 1C). Homogenization produces a protein granules aggregation and a reinforcement of the protein-protein interactions.

1. D. Knorr, B.I.O. Ade-Omawaye and V.Heinz, Proceedings of the Nutrition Society 61 (2002), p. 311.
2. P. Mañas and R. Pagán, Journal of applied microbiology 98 (2005), p. 1387.

A. Aretz, B. Hermanns-Sachweh, J. Mayer (Eds.): EMC 2008, Vol. 3: Life Science, pp. 275–276,
DOI: 10.1007/978-3-540-85228-5_138,

3. R. Sharma, H. Singh and M.W. Taylor, Journal of Food Science 61, 1 (1996), p.28.
4. The authors thank to the Swedish Institute for Food and Biotechnology (Göteborg, Sweden) for supplying the samples studied in this work, the European Commission for the financial support of the project FP6-FOOD-023140 and the "Ministerio de Educación y Ciencia" of Spain for the grant awarded to R. Marco-Molés.

Table I. Technological characteristics of the pulsed electric fields applied. U: electric field strength; t: pulse width; N: number of applied pulses; prf: pulse repetition frequency; R: internal resistance.

	Pulse generator parameters					Temperature (°C)		
Sample	U (kV)	t (μs)	N	prf (Hz)	R (Ω)	In	Out	Increase
HighPEF	32	6	5	250	1000	3.8	32.5	28.7
LowPEF	19	6	5	250	1000	5.4	16.0	10.6

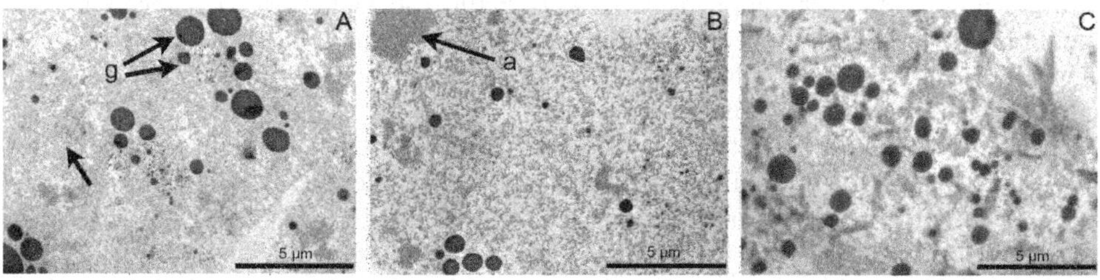

Figure 1. TEM. A: non-treated liquid whole egg. B: LowPEF (19 kV) treated liquid whole egg. C: HighPEF (32 kV) treated liquid whole egg. 1650x. (arrow: lipoprotein matrix; g: protein granules; a: aggregated compounds).

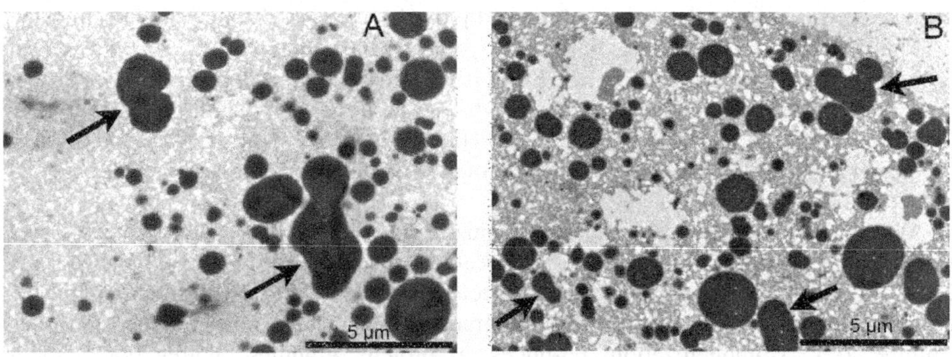

Figure 2. TEM. A: non-treated homogenized liquid whole egg. B: HighPEF (32 kV) treated and homogenized liquid whole egg. 1650x. (arrow: aggregated protein granules).

Microstructural changes in dough treated by Glucose oxidase (GOX) and Transglutaminase (TG), studied by Scanning Electron Microscopy (SEM)

I. Pérez-Munuera, A. Quiles, A. Puig and I. Hernando

Departamento de Tecnología de Alimentos
Universidad Politécnica de Valencia. Camino de Vera s/n 46022, Valencia, Spain

iperez@tal.upv.es
Keywords: Glucose oxidase, Transglutaminase, Scanning Electron Microscopy

The improvement of dough functionality through different enzymes is a new strategy used in the last years for the bakery industry. Some enzymes as glucose oxidase (GOX, EC 1.1.3.4) and transglutaminase (TG, EC 2.3.2.13) modify the protein fraction, mainly the gluten proteins, but is not well-known the influence of these enzymes on the structure of the dough components [1, 2].

The aim of this work has been to study the effect of different combined treatments, based on the GOX and TG treatments, on the microstructure of wheat flour dough.

The samples were fixed (primary fixation with 2.5% glutaraldehyde and secondary fixation with 2% osmium tetroxide) and dehydrated in gradual series of ethanol, 10, 20, 40, 60, 80 and 100% ethanol (ethanol / water, V/V) every 20 min, rinsed in acetone and ultradehydrated by critical point with CO_2 (1100 psi, 31.5 ºC) in a POLARON E3000 instrument (Watford, United Kingdom). Then they were gold-coated using POLARON E6100 Equipment (10^{-4} mbar, 20 mA, 80 s) and observed in a Jeol JSM 6300 Scanning Electron Microscope (SEM) (Jeol, Tokio, Japan) at 15 kV and a working distance of 15 mm.

The control dough structure (Figure 1A) shows a continuous matrix of proteins. The different starch granules populations, from the wheat endosperm, are situated separately in the empty spaces between the protein fibres. Gluten proteins must be the main components of this fibres proteins because of the most of the soluble protein fraction has been eliminated during the sample preparation by SEM. When dough is treated with 1% TG (Figure 1B), reinforcement of the protein-protein interaction is observed; however this enzyme doesn't affect the starch granules-protein interactions. When 0.001% or 0.01% GOX were added to dough (Figures 1C and 1E respectively), a weak reinforcement of the protein matrix is produced too, however the starch granules appear more embedded in the protein matrix. The combined effect of GOX and TG treatments (Figures 1D and 1F respectively) shows that the main dough structural components are interacting each other; protein-protein and protein-starch interactions are reinforced in these samples and these results shows the complementary effect of both enzymes treatments. The combined use of these enzymes could be a good tool to improve the functionality of dough in baked products playing with different dose of both enzymes.

A. Aretz, B. Hermanns-Sachweh, J. Mayer (Eds.): EMC 2008, Vol. 3: Life Science, pp. 277–278,
DOI: 10.1007/978-3-540-85228-5_139,

1. R. Bonet, P.A. Caballero, M. Gomez and C.M. Rosell, Cereal Chemistry 82, 4 (2005), p 425.
2. A. Bonet, C.M. Rosell, P.A. Caballero, M. Gómez, I. Pérez-Munuera and M.A. Lluch, Food Chemistry 99 (2006), p408.
3. The authors thank to the "Ministerio de Educación y Ciencia" of Spain for the financial support of the project AGL 2005-05192-C04-03.

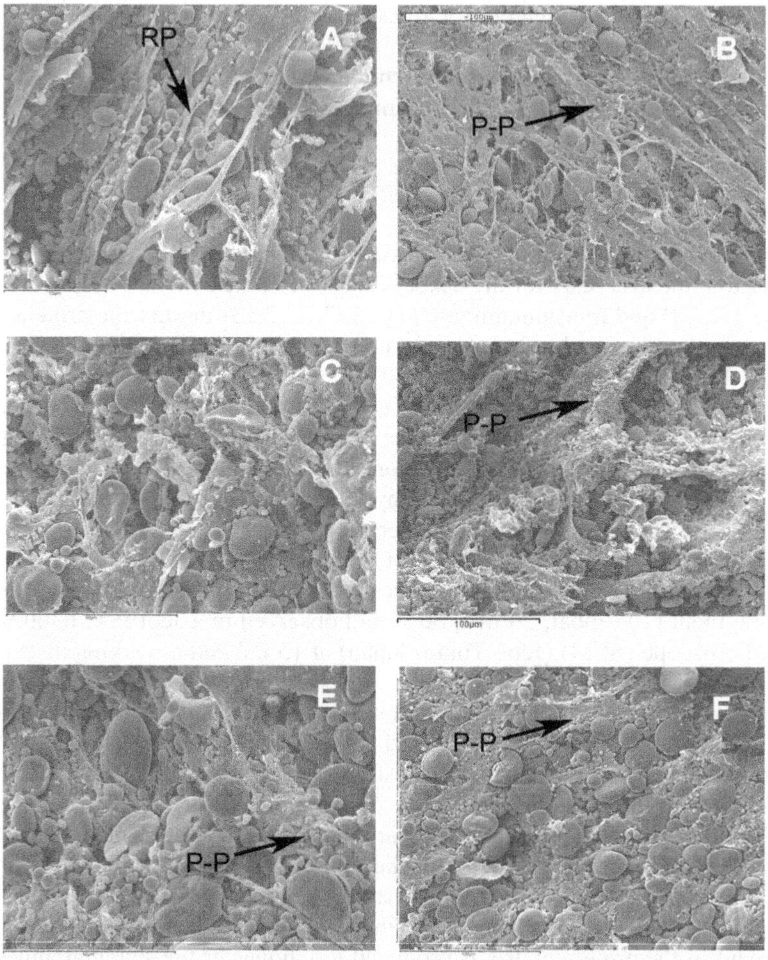

Figure 1. Scanning Electron Microscopy (SEM). **A:** Control dough, **B:** dough with 1% TG, **C:** dough with 0.001% GOX, **D:** dough with 0.001% GOX and 1% TG, E: dough with 0.01% GOX, **F:** dough with 0.01% GOX and 1% TG. (**RP:** protein fibres. **P-P:** protein-protein interactions).

Cytoplasmic K/Na balance in the cardiomyocyte of young and old rats

M.A. Pogorelova, A.G. Pogorelov

Institute of Theoretical and Experimental Biophysics, Russian Academy of Sciences, Pushchino, Moscow province, 142292 Russia

agpogorelov@rambler.ru

Keywords: cardiomyocyte, cytoplasmic potassium, electron probe microanalysis

Age-related changes in the cardiac function in animals are accompanied and/or caused by a decrease in the action potential plateau [1]. This parameter changes with age and depends on the potassium current in the cardiomyocyte [2]. As a result, one should expect a change in the potassium balance in the cell, which would affect key molecular-genetic transformations: DNA recombination [3], gene expression [4], and actin depolymerization [5]. With age, deficient supply of oxygen to tissues develops. This factor causes activation of adaptive mechanisms, which use the potassium and sodium gradients across the cytoplasmic membrane for compensating for hypoxic acidosis and lactosis [6, 7]. The complex of age-related impairments is particularly pronounced in the extreme case of acute deficiency of energy substrate and oxygen.

Therefore, the purpose of this work was to compare the cytoplasmic concentrations of potassium and sodium in the cardiomyocyte of young and old animals in substrate–oxygen deficiency modeled in glucose-free hypoxic perfusion of an solated heart. The investigation was carried out on Wistar rats. The cytoplasmic concentrations of potassium and sodium were measured by electron probe microanalysis.

The response of the isolated heart to hypoxia is most pronounced in glucose-free hypoxic perfusion. Table I summarizes the results of measuring the potassium and sodium concentrations in the cytoplasm of the cardiac muscle cell of young and old animals. For all ages on termination of glucose-free hypoxic perfusion, cellular potassium deficiency develops. Note that, in 40 min of hypoxia, significant potassium accumulation in the cardiomyocyte of young animals is detected. In glucose-free hypoxic perfusion, rats for both ages exhibit similar tendencies, which manifest themselves in sodium accumulation in the cytoplasm and in a maximum of the concentration in the 40th minute.

The results obtained suggests the following conclusions: in the cardiac muscle cell of old animals, an increase in the cytoplasmic sodium concentration without change in the potassium concentration is observed. Oxygen–substrate deficiency induces sodium accumulation and potassium deficiency in the cardiac muscle cell. In acute hypoxic deenergization, the change in the sodium–potassium balance in the cytoplasm of the cardiac muscle cell depends on the age of animal. After electromechanical decoupling in young rats, the cytoplasmic potassium concentration increases. The data obtained are consistent with the hypothesis of the presence of genetically determined age-related changes in the conductance of potassium channels, which occur in old animals against the background of short supply of oxygen and substrate to tissues.

A. Aretz, B. Hermanns-Sachweh, J. Mayer (Eds.): EMC 2008, Vol. 3: Life Science, pp. 279–280,
DOI: 10.1007/978-3-540-85228-5_140,

1. E. P. Anyukhovsky, E. A. Sosunov, A. Plotnikov, et al., Cardiovasc. Res. **54** (2002), p. 462.
2. L. Yue, J. Feng, G. R. Li and S. Nattel, J. Physiol. **496** (1996), p. 647.
3. G. N. Parkinson, M. P. H. Lee and S. Neidle, Nature **417** (2002), p.876.
4. S. Taurin, V. Seyrantepe, S. N. Orlov, et al., Circ. Res. **91** (2002), p.915.
5. W. H. Goldmann, Cell Biol. Int. **27** (2003), p.391.
6. A. G. Pogorelov, V. N. Pogorelova, M. I. Dubrovkin, et. al., Biofizika **47** (2002), p.744.
7. A. G. Pogorelov, V. N. Pogorelova, E. V. Khrenova, et. al., Sechenov Zh. Evol. Biokhim. Fiziol. **40** (2004), p.353.

Table I. Cytoplasmic potassium and sodium concentrations (mM) in the myocyte of the papillary muscle of the heart of young and old Wistar rats at various durations of substrate–oxygen deficiency modeled in glucose-free hypoxic perfusion
Note: Mean values for groups of eight animals are presented. The scatter of results from the mean value in a group does not exceed 10% for potassium and 15% for sodium.

Age	Hypoxic perfusion duration, min							
	0		20		40		60	
	K	Na	K	Na	K	Na	K	Na
3 months	117	37	118	75	155	107	95	88
16 months	120	51	108	82	92	103	82	94

Hyperspectral imaging – a novel concept for marker free chromosome characterization

K. Rebner, T. Merz and R.W. Kessler

Process Analysis and Technology, Institute for Applied Research, Alteburgstrasse 150, 72762 Reutlingen, Germany

karsten.rebner@reutlingen-university.de
Keywords: chemical imaging, karyotyping, cytogenetic

Introduction & Problem statement

Hyperspectral Imaging (HSI) is a powerful and well-established technique. It combines different technologies like optical microscopy, digital imaging and molecular spectroscopy in combination with multivariate data analysis (MVDA) methods [1]. We describe the design and performance of our HSI System for Whiskbroom (single point scanning), Staring (2D wavelength scanning) and Pushbroom (line scanning with spectral dispersion) microspectroscopy. The system is optimized for the UV-VIS-NIR range and for fluorescence. In order to demonstrate the potential of the technique we will focus on the hyperspectral characterization of chromosomes.

Staining techniques like G-banding are routinely used to identify metaphase chromosomes based on their unique banding pattern. Advanced molecular cytogenetic techniques like fluorescence-in-situ-hybridization (FISH) provide a more sensitive tool for complex and small structural aberrations [2]. In case of complex diagnostic findings, the FISH technique is applied on the same slide. These combined methods are expensive and often insufficient because of inconsistent information of the poor dyeing results. We will show that identification and characterization of the spectral signature of the chromosomes can be used as a marker-free screening method.

Materials, Methods and Data Analysis

Human metaphase chromosomes are prepared from peripheral lymphocyte culture and spread on specialized thin-film gold coated slides. Mapping and imaging are carried out in reflectance within the UV-VIS-NIR range with bright- and darkfield illumination. The spectra are analysed by means of a principal component analysis (PCA) using the first derivative spectra.

Selected Results & Discussion

For the sake of simplicity, we present results only from the visible spectral range here. Figure 1 a) and b) show images of unstained chromosomes in darkfield illumination before and after trypsination. As can be seen, due to a decrease in film thickness the colouring of the interference spectra changes from red to green. Figure 1c shows the pattern of chromosome 17 in darkfield illumination.

The spectral signature of the banding structure is representative for the individual chromosome. This signature is due to the superposition of the interference pattern of the different layer thicknesses, the spectral interference of the band pattern, changes in

A. Aretz, B. Hermanns-Sachweh, J. Mayer (Eds.): EMC 2008, Vol. 3: Life Science, pp. 281–282,
DOI: 10.1007/978-3-540-85228-5_141, © Springer-Verlag Berlin Heidelberg 2008

refractive indices along the chromosome axis as well as the absorption of chromophores in different spectral regions of the chromatin condensation.

Figure 2 shows the PCA of the spectra of the trypsinated chromosomes. The PCA extracts latent factors from the spectra which can be used for classification. Principal component 1 (PC 1) represents the thickness pattern of the chromosome and its size. PC 2 may show the interference pattern of the banding. Higher PCs can be used for further discrimination.

The marker free screening of the individual chromosomes by means of their spectral signature e.g. in the VIS in combination with MVDA may open up new potentials for a fast and reliable Karyotyping. Furthermore, applying this new technique, FISH technique can more easily be used as a supplementary feature for fine tuning of the Karyotyping.

1. G. Bearman, R. Levenson, Biomedical Photonics Hand-book, (2003), Kap. 8.
2. J. M. Bayani, J. A. Squire, Cancer Invest 20(3), (2002) p. 373-386.
3. We acknowledge the financial support of Bundesministerium für Bildung und Forschung.

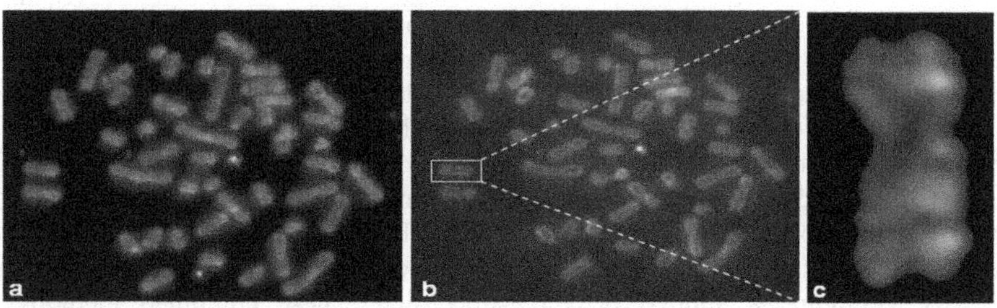

Figure 1. unstained chromosomes in darkfield illumination a) before trypsin digestion b) after trypsin digestion c) with magnification of chromosome 17.

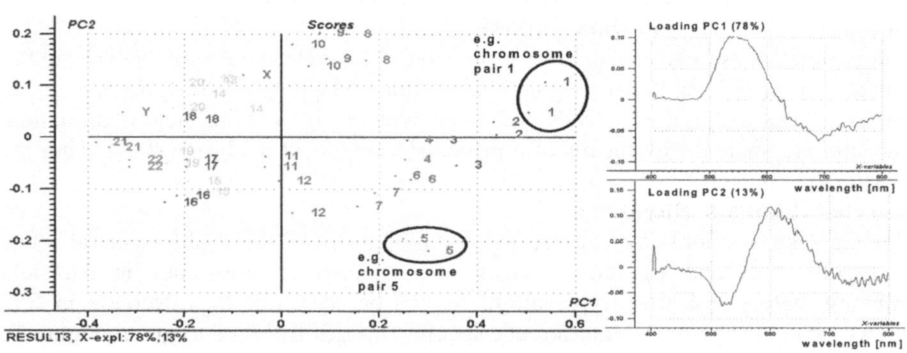

Figure 2. principle component analysis of a darkfield (1c) dataset for the VIS range.

Poxviruses: Morphogenesis of an Orthopoxvirus (CPXV) and an Avipoxvirus (FPXV) in host tissue

S. Richter and P. Wernsdorf

Dept. of Electron Microscopy, Inst. of Veterinary Disease Control – AGES
Robert Kochgasse 17, A-2340 Moedling, Austria

susanne.richter@ages.at
Keywords: poxviruses, morphogenesis, zoonosis

Cowpox (CPX) and Fowlpox (FPX) are highly contagious poxvirus diseases of either public health or veterinary economic importance [1]. Rapid accurate detection in an early stage of infection is necessary for basic control disease measures. Therefore diagnosis at the beginning of virus development is fundamental. This study deals with the virus formation of two different poxvirus species in host tissue.

Electron microscopy of infected tissue and, in the case of CPX, of swab from vesicle fluid was performed by negative staining and tissue sectioning (TEM: Zeiss 906, 80kv). For negative staining all samples solubilised in 4°C-cold PBS (pH 7.3) were cleared by low-speed centrifugation and then sedimented directly onto the grids by airfuge (82000rpm, 20psi). UA and PTA were used as negative stains. Sections of biopsy samples embedded in Epon were contrasted with methanolic UA and lead citrate.

Virus particle concentration in CPX was most significant in the dermis; in FPX histological examination revealed characteristic epidermal and mucosal pox lesions with the presence of characteristic intracellular Bollinger body inclusions [2]. In negative staining preparation, CPXV (Figure 1a) and FPXV (Figure 2a) displayed the characteristic morphology of Orthopoxvirus and Avipoxvirus. In ultrathin section, both virus types (Figures 1b, 2b) revealed the typical common structure of poxviruses consisting of the lipoprotein envelope (le), an intermediate coat (ic) that expands into two lateral bodies (lb) and of the biconcave core (co) composed of the core membrane and the nucleosome. In the case of Cowpox virus (Figure 1), transmission electron microscopy demonstrated developing and mature virus stages in infected human skin cells. Both, cytoplasmic A-type viral inclusions and coarsely granular intracellular B-type inclusions were detected in the host tissue. A-type inclusions (Figure 1c) ranging from less than 1µm to several micrometers consisted of uniformly granular, moderately electron dense material full of immature and mature virions. Virions within these inclusion bodies have gained an additional outer coat; virions surrounding the inclusions were often in varying stages of maturation. In B-type inclusions (Figure 1e), which are also called virus factories, virus development was similar to that shown in infected cell cultures [3]: CPX virus assembly in the human skin tissue began with the appearance of membrane crescents deep within the virus factories. The lipid crescents extended to form spherical immature virions from which the first infectious progeny, the intracellular mature virions, were developed. The spherical immature virions were simple circular structures containing a moderately granular viroplasm. Special interest was put on the ultrastructure and the origin of the membrane of crescents. Like in cell

A. Aretz, B. Hermanns-Sachweh, J. Mayer (Eds.): EMC 2008, Vol. 3: Life Science, pp. 283–284,
DOI: 10.1007/978-3-540-85228-5_142, © Springer-Verlag Berlin Heidelberg 2008

culture [3], section and angular tilt analysis, revealed no continuity between double layered virus crescent membrane and cellular membrane cisternae respectively cellular intermediate compartiment (IC). Crescents were found to form at sites remote from IC membranes – namely within the centre of virus factories. Ultrastructurally, the immense cytoplasmic inclusion bodies of Fowlpox virus (Figure 2b) consisted of large lipid locules filled with mature virions and dense rodlets. Immature virions developed by formation of dense foci of viroplasm surrounded by crescent shaped structures later closing to circular membranes [5]. The final stages of maturation involved the attachment of these virions to small lipid globules during early inclusion formation. Attachment of virions to small lipid globules was also observed in CPX (Figure 1d).

1. S. Essbauer, and M. Pfeffer, Bundesgesundheits-blatt - Gesundheitsforschung 47 (2004), p. 671.
2. D. Boulanger, T. Smith and MA Skinner, J. Gen. Virol. 83 (2000), p. 675.
3. M. Hollinshead, A. Vanderplasschen, G.L. Smith and D.J. Vaux, J. Virol. 73(1999), p. 1503.
4. S. Ch. Weli, Ø. Nilssen and T. Taarvik, Med. Electron Microsc. 37 (2004), p. 225.

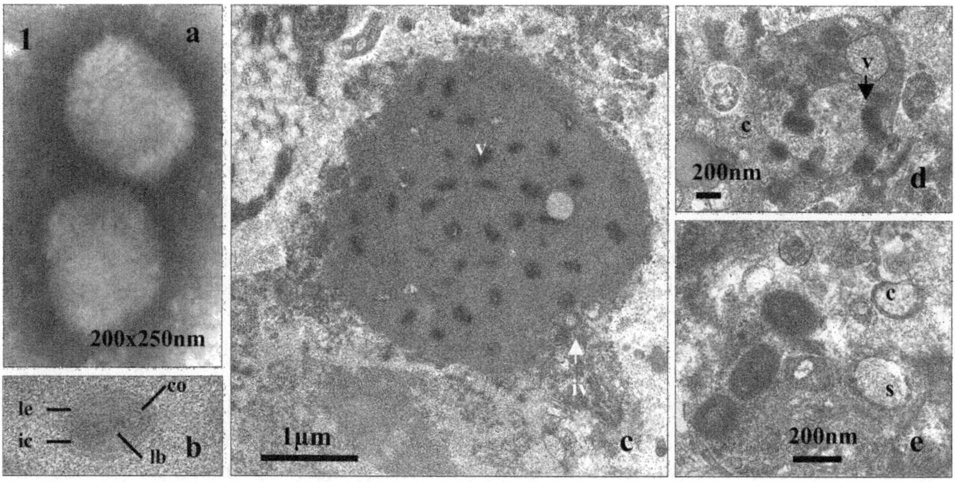

Figure 1. CPXV a) negative staining, b) section through virus, c) A-type inclusions (v = mature virus particle, iv = immature virus particle), d) lipid globules with attached CPXV particles, e) B-type inclusion with cup shaped (c) and spherical lipid crescent (s).

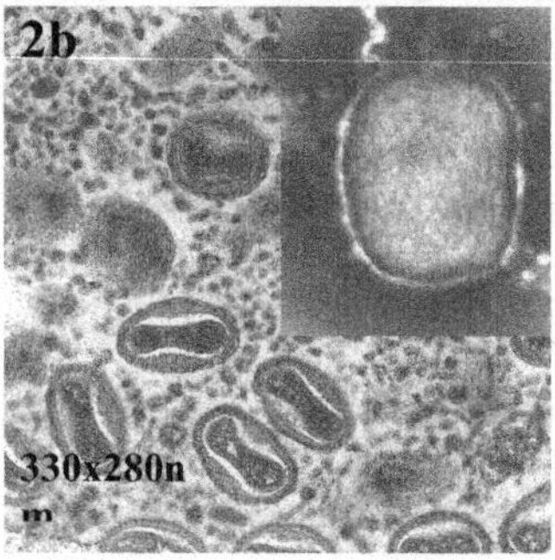

Figure 2. FPXV a) negative staining, b) dense lipid inclusion filled with immature and mature virions and excess coat tubules.

Ultrastructural alterations of methotrexate in mouse kidney

Ayşe Köse Sargın[1], Emin Oğuzhan Oğuz[2], Belgin Can[1], Yüksel Saran[1]

1. Ankara University School of Medicine, Department of Histology and Embryology, Ankara, Turkey.
2. Pamukkale University School of Medicine, Department of Histology and Embryology, Denizli, Turkey

drayse1980@yahoo.com
Keywords: Methotrexate, mouse kidney, ultrastructure

MTX (methotrexate) is a cytostatic anti-neoplastic in the group of folic acid antagonist [1]. It inhibits nucleic acid synthesis by displaying anti-metabolite reaction [2].

In this study, 20 male Albino mice weighing between 25-30 g were used. 15 of the mice were included in the experiment group and 5 in the control group. Experiment animals were randomly divided into three groups. The groups were consecutively injected 20 mg/kg, 80 mg/kg, and 300 mg/kg in single doses by intra-peritoneal way. The control group was injected serum physiologic on the same day. 24 hours later, the kidneys of the both control and experiment groups were excised and the tissues were processed for light and transmission electron microscopy.

The experimental group contained increasing degenerative ultrastructural alterations depending on the doses in the structure of kidney cortex and medulla when compared with sections of the control group. In the electron microscope; enlargement in smooth endoplasmic reticulum, swelling in mitochondria, melting in cristae and electron dense inclusions in mitochondrial matrix were seen. There were many vacuoles adhered to each other in the apical part of the cell. The nucleus borders were in a disordered structure. Condensation and marginalisation of chromatin in nucleus membrane were observed. In tubular lumen, there were necrotic cell pieces.

MTX joins in DNA synthesis, repairment and cellular replication [3, 4]. In this study, it is observed on the ultrastructural level that MTX, which is used in a large dose range in treatment, has a direct toxic effect on kidney.

1. H.T Abelson. , M.T Fosburg. , G.P. Beardsleng., A.M. Goorin, C. Gorka, M. Link, J. Clin Oncol, **1(3)** (1983), p.208.
2. M. G. el-Badavi, M. H. .Amer, N.M.Dahaba, J. A. Fatani, D. M. Sabah, F.A. Mustafa, Chemotherapy, **33(4)** (1987), p.278.
3. M.P. Iqbal, JPMA J Pak Med Assoc **48(11)** (1998), p.341.
4. Şener, G. , Demiralp, E.E. , Çetiner, M. , Ercan, F. , Yeğen, B.Ç. , European Journal of Pharmacology, **542** (2006), p.170.

A. Aretz, B. Hermanns-Sachweh, J. Mayer (Eds.): EMC 2008, Vol. 3: Life Science, pp. 285–286,
DOI: 10.1007/978-3-540-85228-5_143,

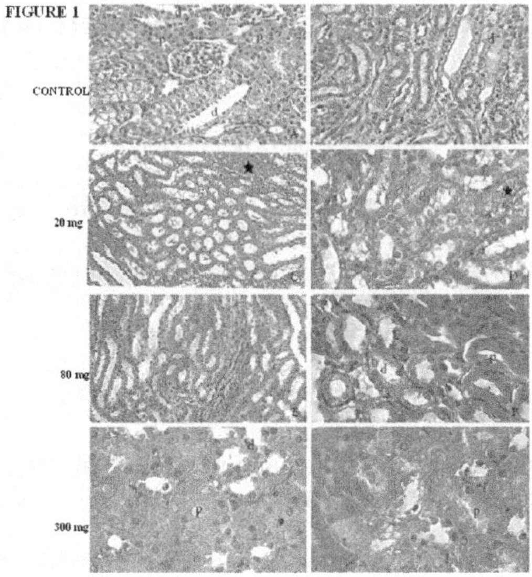

Figure 1. Control, 20 mg/kg, 80 mg/kg, 300 mg/kg MTX administered groups stained by Hematoxylin & Eosin Light Microscope photomicrographs. 1A:X50, 1B:X50, 1C:X25, 1D:X100, 1E:X50, 1F:X250, 1G:X250, 1H:X250. p:Proximal tubule, d:Distal tubule, ✯:Congestion

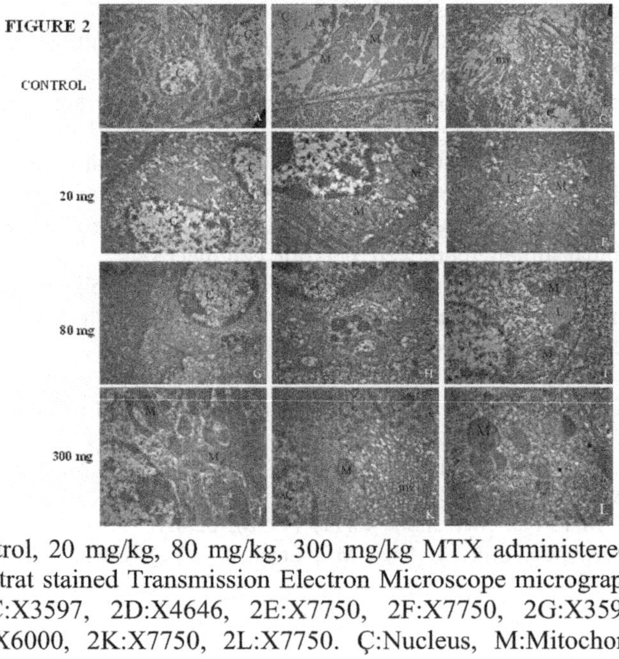

Figure 2. Control, 20 mg/kg, 80 mg/kg, 300 mg/kg MTX administered group Uranil acetate-Lead citrat stained Transmission Electron Microscope micrographs. 2A:X2156, 2B:X3597, 2C:X3597, 2D:X4646, 2E:X7750, 2F:X7750, 2G:X3597, 2H:X7750, 2I:X7750, 2J:X6000, 2K:X7750, 2L:X7750. Ç:Nucleus, M:Mitochondria, L:Lipid, mv:Microvillus

Ultrastructural gadolinium evidence in skin of a patient with Nephrogenic Fibrosing Dermopathy

J.A. Schroeder[1], C. Weingart[2], T. Vogt[3], I. Hausser[4], F. Hofstaedter[1] and B.K. Krämer[2]

Departments of 1. Pathology, 2. Nephrology, and 3. Dermatology, Regensburg University Hospital, F-J-Strauss Allee 11, 93052 Regensburg, Germany
4. Department of Dermatology, Heidelberg University Hospital, Voßstr. 2, 69115 Heidelberg, Germany

josef.schroeder@klinik.uni-regensburg.de
Keywords: gadolinium, nephrogenic fibrosis, EELS

Neprogenic Fibrosing Dermopathy (NFD), now termed Nephrogenic Systemic Fibrosis (NSF), is a recently emerged acquired disorder of the skin and systemic tissues (first description in 2000 by Cowper). It is observed exclusively in patients with renal insufficiency and a history of exposure to gadolinium (Gd) based MRI contrast agents [1]. Gd is widely used in MR imaging as a relatively safe complex with an inorganic linear or macrocyclic chelate. It is distributed equally in the extracellular compartment and is excreted almost exclusively by the kidney, its half-life is increasing dramatically with kidney function deterioration (healthy humans half-life = 1.3 hours versus patients with End-Stage-Renal-Disease = about 120 h) [2]. The pathogenesis of the disorder is not understood, the role of circulating fibroblasts in the fibrosing tissue hypothetical, and the mechanism of the assumed trigger function of Gd elusive [3].

We studied a skin lesion of a 76-y old man with symptoms of a suspected NSF lasting for 5 months; the diagnosis was confirmed by histopathological methods (pronounced fibrosis of corium and subcutis, positive evidence of CD34 and CD68 cells). The EM study of deparaffinized skin samples of this patient, utilizing electron spectroscopic imaging (ESI) and electron energy loss spectroscopic (EELS) microanalysis, revealed numerous Gd deposits (Figures 1, 2) in a wide perivascular zone of the skin vessels. Simultaneously iron was detected in some Gd-deposits as well as in the adjacent tissue [4]. This finding supports the Gd transmetallation hypothesis and its possible direct toxic effect in different tissues and also supports its trigger function for the circulating fibroblasts [5]. We also provide ultrastructural evidence for the ability of lesional fibroblasts to differentiate into myofibroblastic cells [6].

1. W.A. High et al., J Am Acad Dermatol **56** (2007), p. 31-6.
2. T. Grobner and F.C. Prischl, Kidney Int **73** (2007), p. 260-4.
3. S. Swaminathan and S.V. Shah, J Am Soc Nephrol **18** (2007), p. 2636-43.
4. J.A. Schroeder et al., Clin J Am Soc Nephrol (2008), accepted.
5. S.E. Cowper et al., Arthritis Rheum **56** (2007), p.3173-5.
6. B. Eyden, "The myofibrobroblast: a study of normal, reactive and neoplastic tissues, with an emphasis on ultrastructure" (Nuova Immagine Editrice, Siena) (2007).
7. We kindly acknowledge the help of V. Seybold and E. Zellmann (Zeiss Oberkochen), B. Voll and H.I. Siegmund (Central EM-Lab Regensburg).

A. Aretz, B. Hermanns-Sachweh, J. Mayer (Eds.): EMC 2008, Vol. 3: Life Science, pp. 287–288,
DOI: 10.1007/978-3-540-85228-5_144,

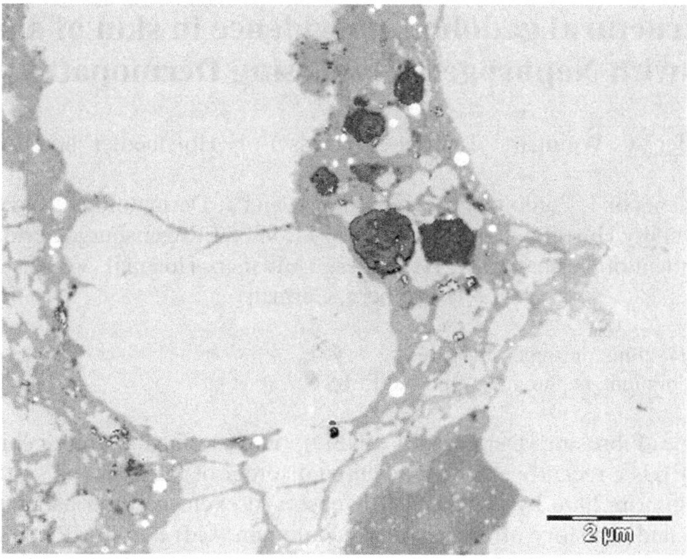

Figure 1. Nephrogenic Fibrosing Dermopathy. Gadolinium and iron mapping in the skin lesion (perivascular area) using ESI. The net Gd (red) and iron (green) signals were combined with the inverted HCI-image (High Contrast Image acquired at 250 eV energy loss in EFTEM-mode) of the non-stained section. Note the tiny iron colocalisation in the intracellular Gd deposits (diameter 100 to 1000 nm) in histiocytes and in the adjacent tissue. Bar 2µm.

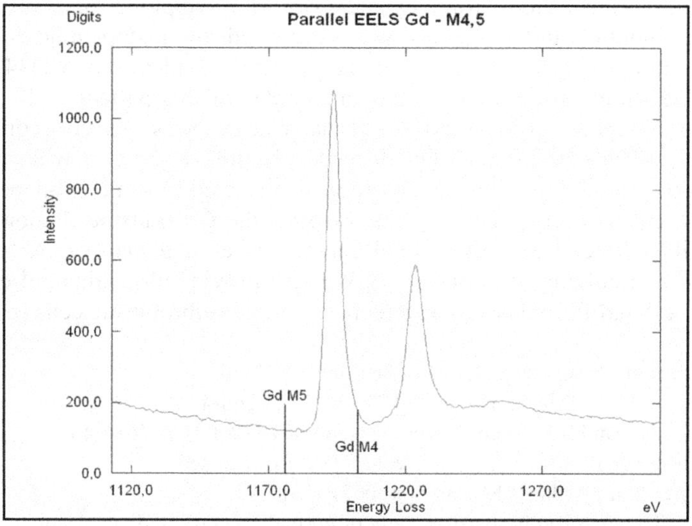

Figure 2. Parallel EELS-spectrum. Record of the measured (EFTEM spot mode) Gd signal (from the large red deposit shown in the Figure 1) displaying the typical "white line" peaks of energy loss at 1,185 eV characteristic for the Gd element.

Nephrotoxicity induced by inorganic Hg(II) and Pb(II): a microscopic and biochemical in vitro study

A. Stacchiotti, F. Morandini, F. Bettoni, L.F. Rodella, P. Grigolato, A. Lavazza and M.F. Aleo

1. Human Anatomy Unit, Dept. Biomedical Sciences and Biotechnology, V.le Europa 11, University of Brescia, I-25123 Brescia, Italy
2. Biochemical Unit, Dept. Biomedical Sciences and Biotechnology, Brescia, Italy
3. 2th Pathological Anatomy, School of Medicine, Brescia University, Italy
4. Istituto Zooprofilattico Sperimentale, Via Bianchi 7/9, I-25100 Brescia, Italy

stacchio@med.unibs.it
Keywords: stress proteins, mercury, lead

Inorganic mercury, Hg(II) and lead, Pb(II), are toxic environmental contaminants and the kidney, mainly cortical proximal tubule, is their important target [1, 2].

Among pathogenic mechanisms of these metals, altered oxidant/antioxidant balance and increased reactive oxygen species (ROS) have been reported.

Stress proteins are a fundamental class of molecules used by cells as chaperones to counteract heavy metals toxicity. They comprise heat shock proteins (HSPs), glucose-regulated-proteins (GRPs) and metallothioneins (MTs) [3].

The aim of this study was to compare the ability of $HgCl_2$ or $PbCl_2$ to activate a stress response in normal rat proximal tubular cells (NRK-52E cell line) and to correlate this reaction to the oxidative damage.

Qualitative and quantitative expression of four stress proteins (HSP25, GRP78, HSP72 and MTs) was analyzed by immunohistochemistry and Western blotting (WB), after 24h, 48h and 72h from the beginning of Hg(II) or Pb(II) treatments; moreover, the transcription of MT1, an early marker of oxidative damage, was tested by RT-PCR at 3h and 24h. The dose-dependent morphological effects of two metals were also checked using light and transmission electron microscopy. To assess the relationship between the stress response and oxidative damage, the intracellular ROS amount, the total glutathione (GSH) levels and Glutathione S-transferase (GST) activity were further estimated.

Immunohistochemistry demonstrated that, at 24h, 20μM $HgCl_2$ was able to induce HSP72 that was instead undetectable in untreated and 60μM or 300μM $PbCl_2$ treated cells. HSP25 enhanced in Hg(II) exposed cells, but it translocated inside nucleus after Pb(II) treatments (Figure 1). Differently, GRP78, the main endoplasmic reticulum chaperone, was more intense and granular after 60μM or 300μM $PbCl_2$, but less evident and diffuse in the cytoplasm after 20μM $HgCl_2$. This trend was quantitatively confirmed by WB and RT-PCR where MT1 mRNA was early transcripted (3h) only after Hg(II) exposure.

Light microscopy studies showed necrosis and vacuolization after 20μM Hg(II) treatment just at 24h, while Pb(II) induced nuclear dense inclusions at 48h. Moreover,

A. Aretz, B. Hermanns-Sachweh, J. Mayer (Eds.): EMC 2008, Vol. 3: Life Science, pp. 289–290, DOI: 10.1007/978-3-540-85228-5_145,

as shown by TEM microscopy, Hg(II) administration to NRK-52E cells induced mitochondrial damage, while Pb(II) caused nucleolar segregation in a dose-dependent manner (Figure 2). As regards $HgCl_2$ treatment, microscopical evidences well correlated with biochemical data that showed marked ROS production at 48h. Remarkably, $PbCl_2$ did not increase ROS levels at every dose and time tested.

In conclusion, the different stress response of these metals support the idea that Pb(II) nephrotoxicity may be due to another mechanism rather than to direct oxidative damage. Our in vitro approach could be useful to test the efficacy of chelating agents or antioxidants against environmental toxic metals.

1. R.K.Zalups, Pharmacol Rev **52** (2000), p.113-143.
2. M.Logham-Adham, Environ Health Perspect **105** (1997), p.928-938.
3. F.X.Beck, W.Neuhofer, E.Muller, Am J Physiol Renal Physiol **279** (2000), p.F203-F215.
4. We kindly acknowledge the expertise of Bertocchi G. for digital imaging, Cadei M. for FACS analysis and Pedretti N. for sectioning TEM samples.

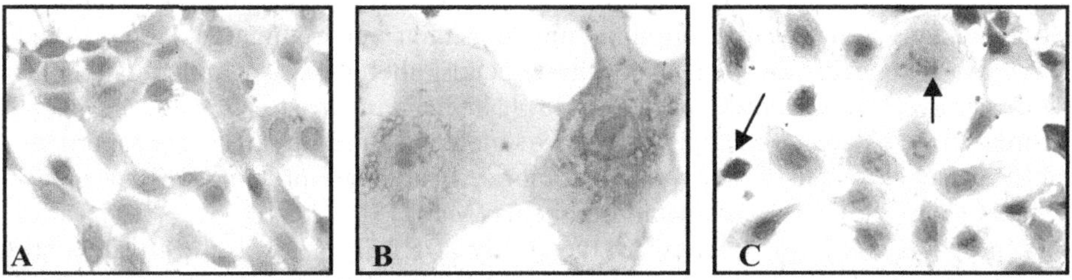

Figure 1. Anti-HSP25 immunostaining in untreated NRK-52E cells (A), after 20μM $HgCl_2$ (B) or 60μM $PbCl_2$ (C) at 24h. Arrows: nuclear positivity. Original magnification: 400X (A,C) and 1,000X (B)

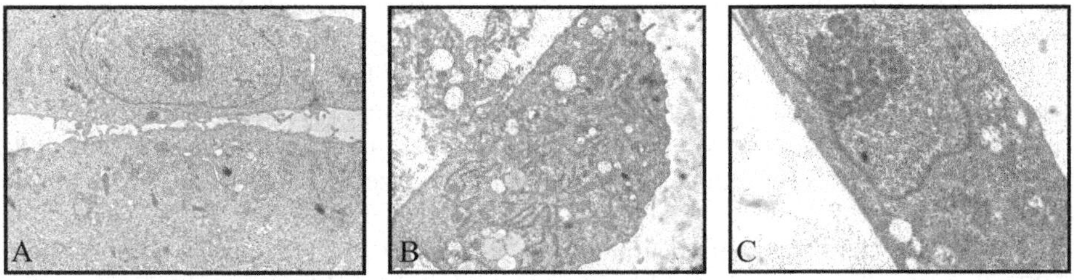

Figure 2. NRK-52E untreated cells (A) or treated with 20μM $HgCl_2$ (B) or 60μM $PbCl_2$ (C) at 24h. Normal ultrastructure in controls, but necrosis or nucleolar segregation after nephrotoxic metals. TEM magnification: 5,200X (A, B) and 6,600X (C)

Tissue distribution of peroxisomes in zebrafish

M. Stevens, O. Krysko , M. Baes, M. Espeel

1. Dep. Anatomy, Embryology, Histology and Medical physics, University of Gent, De Pintelaan 185, 9000 Gent, Belgium
2. Lab of Cell Metabolism, Faculty of Pharmaceutical sciences, Katholieke Universiteit Leuven, Herestraat 49, 3000 Leuven, Belgium

Mieke.Stevens@Ugent.be
Keywords: peroxisomes, zebrafish, metabolic disorders

Peroxisomes are single membrane-bound cell organelles present in almost all eukaryotic cells; they are in particular abundant and large in mammalian liver and in the proximal tubules of the kidney. Peroxisomes have a fine granular matrix containing oxidases, catalase and other enzymes which are involved in VLCFA breakdown, catabolism of purines and polyamines, inactivation of toxic oxygen species and in plasmalogen synthesis [1]. The importance of these metabolic functions is emphasized by the existence of human peroxisomal disorders (Zellweger syndrome, neonatal adrenoleukodystrophy,...). In most cases patients display severe neurological dysfunctions and craniofacial malformations. Sofar, the pathogenic mechanism of these inherited metabolic disorders remains largely unexplained. In order to elucidate the pathogenesis, several mouse models have been generated [2]. In recent years zebrafish models for metabolic human diseases have been developed as an alternative for mouse models [3]. We are exploring the suitability of zebrafish as a model organism for peroxisomal biogenesis disorders. We report here our first observations on the occurrence of peroxisomes in major organs of zebrafish by DAB cytochemistry for catalase activity – catalase is the peroxisomal marker enzyme - and immunostaining against a peroxisomal membrane protein (PMP70).

In the liver of 96 hpf (hours post fertilization) embryos, peroxisomes have a predominantly spherical shape, and they are scattered in the cytoplasm of the hepatocytes (fig 1); in adult liver they are numerous and organized in clusters near the bile canaliculi (BC) (fig 1). In the pronephros of 96 hpf embryos, spherical peroxisomes are prominent, they are concentrated at the apical to perinuclear region of the pronephric duct cells. In the mesonephric tubules of the adult zebrafish, large peroxisomes are seen in the proximal tubules (fig 2). Sofar, no cylindrical elongated peroxisomal profiles, which are found in mouse distal tubules were observed. In the intestinal bulb of 96 hpf embryos no peroxisomes can be identified after DAB incubation. However, in the posterior parts of the intestine numerous small peroxisomes can be seen in the epithelial cells. The predominantly spherical peroxisomes are mostly located at the apical part of the epithelial cells. In adult zebrafish DAB reactive spherical peroxisomes are clearly visualized in the epithelium of all regions of the digestive tract (fig 3). They are mainly present at the apical part of the folds. Compared to the hepatic peroxisomes they are smaller in size. Clusters of peroxisomes are not seen in the intestine.

A. Aretz, B. Hermanns-Sachweh, J. Mayer (Eds.): EMC 2008, Vol. 3: Life Science, pp. 291–292,
DOI: 10.1007/978-3-540-85228-5_146,

We conclude that, in accordance to the data in mouse and men, peroxisomes in zebrafish are most abundant in the liver and the proximal tubules of the kidney (mesonephros). Moreover and in agreement with the situation in mammals, their size and number vary between different organs and within a single organ. Based on these main similarities we suggest that zebrafish might be a suitable alternative for the current expensive and labour-intensive mouse models for peroxisomal diseases; we intend to further explore this alternative model.

1. R.J.A. Wanders and H.R. Waterham, Annu. Rev. Biochem. 75 (2006), p. 295-332.
2. M. Baes and P.P. Van Veldhoven, Biochim. Biophys. Acta 1763, (2006), p. 1785 – 1793.
3. K.N. Baden *et al.* , J.Biol.Chem 282 (2007), p. 34839-34849 .

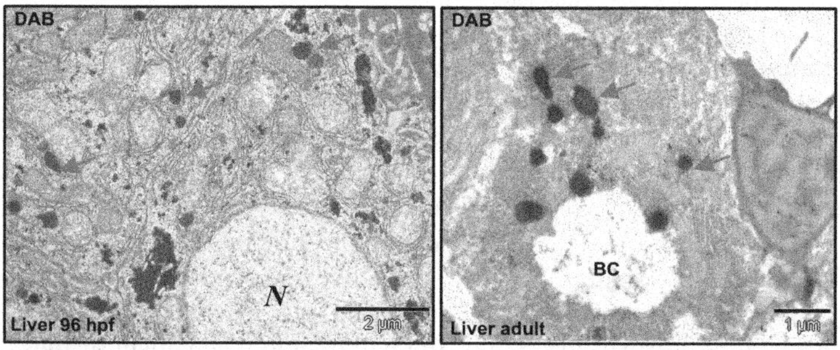

Figure 1. EM photographs of peroxisomes (↓)in the liver of a 96 hpf embryo (left) and adult (right) zebrafish after DAB incubation for catalase activity.

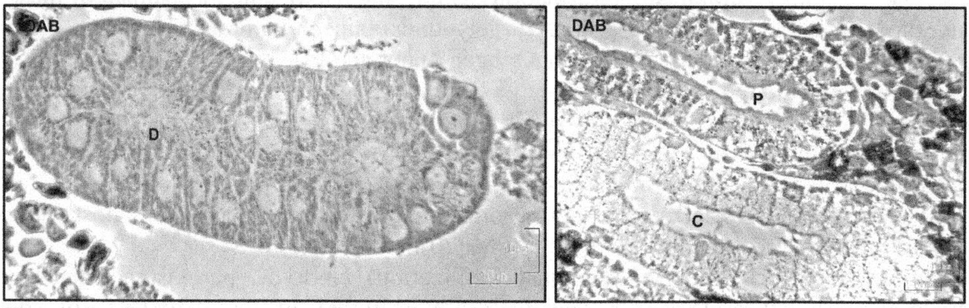

Figure 2. phase contrast of a distal tubule (D, left), a proximal tubule (P) and collecting duc (C; right) of an adult zebrafish after DAB incubation for catalase activity.

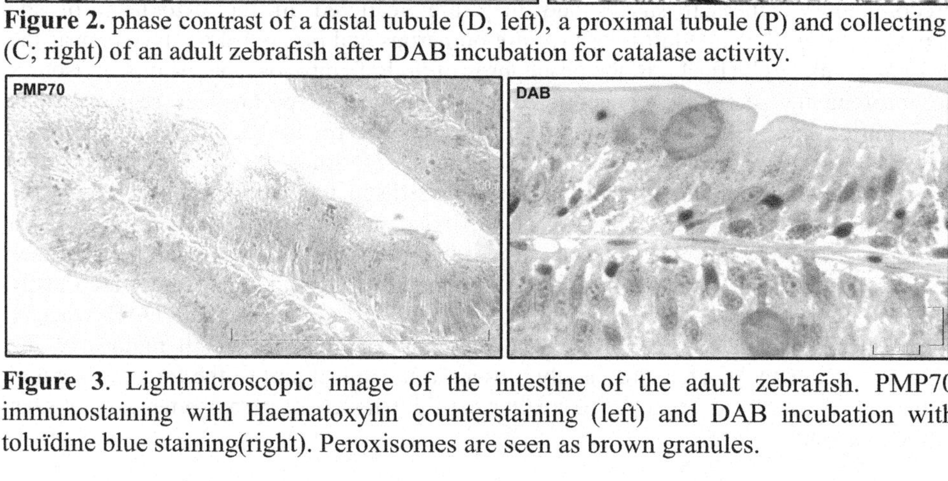

Figure 3. Lightmicroscopic image of the intestine of the adult zebrafish. PMP70 immunostaining with Haematoxylin counterstaining (left) and DAB incubation with toluïdine blue staining(right). Peroxisomes are seen as brown granules.

An analysis of the ultrastructure of matured HIV

I. Takahashi[1], M. Takama[1] and A.M. Ladhoff[2]

1. Lab EM, Teikyo Univ. Sch. Med., Tokyo, Japan
2. Charite Clinic Humboldt Univ. Berlin, Germany

ich@med.teikyo-u.ac.jp
Keywords: HIV matured model, ultrastructure, negative stain technique

The HIV model proposed by L.Henderson and L.Arthur (NIH, USA) or by United States Department of Energy is now well-known as a standard structural model. However from ultrastructural points of view, the features of p17 as the matrix protein were not described in their model. In this research, we proposed a further HIV model based on the ultrastructures and the localizations of p17 by detergent-treated /immuno-labeled negative stain techniques.

HIV-producing H9-clones delived from non-infected HT-lymphocyte lines were cultured for 5 days, 37 degrees centigrade to obtain HIV-1 particles (LAV/HTLV-III strains). Before fixed with formaldehyde-glutaraldehyde mixed solution, part of samples were detergent-treated by saponin and then immuno-labeled with anti-p17/gp41-MoAb. Negative-stained images of these samples were compared to those of no detergent-treated/immuno-labeled samples to reveal the fine structures of p17 and the mutual relations between p17 and gp41.

Negative-stained images of usual HIV particles revealed large amounts of fibrous structures, that seemed to be connected to viral envelopes and cores tightly. To clear the components of these meshwork-like structures, viral envelopes were dissolved by detergent treatments, and then immuno-label procedures with anti-p17/gp41-MoAb were done after detergent treatments to obtain further results. Figure 1 showed a saponin-treated negative-stained image of fibrous structures and Figure 2 showed the localization of p17 on these fibrous structures. Furthermore from the immuno-labeled observations of gp41 that were localized distinctly on the rim of p17 structures, the mutual connections between p17 and gp41 were suggested. Also we obtained another images of p17 that connected to viral cores. According to these results, it was suggested that p17 might play a role as the matrix protein in HIV particles. Hitherto the structures of p17 were considered as thin membranes under the envelopes, however by detergent-treatment the fibrous meshwork-like structures of p17 were revealed. Based on these results, we proposed a modified HIV model in this report.

1. Under the proposal of Dr. I.Takahashi, these experiments were supported by the cooperation with the group of Dr. Hans Gelderblom (Koch Institute. Berlin). We wish to express our gratitudes for all their cooperations.

A. Aretz, B. Hermanns-Sachweh, J. Mayer (Eds.): EMC 2008, Vol. 3: Life Science, pp. 293–294,
DOI: 10.1007/978-3-540-85228-5_147, © Springer-Verlag Berlin Heidelberg 2008

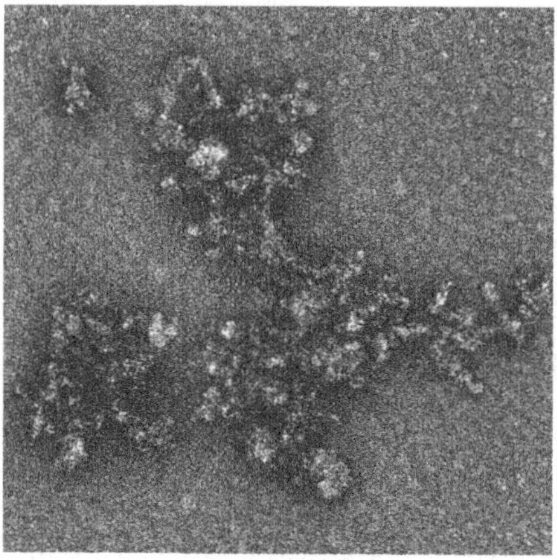

Figure 1

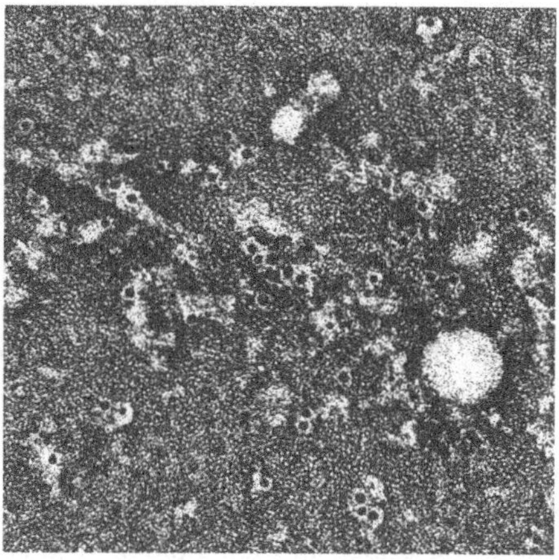

Figure 2

Trichinella spiralis and *Trichinella britovi* cuticle and hypodermal glands Ultrastructure

Tosti M.

Dip. Sc. Biopat. Ig. Prod. An. Alim. – Sez. Parassitologia - Via San Costanzo, n° 4 – Perugia (I) –

parassit@unipg.it
Keywords:TEM, SEM, Trichinella

Cuticles and hypodermal glands of *Trichinella spiralis* and *T. britovi* (Larvae L1) have studied by TEM.

Trichinella spiralis cuticle (Photo No. 2) has produced by an epithelial fine undercuticle or hypoderm, that with the undermuscular layer forms the muscle cutaneous sack. Hypoderm is covered inside by a single muscular cell layer, the contractile fibrillae of which extend up to connect directly with the nervous cords. *T. britovi* cuticle (Photo No. 4) has stratified and presents a similar-papillar disc structure which integrates solidly to the second layer going in connection, first, with the muscular tissue, then, with nervous cords which consent the sensitive function; an excretive duct appears. *T. spiralis* lipidic drops (Photo No. 3) have an oval shape, vague contents and a 200 um diameter; *T. britovi* drops (Photo No. 5) have a round shape, electrondense contents and 300 um diameter. The study of *T. spiralis* and *T. britovi* L1 larvae cuticle and hypodermal glands by SEM and TEM provide significant items useful on taxonomic, biochemical and pharmacological investigations aiming to isolate new anthelminthic molecules.

A. Aretz, B. Hermanns-Sachweh, J. Mayer (Eds.): EMC 2008, Vol. 3: Life Science, pp. 295–296,
DOI: 10.1007/978-3-540-85228-5_148,

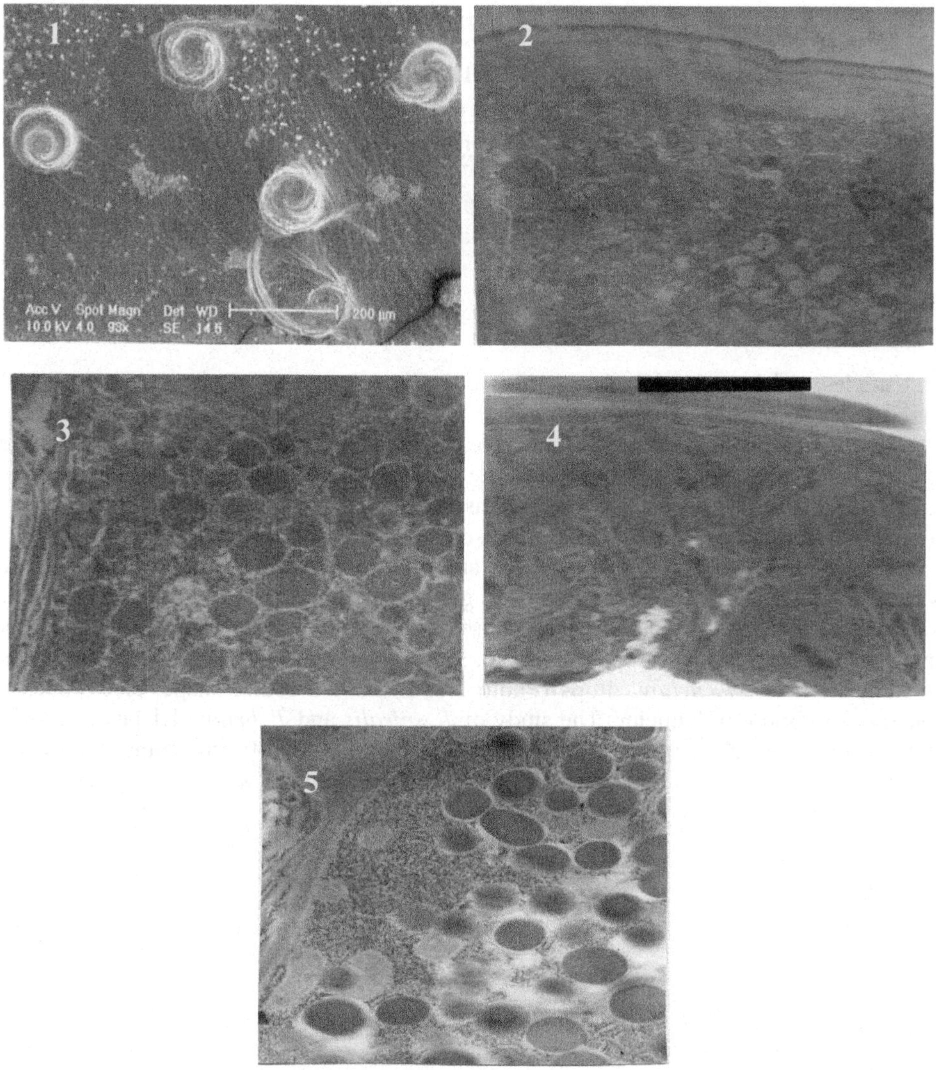

Figure 1. *T. britovi* to SEM, 93x - **Figure 2, 3.** *T. spiralis* to TEM, 10.000x, 12.000x - **Figure 4, 5.** *T. britovi* to TEM, 16.000x, 25.000x -

Ultrastructural analysis of lysosomal storage diseases: effects of therapy

Marius A van den Bergh Weerman[1], Josanne Cox-Brinkman[2], Anouk C Vedder[3], Frits A. Wijburg[2], Carla E.M. Hollak[3], Sandrine Florquin[1]

1. Department of Pathology, Academic Medical Center, Amsterdam, The Netherlands
2. Department of Pediatrics, Academic Medical Center, Amsterdam, The Netherlands
3. Department of Internal Medicine; Academic Medical Center, Amsterdam, The Netherlands

m.a.vdberghweerman@amc.uva.nl
Keywords: electron microscopy, therapy, lysosomal storage disease.

There are more than 50 lysosomal storage diseases. Of these,the diseases of Fabry, mucopolysacharidosis type I (MPS-I), Pompe, Gaucher, Nieman-Pick, Metachromatic leukodystrophy, Krabbe and Neuronal ceroid lipofuscinoses are the most frequent.

Due to an enzyme deficiency, accumulation of abnormal materials in different cell types results in a unique image in electron microscopy.

Bone marrow transplantation, enzyme replacement therapy, substrate reduction therapy and hematopoietic stem cell therapy aim to compensate the enzyme deficiency and restore tissue morphology and function.

In MPS-I, glycosaminoglycans accumulate in different cell types, causing characteristic vacuolization. The accumulation is due to a deficiency of the enzyme α-L-iduronidase. Bone marrow transplantation as well as enzyme replacement therapy aim to restore tissue morphology by delivering α-L-iduronidase to the deficient cells.

We investigated the effect of both therapies on the morphology of dermal fibroblasts in twelve patients by electron microscopy. Skin biopsies were taken before and during two years of therapy.

The same was done for patients with Fabry disease, who lack the enzyme α-galactosidase A, and as a result of which accumulate glycosfingolipids. We scored the accumulation of glycosfingolipids in endothelial cells and pericytes of capillaries in skin biopsies of 17 patients before and during two years of therapy.

Accumulation was rated according to a semi-quantitative scoring system.

Electron microscopy for the diagnosis of lysosomal storage diseases as well as for the evaluation of the efficiency of therapy will be discussed.

A. Aretz, B. Hermanns-Sachweh, J. Mayer (Eds.): EMC 2008, Vol. 3: Life Science, pp. 297–298,
DOI: 10.1007/978-3-540-85228-5_149,

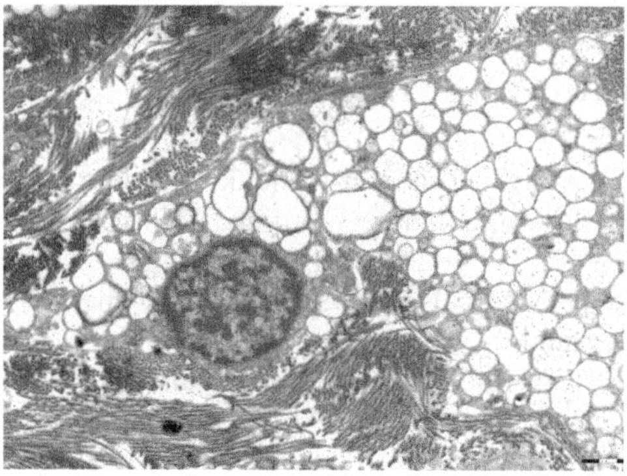

Figure 1. MPS-I patient. Dermal fibroblas, fully occupied by vacuoles. Original magn.: 20.000X

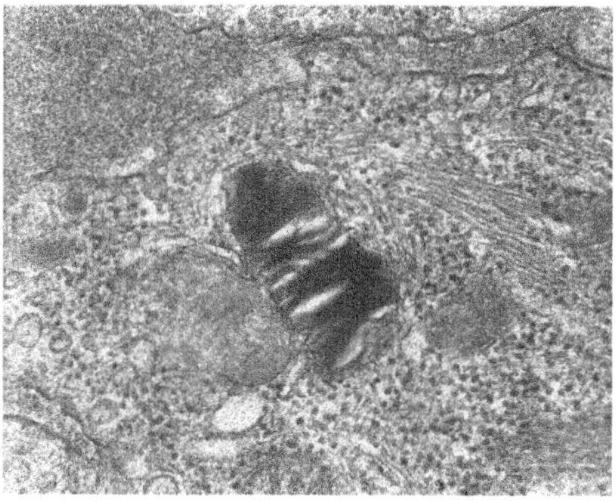

Figure 2. Fabry patient. Dermal capillary: characteristic zebra body in the endothelial cell Original magn.: 100.000X

Immunolabeling for Electronmicroscopy

G. Posthuma

Cell Microscopy Center, Department of Cell Biology, University Medical Center Utrecht, Utrecht, The Netherlands, HTTP://www.cmc-utrecht.nl.

g.posthuma@umcutrecht.nl
Keywords: immunolabeling, electron microscopy, rehydration method

As long as template matching[1] is not a standard technique that allows researchers to generate a template by simple means, immuno electron microscopy is the method of choice to localize interesting molecules at a sub cellular level. In its most simple application the specimen is bound to a grid, labeled with an antibody and a marker (usually a gold particle), fixed, stained, dried and examined in the electron microscope. This is particularly useful for small organisms like bacteria [2] (Figure 1). However for larger specimens more complex procedures are required due to the fact that intracellular epitopes cannot be reached by the antibodies and markers and the thickness of the specimen is limited by its "transparency" for electrons. Whether or not epitopes can be visualized depends on two factors: the actual number of epitopes present in the section and the labeling efficiency. In Table I a theoretical example is given of the amount of marker particles that can be expected after labeling. It is obvious, that even when all the epitopes in a section can be marked (100% efficiency), a substantial number of molecules is necessary to generate enough marker particles.

Table I: Expected number of marker particles in two different locations in a 50 nm section of a standard 10x10x10 µm sized cell labeled with different efficiencies.

	Outer membrane		Cytoplasm	
Molecules in the cell	100% efficiency	10% efficiency	100% efficiency	10% efficiency
1,000	3	0.3	5	0.5
10,000	33	3.3	50	5
100,000	333	33	500	50

Several techniques have been developed to visualize as many epitopes as possible of which the Tokuyasu technique[3] is one of the most successful because it does not use denaturizing solvents nor resin embedding. The Tokuyasu technique consists of 4 major steps: chemical fixation, sectioning, immunolabeling, contrast enhancement. Chemical fixation is one of the least understood parts of the procedure. It takes fixatives (often aldehydes) at least seconds to enter the cells and each and every molecule is affected in a different way. Additionally the size and shape of some organelles is altered as was shown by Murk et al. [4]. Using cryo-immobilization by ultra rapid freezing possible changes in morphology can be minimized. Typically a specimen is frozen in 10 milliseconds, with freezing rates of > 20.000 K/sec. The high freezing rate ensures that ice crystal formation is minimal or even absent. We explored the possibility to perform

A. Aretz, B. Hermanns-Sachweh, J. Mayer (Eds.): EMC 2008, Vol. 3: Life Science, pp. 299–300, DOI: 10.1007/978-3-540-85228-5_150,

immuno-labeling on sections of cryo-immobilized material. Two different approaches were chosen: the section fixation method [5] and a novel approach: the rehydration method [6] in which cryo-immobilized material after freeze substitution with acetone and low temperature fixation with combinations of glutaraldehyde, osmium and uranylacetate was rehydrated, infiltrated with cryo-protectant (2.3 M sucrose in buffer) and frozen again and treated as a standard Tokuyasu specimen. Directly fixed sections can be immuno-labeled but only 5-15 % of the section surface displayed a good morphology. The rehydration method gives far better results but requires an elaborate and delicate freeze substitution procedure and low temperature fixation. The morphology and labeling capabilities are as good as a normal Tokuyasu section (Figure 2).

In conclusion: When a superb morphology with high labeling efficiency is an absolute requirement, the rehydration method is an outstanding, but laborious and difficult method. However in many case the standard Tokuyasu method yields excellent results, as long as one is aware of the possible morphological changes due to the chemical fixation.

1. C. Best, S. Nickell, and W. Baumeister, Methods Cell Biol **79** (2007), p. 615.
2. W.J. van Wamel, A.P. Hendrickx, M.J. Bonten, J. Top, G. Posthuma, R.J. Willems, Infect Immun **75**(2) (2007), p. 924.
3. J.W. Slot and H.J. Geuze, Nat Protoc **2**(10) (2007), p. 2480.
4. J.L. Murk, G. Posthuma, A.J. Koster, H.J. Geuze, A.J. Verkleij, M.J. Kleijmeer, B.M. Humbel, J Microsc, **212** (2003), p. 81.
5. W. Liou, H.J. Geuze, and J.W. Slot, **106** (1996), p. 41.
6. E.van Donselaar, G. Posthuma, D. Zeuschner, B.M. Humbel, J.W. Slot., Traffic **8**(5) (2007), p. 471

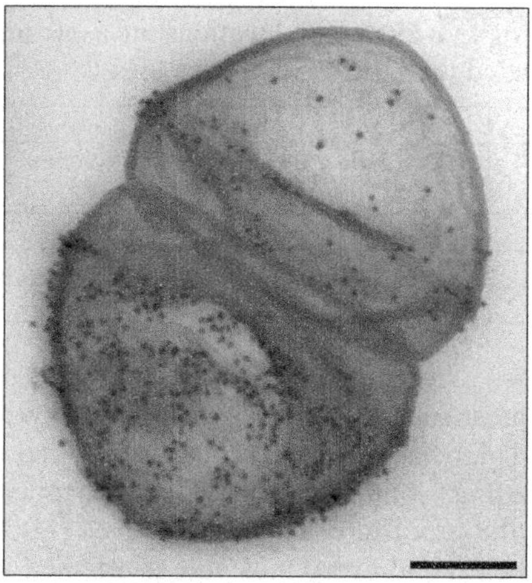

Figure 1. Whole mount immuono-labeling of an Enterococcus surface protein. Bar = 200 nm

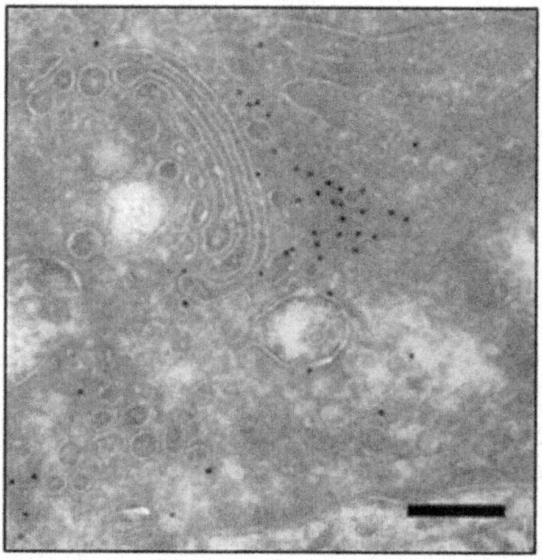

Figure 2. Section made with the rehydration method and labeled for β COP -10 nm protein A-gold particles. Bar = 200 nm

Molecular Organisation of Cadherins in Intercellular Adhesion Junctions by Cryo-Electron Tomography of Vitreous Sections

Ashraf Al-Amoudi and Achilleas Frangakis

European Molecular Biology Laboratory, Meyerhofstrasse 1, 69117 Heidelberg, Germany

alamoudi@embl.de
Keywords: cadherins, cryo-electron tomography of vitreous sections, desmosomes

Intercellular adhesion junctions are fundamental for the function and development of multi-cellular organisms. Desmosomes and adherens junctions (AJs) represent major categories of these junctions. Both desmosomes and AJs are present in epithelia but desmosomes are more abundant in tissues prone to mechanical stress such as heart and skin. The intercellular space of desmosomes and AJs relies on the associations between members of Ca^{2+}- dependent adhesion molecules called cadherins which share high sequence similarity among them. Recently, the first crystal structure of the full extracellular domains of C-cadherin (a representative of classical cadherins) showed that the cadherin molecules adopt a stable, curved conformation [1]. In this crystal structure, the neighbouring molecules are oriented in antiparallel and engaged through a mutual exchange of the tryptophan 2 (Trp2) forming a W-like shape. Such Trp2 *trans*-interactions were supported by mutagenesis data and cell-adhesion assays and are now considered to be physiologically relevant [2].

A recent study based on freeze-substitution of neonatal mouse epidermis and electron tomography showed that the extracellular region of desmosomes form a disorganized three-dimensional (3D) array of cadherin molecules [3]. By fitting the crystal structure of C-cadherins ectodomain [1] into the resulting tomographic maps showed that the cadherins interact via their N-terminal tips. The fitting reproduced the W-shape of C-cadherins [1] but it also revealed a predominant S-shape which results if the N-terminal of one cadherin rotates leaving its Trp2 free without inserting it into the hydrophobic pocket of the neighbouring molecule. This produces a highly flexible cadherin system in which the molecules are arranged in a stochastic manner. In contrast to freeze substitution, we have previously shown by cryo-electron microscopy of vitreous sections (CEMOVIS) [4] and later by cryo-electron tomography of vitreous sections [5] that the extracellular domains of desmosomal cadherins protrude straight from the cell membranes forming a densely packed periodic array. Quantitative analysis using sub-tomogram averaging revealed two predominant *cis*- and *trans*- interactions alternating in a periodic manner similar to the arrangements observed in the linear zipper of the crystal structure of N-cadherins [6]. In addition, the resulting molecular model explains previous two dimensional images observed with CEMOVIS at various orientations and yields important insights into the assembly of cadherin-based intercellular junction.

A. Aretz, B. Hermanns-Sachweh, J. Mayer (Eds.): EMC 2008, Vol. 3: Life Science, pp. 301–302,
DOI: 10.1007/978-3-540-85228-5_151,

1. T.J. Boggon, J. Murray, S. Chappuis-Flament, E. Wong, B.M. Gumbiner and L. Shapiro, Science. **296** (2002) 1308.
2. S. Troyanovsky, Eur J Cell Biol **84** (2005), 225.
3. W. He, P. Cowin and D.L. Stokes, Science. **302** (2003) 109.
4. A. Al-Amoudi, , L. Norlén and J. Dubochet, J. Struct.. Biol.**124** (2005), 764.
5. A. Al-Amoudi, D. C. Díez, M. Betts, A. S. Frangakis**,** Nature **450** (2007), 832
6. L. Shapiro, A.M. Fannon, P.D. Kwong, A. Thompson, M.S. Lehmann,G. Grubel, J.F. Legrand, J. Als-Nielsen, D.R. Colman and W.A. Hendrickson, Nature. **374** (1995) 327.

Ultrastructural observations of indium in the lactating mammary gland cells

Ayadi Ahlem[1], Maghraoui Samira[1], El Hili Ali[2] and Tekaya Leila[1]

1. Laboratoire de Physiologie, Faculté de Médecine de Tunis, 15 rue Jebel Lakhdhar , 1007 Bab Saâsoun, Tunis, Tunisie
2. Laboratoire de Microscopie Electronique, Faculté des Sciences de Tunis, Campus Universitaire 2092, El Manar II Tunis Tunisie

ayadi.ahlem@yahoo.fr
Keywords: aluminium, kidney, microscopy

It becomes more and more important to explore the behaviour into the organism of some mineral elements such as IIIA group elements which are increasingly used in both modern industries and medicine. Some of them like indium are known to be very toxic.

The aim of this work is to study the intracellular localization of this element in the lactating mammary gland cells after its intraperitoneal injection to female rats using the conventional transmission electron microscopy.

The ultrastructural study of the ultra-thin sections from lactating mammary glands has demonstrated the presence of very dense granules in the lysosomes of the glandular epithelial cells (figure 1).

No inclusions were observed in the cell lysosomes of the control rats (figure 2).

Previous studies have attempted to research the intracellular localization of indium after its intraperitoneal injection. These results demonstrated that this element was precipitated in the cell lysosomes of the kidney associated with phosphorus [1; 2; 3].

Another element of the same group, the gallium has been studied. This element has been found, associated with phosphorus, in the lysosomes of the epithelial glandular cells [4].

These data show that the deposits observed in the cell lysosomes may probably be composed of an insoluble salt of indium phosphate. The lysosome seems to be the main organelle in witch are sequestered foreign or toxic elements.

1. J.P. Berry. *The role of lysosomes in the selective concentration of mineral elements. A microanalytical study*. Cell. Mol. Biol. **42 (3)** (1996) 395- 41
2. Galle P. Mécanisme d'élimination rénale de deux éléments du groupe IIIA de la classification périodique: l'aluminium et l'indium. C.R. Acad. Sci. Paris. **292**(1981).91-96.
3. Galle P. Levi-Setti R., Lamperti A., Bourahla K., Escaig F. Subcellular localization of aluminum and indium in the rat kidney. Appl. Surf. Sci. **231-232**(1981). 475-478.
4. J.P. Berry, M.F. Poupon, S. Galle and F. Escaig *Role of lysosomes in gallium concentration bymammalian tissues*. Biol. Cell. **51** (1984) 43-52.

A. Aretz, B. Hermanns-Sachweh, J. Mayer (Eds.): EMC 2008, Vol. 3: Life Science, pp. 303–304, DOI: 10.1007/978-3-540-85228-5_152,

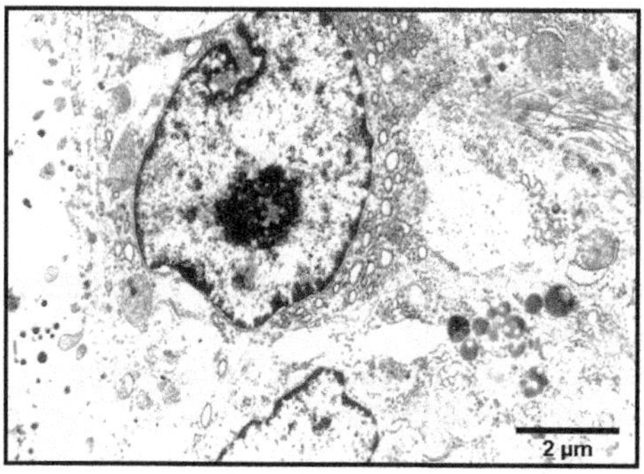

Figure 1. Treated mammary glandular epithelial cell. Altered mitochondria (m) and numerous charged lysosomes (L) are observed in the cytoplasm.

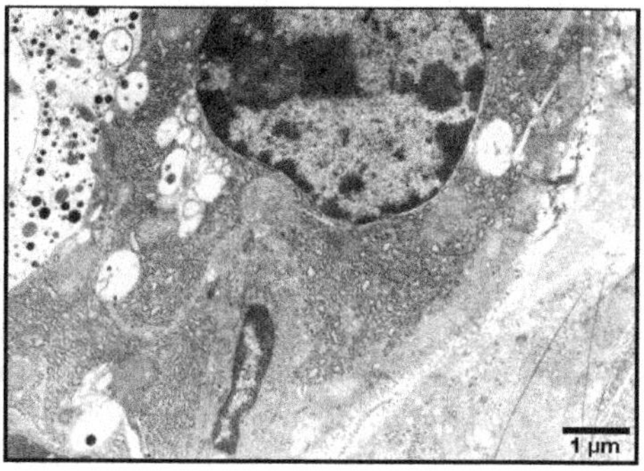

Figure 2. Control mammary glandular epithelial cell. No alterations were observed in lysosomes

More sensitive analysis, such as secondary ion mass spectrometry and electron microprobe, have to been carried out to precise the chemical nature of the intralysosomal deposits.

An active mechanism flanks and modulates the export of the small ribosomal subunits

B. Cisterna [1], M. Malatesta [2], J. Dieker [3], E. Prosperi [4], S. Muller [3], M. Biggiogera [1,4]

1. Laboratorio di Biologia Cellulare e Neurobiologia, Dipartimento di Biologia Animale, Università di Pavia, Italy
2. Dipartimento di Scienze Morfologico-Biomediche, Sezione di Anatomia e Istologia, Università di Verona, Italy
3. Centre National de la Recherche Scientifique (CNRS), Unité Propre de Recherche 9021, Institut de Biologie Moléculaire et Cellulaire, Strasbourg, France
4. Istituto di Genetica Molecolare del CNR, Sezione di Istochimica e Citometria, Università di Pavia, Italy

barbara.cisterna@unipv.it
Keywords: Nucleolus; Ribosomal subunits; Nucleus-Cytoplasm export; Motor proteins; ATP depletion; Electron microscopy; Immunocytochemistry

Ribosome biogenesis is a multi-step process taking place in the nucleolus, and which can be followed through the morphologically well-defined nucleolar components, known as dense fibrillar components (DFC), fibrillar centers (FC) and granular component (GC). DFC and FC are active in the transcription and maturation process of the ribosomal subunits, whereas the latter is the site of later assembling and storage of the pre-ribosomes.

The modalities of movement of the subunits from the nucleolus to the nuclear pores have been only partially clarified, and the idea of a diffusive mechanism is normally favoured, as recently suggested for other RNA-containing structures [1]. However, we have recently suggested the existence of an active, motor protein madiated mechanism involved in the transport of a limited subset (about 10%) of small ribosomal subunits (SSU) [2].

In order to confirm this hypothesis, we give further evidence for the existence of such a mechanism by means of different approaches, such as blocking the *de novo* synthesis of ATP and *in vivo* incorporation (*via* lysophosphatidylcholine, LPC) of antibodies against motor proteins.

Our data can be summarized as follows:

- Both the blockade of ATP synthesis and the incorporation of anti-NMI or anti-actin antibodies modify the distribution and the relative content of the nucleolar components. We observed, by means of an ultrastructural morphometrical analysis: i) a decrease in the area occupied by FC and DFC, active in the rRNA transcription/processing; ii) an increase of GC, site of ribosomal subunits maturation/storage; iii) an increase of the nucleolus associated chromatin (NAC). Figure 1
  The significant reduction of FC and DFC, suggestive of a decrease in the pre-ribosomes production, strongly indicate a decrease in nucleolar activity.

A. Aretz, B. Hermanns-Sachweh, J. Mayer (Eds.): EMC 2008, Vol. 3: Life Science, pp. 305–306,
DOI: 10.1007/978-3-540-85228-5_153,

Intriguingly, the decrease in the production of ribosomal subunits is accompanied by an increase of the GC, thus indicating the presence of an unexpected accumulation of subunits. This suggests that the reduced transcription rate, induced by ATP depletion, would lead to a reduced amount of ribosomal subunits that, however, are not actively exported from GC, but are retained enlarging the latter.

- ATP blockade and temperature lowering increase the amount of RNA retained in the nucleus, as shnown by the cytofluorimetry, and the nuclear S6 (ribosomal protein, marker of the SSU) signal increases after ATP depletion, as observed in western im munoblotting analysis.

It is interesting that the NMI- and actin-blockade induces the same nucleolar reorganization as ATP depletion, thus suggesting for the nuclear motor proteins a role in the transport of ribosomal subunits, as already proposed in [2].

Therefore, the increase in nucleolar GC suggests that ATP depletion, by affecting NMI and actin, can influence not only transcription but also the motor protein-dependent export of the SSU. The existence of this secondary, active mechanism might be utilized when a more rapid and directional export is needed.

1. J.C. Politz, R.A. Tuft, K.V. Prasanth, N. Baudendistel, K.E. Fogarty, L.M. Lifshitz, J. Langowski, D.L. Spector, T. Pederson. Mol. Biol. Cell. **17** (2006) 1239-1249.
2. B. Cisterna, D. Necchi, E. Prosperi, M. Biggiogera. Faseb J. doi:10.1096/fj.05-5278fje.

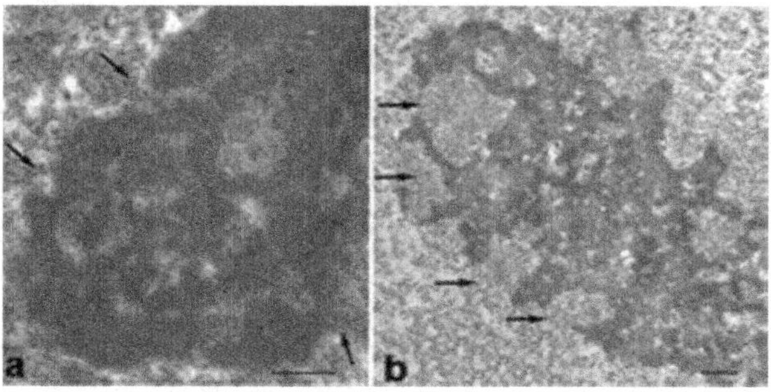

Figure 1. Ultrastructural image of a typical normal nucleolus (a) and a nucleolus of cell treated for the ATP depletion (b), characterized by a reorganization of its components and an increase of NAC (arrows).

New methods for micro-domain detection in bacterial sacculi

M.A. de Pedro[1] and H. Schwarz[2]

1. Centro de Biología Molecular "Severo Ochoa", Nicolás Cabrera 1, 28049 Madrid, Spain
2. Max-Planck Institut für Entwicklungsbiologie, Spemanstrasse 35, 72076 Tübingen, Germany

madepedro@cbm.uam.es
Keywords: sacculi, morphogenesis, immunolabeling.

Bacterial cell shape is defined and maintained by the peptidoglycan (murein) sacculus, a covalently closed, bag-like macromolecule which surrounds the cell on the external side of the cytoplasmic membrane. Because of its covalent net-like structure the shape and dimensions of the sacculus dictate the form and size of the cell [1,2].

To understand how the bacterial sacculus is constructed and how cell shape is determined it is essential to know when and where new subunits and auxiliary components (as Brauns's lipoprotein) are inserted into the pre-existing cell wall structure. While incorporation of new material is clearly necessary to enlarge the sacculus, it is the geometric distribution of these insertion events that is believed to affect the final shape of the bacterium. Mapping these sites should provide a snapshot of the distribution, number and geometry of metabolically active zones at specific times during the cell cycle. Therefore it is important to develop molecular tools for visualizing metabolic activities on the bacterial sacculus.

We will report on the development and early results of two methods aimed to precisely locate active growth sites and murein bound lipoprotein in purified sacculi from Escherichia coli by means of immuno-electron microscopy methods.

Detection of active growth sites in E. coli sacculi is based on the fact that in many bacterial species, including E. coli, the C-terminal D-ala present in murein precursors is cleaved off during the transpeptidation reaction, which cross-links the incoming precursors to the sacculus, or is removed shortly afterwards by the activity of DD-carboxypeptidases [1]. As a consequence, only trace amounts of muropeptides in the sacculus retain the characteristic D-ala-D-ala terminal dipeptide. It is generally accepted that this residual fraction defines insertion sites as the only locations where sizable amounts of D-ala-D-ala terminated muropeptides accumulate. The antibiotic vancomycin has a very high affinity to D-ala-D-ala containing muropeptides and potentially makes a good affinity label for regions enriched in D-ala-D-ala [3]. Vancomycin in turn can be immunolabeled with a nanogold conjugated probe followed by silver enhancement. We are developing the method further to perform "flash-inhibition" of DD-carboxypeptidase activities to allow for controlled periods (down to 3-4 minutes) of D-ala-D-ala accumulation under selected growth conditions. The

A. Aretz, B. Hermanns-Sachweh, J. Mayer (Eds.): EMC 2008, Vol. 3: Life Science, pp. 307–308,
DOI: 10.1007/978-3-540-85228-5_154,

outcome of such experiments should provide new information on the dynamics of growth sites under variable conditions.

A second topic of interest to us is to find out when newly inserted precursors become attached to Braun's lipoprotein (Lpp) [1,4]. Lpp seems to play a key role in anchoring the outer membrane to the sacculus, therefore it is relevant to know when the incoming new materials become "competent" as anchoring sites, in particular at division sites, where the different layers of the cell envelope become temporarily detached. Detection of murein bound Lpp poses specific problems (difficulty to purify Lpp-containing sacculi, lack of appropriate antibodies, etc.). We are developing a new protocol which takes advantage of the presence of a tyrosine residue close to the Lpp C-terminal, right after a thermolysin cleaving site [4]. This combination makes it possible to purify sacculi which instead of the complete Lpp molecule retain the C-terminal hexapeptide bound to the murein. Sacculi can then be specifically labeled at the tyr residue by a modification of the tyramide amplification reaction, and analysed by immuno-electron microscopy methods using nanogold-conjugates and silver enhancement for maximum labeling density. Regions impoverished in Lpp should show no or very reduced labeling in comparison with the old murein which should label heavily because of the high abundance of Lpp.

1. J.-V. Höltje, Microbiol. Mol. Biol. Rev. **62** (1998), 181-203
2. T. den Blaauwen, M.A. de Pedro, M. Nguyen-Disteche, and J.A. Ayala. FEMS Microbiol. Revs. **32** (2008), 321-344
3. P.J. Loll, and P.H. Axelsen. Annu. Rev. Biophys. Biomol. Struct. **29** (2000), 265-289.
4. V. Braun, and V. Bosch. Eur. J. Biochem. 28 (1972), 51-69.

Subcellular localization of Myogenic Regulatory Factors along skeletal muscle development

E. Falcieri[1,3], S. Burattini[1], L. Biagiotti[1], P. Ferri[1], A. D'Emilio[1], M. Guescini[2], E. Barbieri[2], V. Stocchi[2]

1. Istituto di Scienze Morfologiche Università degli Studi di Urbino "Carlo Bo"
2. Istituto di Ricerca sull'Attività Motoria, Università degli Studi di Urbino "Carlo Bo"
3. Istituto di Genetica Molecolare,CNR, Istituti Ortopedici Rizzoli, Bologna

elisabetta.falcieri@uniurb.it
Keywords: skeletal muscle cells, MRF localization, Western Blot, PCR-real time

Myogenesis is a multistep process controlled by a transcriptional cascade in which myogenesis regulatory factors (MRFs) have a crucial role [1]. MRFs are a family of basic helix-loop-helix proteins, composed by Myo-D, Myf-5, Myogenin and MRF-4 [2]. They are not activated at the same time: in fact, Myo-D and Myf-5 induce differentiation at earlier step, while Myogenin and MRF-4 are important for myotube formation and for myogenic lineage maintenance [2,3]. C2C12 are murine myoblasts derived from satellite cells and their behaviour corresponds to that of progenitor lineage. They are a subclone of C2 myoblast which spontaneously differentiates in culture after serum removal [5].

For their characteristics, C2C12 murine cell line is considered a reliable model to study skeletal muscle development, both in terms of morphogenesis and expression of regulative molecules.

C2C12 adherent myoblasts were grown and induced to differentiate as previously described [6]. MRFs were investigated by different approaches. The PCR-real time and Western blot techniques were used to quantify mRNA and protein expression levels (see figure), whereas protein subcellular localization was investigated by immunofluorescence (IF), also observed at confocal microscope. Moreover, MRF protein distribution was studied by means of immunogold (IG) technique and investigated by transmission electron microscopy, on LRW thin sections from paraformaldehyde-fixed cells. Undifferentiated, early and late differentiation stage were investigated for all MRFs and the behaviour of MRF-4 is displayed in the figure, as an example.

Nucleus and cytoplasm of undifferentiated cells are positive to MRF-4; in particular, discrete positive patches are shown, throughout cytoplasm, both by IF (A) and IG (B), particularly in perinuclear areas. At early differentiation stage, the positivity is evident both in the nucleus, were the nucleoli are negative, and, more, in the cytoplasm (C,D). Myotubes present a strong staining, more intense throughout the cytoplasm (E,F).

PCR-real time (G) and Western Blotting (H) results are in agreement with microscopic research (IF, IG). To conclude, not only the quantitative time-correlated MRF expression, but also their subcellular localization ,which changes along the differentiation, plays a significant role in skeletal muscle differentiation.

A. Aretz, B. Hermanns-Sachweh, J. Mayer (Eds.): EMC 2008, Vol. 3: Life Science, pp. 309–310,
DOI: 10.1007/978-3-540-85228-5_155,

1. Travaglione et al., Cell Death Diff, 12:78, 2005;
2. Kataoka et al., J Biol Chem, 45:44178, 2003;
3. Sun et al., J Biol Chem, 280:26448, 2005;
4. Kassar-Duchossoy L. et al. Nature, 431:466-471, 2004;
5. Burattini et al. Eur J Histochem, 48:223, 2004;
6. Curci et al. Micron 2008, in press

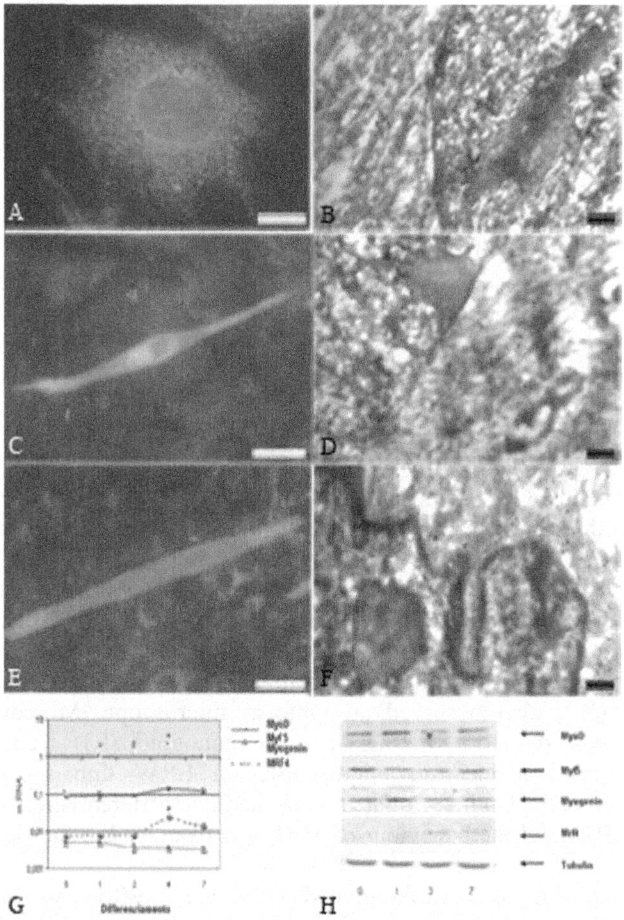

Figure 1. MRF-4 localization in C2C12 cells at 0 days (A,B), 3 days (C,D) and 7 days (E,F) after IF (A,C,E) and IG (B,D,F). At 0 days nucleus and cytoplasm, with a particular spot-like distribution, are positive; at 3 days the positivity appears both in the nucleus and in the cytoplasm, more evidently around nucleus. Myotubes present a strong labelling, diffusely in the cytoplasm (E,F).
G, PCR-real time and H, Western Blotting, both referred to all MRFs and to the complete differentiation protocol. A, bar=0,5μm; C,E, bar=20 μm; B,D,F, bar=0,2 μm.

Immunocytochemical Strategies: LR Resins or Lowicryl, Gold or Peroxidase – Which is Better?

J. A. Hobot and C.J. von Ruhland

Medical Microscopy Sciences, School of Medicine, Cardiff University. Heath Park, Cardiff CF14 4XN, UK.

hobot@cf.ac.uk
Keywords: Immunocytochemistry, resins, tomography

On-section immunocytochemistry is divided into two parts: (i) processing of biological tissue in order to obtain sections for microscopy, and (ii) immunolabelling of sections. Many of the more successful microscopical methods employ delicate aldehyde fixation of biological tissue followed by "sympathetic" processing into an acrylic resin. Processing regimens do not have to be complicated. Simple and cost effective room temperature protocols utilising partial dehydration have been devised, and they can be as effective as the more complex low temperature techniques in preserving both ultrastructure and antigenic reactivity [1,2]. The embedded material can be investigated by either light or electron microscopy. Frozen sections can be cut and immunolabelled but only if the tissue is chemically fixed first, as in resin embedding. Fixation with low concentrations of aldehyde (< 0.2%) will normally better preserve tissue immunoreactivity but this may be at the expense of good ultrastructure with these protocols. If so, low temperature resin embedding methods or rapid freezing and cryosubstitution can be tried. The choice of processing protocol will determine which acrylic resin to use, as will the preference for subsequent immunolabelling with either gold (Figure 1a) or peroxidase/diaminobenzidine (DAB) (Figure 1b). Both types of labelling system offer advantages to localisation studies and can be used in combination for double or even triple labelling. Silver enhancement of the colloidal gold or DAB allows for improved observation by light microscopy.

Further, as antibodies can be shown to penetrate into acrylic resin sections, markers that also penetrate into the resin section can be used to immunolabel antigenic sites *within* the section. This is certainly the case with the peroxidase/DAB system but not colloidal gold, where the gold only sits on the surface of the section [3]. The advantages for 3D electron tomography should become obvious. Whilst 3D tomography is an important tool for furthering our understanding of cellular structures, current applications of this technology for immunolabelling utilising just colloidal gold are very limited. The gold simply sits on the section surface, whilst all the antigenic sites *within* the section stay unlabelled. Thus the 3D cellular structure is revealed, but no new information about the distribution of antigenic sites is uncovered. By accepting the evidence for antibody penetration into sections and using peroxidase/DAB, or by developing novel penetrable markers, 3D electron tomography would be able to offer a whole new range of exciting possibilities to biology.

A. Aretz, B. Hermanns-Sachweh, J. Mayer (Eds.): EMC 2008, Vol. 3: Life Science, pp. 311–312,
DOI: 10.1007/978-3-540-85228-5_156,

1. J.A. Hobot and G.R. Newman, Scanning Microsc Suppl **5** (1991) p. S27.
2. G.R. Newman and J.A. Hobot "Resin Microscopy and On-Section Immunocytochemistry" (2nd Ed. Springer-Verlag, Heidelberg) (2001).
3. G.R. Newman and J.A. Hobot, J. Histochem Cytochem **35** 971 (1987).

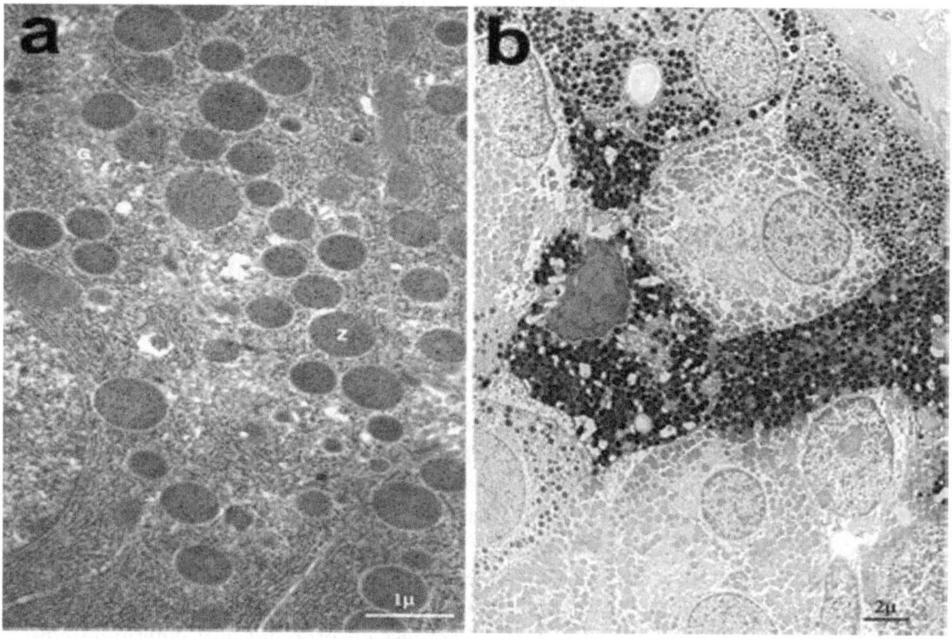

Figure 1. Immunolabelling of thin sections with colloidal gold (a) or DAB/peroxidase. In (a) rat pancreas has been perfusion fixed with 1% glutaraldehyde, embedded in Lowicryl HM20 by the progressive lowering of temperature technique, and labelled for rat anionic trypsin with goat anti-rabbit IgG conjugated to 10nm colloidal gold. Sections counterstained with uranyl and lead acetate. In (b) pituitary of human has been immersion fixed with buffered glutaraldehyde picric acid, embedded in LR White following partial dehydration at room temperature, and treated with DNP-labelled rabbit anti-ACTH followed by treatment with DAB/peroxidase and gold chloride. No counterstaining.

Elemental Analysis in Electron Microscopy for Medical Diagnostics

L. Jonas[1] and S. Schäd[2]

1. Electron Microscopic Centre in the Department of Pathology, Medical Faculty, University of Rostock, Strempelstr. 14, D-18055 Rostock, Germany
2. Department of Dermatology and Venerology, Medical Faculty, University of Rostock, Augustenstr. 80, D-18055 Rostock, Germany

ludwig.jonas@med.uni-rostock.de
Keywords: Medical Diagnostics, elemental analysis, EDX/EELS

The study of tissues for medical diagnostics in electron microscope gives not only information about the ultrastructure, but also indices about the elemental composition of samples, when you have an analytical instrument. When an electron beam irradiates a specimen, a number of different interactions occur. Besides others X-rays and inelastically scattered electrons carry information about the elemental composition of the specimen in the region that is being radiated. Analytical electron microscopes are in most cases combined with an Energy Dispersive X-rax (EDX) system [1]. In transmission electron microscopy there is an additional possibility for elemental analysis: Electron Energy Loss Spectroscopy (EELS) by filtering the inelastically scattered electrons [2].

In the Electron Microscopic Centre (EMZ) Rostock, there are a SEM DSM 960 A (Zeiss) with an EDX system (KEVEX) and two TEMs: EM 902 A (Zeiss) with EELS and Libra 120 with EELS and EDX (EDAX). All three analytical instruments are used for scientific and diagnostic purpose.

Here, we give three examples of usefulness of elemental analysis to differentiate skin discolorations and skin diseases: 1. Argyria (Figure 1) after abusus of silver containing drugs [3], 2. Chrysosis (Figure 2) after long-term administration of colloidal gold for medication of rheumatoid arthritis 3. Gadolinium mediated nephrogenic systemic fibrosis with Scerodermia (Figure 3) after several Gadolinium containing contrast material injections for MRT diagnostics. Several other cases were published, where we used elemental analysis in TEM and SEM for medical diagnostics [4, 5, 6, 7].

1. A. Warley: X-ray microanalysis for biologists. In: Practical methods in electron microscopy (ed by A.M. Glauert), Portland Press London 2002
2. RRR.F. Egerton: Electron energy-loss spectroscopy in the electron microscope. Plenum Press new York, London 1996
3. L. Jonas et al. Ultrastruct Pathol 31 (2007), 379-384.
4. L. Jonas et al. Ultrastruct Pathol 25 (2001), 111-118.
5. L. Jonas et al. Ultrastruct Pathol 26 (2002), 23-26.
6. L. Jonas et al. Ultrastruct Pathol 26 (2002), 323-329.
7. L. Jonas et al. Ultrastruct Pathol 29 (2005), 405-413.

A. Aretz, B. Hermanns-Sachweh, J. Mayer (Eds.): EMC 2008, Vol. 3: Life Science, pp. 313–314,
DOI: 10.1007/978-3-540-85228-5_157,

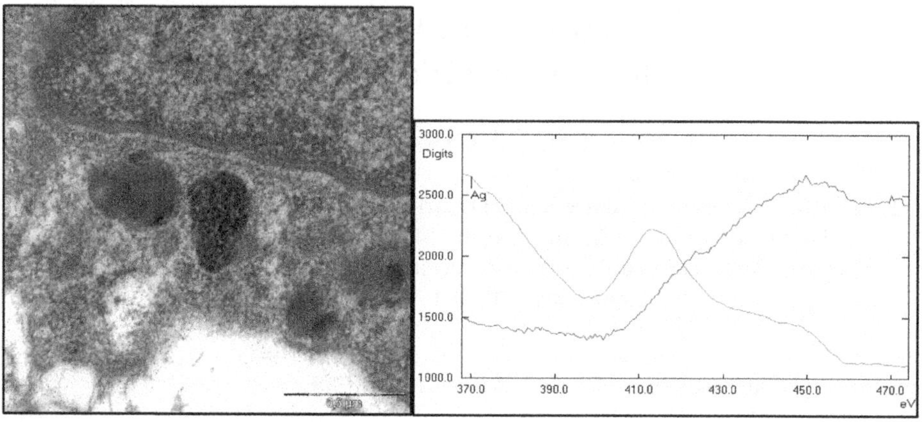

Figure 1. Argyria, silver accumulation in skin histiocyte, Ag, detection by EELS in EM 902 A

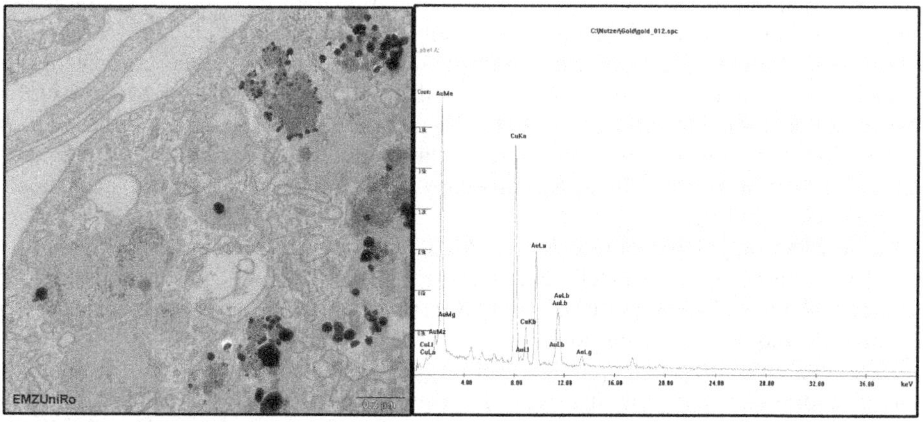

Figure 2. EDAX

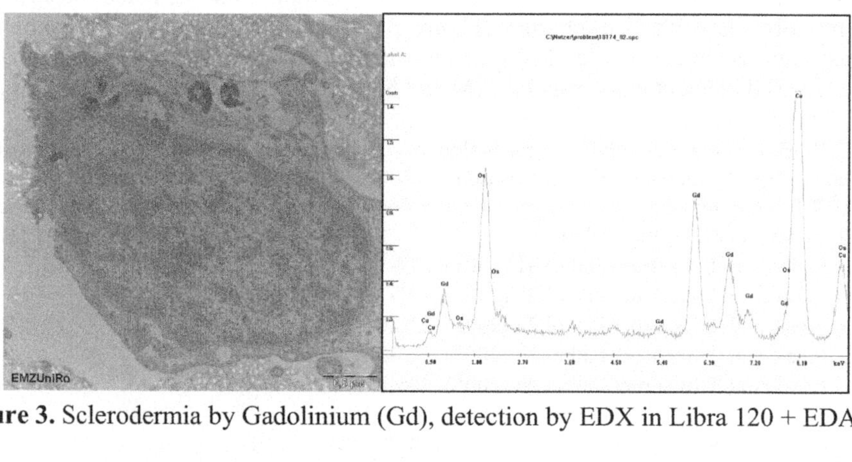

Figure 3. Sclerodermia by Gadolinium (Gd), detection by EDX in Libra 120 + EDAX

Synaptic localization of KCa1.1 potassium channels in central neurons revealed by postembedding immunogold and SDS-digested freeze-fracture replica labelling

W.A. Kaufmann[1], R. Shigemoto[2], H.-G. Knaus[3] and F. Ferraguti[1]

1. Dept. Pharmacology, Innsbruck Medical University, 6020 Innsbruck, Austria
2. Div. Cerebral Structure, Nat. Inst. Physiol. Sciences, Okazaki 444-8787, Japan
3. Div. Molecular and Cellular Pharmacology, Innsbruck Medical University, 6020 Innsbruck, Austria

walter.kaufmann@i-med.ac.at
Keywords: immunogold, subcellular, microdomain

The physiological impact of each ion channel strongly depends on the site where it is expressed: in which type of neurons and exactly in which subcellular domain within each neuron. This study addressed the precise subcellular localization of calcium-activated potassium channels of the BK type (KCa1.1 or BK channels) in hippocampal pyramidal and cerebellar Purkinje cells. Affinity purified rabbit antisera directed against the principal subunit alpha1 were used for immunodetection. The ultrastructural distribution was studied by means of preembedding immunoperoxidase, postembedding immunogold and SDS-digested freeze-fracture replica labelling (SDS-FRL). Specificity of immunoreaction was controlled and confirmed on samples from BK null mice (kindly provided by P. Ruth, University of Tuebingen, Germany).

In CA1 stratum radiatum, BK channels were found enriched at the active zone of axon terminals forming asymmetric synapses. They were also detected in dendritic spines of CA1 pyramidal cells localized to the extrasynaptic membrane of spine head and neck. In the cerebellar cortex, BK channels were present in postsynaptic domains of Purkinje cells. Scattered and clustered pools of channels were found in the extrasynaptic plasma membrane of dendritic and somatic sites (Figure 1). Parallel fibre inputs were immunonegative. In both, hippocampal pyramidal as well as cerebellar Purkinje cells, BK immunoreactivity was associated with excitatory, asymmetric synapses but not with inhibitory synaptic junctions (Figure 2). This was revealed by co-localization experiments of BK channels with domain-specific marker proteins such as GABA-A alpha1, NMDA-R1 or mGluR1alpha.

Our data provide the first clear evidence that BK channels do reside at both pre- and postsynaptic domains of excitatory synapses in hippocampus and cerebellar cortex. However, labelling at both the pre- and postsynaptic elements of the same synapse was never detected.

A. Aretz, B. Hermanns-Sachweh, J. Mayer (Eds.): EMC 2008, Vol. 3: Life Science, pp. 315–316,
DOI: 10.1007/978-3-540-85228-5_158,

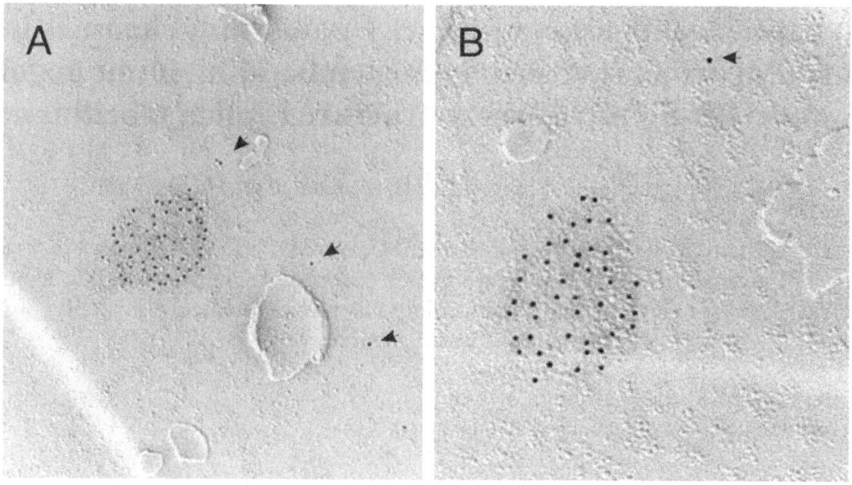

Figure 1. Localization of BK channels (10nm gold) at the plasma membrane P-face of cerebellar Purkinje cells. Antibodies directed against sequence 913-26 (shown in A) and 1118-32 (shown in B) yielded identical results. Two pools of channels were detected, one being clustered and the other scattered over the membrane (indicated with arrows).

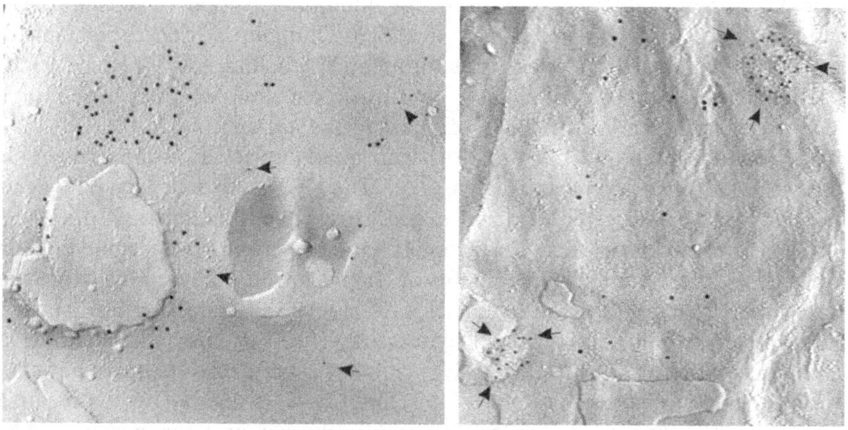

Figure 2. Distribution of BK channels and $GABA_A$ receptor alpha1 subunits at the Purkinje cell plasma membrane: BK channels (10nm gold particles, indicated with arrows) are not co-localized with GABA-A receptor alpha1 (15nm gold particles) indicating that both the scattered (image to the left) and the clustered pool of BK channels (image to the right) are absent from postsynaptic membrane specializations of inhibitory synapses.

Ultrastructural studies of rod photoreceptor nuclei from SCA7 mouse

C. Kizilyaprak, D. Spehner, D. Devys, L. Tora, P. Schultz

1. Department of structural biology and genomics IGBMC 67404 Illkirch France
2. Department of functional genomics IGBMC 67404 Illkirch France

caroline.kizilyaprak@igbmc.u-strasbg.fr
Keywords: SCA7, TFTC/STAGA, high pressure freezing, cryo-substitution, electron microscopy.

The spinocerebellar ataxia SCA7 is a polyglutamine (polyQ) expansion disease in which the retina is affected and results in blindness. Ataxin-7 (ATXN7), the protein mutated in SCA7, is a subunit of the hSAGA complex. The hSAGA coactivator is known to regulate transcription, notably through histone acetylation mediated by the GCN5 histone acetyltransferase. In SCA7 mouse models the overexpression of polyQ-expanded ATXN7 in rod photoreceptors induces a progressive retinal phenotype resulting in rods dysfunction with electrophysiological, histological and transcriptional abnormalities leading to a dramatic alteration of the rod nuclear architecture [1].

Analysis of SCA7 mouse shows an accumulation of poly-Q expanded ATXN7 in rod photoreceptors nuclei leading to the formation of insoluble nuclear inclusions, a hallmark of poly-Q disease. Nuclear accumulation of mutant ATXN7 is an important step in pathogenesis [2] although the mechanisms underlying polyQ toxicity remain unclear [3].

In a first step an ultrastructural investigation of rod photoreceptors nuclei was performed in order to better understand the pathogenic mechanism. Normal and mutant retina samples were prepared by high pressure freezing to preserve optimally the structure of the nuclei. The sample was cryo-substituted, sectioned and observed in electron microscopy. The good quality of the ultrastructural preservation of the rod photoreceptor nuclei was assessed by the absence of ice crystal induced segregation. Wild-type nuclei show a particular organization of the chromatin in which a thin border of euchromatin surrounds a large central heterochromatin territory (Figure 1). In SCA7 nuclei, chromatin is dramatically disorganized with an apparent decondensation evidenced by the appearance of internal euchromatin patches (Figure 2).

In a second step, cryo-sections were produced in order to localize polyQ aggregates by immuno-staining. The Tokuyasu method [4] was used to perform both immuno-fluorescence and immuno-electron microscopy experiments on the same sample. At the optical microscopy level, ultra-thin cryo-sections revealed a single poly-Q aggregate per nucleus (Figure 3). Electron microscopy inspection showed that these inclusions are exclusively located in the euchromatin compartment and present a fibrillar structure (Figure 4).

Future electron tomography analysis will allow us to describe in more details the interaction of chromatin with polyQ aggregates and investigate their possible influence on the deregulated gene expression that is associated with the disease.

A. Aretz, B. Hermanns-Sachweh, J. Mayer (Eds.): EMC 2008, Vol. 3: Life Science, pp. 317–318,
DOI: 10.1007/978-3-540-85228-5_159,

1. D.Helmlinger, S.Hardy, G. Abou-Sleymane, A. Eberlin, AB Bowman, A Gansmu S, S. Picaud, HY. Zoghbi, Y. Trottier, L. Tora and D. Devys, PLoS Biol **4** (2006).
2. G. Yvert, D.Lindenberg, D. Devys, D.Helmlinger, GB. Landwehrmeyer, JL. Mandel, Hum Mol Genet **10** (2001).
3. HT. Orr and HY. Zoghbi, Annu. Rev. Neurosci **30** (2007)
4. KT. Tokuyasu, the Journal of Cell Biology **57** (1973).

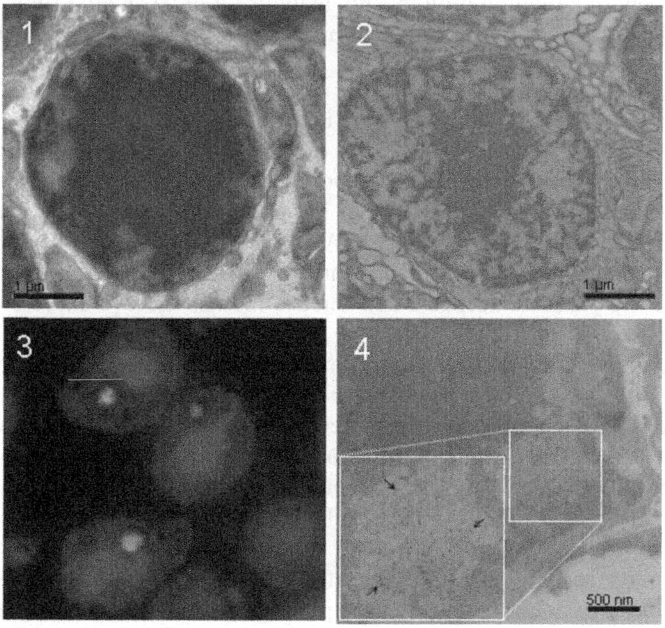

Ultrastructure of rods photoreceptors nuclei:

Figure 1. Electron micrograph of an ultrathin (100nm) section showing a nucleus of rod photoreceptor from wilt-type, 13 weeks old, mice after high pressure freezing, cryo-substitution and Epon embedding. The sample appears well preserved with a thin border of euchromatin surrounding a large heterochromatin territory.

Figure 2. Electron micrograph of 100nm thick section of rod nucleus from SCA7 mice (prepared as in Figure 1) showing a dramatic chromatin disorganization.

Figure 3. Fluorescence micrograph showing a 100nm cryo-section of rods nucleus from SCA7 mice labelled for ATXN-7 and detected with Alexa-488.

Figure 4. Electron micrograph showing the same immuno-labelling detected with 10-nm gold coupled protein A. Both detections yield similar labelling patterns. ATXN-7 forms fibrillar nuclear inclusion in euchromatin territory (arrows).

Application of electron energy loss spectroscopy for heavy metal localization in unicellular algae

U. Lütz-Meindl, M. Milla, B. Stöger

Plant Physiology Division, Cell Biology Department, University of Salzburg, Hellbrunnerstrasse 34, 5020 Salzburg, Austria

ursula.meindl@sbg.ac.at
Keywords: algae, EELS, heavy metals

Fresh water algae may be exposed to various heavy metals due to agricultural practices, industrial waste waters and atmospheric pollution. Whereas numerous studies have investigated heavy metal impact on physiological and molecular responses of algae, or on biomass production, only little information is so far available on subcellular localization of heavy metals in intra- or extracellular compartments. This lack in data is at least in part due to limited employment of analytical transmission electron microscopy, like electron energy loss spectroscopy (EELS) in biological sciences and particularly in plant biology [1].

In the present study the fresh water green alga *Micrasterias* (Desmidiaceae) which has been used as cell biological model system in numerous investigations is experimentally exposed to salts of the heavy metals copper, cadmium, aluminium and zinc in order to analyse the response of the algae to their impact on growth, development, cell division rate and ultrastructural changes of organelles, secretion pathway or cell walls. Intracellular distribution of the heavy metals is investigated by EELS and element mapping (electron spectroscopic imaging, ESI) on ultrathin sections of high-pressure frozen and cryo-substituted *Micrasterias* cells (for methods see [2,3]) to obtain insight into possible defence or detoxification mechanisms. This is of high ecological relevance as the natural habitats of this group of algae are acid oligotrophic peat bogs which are among the last intact ecosystems in Europe.

The results show that when applied during cell development aluminium, zinc and cadmium induce distinct changes in cell growth ranging from reduction in cell size to severe alterations in cell morphology and symmetry of the cell pattern (Figures 1-3). At an ultrastructural level aluminium and copper lead to malformations of the primary cell wall which forms abnormal pulvinate accumulations of wall material in the case of aluminium (Figure 4) respectively appear unusual inhomogenous in the case of copper (Figure 5). Dictyosomal morphology and secretion activity is affected by aluminium and copper but not by zinc and cadmium. Vacuolar precipitations are found after exposure to aluminium and copper.

EELS measurements revealed accumulation of aluminium in the cell wall already after short term treatment (4h) whereas copper could be measured in the cell wall only after long-term (3 weeks) exposure (Figures 6,7). Additionally copper was measured in mucilage vesicles and in the chloroplast. As in plant cells extracellular polysaccharides like pectins which are main constituents of the cell wall or the adjacent mucilage layer

A. Aretz, B. Hermanns-Sachweh, J. Mayer (Eds.): EMC 2008, Vol. 3: Life Science, pp. 319–320,
DOI: 10.1007/978-3-540-85228-5_160, © Springer-Verlag Berlin Heidelberg 2008

may adsorb heavy metals due to negative charges, ultrathin sections of control cells were pre-treated with heavy metal solutions, rinsed and investigated by EELS. This reveals that e.g. copper and cadmium are specifically bound by cell wall, mucilage and the corresponding vesicles, but not by any other cell component. These results are in good correlation with recent investigation on the calcium binding capacities of the extracellular sheath in *Micrasterias* [4].

1. U. Lütz-Meindl, Micron **38** (2007), p.181
2. U. Lütz-Meindl, N. Aichinger, Protoplasma **223** (2004), p. 155
3. U. Lütz-Meindl, C. Lütz, Micron **37** (2006), p. 452
4. M. Eder, U. Lütz-Meindl, J. Microscopy (2008), in press

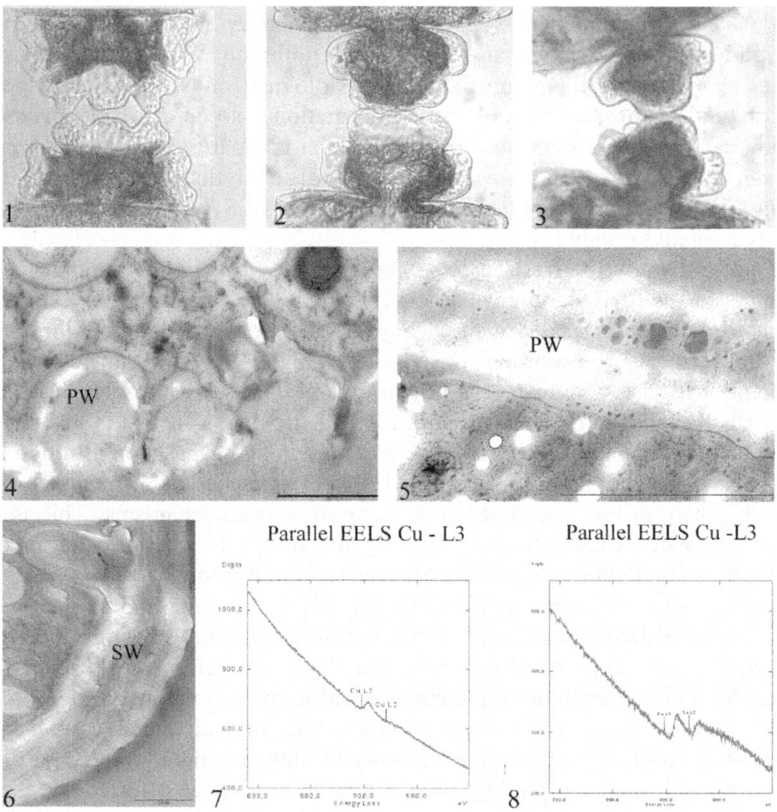

Figure 1-8. Semicells of *Micrasterias denticulata* treated with 30µ $ZnSO_4$ (1), 5 µM $CdSO_4$ (2) and 20µM $AlSO_4$ (3) for 4 h. Malformations of the primary cell wall after exposure to 20µM $AlSO_4$ (4) and 400nM $CuSO_4$ (5). EELS of Cu-$L_{2,3}$ edge (7) measured in the secondary wall (6) after 3 weeks exposure. Cu-$L_{2,3}$ edge after pre-treatment of section with 10mM $CuSO_4$ measured in a mucilage vesicle (8). Scale bar 1µm (4), 2µM (5), 0.5µM (6)

Localization of "flagella" proteins in the Archaeon *Ignicoccus hospitalis*

C. Meyer[1], D. Müller[1], S. Gürster[1], H. Huber[1], R. Wirth[1], R. Rachel[1]

1. Dept of Microbiology / Centre for Electron Microscopy, University of Regensburg, Universitaetsstr 31, D-93053 Regensburg, Germany

carolin.meyer@biologie.uni-regensburg.de
Keywords: flagella, transmission electron microscopy, immun-localization, cryo-preparation

Ignicoccus hospitalis is a strain of hyperthermophilic Archaea, which grow as strict chemolithoautotrophs, using molecular hydrogen as electron donor and elemental sulphur as acceptor [1, 2]. All strains of this genus were described as possessing flagella, although no motility is detectable by light microscopy. The structure, biochemistry and function of these "flagella" are focus of our studies.

"Flagella" were enriched from supernatants of fermenter-grown cells by PEG/NaCl induced precipitation, followed by CsCl gradient centrifugation and dialysis. Transmission electron micrographs of negatively stained samples show filaments of uniform width, 14 nm, with a length of up to 10 µm. In contrast to the flagella of most other Archaea, *Ignicoccus* "flagella" are made of only one type of protein, with a relative mass of 33 kDa. Antibodies were raised in rabbits, which were shown by Western blot analysis to be highly specific. First immun-labelling experiments of the "flagella" for electron microscopy were difficult to interpret: the distribution of the gold clusters was restricted to both ends of the flagella only, while the greatest part of the filaments remained unlabelled (Fig 1A).

We modified our labelling protocol, by first growing the cells directly adhering on carbon-coated EM grids, resulting in a higher number of flagella amenable for the immun-labelling. Secondly, before applying the first antibody solution, the grids with the "flagella" were completely air-dried or heat-treated (up to 140°C). The result was that "flagella" were labelled in their whole length (Fig 1B); gold clusters indicating bound antibodies were found in a much higher density than without heat-treatment.

In order to identify the distribution of "flagella" protein in the intact cell, immun-labelling was also performed on ultrathin sections of high-pressure frozen, freeze-substituted, resin embedded cells. Previous studies showed that this procedure is necessary in order to preserve the delicate architecture of *Ignicoccus* cells [2, 3]; these cells were successfully used for immun-localization of proteins [4]. Surprisingly, the gold clusters are not only found on few distinct foci, i.e. at the spots where the "flagella" are anchored in the cell body. In fact, labels were found scattered on the inner and outer membrane of the cells, in a rather statistical distribution and surprisingly high density (Fig 2). This means that the present concept how these "flagella" are assembled in the cell / cell envelope needs to be re-evaluated.

A. Aretz, B. Hermanns-Sachweh, J. Mayer (Eds.): EMC 2008, Vol. 3: Life Science, pp. 321–322,
DOI: 10.1007/978-3-540-85228-5_161,

1. H Huber et al., IJSEM **50** (2000), 2093.
2. W. Paper et al., IJSEM **57** (2007), 803.
3. R. Rachel et al. Archaea **1** (2002), 9.
4. T. Burghardt et al., Mol. Microbiol. **63** (2007), 166.
5. This work is supported by a grant of the DFG to RW and RR.

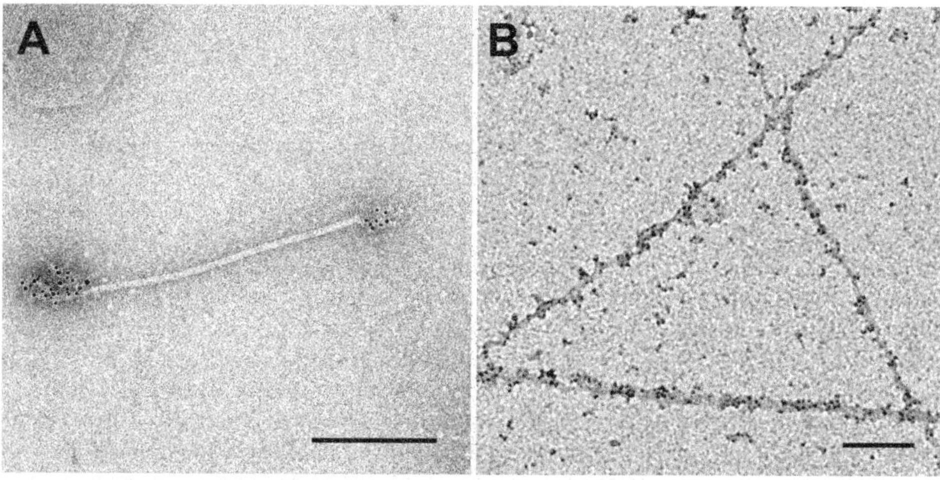

Figure 1. Transmission electron micrographs of *Ignicoccus* "flagella", labelled with specific antibodies directed against the main protein. A: native "flagella"; B: "flagella" after heat-treatment at T=140°C. Bar, 250 nm.

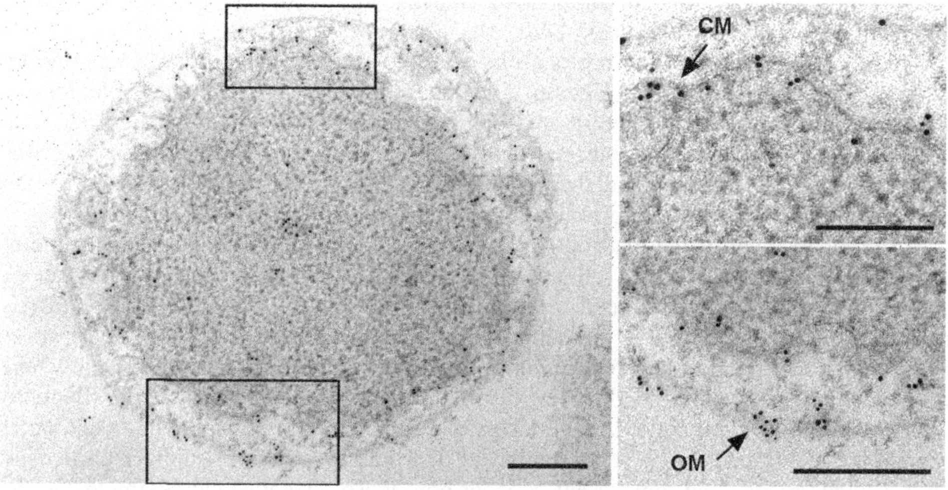

Figure 2. Transmission electron micrograph of an ultrathin section of *Ignicoccus* cells, prepared as described (see also: [4]), and labelled with antibodies directed against the "flagella" protein. Boxed areas are shown at higher magnification on the right. CM, cytoplasmic membrane; OM, outer membrane. Bar, 200 nm.

Detection of gold nanoparticles by Autrata YAG detector in FE SEM working in cryo mode

J. Nebesarova[1,2], M. Vancova[1,3], S. Hucek[1,4] and J. Vanecek[1]

1. Biology Centre of ASCR, Branisovska 31, CZ-37005 Ceske Budejovice, Czech Republic
2. Faculty of Science, Charles University in Prague, Vinicna 7, CZ-12843 Prague, Czech Republic
3. Faculty of Science, University of South Bohemia, Branisovska 31, CZ-37005 Ceske Budejovice, Czech Republic
4. Pedagogical Faculty, University of South Bohemia, Jeronymova 10, CZ-37115 Ceske Budejovice, Czech Republic

nebe@paru.cas.cz
Keywords: Autrata BSE YAG detector, cryo field emission SEM, colloidal gold nanoparticles

In lens and semi-in-lens cold field emission scanning electron microscopes offer high resolution imaging. In connection with a high sensitive detection of backscattered electrons they would present an effective tool for observing of colloidal gold markers on a surface of biological specimens.

The Autrata detector with the yttrium aluminium garnet (YAG) scintillator activated by cerium [1] belongs to the family of detectors with the highest resolution allowing the observation of colloidal gold labels. In addition it can work even at low energies of primary electrons. Erlandsen et al. [2, 3] proved that 10 nm gold particles could be detected by means of this detector at FESEM working at room temperature with accelerating voltages of 2-5 keV. In our study we would like to find out, if it is possible to use this type of detector at FESEM working at cryo-mode for an observation of gold nanoparticles and to determine optimal conditions for its using in this mode.

FESEM JEOL 7401 equipped by Autrata YAG BSE detector and cryo-attachment ALTO 2500 (Gatan) was used in this study. The observation of colloidal gold particles was performed on specimens prepared by the following ways: 1/ 10 % gelatine was dropped on an aluminum target, then goldsol 15 nm (Aurion) was added on the surface of gelatine, after few minutes an excess of solution was removed, the target was immersed into liquid nitrogen and the frozen specimen was transported into the cryo-chamber of ALTO 2500. Before inserting the specimen into a cryo-stage of the microscope we sublimated 15 min at -95° C an ice contamination and its surface was coated by Pt. 2/ goldsol 10 nm (Aurion) was directly added into 10 % gelatine and the solution was dropped on the target. The specimen was frozen also by immersion into liquid nitrogen and fractured in the cryo-chamber of ALTO 2500. The temperature of cryo-stage during the observation was -135° C.

Our results prove that Autrata YAG BSE detector can be used in FESEM working in cryo-mode for the detection of colloidal gold particles with the size around 10 nm. A lot of factors influence the quality of imaging, e.g. accelerating voltage, working distance,

A. Aretz, B. Hermanns-Sachweh, J. Mayer (Eds.): EMC 2008, Vol. 3: Life Science, pp. 323–324,
DOI: 10.1007/978-3-540-85228-5_162,

working mode of microscope and type of specimens. The threshold for BSE detection in the microscope working at basic mode was the accelerating voltage 4 kV at the smallest working distance. With the help of the gentle beam method, which slows electrons just before their impact on the specimen, we obtained the sufficient BSE signal at accelerating voltage around 1 keV at the smallest working distance. The reducing of primary beam energy is inevitable in case of frozen specimens like frozen gelatine, which was very unstable under the electron beam. The high attention also has to be paid to specimen preparation. In Figure 2, gold nanoparticles on the fracture are worse distinguishable in comparison with particles located on the specimen surface (Figure1).

1. R. Autrata, EMSA Bull. 22 (1992), p. 54-58.
2. S.L. Erlandsen, P.T. Macechko and C. Frethem, Scanning Microsc. 13 (1999), p. 43-54.
3. S. Erlandsen, Y. Chen, C. Frethem, J. Detry and C. Wells, J. of Microscopy 211, p. 212-218.
4. Supported by grant project of Academy of Sciences of Czech Republic No. KAN200520704 and Z60220518.

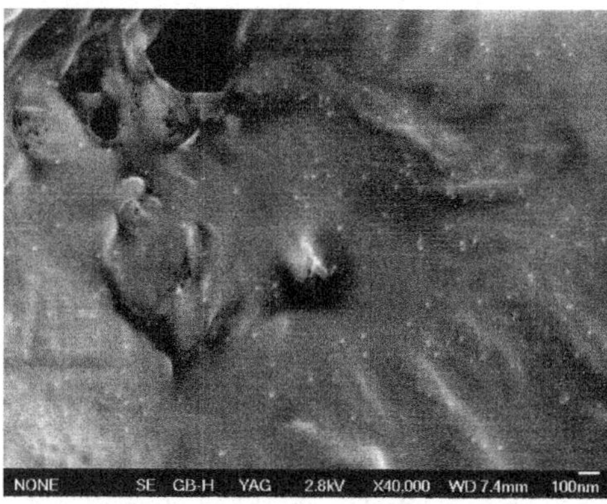

Figure 1. 10 nm colloidal gold particles on the surface of frozen gelatine observed at 2,8 keV in cryo FESEM JEOL 7401 working in GB-H mode. The specimen was coated 10 s by Pt.

Figure 2. 15 nm colloidal gold particles on the fracture of frozen gelatine observed at 1 keV in cryo FESEM JEOL 7401 working in GB-H mode. The specimen was coated 20 s by Pt.

Correlative cryo-fluorescence and electron microscopy

A. Rigort[1], A. Leis[1], M. Gruska[1], A. Sartori[2], W. Baumeister[1] and J. M. Plitzko[1]

1. Max-Planck-Institut für Biochemie, Abt. Molekulare Strukturbiologie, Am Klopferspitz 18, D-82152 Martinsried, Germany
2. Institut Pasteur, Imagopole, Dynamic Imaging Platform (PFID) and Ultrastructural Microscopy Platform (PFMU) 25-28 rue du Dr. Roux, F-75015 Paris, France

rigort@biochem.mpg.de
Keywords: correlative microscopy, cryo-fluorescence microscopy, cryo-electron microscopy

Cryo-electron tomography (CET) can provide 3-dimensional information about the structural basis of complex cellular processes on the scale of a few nanometres. However, a main drawback in locating features of interest at the level of magnification offered by electron microscopy results from the fact that biological material embedded in amorphous ice has intrinsically low contrast and is sensitive to irradiation by the electron beam. Due to these circumstances, it is complicated to retrieve areas of interest without inducing *a priori* damage by the search process.

Correlative microscopy [1] can greatly facilitate the retrieval of regions of interest by allowing pre-screening of vitrified specimens at the light microscopic level prior to an in-depth analysis at higher magnifications by means of CET. For this purpose, a cryo-stage adapted to an inverted motorised fluorescence microscope was developed, which allows imaging of fluorescent samples after vitrification. The frozen grid can be scanned using multifield imaging with automated stitching routines. Grid maps can be generated where coordinates of specific features can be defined by the user. From the same sample, an additional grid map is recorded at the electron microscope via the "gridscan" routine of TOM toolbox [2,3] implemented in MatLab. Subsequently, the coordinates defined using cryo-light microscopy are recovered on the electron microscopy grid map. With the correlative approach, the total electron dose spent on locating areas of interest is minimized, search efficiency is increased, and an independent, specific signal can be used to confirm the identities of features, where identification via an EM density map might be ambiguous or even impossible.

The entire experimental setup was tested successfully on primary human keratinocytes grown on holey carbon coated gold finder-grids. Due to their unique morphology with long, flat shaped cellular extensions, primary keratinocytes are an ideal system to investigate the molecular structure of cell-substrate adhesion points. This is especially true for investigations by CET where thin flat samples (a few hundred nanometres but less than 500 nm) are mandatory for a high resolution structural analysis.

Where cells are too thick for CET, vitreous cryosections must be cut from high-pressure frozen material. Using murine cardiomyocytes prepared in this way, we could also demonstrate an unambiguous signal from fluorescently stained but chemically unfixed mitochondria (Figure 1) [4].

A. Aretz, B. Hermanns-Sachweh, J. Mayer (Eds.): EMC 2008, Vol. 3: Life Science, pp. 325–326,
DOI: 10.1007/978-3-540-85228-5_163,

1. A. Sartori, R. Gatz, F. Beck, A. Rigort, W. Baumeister & J. M. Plitzko. Correlative microscopy: bridging the gap between fluorescence light microscopy and cryo-electron tomography. J. Struct. Biol., **160** (2007), 135-45.
2. S. Nickell, F. Förster, A. Linaroudis, W. Del Net, F. Beck, R. Hegerl, W. Baumeister, J.M. Plitzko. TOM software toolbox: acquisition and analysis for electron tomography. J. Struct. Biol., **149** (2005), 227-34.
3. S. Nickell, F. Beck, A. Korinek, O. Mihalache, W. Baumeister & J. M. Plitzko. Automated cryoelectron microscopy of "single particles" applied to the 26S proteasome. FEBS Lett., **581** (2007), 2751-6.
4. M. Gruska, O. Medalia, W. Baumeister & A. Leis. Electron tomography of vitreous sections from cultured mammalian cells. J. Struct. Biol., **161** (2008), 384-92.

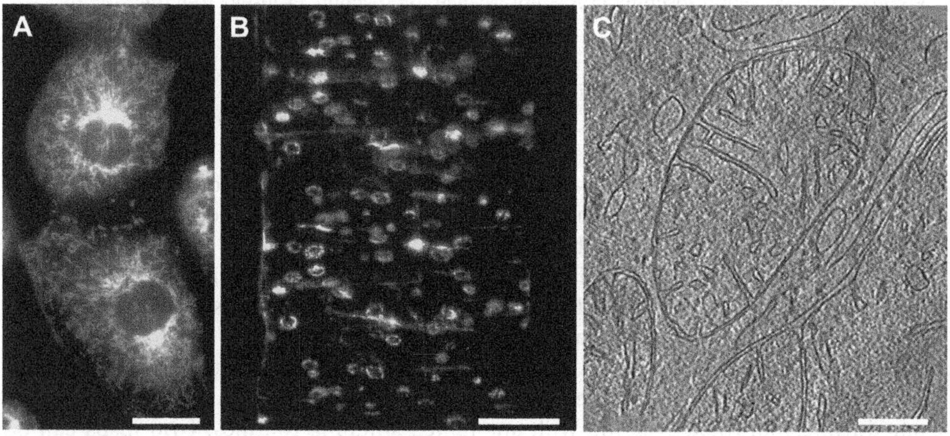

Figure 1. (a) Adherent murine cardiomyocytes viewed by epifluorescence microscopy at 37°C after staining with Mitotracker Green FM. (b) Vitreously sectioned, detached cardiomyocytes as shown in (a), viewed using cryo-fluorescence microscopy. Nominal section thickness: 150 nm. (c) Digital slice from a cryo- electron tomogram of a vitreously sectioned cardiomyocyte, showing several mitochondria in cross-section. Nominal section thickness: 85 nm. Scale bars: A = 20 μm, B = 100 μm, C = 200 nm.

The effect of hypometabolizing molecules on transcription as shown by a new two-step pulse-chase method

A. Spedito[1], M. Malatesta[2] and M. Biggiogera[1]

1. Dipartimento di Biologia Animale, Lab. Biologia Cellulare e Neurobiologia, University of Pavia, Piazza Botta 10, Pavia, Italy
2. Dipartimento di Scienze Morfologico-Biomediche, Sez. Anatomia e Istologia, University of Verona, Strada Le Grazie 8, Verona, Italy

alessandro.spedito@unipv.it
Keywords: transcription, hypometabolism, RNA precursors

Hibernation is an adaptive strategy developed by some animal species when ambient temperature and food decrease during winter. In this period, the animals lower drastically their metabolic activity. We have been studying for many years the effect of two synthetic enkephalins, DADLE and DALE, which can induce a hibernation-like state in HeLa cells. The main aim of our work is to understand the modifications occurring to the transcription process during hibernation. In several morphological [1] or immunohistochemical [2,3] studies, we have described the decrease of the transcriptional activity during hibernation

as well as after DADLE and DALE treatments. We propose a new method to evaluate transcription by means of incorporation of two RNA precursor: iodouridine and chlorouridine. First, Iodouridine is added to the colture medium for 15 min, the medium is then replaced with fresh medium for 15 min. Finally, chlorouridine is added for an additional 15 min. Cells are then embedded and the precursors revealed by immunocytochemistry [4] on thin section and observed in a Zeiss EM 900 electron microscope, The aim of this experiment is to follow the labelling on perichromatin fibrils (morphological counterpart of RNA) in different condition: control cells; after 48h treatment with DADLE; after 48h treatment with DADLE plus 48h recovery; after 48h treatment with DALE; after 48h treatment with DALE plus 48h recovery. PF show 3 different patterns: labelling for one, two precursors or no labelling. If a single fibril is transcribed within 16 min, one expects a labelling for 0 or 1 precursor; if the transcription takes place between 16 to 30 min, the labelling should mark 1-2 precursors and, in the case of synthesis occurring after 31 min up to 45 min both precursors should be labelled "Figure 1". We have observed that after DADLE and DALE treatments for 48 h the amount of double labelled fibrils increase. After DADLE + Rec the amount of double labelled fibrils reaches the level of control cells, while after DALE + rec there is no change in comparison with DALE 48h "Figure 2". These data can be explained considering previous studies on DADLE and DALE. The only difference between the two enkephalin is a L-Leucine present in DADLE and a R-Leucine in DALE. Due to this, the cells quickly degradate DADLE, but not DALE. After DALE recovery, we have observed perichromatin granules accumulation, a possible evidence for cellular damage. The decrease of the double labelled fibrils after DADLE and DALE can be

A. Aretz, B. Hermanns-Sachweh, J. Mayer (Eds.): EMC 2008, Vol. 3: Life Science, pp. 327–328,
DOI: 10.1007/978-3-540-85228-5_164,

explained in different ways. Probably DADLE causes a decrease in RNA transcription without damage, because after recovery the quantity of double labelled fibrils is comparable to control cells. We can suggest this explanation considering previous result too. The decrease in double labelled fibrils after DALE does not revert to the control levels, but is accompanied by a perichromatin granule accumulation. Generally a such accumulation is associated with cellular damage, but is present during ageing too. Consequently, it is not clear whether the observed decrease of the amount of double labelled fibrils is due to the hypometabolical effect of the peptide or to a cellular damage. Further analysis should be carried out to understand if DALE could be harmful.

1. M. Malatesta, S. Battistelli, M.B. Rocchi, C. Zancanaro, S. Fakan and G. Gazzanelli, Cell. Biol. Int. **25(2)** (2001), p. 131-138.
2. L. Vecchio, C. Soldani, M.G. Bottone, M. Malatesta, T.E. Martin, L.I. Rothblum, C. Pellicciari and M. Biggiogera, Histochem. Cell. Biol. **125(1-2)** (2006), p. 193-201.
3. M. Biggiogera, B. Cisterna, A. Spedito, L. Vecchio and M. Malatesta, Differentiation **76(1)** (2008), p. 57-65.
4. L. Vecchio, L. Solimando, M. Biggiogera and S. Fakan, J. Histochem. Cytochem 56(1) (2008), p. 45-55

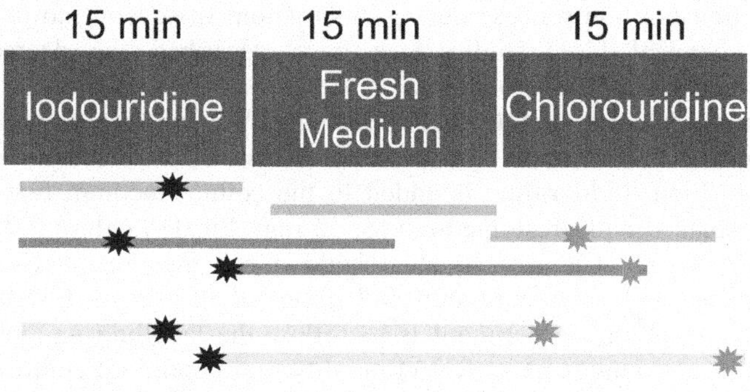

Figure 1. Scheme of the experiment

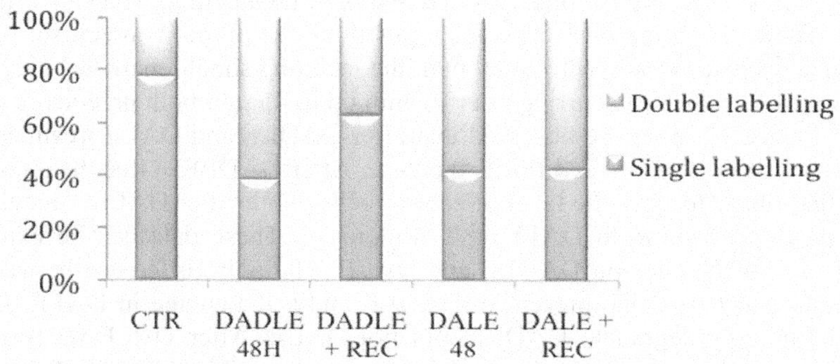

Figure 2. The graph show the amount of double labelling compared with single labelling.

Cryofixation and Freeze-Substitution Combined with Tokuyasu Cryo-section Labelling

Y.-D. Stierhof[1] and H. Schwarz[2]

1. Zentrum für Molekularbiologie der Pflanzen ZMBP, Universität Tübingen, Auf der Morgenstelle 5, 72076 Tübingen, Germany
2. Max-Planck-Institut für Entwicklungsbiologie, Spemannstr. 35, 72076 Tübingen, Germany

york.stierhof@zmbp.uni-tuebingen.de
Keywords: high-pressure freezing, immunolabelling, Quantum dot markers

In biological electron microscopy, fixation techniques based on cryofixation have found widespread application not only in visualization techniques for frozen-hydrated molecules, molecule complexes, sections and thin regions of cells but also in resin embedding of tissue for structural analysis and immunocytochemistry. Recently it has been shown that for thawed cryo-section labelling according to Tokuyasu [1] it is also possible to use cryofixation instead of chemical fixation [2,3,4]. In this case, the cryofixed sample has to be dehydrated and chemically fixed during freeze-substitution and therafter rehydrated, in order to infiltrate the samples prior to conventional freezing and cryosectioning with aqueous cryoprotectants like sucrose (Figure 1).

The advantages of this technique are severalfold: (i) tissues like plant, insect embryo and nematode tissue, which are difficult to fix by conventional chemical fixation at ambient temperature can be better preserved and reliably prepared for thawed cryo-section labelling according to Tokuyasu [3,4], (ii) fixation-sensitive antigens can be better preserved after cryofixation and chemical fixation during freeze-substitution [3,4], (iii) a wider range of fixatives is available and (iv), the overall appearance of cell membranes and organelles resembles more a cryofixed, freeze-substituted and resin-embedded sample with unwrinkled membranes and reduced shrinkage [3,4]. The main disadvantages are the limited sample size, the time consuming preparation procedure and the treatment with solvents during freeze-substitution.

There are some unexpected observations when this technique is applied to different tissues: Chemical fixation during freeze-substitution, below -30°C, e.g., with e.g., 0.5% glutaraldehyde, 0.1% osmium tetroxide and 0.2% uranyl acetate, does not seem to be sufficient to protect the sample against deleterious effects caused by rehydration, cryoprotectant infiltration and immunolabelling [2,3] with the consequence that a postfixation step has to be introduced. In addition, the typical appearance of the membrane`s bilayer can be achieved only when water is added to the freeze-substitution cocktail at -90°C, the amount depending on the type of tissue [2,3].

Here, we have investigated the usefulness of different fixatives (osmium tetroxide, glutaraldehyde, formaldehyde, uranyl acetate, tannic acid) during freeze-substitution and postfixation and the effects of water added during freeze-substitution in more detail. In addition we have evaluated the possibility of labelling semithick cryo-sections after

A. Aretz, B. Hermanns-Sachweh, J. Mayer (Eds.): EMC 2008, Vol. 3: Life Science, pp. 329–330, DOI: 10.1007/978-3-540-85228-5_165,

cryofixation, freeze-substitution/chemical fixation and rehydration/chemical fixation with gold and quantum dot markers throughout the whole section thickness, a prerequisite for reliable `immuno-electron tomography´ [6].

1. K.T. Tokuyasu in "Cells: A Laboratory Manual", ed. D.L. Spector, R.D. Goldman and L.A. Leonwand, (Cold Spring Harbour Laboratory Press, Cold Spring Harbour, NY) (1997), 131.1.
2. E. van Donselaar, G. Posthuma, D. Zeuschner, B.M. Humbel and J.W. Slot. Traffic **8** (2007), 471.
3. D. Ripper, H. Schwarz and Y.-D. Stierhof. Biol Cell., published online September 28 (2007), DOI:10.1042/BC20070106 .
4. I. Reichardt, Y.-D. Stierhof, U. Mayer, S. Richter, H. Schwarz, K. Schumacher and G. Jürgens. Curr. Biol. **17** (2007), 2047.
5. P. Walther and A. Ziegler. J. Microsc. **208** (2002), 3.
6. D. Zeuschner, W.J. Geerts, E. van Donselaar, B.M. Humbel, J.W. Slot, A.J. Koster J. Klumperman. Nat. Cell Biol. **8** (2006), 377.
7. This work was supported by Deutsche Forschungsgemeinschaft through SFB 446.

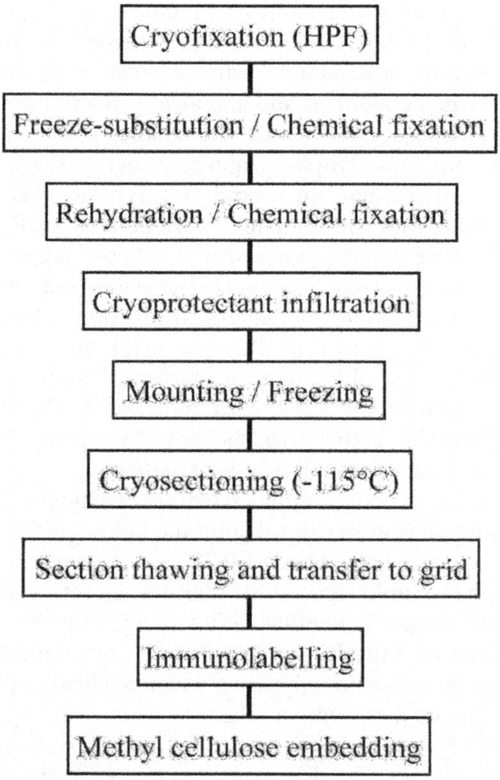

Figure 1. Flow chart of the Tokuyasu thawed cryo-section labelling technique modified by cryofixation / freeze-substitution.

The *Drosophila* active zone architecture: combining confocal, STED and transmission electron microscopy

C. Wichmann[1,2], W. Fouquet[1], S. Mertel[1], D. Owald[1], M. Dyba[3], S. Eimer[2], and Stephan J. Sigrist[1]

1. Rudolf Virchow Center, DFG Research Center for Experimental Biomedicine, University of Würzburg, Germany
2. European Neuroscience Institute Göttingen, Grisebachstr. 5, 37077Göttingen, Germany
3. Leica Microsystems, Mannheim, Germany

cwichma@gwdg.de
Keywords: *Drosophila*, active zone architecture, high pressure freezing, STED

Synapses form by an asymmetric association of highly specialized membrane domains: transmitter receptors at the postsynaptic density sense the neurotransmitter signal, while the synaptic vesicles release their content at the presynaptic active zone. Here, macromolecular specializations have evolved that are visible in the transmission electron microscope and therefore are referred to as electron dense projections or dense bodies. Dependent on their transmission kinetics, they show a high variation in shape and size. Many components of the release machinery have already been identified. However, the molecular composition of such dense bodies is in most cases still completely unknown.

Currently, we could show that in the absence of the *Drosophila* CAST/ERC/ELKS homolog Bruchpilot (BRP), the *Drosophila* dense bodies (T-bars) are entirely lost [1, 2]. This is accompanied by a severely impaired evoked vesicle release, due to a defect in the clustering of presynaptic Ca^{2+} channels [2]. Using STED (stimulated emission depletion) microscopy, a novel form of light microscopy breaking the diffraction limit, we previously reported that BRP shows a donut-shaped distribution centered at the active zone [2, Figure 1].

Here, we combine confocal, STED and transmission electron microscopy in order to analyse the structural organization of active zone components like BRP at the presynapse. We demonstrate the spatial arrangement of the BRP-molecule using two different anti-BRP antibodies. Moreover, we give an overview of the T-bar ultrastructure at the *Drosophila* neuromuscular junction (Figure 2) prepared with high pressure freezing and freeze substitution.

1. D.A Wagh, T.M. Rasse, E. Asan, A. Hofbauer, I. Schwenkert, H. Durrbeck, S. Buchner, M.C. Dabauvalle, M. Schmidt, G. Qin, C. Wichmann, R. Kittel, S.J. Sigrist, and E. Buchner. Bruchpilot, a protein with homology to ELKS/CAST, is required for structural integrity and function of synaptic active zones in *Drosophila.* Neuron **49** (2006), 833-44.

A. Aretz, B. Hermanns-Sachweh, J. Mayer (Eds.): EMC 2008, Vol. 3: Life Science, pp. 331–332,
DOI: 10.1007/978-3-540-85228-5_166,

2. R.J. Kittel, C. Wichmann, T.M. Rasse, W. Fouquet, M. Schmidt, A. Schmid, D.A. Wagh, C. Pawlu, R.R. Kellner, K.I. Willig, S.W. Hell, E. Buchner, M. Heckmann, and S.J. Sigrist. Bruchpilot promotes active zone assembly, Ca^{2+} channel clustering, and vesicle release. Science **312** (2006), 1051-4.

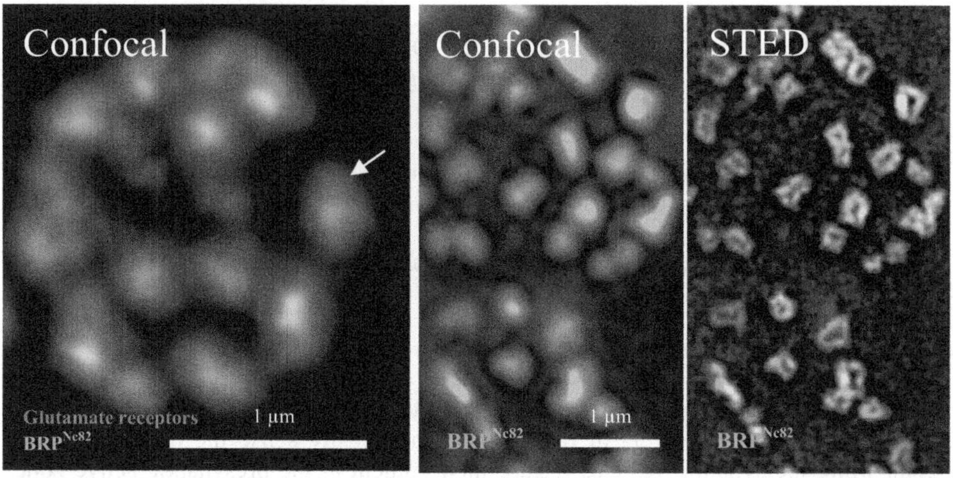

Figure 1. Bruchpilot (BRP) localization at the active zone. Left panel: Confocal image of a typical *Drosophila* larval presynaptic terminal of the neuromuscular junction with individual synapses (white arrow). The postsynaptic density is labeled for glutamate receptors and the presynapse with the active zone marker BRP. The middle and right panel illustrate the resolution difference of the BRP signal with conventional confocal compared to STED microscopy [modified from 2].

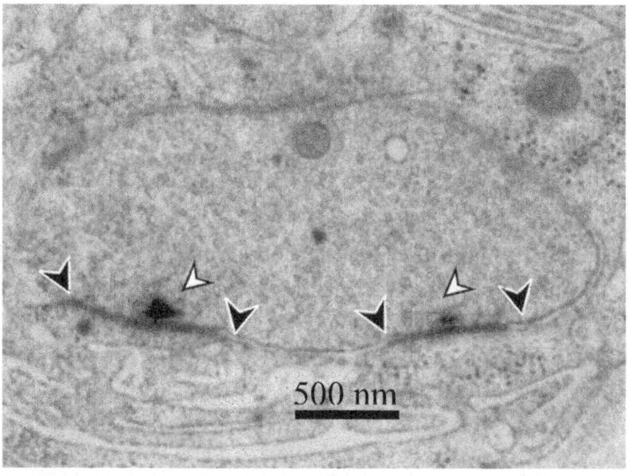

Figure 2. High pressure frozen presynaptic terminal of the *Drosophila* larval neuromuscular junction. Individual synapses (black arrowheads) are decorated with dense bodies (T-bars, white arrowheads).

Distribution and function of amorphous $CaCO_3$ and calcite within the tergite cuticle of *terrestrial isopods* (Crustacea)

A. Ziegler and S. Hild

Central Facility for Electron Microscopy; University of Ulm, Albert-Einstein-Allee 11, 89069 Ulm

andreas.ziegler@uni-ulm,de
Keywords: Biomineralisation, Moult, Raman spectroscopy

The mineralized exoskeleton (cuticle) of terrestrial isopods contains crystalline $CaCO_3$ in the form of calcite and amorphous calcium carbonate (ACC) [1]. The cuticle is subject to a unique biphasic molting process in which the isopod sheds off first the posterior and then the anterior part of the cuticle. Before the posterior molt the cuticle is decalcified and the mineral is stored within sternal ACC-deposits within the ecdysial gap of the anterior sternites and at the cuticle of the anterior tergites. Between the posterior and the anterior moult mineral from the sternal deposits and the anterior tergites is used to mineralize the new larger posterior cuticle, synthesized before shedding [2]. In order to understand the reason for the presence of two $CaCO_3$ phases we analysed the ultrastructure of the mineralized cuticle using field emission electron microscopy (Figure 1) and the distribution of ACC and calcite employing micro-Raman spectroscopy and confocal micro-Raman spectroscopic imaging. In *Armadillidium vulgare, Porcellio scaber* and many other species we found that calcite is restricted to the distal area of the cuticle, whereas ACC is localized proximally having only little overlap with the calcite layer (Figure 2). ACC is the least stable form of the known phases of calcium carbonate, and is thought to be a precursor for crystalline modifications and, because of its high solubility [3], serves as transient calcium carbonate reservoir. In order to test if ACC is preferentially resorbed over calcite during mineral resorption from the cuticle and if ACC has a precursor function for calcite during mineralisation of the distal layers, we analysed the mineral phase distribution within the exuvia and the anterior and posterior cuticle during various molting stages of *P. scaber.* We show that the protective outer calcite layer is shed away during each molt, while ACC is recycled to quickly re-establish the protective calcite layer in the new cuticle. During early stages of mineralization ACC occurs in the distal areas of the cuticle suggesting that calcite is formed by crystallization of ACC.

1. F. Neues, A. Ziegler, and M. Epple,. Cryst. Eng. Comm. **9**, (2007) p. 1245.
2. A. Ziegler, M. Hagedorn, G.A. Ahearn, and T.H. Carefood, J. Comp. Physiol. B **177** (2007) p. 99
3. L. Brecevic, and A.E. Nielson, J. Cryst. Growth **98** (1989) p. 504.
4. This work was supported by the Deutsche Forschungsgemeinschaft within the research program „Principles of Biomineralisation“ SPP 1117 (Zi 368/4-3).

A. Aretz, B. Hermanns-Sachweh, J. Mayer (Eds.): EMC 2008, Vol. 3: Life Science, pp. 333–334,
DOI: 10.1007/978-3-540-85228-5_167,

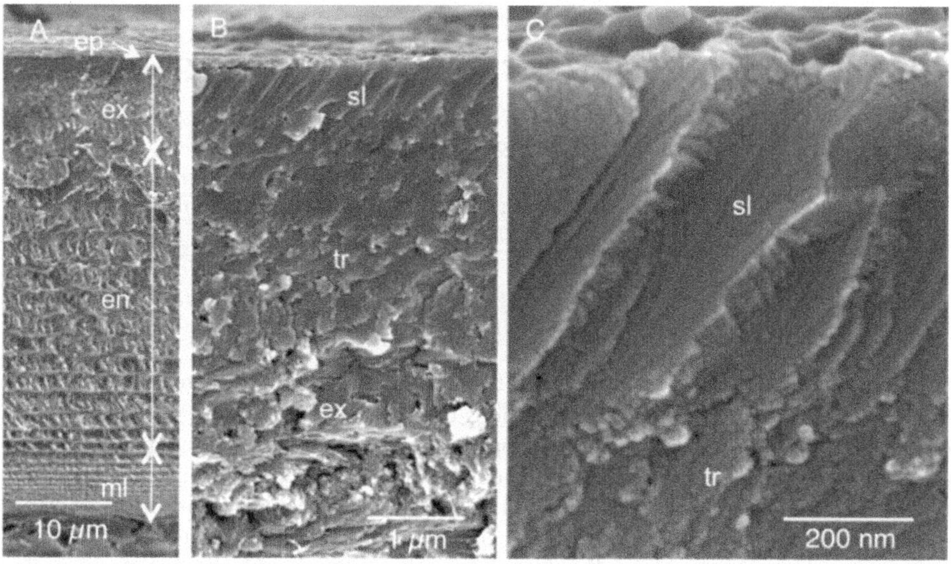

Figure 1. Field-emission SEM of mineralized tergite cuticle of *A. vulgare*. (A) Overview showing the distal epicuticle (ep), exocuticle (ex), endocuticle (en) and the membranous layer. (B) Distal region of the exocuticle showing a distal smooth layer (sl) and a transition zone that appears rough (tr). (C) Detail of the smooth layer showing oblique sheets composed of rows of 20 nm thick granules (arrows).

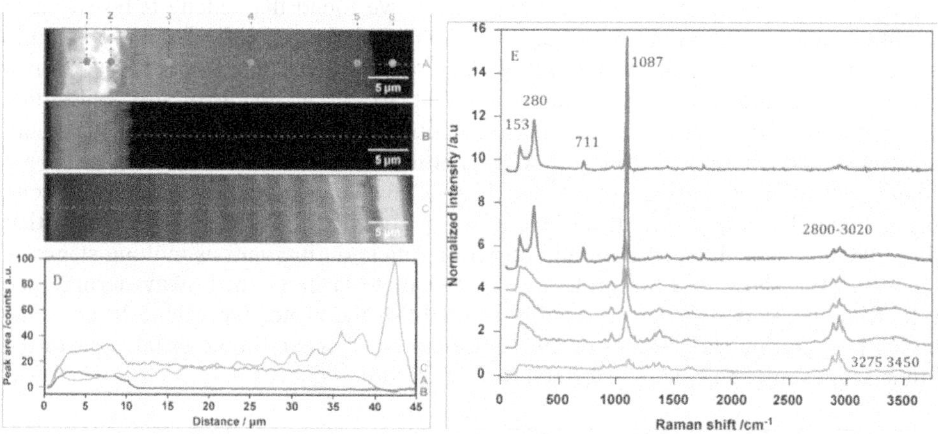

Figure 2. Raman spectroscopic images of the tergite cuticle of *A. vulgare* (A-C, line scans (D) and spectra (E) recorded from a microtome-polished surface of the tergite cuticle of *A. vulgare* showing the distribution of calcium carbonate (A) calcite (B) and organic material (C).

A symbiosis: tracking cell signaling with expression probes, quantum dots and a programmable array microscope (PAM)

M. G. Botelho[1], W. Caarls[1], S. Kantelhardt[2], A. Cambi[3], G. Hagen[4], D. J. Arndt-Jovin[1] and T. M. Jovin[1]

1. Laboratory of Cellular Dynamics, Max-Planck Institute for Biophysical Chemistry, 37077 Göttingen, Germany
2. Department of Neurosurgery, Georg-August-University of Göttingen, 37075 Gottingen, Germany
3. Department of Tumor Immunology, Nijmegen Centre for Molecular Life Sciences, Radboud University Nijmegen Medical Centre, 6500HB Nijmegen, The Netherlands
4. Department of Cell Biology,Institute of Physiology, Charles University in Prague, Czech Republic

mgralle1@gwdg.de
Keywords: single-molecule imaging, EGF receptor, dendritic cells

Quantum dots (QDs) are colloidal inorganic semiconductor nanocrystals composed typically of a CdSe, CdS or CdTe core and a ZnS shell. There are many advantages in the use of QDs as fluorophores: they can be excited over a broad spectral range and they have narrow emission bands that can be tuned from ultraviolet to infrared by adjusting size and composition. Their bright emission fluorescence and resistance to photobleaching make QDs ideal for single-particle detection and permit imaging over prolonged time periods. Because of these advantages, QDs are finding increasing use in *in vivo* and *in vitro* studies [1].

Activation of the erbB receptor tyrosine kinases (erbB1, 2, 3 & 4) induced by the extracellular binding of peptide ligands triggers signaling cascades responsible for cellular motility, cell division, and differentiation. We have genetically tagged the ErbB proteins with fluorescent proteins and/or the acyl carrier protein (ACP) sequence [2]. QDs have been targeted to receptors on the external cell surface through the growth factor receptor, EGF, or by covalently linking to the ACP tag allowing the visualization in living cells of individual receptors, the diffusion of which has been determined on different cell types. We have also used them to detect dimerization and activation of the transmembrane erbB proteins upon ligand binding. These reagents have revealed a new mode of retrograde transport of the activated receptor from filopodia to the surface of the cell [3,4]. The process is linked to treadmilling of actin filaments. This phenomenon acts as a biosensor, in that receptors are transferred from remote sites of detection/activation to the transduction mechanisms in the cell body.

Results from basic research studies of erbB tyrosine kinase receptors have led to the application of QD probes in delineating glioblastoma tumors in normal brain in collaboration with neurosurgeons. The objective of these studies is to facilitate the localization of tumor margins during surgery, thereby facilitating the accurate resection of the tumor with minimal loss of normal brain tissue.

A. Aretz, B. Hermanns-Sachweh, J. Mayer (Eds.): EMC 2008, Vol. 3: Life Science, pp. 335–336,
DOI: 10.1007/978-3-540-85228-5_168,

Specialized cells called dendritic cells prime the immune system when the body is confronted with foreign proteins or pathogens. Dendritic cells internalize antigens for degradation and presentation onto MHC molecules utilizing a C-type lectin called DC-SIGN. Ligands of DC-SIGN, such as the complex carbohydrate, LewisX, or the HIV-1 envelope protein gp120, when complexed to QDs mimic pathogens and can be used to study binding, internalization and sorting pathways in dendritic cells [5].

The detection and tracking of single molecules, and elucidation of fast, short-lived physiological processes, such as dimerization of proteins on the surface of live cells, require rapid and sensitive optical sectioning microscopy. For that purpose, we have devised an instrument denoted as the Programmable Array Microscope (PAM) [6]. The PAM utilizes a spatial light modulator (SLM) placed at the primary image plane to construct a pattern of conjugate illumination and detection. By integrating over a number of such patterns, an optically sectioned image can be quickly generated. Possible patterns include point scanning, line scanning, and pseudorandom (Sylvester) sequences. A non-conjugate image - consisting of light collected from pixels not used for illumination - serves to reduce the background usually associated with multiple pinhole systems, e.g. spinning disk confocal microscopes, and to permit the use of the very light efficient Sylvester sequences.

1. A.P. Alivisatos, W. Gu and C. Larabell, Annu. Rev. Biomed. Eng. **7** (2005), p. 55.
2. N. Johnsson, N. George and K. Johnsson, Chembiochem. **6** (2005), p. 47.
3. D.S. Lidke, P. Nagy, R. Heintzmann, D.J. Arndt-Jovin, J.N. Post, H.E. Grecco, E.A. Jares-Erijman and T.M. Jovin, Nat. Biotechnol. **22** (2004), p. 198.
4. D.S. Lidke, K.A. Lidke, B. Rieger, T.M. Jovin and D.J. Arndt-Jovin, J. Cell Biol. **170** (2005), p. 619.
5. A. Cambi, D.S. Lidke, D.J. Arndt-Jovin, C.G. Figdor and T.M. Jovin, Nano Lett. **7** (2007), p. 970.
6. G.M. Hagen, W. Caarls, M. Thomas, A. Hill, K.A. Lidke, B. Rieger, C. Fritsche, B. van Geest, T.M. Jovin and D.J. Arndt-Jovin, Proc. SPIE **6441** (2007), p. S1.
7. G.M.Hagen et al. in "Single Molecule Dynamics", eds. Ishii, Y. & Yanagida, T. (Wiley, Orlando) (2008), in press.

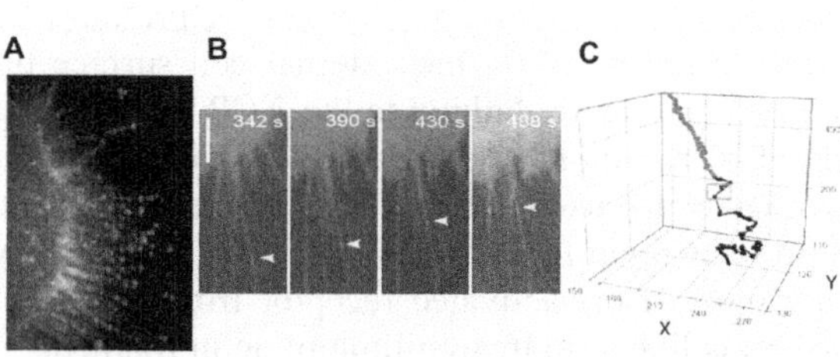

Figure 1. (A) Confocal micrograph from the PAM showing distribution of EGFR on A431 cells expressing EGFR-eGFP (green). QD655-EGF (red) bind to the EGFR on the filopodia and the cell membrane. (B) Time series and particle track of QD-EGF-EGFR on an A431-eGFP cell. Bar, 5 μm. (with a Zeiss LSM 510). (C) Trajectory of the indicated QD-EGF-EGFR complex (B, arrowhead) exhibits random diffusional movement (black) until the addition of free EGF (green box), after which the complex begins active retrograde transport (red). Taken from [7].

Golgi twins in mitosis revealed by genetically encoded tags for live cell imaging and correlated electron microscopy

Ben N. G. Giepmans[1,2,3], Guido M. Gaietta[1], Thomas J. Deerinck[1], Stephen R. Adams[2], Roger Y. Tsien[2,4] and Mark H. Ellisman[1]

1. National Center for Microscopy and Imaging Research, Center for Research in Biological Systems and Dept. of Neurosciences;
2. Dept. of Pharmacology;
4. Howard Hughes Medical Institute, University of California, San Diego; La Jolla, CA USA.
3. Current address: Molecular Imaging & EM, Dept of Cell Biology, University Medical Center Groningen, 9713 AV-1 Groningen, The Netherlands

b.n.g.giepmans@med.umcg.nl
Keywords: Golgi apparatus, mitosis, correlated microscopy

The dynamic behaviour of cells is a consequence of the coordinated and elaborate interactions between complexes of macromolecules that constitute their formed structures or organelles. Our understanding of basic cell structure and function has been greatly aided by the identification of proteins at the ultrastructural level. Advances in molecular biology, organic chemistry, and materials science have recently led to the creation of several new classes of fluorescent probes for live cell imaging and correlated light and electron microscopy (EM; Fig. 1; [1,2]). Here we show the application of (i) small organic fluorescent dyes, (ii) nanocrystals ("quantum dots"), (iii) fluorescent proteins, including the optimized mCherry, and (iv) genetic encoded tetracysteine tags complexed with biarsenical dyes (FlAsH and ReAsH), to determine protein dynamics and ultrastructural localization of proteins and organelle structures.

Combinatorial tags visible at the light and electron microscopes become particularly advantageous in the analysis of dynamic cellular components like the Golgi apparatus. This organelle disassembles at the onset of mitosis and, after a sequence of poorly understood events, reassembles after cytokinesis. The precise location of Golgi membranes and resident proteins during mitosis remains unclear, partly due to limitations of molecular markers and the resolution of light microscopy. We generated a fusion consisting of α-mannosidase II tagged with a fluorescent protein and a tetracysteine motif. The mannosidase component guarantees docking into the Golgi membrane, with the tags exposed in the lumen. The fluorescent protein is optically visible, whereas the tetracysteine tag can be labeled with ReAsH, monitored in the light microscope, and used to trigger the photoconversion of diaminobenzidine, allowing 4D optical recording on live cells and correlated ultrastructural analysis (Fig. 2; [3]). These methods reveal that Golgi reassembly is preceded by the formation of four co-linear clusters at telophase, two per daughter cell. Within each daughter, the smaller cluster near the midbody gradually migrates to rejoin the major cluster on the far side of the nucleus and reconstitutes a single Golgi. Our studies provide new insights into Golgi

A. Aretz, B. Hermanns-Sachweh, J. Mayer (Eds.): EMC 2008, Vol. 3: Life Science, pp. 337–338, DOI: 10.1007/978-3-540-85228-5_169,

reassembly during mitosis and illustrate the potential of the new fluorescent probes to follow protein distribution in 4D imaging and correlated high-resolution analysis.

1. Giepmans BNG, Adams SR, Ellisman MH & Tsien RY (2006). The fluorescent toolbox for assessing protein location and function. *Science 312*, 217-224.
2. Sosinsky GE, Giepmans BNG, Deerinck TJ, Gaietta GM & Ellisman MH (2007). Markers for correlated light and electron microscopy. In: Cellular Electron Microscopy, ed. McIntosh JR, San Diego: Academic Press, Elsevier Inc. *573-589.*
3. Gaietta, GM, Giepmans BN, Deerinck TJ, Smith WB, Ngan L, Llopes J, Adams SR, Tsien RY & Ellisman MH (2006). Golgi twins in late mitosis revealed by genetically encoded tags for live cell imaging and correlated electron microscopy. *PNAS 103, 17777-82.*

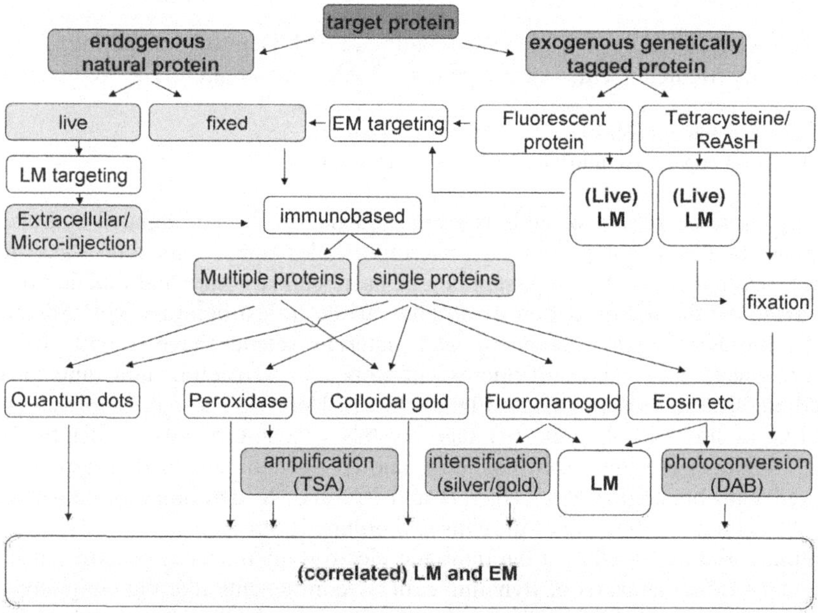

Figure 1. Flowchart with tools for correlated microscopy (adapted from [2]).

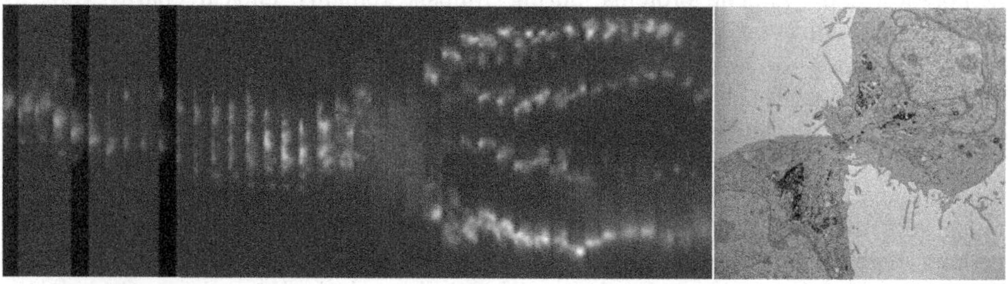

Figure 2. Dynamic (left) and high-resolution (right) imaging of the Golgi during mitosis using a combinatorial tag of green fluorescent protein and the tetracysteine tag stained with ReAsH on a Golgi-resident enzyme (detailed in [3]).

Understanding the 3D architecture of organelle bound protein complexes using cryo-electron tomography of frozen hydrated sections and immunogold; our next great challenge

Peter J. Peters, Jason Pierson, Diane Houben, Musa Sani, Sue Godsave, Pekka Kujala, Matthijn Vos, Erik Bos, Hans Jansen, Maaike van Zon and Nicole Van der Wel

Netherlands Cancer Institute - Antoni van Leeuwenhoek Hospital
Plesmanlaan 121 - H4, 1066CX Amsterdam, The Netherlands

p.peters@nki.nl
Keywords: protein complex, cryo-electron tomography

One of the focal points of our group is to reveal the macromolecular organisation of cells by means of cryo-electron tomography. This is the only method that can be used to obtain molecular resolution information of intact cells constituencies in a near-native situation. The tomogram contains a 3D map of the cellular proteome and we are just beginning to explore its potential after we worked for a few years on some technical hurdles of specimen preparation. The other central point of our group is to localize gene product in cellular structures by EM at the highest resolution with gold probes on cryo-sections. Our main focus is on understanding the molecular machinery and organization within the endomembrane system. We concentrate our work on the transport mechanism within the endocytic pathway and the link to pathogenesis. Cryo-electron tomography of unfixed frozen hydrated cryo-sections and cryo immunogold-EM of aldehyde fixed cells are our main techniques, which allow the subcellular detection and visualization of molecular machines at the highest resolution.

Bridging the gap between in situ EM and single particle EM and X-ray

Large macromolecular protein complexes (molecular machines) perform there functions in a crowded cellular environment. The technique of cryo preservation for electron microscopy assures that cells and tissues are observed in a close to native state, absent of fixation, dehydration and other related artifacts. Combined with tomography, a three-dimensional map of cells and cellular components can be obtained. We have implemented novel techniques to facilitate tomography of vitreous sections. Furthermore with improved techniques for attachment of cryo-sections and we have produced high resolution tomograms that reveal macromolecules within their cellular environment. By using automatic particle estimation and volumetric averaging within the vitreous sections we have produced high-resolution density maps of a molecular machine: the 80s S.cerevisiae ribosome, unaltered and within a cellular environment. With a resolution of 2.5 nm we are currently template matching the X-ray crystal structure with the ultimate goal to explain the structure of our averaged density maps. With the ribosome as a validation step, large eukaryotic cellular macromolecules of living cells can now be imaged in their natural environment using our technique of

A. Aretz, B. Hermanns-Sachweh, J. Mayer (Eds.): EMC 2008, Vol. 3: Life Science, pp. 339–340,
DOI: 10.1007/978-3-540-85228-5_170,

tomography of vitreous sections, bridging the gap between single particle EM, x-ray crystallography, NMR and EM of sectioned cells. More details on the abstract of Jason Pierson at this meeting.

Mycobateria

I will also present an update of our work (Cell. 2007;**129**:**1287**) on the fate of virulent mycobacteria **using EM studies**. After prolonged infection, mycobacteria relocated from phago-lysosomes to the cytosol. The detection of cytosolic mycobacteria coincides with replication of mycobacteria in human cells. The attenuated M. bovis BCG strain failed to escape from the phago-lysosome in human cells and this provide an explanation of the specific pathogenesis. Over the last two years we have also been applying cryo-electron tomography of vitreous sections to study the spatial organization of the mycobacterial cell wall with the aim of refining our molecular view of its architecture. We have found evidences for the presence of an asymmetrical outer membrane bilayer in addition to the already known symmetric cytosolic bilayer. These results are in agreement with the recent publication of Hoffmann et al., PNAS; 2008;**105**:**3963** that also showed similar results. These exciting data will no doubt lead to substantial revision of the current views of mycobacterial cell wall architecture.
If you have interest in working with us please contact p.peters@nki.nl

Collaborators:
Jose Jesus Fernandez[1], Jose l. Carrascosa[1], Bram Koster[2], Bennie Bel[3], Helmut Gnaegi[4], Henny Zandbergen[5], Michael Brenner[6], Hans Clevers [7], Peter Frederik[8] and Nico Sommerdijk[8] Wilbert Bitter[9]

1. Centro Nacional de Biotechnologia-Conejo Superior de Investigaciones Cientificas, Madrid
2. Department of Molecular Cell Biology, Leiden University Medical Center, Leiden
3. Simco Nederland B.V., Lochem, the Netherlands
4. Diatome Ltd, Biel, Switzerland
5. Kavli Institute of Nanoscience, Technical University Delft
6. Harvard Medical School Smith Building One Jimmy Fund Way, Boston
7. Hubrecht laboratory, Utrecht,
8. Technical University, Eindhoven
9. Free University, Amsterdam

Cryo-electron microscopy of vitreous sections

B. Zuber[1], M. Chami[2], C. Houssin[3], J. Dubochet[4], G. Griffiths[5] and Mamadou Daffé[6]

1. MRC Laboratory of Molecular Biology, Hills Road, Cambridge CB2 0QH, United Kingdom
2. M.E. Müller Institute for Structural Biology, Biozentrum, University of Basel, Klingelbergstrasse 70, 4056 CH. Basel, Switzerland
3. Institut de Génétique et Microbiologie, Université Paris-Sud, F-91405 Orsay, France and Centre National de la Recherche Scientifique, F-91405 Orsay, France
4. DEE, University of Lausanne, Biophore Building, 1015 Lausanne, 20 Switzerland.
5. EMBL, Postfach 102209, 69117 Heidelberg, Germany
6. Université Paul Sabatier (Toulouse III), Institut de Pharmacologie et de Biologie Structurale (IPBS), 205 Route de Narbonne, 31077 Toulouse cedex 04, France and Centre National de la Recherche Scientifique (CNRS), IPBS, Département des Mécanismes Moléculaires des Infections Mycobactériennes, 205 Route de Narbonne, 31077 Toulouse cedex 04, France

bzuber@mrc-lmb.cam.ac.uk
Keywords: cryo-electron microscopy, Mycobacterium, Corynebacterineae

The resolution at which biological specimens can be imaged by electron microscopy is often limited by the need to dehydrate specimens prior to their insertion in the vacuum of the microscope. In the early 1980s the introduction of cryo-electron microscopy solved this problem [1]. Specimens are vitrified (i.e. rapidly frozen without ice crystal formation) and observed in their hydrated, close-to-native state at about -180 °C, a temperature at which water does not evaporate in the microscope. However, the commonly used thin-film plunge-freezing technique allows the vitrification and the observation of specimens no thicker than about 0.5 μm, i.e. macromolecule suspensions, viruses, thin bacteria and areas of flat eukaryotic cells. More recently, cryo-electron microscopy of vitreous sections (CEMOVIS) has extended the optimal structural preservation of cryo-electron microscopy to specimens of several hundreds-μm in thickness, i.e. most eukaryotic cells and tissue biopsies [2,3]. The sample is vitrified by high-pressure freezing and cut in thin sections at -140 °C. These sections can then be observed in a cryo-electron microscope. In the talk, the technique will be presented as well as an application to the study of the structure of the cell envelope of mycobacteria. These bacteria belong to an atypical group of Gram-positive bacteria, whose cell envelope has been particularly difficult to preserve by conventional (i.e. non-cryo) electron microscopy. CEMOVIS reveals the presence of an outer membrane similar to the one found in Gram-negative bacteria, consistent with the very low permeability to hydrophilic molecules of mycobacterial cell envelope.

1. M. Adrian, J. Dubochet, J. Lepault and A.W. McDowall, Nature **308** (1984), pp. 32–36.
2. B. Zuber, I. Nikonenko, P. Klauser, D. Muller, and J. Dubochet. PNAS **102** (2005), pp. 19192-19197.
3. A. Al-Amoudi, D.C. Diez, M.J. Betts, and A.S. Frangakis. Nature **450** (2007), pp. 832-837.

A. Aretz, B. Hermanns-Sachweh, J. Mayer (Eds.): EMC 2008, Vol. 3: Life Science, pp. 341–341, DOI: 10.1007/978-3-540-85228-5_171, © Springer-Verlag Berlin Heidelberg 2008

The visualization of viruses in the low-voltage electron microscope

Bielniková H., Nebesářová J., Vancová M.

Institute of Parasitology, Biology Centre of ASCR, Branišovská 31, České Budějovice, Czech Republic
Faculty of Science, University of South Bohemia, Branišovská 31, České Budějovice, Czech Republic

hana.bielnikova@email.cz, nebe@paru.cas.cz
Keywords: low voltage transmission electron microscope, coating method, carbon films

Low voltage electron microscope (LV TEM; Delong Instruments s.r.o., Brno, Czech Republic) is a microscope operating at accelerating voltage 5 kV. The LV TEM is composed of two basic parts; first is the miniaturised electron optical part containing electron lenses with permanent magnets and electrostatic lenses with a relatively low magnification, and second part is the standard light microscope for the observation and recording of the light image on the structureless YAG fluorescent screen.[1,2] In our previous study we showed, that the LV TEM provide high contrast for biological specimens composed from low atomic number elements without an addition of any contrasting agents and reach a resolution of 2.5 nm [2,3]. In comparison to the high voltage (HV) TEM operating at accelerating voltage 100 kV, the LV TEM can be used for the observation of specimens thinner than 20 nm, i.e., viral particles, replicas, ultrathin resin sections, and negatively stained particles.

The aim of the study was to examine an influence of contrasting agents on imaging of viral particles in both the LV TEM and the HV TEM. We optimized a standard negative staining method of specimen preparation for the HV TEM. Material used for the observation was prepared with and without negative stains such as uranyl acetate, ammonium molybdate and others. We compared the ultrastructural appearance of viral particles in the presence of fixative/contrasting agent osmium tetroxide as well. We used purified suspensions of four kinds of viruses; tobacco mosaic virus, tobacco necrosis virus, tick-borne encephalitis virus and filamentary plant virus.

Viral particles were allowed to adhere onto a support carbon films prepared with an indirect and a direct coating method. The thickness of the carbon film in the LV TEM was required in range 3-10 nm. The carbon film was not susceptible to radiation damage and despite of its low atomic number composition demonstrated own granular structure. Carbon films prepared with the indirect coating method was more homogenous and without contaminations, compared to films prepared with the direct coating method. According our results, the surface of carbon films remained hydrophilic about 14 to 20 days. Hydrophilic character of carbon films allowed satisfactory adherence of viral particles onto carbon film surface. The adherence was diminished after that time.

A. Aretz, B. Hermanns-Sachweh, J. Mayer (Eds.): EMC 2008, Vol. 3: Life Science, pp. 343–344,
DOI: 10.1007/978-3-540-85228-5_172,

In the LV TEM, we observed viral particles of different range of thickness from 18 nm viral particles (tobacco mosaic virus) to 40-60 nm (tick-borne encephalitis virus). Viruses without an addition of contrasting agents exhibited high contrast in the LV TEM “Figure 1” in comparison to low contrast in the HV TEM “Figure 2”. High contrast allowed to omit staining procedures and gave us the possibility to observe these specimens closer to their native state. The LV TEM seems to be a very promising tool for the observation of viral particles even without an addition of any chemicals.

1. Delong A., Electron Microscopy **1** (1992), p. 79-82.
2. Delong A., Hladil K., Kolařík V., Microscopy & Analysis **27** (1994), p. 13-15.
3. Drummy F.L., Yang J., Martin C.D., Ultramicroscopy **99** (2004), p. 247-256.
4. Acknowledgements: This work was supported by grant project of Academy of Sciences of Czech Republic No. 1QS600220501 and Z60220518

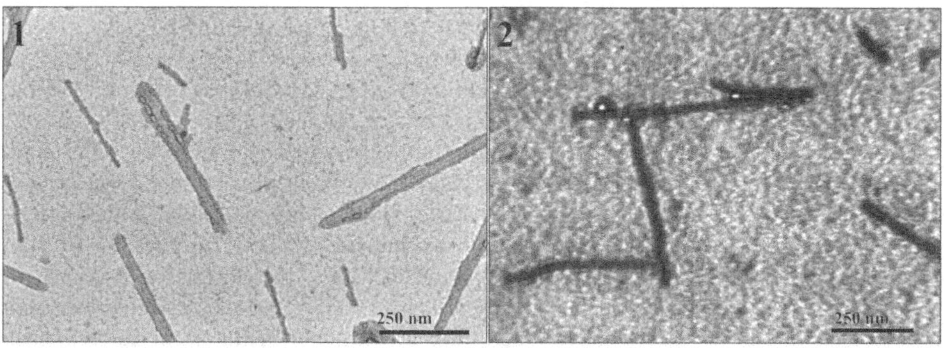

Figure 1. Isolated tobacco mosaic virus without an addition of contrasting agents, TEM JEOL 1010, 80 kV.
Figure 2. Tobacco mosaic virus without an addition of contrasting agents. LVEM 5 Delong instruments, 5 kV.

Exploring HPF/FS methods for sensitive immunogold labeling on resin sections

Hozák P.[1], Philimonenko A.A.[1], Janáček J.[2], and Janda P.[3]

1. Institute of Molecular Genetics, v.v.i., ASCR, Vídeňská 1083, 142 20 Prague, Czech Republic
2. Institute of Physiology v.v.i., ASCR, Vídeňská 1083, 142 20 Prague, Czech Republic.
3. J. Heyrovský Institute of Physical Chemistry, v.v.i., ASCR, Dolejškova 2155/3, 182 23 Prague 8, Czech Republic

hozak@img.cas.cz
Keywords: immunogold labeling, ultrastructure, cryomethods, spatial statistics

Various means for obtaining high-yield immunogold labelling together with fine ultrastrure details and spatial statistics evaluation of immunogold labelling will be presented and discussed.

A protocol for high pressure freezing and LR White embedding of mammalian cells suitable for fine ultrastructural studies in combination with immunogold labelling is presented. HeLa S3 cells in 0.5% low gelling temperature agarose were high-pressure frozen, freeze-subtituted in acetone, and embedded in LR White at 0°C. The morphological and antigenic preservation of such cells was excellent in comparison with chemically fixed cells embedded in the same resin. The ultrastructural preservation was excellent, and the immunolabelling signal was ~4-13 times higher in high pressure frozen than in chemically fixed cells, with additional benefit of low resin toxicity. We conclude that the LR White resin in combination with high pressure freezing and freeze substitution can be successfully used for fine ultrastructural immunocytochemistry.

Protocols using various resins for cryosubstitution will be critically evaluated, including a discussion about the surfaces of sections based on atomic force microscopy evaluation.

The ultrastructural localization of various antigens in a cell using antibodies conjugated to gold particles is a powerful instrument in biological research. However, statistical or stereological tools for testing the significance of non-random location of gold particles are missing. We have therefore developed methods which allow one

(1), to detect clustering or colocalization of antigens/gold particles using the distribution of distances between them

(2), to delineate the borders of cellular compartments, which are defined by immunogold labelling of specific molecules even if they are morphologically inconspicuous.

Furthemore, we have developed plug-ins for the Ellipse program (www.ellipse.sk) that form a set of tools for processing and evaluation of immunogold labeling results. They allow one to detect reliably gold particles in EM images, to evaluate statistically the observed immunogold labelling patterns (clustering, colocalization, compartments), and to produce a convenient output for publications of results. The plug-ins are useful

A. Aretz, B. Hermanns-Sachweh, J. Mayer (Eds.): EMC 2008, Vol. 3: Life Science, pp. 345–346,
DOI: 10.1007/978-3-540-85228-5_173,

addition for any image analysis software which accompanies most of the modern CCD cameras for electron microscopes. These plug-ins are available at our web-site http://nucleus.img.cas.cz/gold. Further progress in 3D analysis of ultrastructural tomography data will be presented.

1. Acknowledgements: This work was supported by the Academy of Sciences of the Czech Republic (reg. no. KAN200520704), grants LC545 and 2B06063 of the MŠMT ČR, and by the institutional grants no. AV0Z50520514.

Low contrast of the ER-membranes in the high pressure frozen, freeze substituted specimens

M. Lindman, H. Vihinen and E. Jokitalo

Institute of Biotechnology, Electron Microscopy Unit, University of Helsinki, Finland

mervi.lindman@helsinki.fi
Keywords: TEM, membrane staining, freeze substitution, endoplasmic reticulum

To preserve the structures of cells in their most natural state the top-rated method is cryofixation. In the ultrarapid freezing methods like plunge and high pressure freezing (HPF) the structures of organelles and macromolecules are frozen in the time scale of milliseconds [1]. To ease the preparing of thin-sections as well as the collection of electron tomographic tilt series we freeze substituted the high pressure frozen specimens by an organic solvent containing chemical fixatives prior plastic embedding.

In vast majority of our specimens (*i.e.*, monolayers grown on sapphire cover slips of NRK-52E and BHK cells) the cell structures were flawlessly cryofixed without notable ice crystal formation by using HPF. However, the detailed investigation of ER-membrane structures was infeasible due to the poor staining of the membranes (Figure 1). We have tested several freeze substitution protocols with different substitution time and rate, medium and fixatives and have consistently found poor contrast on the ER membranes whereas good contrast on other membranes *e.g.*, mitochondria and lysosomes, was detected (Figure 1). It seems that the chemistry of the staining at low temperatures is still poorly understood and that there is no valid, common method to prepare a specimen with well stained ER-membranes [2, 3].

We wish to raise discussion on this matter and to find out whether others have had similar experiences, and in hope to get good suggestions how to improve the situation.

1. R. Dahl and L.A. Staehelin, J. Electron Microsc. Tech. **13** (1989), p. 165.
2. T.H. Giddings, J. Micros. **212** (2003), p. 53.
3. B.S. Donohoea, S. Mogelsvanga and L.A. Staehelin, Methods **39** (2006), p. 154.

A. Aretz, B. Hermanns-Sachweh, J. Mayer (Eds.): EMC 2008, Vol. 3: Life Science, pp. 347–348,
DOI: 10.1007/978-3-540-85228-5_174,

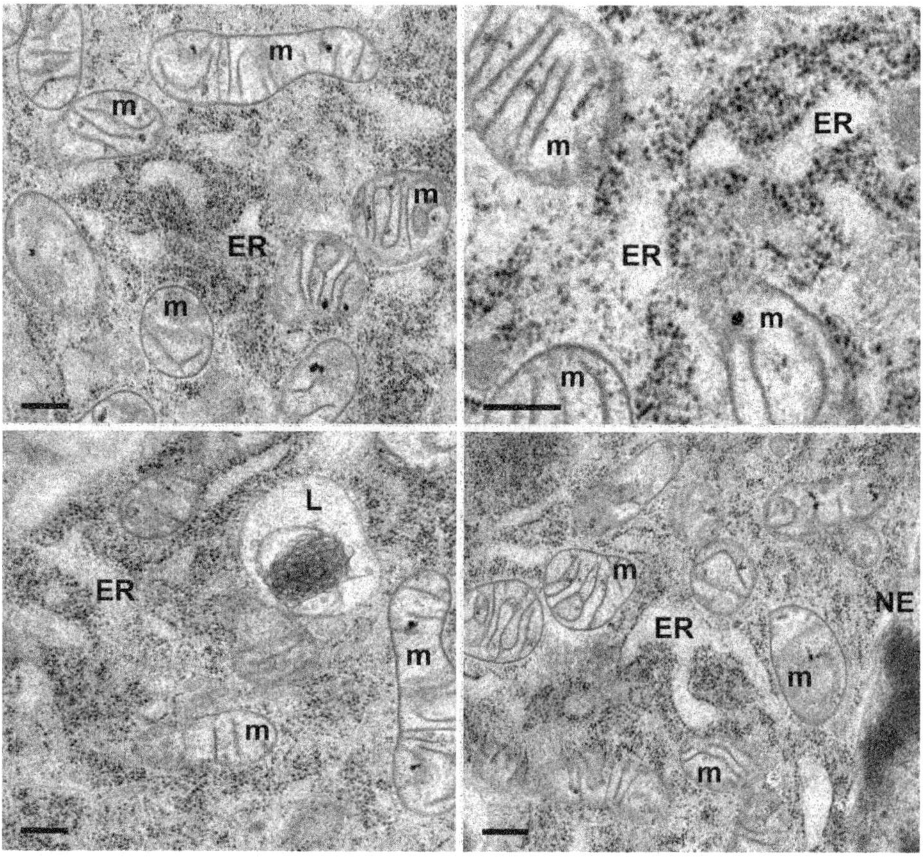

Figure 1. NRK-52E cells were high pressure frozen and freeze substituted in 2% osmiumtetroxide and 0.5% water in acetone prior embedding in Epon. An 80-nm-section was imaged at 80 kV after contrasting with uranyl acetate and lead citrate. Mitochondrial and lysosomal membranes were stained whereas the ER and NE membranes remained unstained. M, mitochondria; ER, endoplasmic reticulum; L, lysosome; NE, nuclear envelope. Bars 200 nm.

HPF of cultured cell monolayers: towards a standard method for high pressure freezing, freeze substitution and electron tomography acquisition

C. López-Iglesias[2], J. Fontana[1], S. Ruíz[2] and C. Risco[1]

1. Laboratorio de Estructura Celular, Centro Nacional de Biotecnología, CSIC, Campus Universidad Autónoma, Cantoblanco, 28049 Madrid, Spain.
2. Unidad de Microscopía Electrónica y Reconocimiento Molecular *in situ*, SCT, Universidad de Barcelona, 08028 Barcelona, Spain

clopeziglesias@ub.edu
Keywords: high pressure freezing – freeze substitution – electron tomography

Electron tomography allows the structures such as individual macromolecules, viruses and parts and even whole cells, to be reconstructed in their near-native state in three dimensions. In electron tomography, the transmission electron microscope is used to obtain the projection images of a sample from multiple angles, and those images are back-projected to reconstruct the original object in three dimensions.

The near-native state is reached by cryoimmobilization of the specimen, either plunge-, impact- or high-pressure freezing (HPF). Any method is valid if the result is the vitrification of the water in the sample.

Two ways to analyze by electron tomography cryoimmobilized samples are possible: they can be directly cryotransferred to the cryoEM, or if they do not have the appropriated thickness for a determined study they can be sectioned. In this case, the sample can be cryo-sectioned or resin embedded and sectioned after freeze-substitution. The advantages of the second method are the minor sensitivity to the radiation, the possibility to use thicker sections and to have more contrast, even without chemical staining. Several authors have used freeze-substitution and resin embedding for electron tomography [1, 2, 3, 4, 5]. In all cases, acetone is the solvent used during the freeze-substitution, different fixatives are added to the acetone, epoxy resins are used for embedding and uranyl acetate and lead citrate are used as staining agents. Moreover, we use these conditions for ultrastructural studies too.

In this study our goal is to find a standard method for HPF, freeze-substitution and electron tomography acquisition for cultured cell monolayers, avoiding the addition of support plastics and post-staining, minimizing chemical fixatives and looking for the maximum stability of the sections during the acquisition.

As a first approach we have worked to find the best protocol to freeze cells by HPF in a Leica EMPact. For that, we compare three ways: Cells seeded onto 1.5 mm sapphire discs, cells grown onto Formvar-gelatin-coated golden grids [3] and enzyme detached cells. We found that grids in the EMPact are more difficult to handle, so we decided to use trypsin detachement and sapphire discs. We tried the 200 µm flat specimen carriers and membrane carriers and we had good results with both of them.

A. Aretz, B. Hermanns-Sachweh, J. Mayer (Eds.): EMC 2008, Vol. 3: Life Science, pp. 349–350,
DOI: 10.1007/978-3-540-85228-5_175,

The second step was to determine the composition of the freeze-substitution medium and resin. We compare different fixatives, osmium tetroxide, uranyl acetate and glutaraldehyde in different combinations in acetone followed by Epon embedding. On the other hand, we compare acetone and methanol with uranyl acetate, followed by Lowicryl HM23 resin embedding. We used this type of resin because it is the most transparent one to the electron beam. Our conclusion in this step was that 0.5% uranyl acetate in acetone followed by Lowicryl HM23 embedding is the most adequate for our tomographic purpose.

The last step was to determine the support for the sections and if they need post-staining. In order to avoid more plastic in the preparation, we replace the standard Formvar coated grids by Quantifoil® grids, that consist in a holey carbon film. In fact, this type of support allowed us to use Lowicryl HM23 as resin, because these grids provided an adequate stability of Lowicryl sections during tomography acquisition. Finally, post-staining was not necessary, because contrast was optimal.

All together, these results show that non-stained sections on Quantifoil® grids of high pressure frozen trypsin detached cells, freeze-substituted in 0.5% uranyl acetate in acetone and Lowicryl HM23 embedded, are very convenient for electron tomographic studies. With these conditions, we are now ready to face the study of a variety intracellular structures in large eukaryotic cells by electron tomography.

1. J.G. Duman, N.J. Pathak, M.S. Ladinsky, K.L. McDonald and J.G. Forte, J. Cell Sci. 115 (2002), p. 1251.
2. M.S. Otegui, R. Capp and L.A. Staehelin, Plant Cell 14 (2002), p. 1311.
3. J.L.A.N. Murk, G. Posthuma, A.J. Koster, H.J. Geuze, A.J. Verkleij, M.J. Kleijmeer and B.M. Humbel, J. Microsc. 212 (2003), p. 81.
4. J.D. Baines, C-E. Hsieh, E. Wills, C. Mannella and M. Marko, J. Virol. 81 (2007), p. 2726.
5. A.B. Noske, A.J. Costin, G.P. Morgan and B.J. Marsh, J. Struc. Biol xxx (2007), p. xxx

Life-like physical fixation of large samples for correlative microscopy

M.S. Lucas[1], M. Wölfel[1], F. Lucas[1,2] and R. Wepf[1]

1. Electron Microscopy ETH Zurich (EMEZ), ETH Zurich, 8093 Zurich, Switzerland
2. Leica Microsystems, Balzers, Principality of Liechtenstein

miriam.lucas@emez.ethz.ch
Keywords: High-pressure freezing, increased sample size, correlative microscopy

High pressure freezing (HPF) is a well-established method for immobilization of hydrated organic and biological samples for electron microscopy (EM). Correlative microscopy involves the effective application of a variety of microscopy techniques to a single sample. But different microscopy modalities require different sample supports and the ability to process large samples of varying geometries. Commercially available HPF systems share the principle of a rigid sample container consisting of metal sandwich parts. Until now, the maximum sample size that could be adequately frozen was 2 mm in diameter and a few hundred µm in thickness [1] using e.g. the Bal-Tec HPM 010 high pressure freezer (Balzers, Liechtenstein).

The new Leica EM HPM 100 high pressure freezer was designed to utilize sample holders that vary in shape, size and material properties (Fig. 1B). It is the first HPF device to offer the capability to simply and inexpensively modify the sample holders to accommodate material of various sizes and geometries. It can freeze samples up to 6 mm in diameter with the same excellent ultrastructure as the smaller, conventionally high pressure frozen samples (Fig. 1D).

This extended sample size opens up new possibilities for correlative microscopy without the drawbacks of chemical fixation [2]. By introducing fluorescent stains during standard freeze-substitution and embedding for EM, it becomes possible to make a true pre-selection of a region of interest (ROI) from large sample areas using confocal laser scanning microscopy (CLSM) [3]. It is then possible to cut sections of the very same ROI for investigation of the ultrastructure by EM (Fig. 1C & D).

The new HPF device also incorporates a microscope stand to view and manipulate samples directly inside the cavity in which they will be frozen (Fig. 1A). This makes it possible to investigate time-dependent processes by instantly freezing as soon as the desired process is detected. In addition this open sample preparation platform allows retrofitting of an inverse light microscope set-up for high resolution *in vivo* observations of dynamic events prior to freezing. These techniques are equally suitable for investigation of cell culture as well as tissues.

The physical fixation of large samples (diameter > 2 mm) without cryo-protectant or chemical fixative has become possible and results in a life-like preservation of larger volumes that are suitable for statistical and histological investigations. *In vivo*, *in situ* and *en-bloc* correlative microscopy with conventional light, confocal laser scanning and electron microscopy are all realized with the new high pressure freezer.

A. Aretz, B. Hermanns-Sachweh, J. Mayer (Eds.): EMC 2008, Vol. 3: Life Science, pp. 351–352,
DOI: 10.1007/978-3-540-85228-5_176, © Springer-Verlag Berlin Heidelberg 2008

1. U. Riehle, Diss., ETH Zurich (1968).
2. J. L. Murk et al., J. Microsc. 212, **81** (2003), p. 81.
3. S. S. Biel et al., J. Microsc. 212, **91** (2003), p. 91.

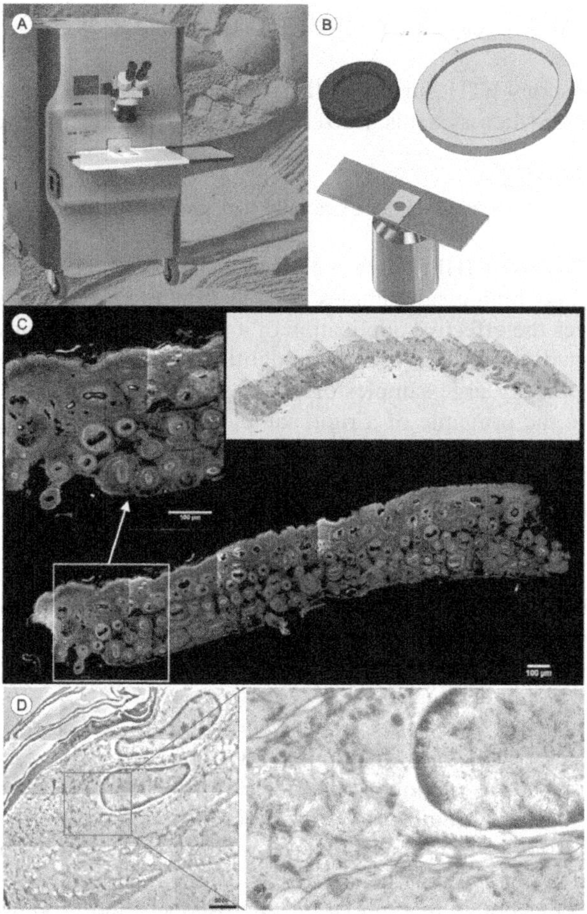

Figure 1. (A) The new Leica EM HPF 100 incorporates a microscope stand to view and manipulate samples directly in the sample holder. (B) Sample holders of various sizes, geometries and material properties are available and samples of up to 6 mm in diameter can be fixed. (C) Biopsy of mouse skin (diameter 6 mm) frozen with the HPM 100. The sample was freeze-substituted in acetone containing uranyl acetate and the fluorescent dye Safranin O (Merck). The embedded sample was bisected, excess resin was removed and a smooth surface was generated using diamond knife (Diatome). This blockface was investigated by CLSM previous to sectioning for LM (inset, stained with Toluidine blue, Merck) and EM investigation. (D) Electron micrograph of the same sample demonstrating the excellent preservation of ultrastructure.

Integrating automated single, double and triple *in situ* molecular detection with imaging for the analysis of inter- and intracellular events in the seminiferous tubule

K. Manova, Y. Romin, A. Barlas, S. Gonzalez, M. Turkekul, T. Tong, S. Suh, V. Gueorguiev and Z. Lazar

Molecular Cytology Core Facility, Sloan Kettering Institute of Cancer Research, Memorial Sloan Kettering Cancer Center 1275 York Avenue, New York, NY 10065, USA

k-manova@ski.mskcc.org
Keywords: optical microscopy, *in situ* molecular detection, seminiferous tubule

The seminiferous tubule of the testis represents a unique organoid. The dynamics of its microenvironment are strictly controlled, where somatic cells (peritubular myoid and Sertoli cells) provide for the needs of the germ cells. All cellular components work together synchronously to execute spermatogenesis and sperm genesis. As a result of the precisely controlled partnership of the cellular components of the tubule during the adult life, strictly programmed mitotic divisions generate undifferentiated spermatogonia. The majority of these cells progress along differentiation, before acquiring competency to enter meiosis and giving rise to millions of mature spermatozoa, constantly produced and released.

Surprisingly, in the era of the molecular revolution in biology, the description of the spermatogenesis process, schematized in 12 stages, is still based mostly on morphological characteristics [1]. Data provided in recent years have pointed to the seminiferous tubule as an important site for generating stem cells in juvenile but also in adult testis [2,3]. In addition to spermatogonial stem cells, responsible for maintenance of spermatogenesis, it has been shown, that some cells produced in the seminiferous tubule carry pluripotency [4,5,6]. Independently of the discovery of numerous molecules and clarification of the roles they play in the seminiferous tubule, most frequently links between functions and exact cellular/subcellular localization of are still missing, neither undisputable markers for stem, undifferentiated and different types of differentiating spermatogonia are being uncovered.

Our study focuses on the *in situ* localization of molecular markers with important functions in major cellular and intercellular processes occurring in the seminiferous tubule, using optical microscopy, image processing, analysis and 3D rendering technology. High efficiency was achieved by detecting different markers in the same tissue sections by automated double and triple immunofluorescence. We have generated data for the expression of more than 50 molecular markers in the testis. Data, obtained from such approach provides wealth of information. The molecular signatures are useful for distinguishing stem cells from the rest of the cells, allow spatial detection of functional activities and are useful for precise staging of the spermatogenesis process.

A. Aretz, B. Hermanns-Sachweh, J. Mayer (Eds.): EMC 2008, Vol. 3: Life Science, pp. 353–354, DOI: 10.1007/978-3-540-85228-5_177,

Normal and abnormal testis could be easily compared. Our experimental approach has a great potential for expansion.

We have found different subpopulations of spermatogonia, expressing PLZF. Some of these cells are also expressing c-Kit, considered a marker for differentiating spermatogonia (Fig.1). pMAPK is found mostly in Sertoli cell nuclei with prominent waves of strong cytoplasmic expression (Fig 2). Only when localized in close proximity to the interstitial tissue, PLZF – immuno-labeled spermatogonia are very active metabolically, expressing high levels of P-S6 ribosomal protein (Fig 3). Although the vast majority of PLZF expressing spermatogonia are also positive for PCNA, they do not express another cell proliferation marker Ki-67. The expression of DAZL, FAK, β– catenin, ZO1 is dynamic and the patterns are spermatogenesis- stage specific. We are in a process of creating a map of molecular signatures of different stages of the germ and somatic cells of the testis seminiferous tubule.

1. L.D. Russell et al., "Histological and histopathological evaluation of the testis", (Cache River Press, St. Louis) (1990), p. 120-161.
2. D.G. de Rooij, Reproduction **121** (2001), 347-354.
3. A. Simon and J. Frisen, Cell **128** (2007), 825-826.
4. S. Baba et al., Stem Cells **25** (2007), 1375- 1383.
5. K. Guan et al., Nature **440** (2006), 1199-1203.
6. M. Seadel et al., Nature **449** (2007), 346-350.

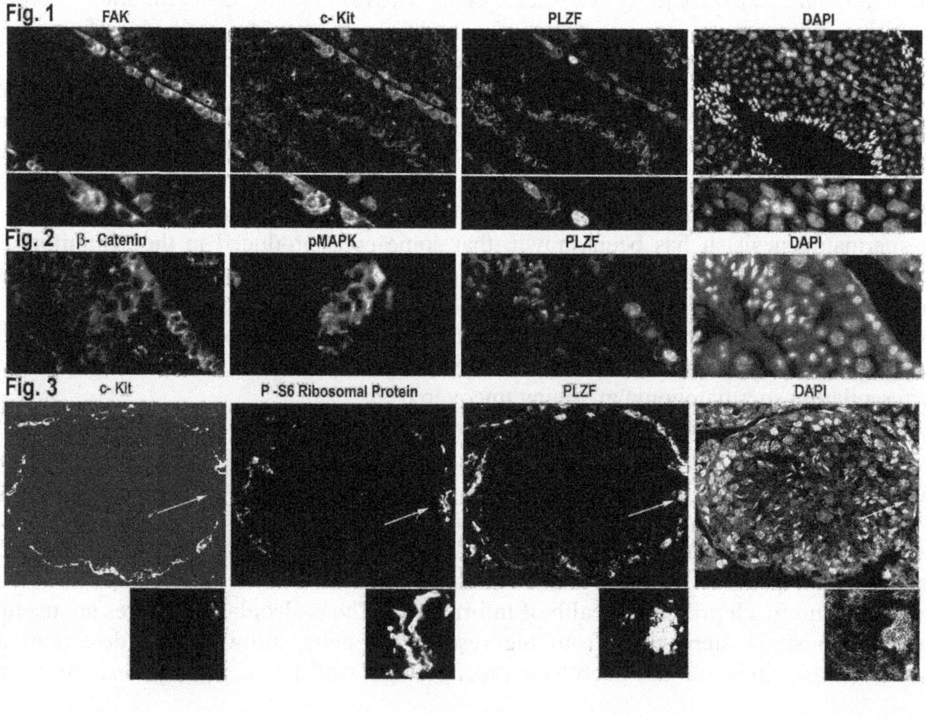

Freeze-substitution in Epon: An attempt to combine immunolabeling and improvement of structural preservation

E. M. Schraner[1], S.Leisinger[1], M. Müller[2*] and P. Wild[1]

1. Electron Microscopy, Institutes of Veterinary Anatomy and Virology, University of Zürich, CH-8057 Zürich,
2. Electron Microscopy Centre, Swiss Federal Institute of Technology, CH-8093 Zürich, Switzerland

elisabeth.schraner@access.uzh.ch
Keywords: freeze-substitution, epon, immunolabeling, herpesvirus

Epoxy resins solved in acetone have been successfully used for freeze-substitution of nematodes (1). Understanding of intracellular transportation of viruses, such as herpesviruses, requires imaging of infected cells at high temporal and high spatial resolution that can be achieved by high-pressure freezing followed by freeze-substitution (3,4). To locate exactly proteins involved in viral formation and/or transportation immunolabeling must be performed on equally well preserved cells as for ultrastructural studies. Therefore, we searched for a protocol allowing immunolabeling at high accuracy because the number of viral proteins in a given section plane is generally low.

Cells grown on sapphire disks were infected with herpesviruses. After incubation for 17 to 24 h cells were high-pressure frozen and freeze-substituted in acetone containing 20% epon at -88°C for 17 h: The temperature was slowly raised to 4°C, the cells further infiltrated wit 50% epon (in acetone) for 6 hours and embedded in epon, and polymerized for 2½ days at 60°C.

Membranes of cells substituted in epon are not stained (1). To improve contrast we treated sections with ethanol, glutaraldehyde or osmiumtetroxyde prior to stainig with uranyle acetate and lead citrate. Ultrastructural details of epon substituted cells were well preserved. Pretreatment with ethanol resulted in reasonable visibility of cell membranes. Cell membranes, however, were clearly recognisable when decorated by ribosomes or by viral proteins (Fig. 1). To identify the nature of viral proteins decorating Golgi membranes or folded nuclear membranes (Fig.2) cells were immunolabeled using antibodies against viral proteins such as the tegument protein VP16. Prior to immunolabeling sections were etched either with ethanol or 2% Na-perjodate for 1-2 min. To block unspecific bindings to epon sections were treated with 3% bovine serum albumin, 3% casein, 3% Skim milk, 3% Fraction V bovine serum albumin or 3% Aurion bovine serum albumin-C. It turned out that each antibody used required its own blocking solution. Incubation with viral antibodies led to distinct labeling of viral proteins decorating membranes but to much less distinct labeling of proteins on virions. It has to borne in mind that the intensity of labelling on virions is very low (4).

1. N.Matsko, and M. Mueller, Journal of Structural Biology **152** (2005), 92-103.
2. H.Leuzinger, U. Ziegler, et al., Journal of Virology **79** (2005), 13047-13059.

A. Aretz, B. Hermanns-Sachweh, J. Mayer (Eds.): EMC 2008, Vol. 3: Life Science, pp. 355–356,
DOI: 10.1007/978-3-540-85228-5_178,

3. P.Wild, M. Engels, et al., Journal of Virology **79** (2005) 1071-83.
4. R.Naldinho-Souto, H. Browne, et al., Journal of Virology **80** (2006) 2582-2584.

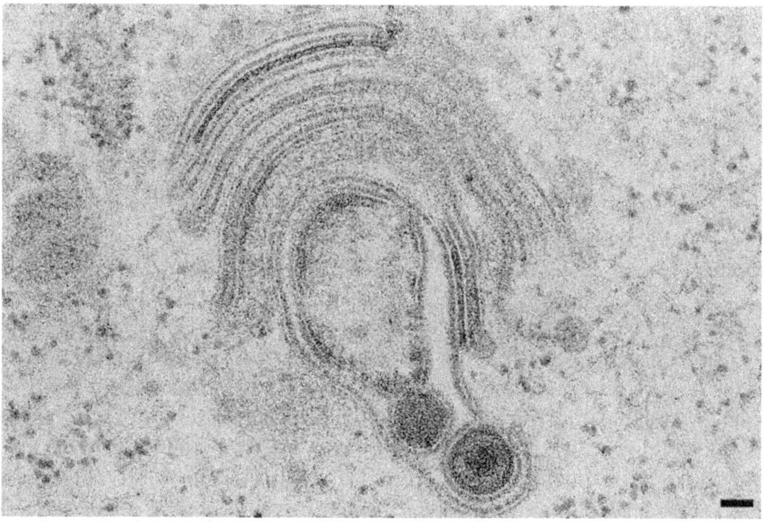

Figure 1. Golgi complex of herpesvirus infected MDBK cell after freeze-substitution in aceton/epon.Bar: 100 nm

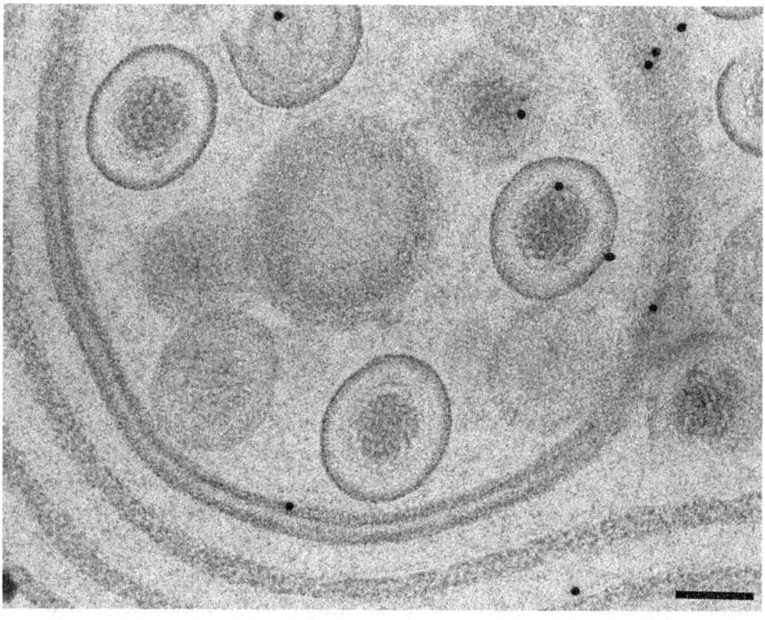

Figure 2. Folds of nuclear membranes of herpesvirus infected cells after immuno-labeling with anti vp16. Bar: 100 nm

*Present Address: Domaine du Parc St. Georges, F-30200 Venejan, France

The Shape of Caveolae is Omega-like after Chemical Fixation and Cup-like after Cryofixation

M. Westermann[1], F. Steiniger[1], S. Nietzsche[1], G. Hause[2], C. Lemke[3], W. Richter[1]

1. Elektronenmikroskopisches Zentrum, Klinikum der Friedrich-Schiller-Universität Jena, Ziegelmühlenweg 1, D-07740 Jena, Germany
2. Biozentrum der Martin-Luther-Universität Halle, Weinbergweg 22, D-06120 Halle/Saale, Germany
3. Institut für Anatomie I, Klinikum der Friedrich-Schiller-Universität Jena, Teichgraben 7, D-07740 Jena, Germany

martin.westermann@uni-jena.de
Keywords: Caveolae, Structure, Fixation

Caveolae were first observed by electron microscopists in the 1950s [1]. They were defined as flask- or omega-shaped plasma membrane invaginations, abundant in adipocytes, fibroblasts, endothelial and smooth muscle cells. The major protein component of caveolar membranes is an integral membrane protein named caveolin (18-24 kDa).

In recent studies we analysed the distribution of caveolin at caveolae [2] using the technique of SDS freeze-fracture replica labelling (SDS-FRL) [3]. In SDS-FRL, chemically unfixed, quick-frozen cells are freeze-fractured and replica-attached membrane proteins can be labeled immunocytochemically.

Comparing the freeze-fracture shape of caveolae in glutaraldehyde fixed and cryofixed 3T3 mouse fibroblast cells (Figure 1) we found significant differences. In chemically fixed cells almost all caveolae were cross-fractured through their pore (Figure 1A, inset, dashed line) and very few caveolar membranes were completely replicated (Figure 1A, arrow). We found the reverse situation in quick-frozen cells. Most caveolar membranes were completely replicated (Figure 1B, inset, dashed line) and only a small number was cross-fractured (Figure 1B, arrow). In ultrathin sections of chemically fixed heart endothelial cells caveolae exhibit the classical flask-like shape (Figure 2A), whereas, after near to life preparation by high-pressure freezing, freeze substitution and low temperature embedding the caveolae in heart endothelial cells showed in most cases a cup-like shape without constriction (Figure 2B).

These findings indicate that in most cases the constricted "neck" region of caveolae is a glutaraldehyde induced artefact and that caveolae *in vivo* show all degrees of invagination from nearly flat via deeply invaginated to flask-like. It seems that glutaraldehyde cannot preserve the native configuration of the lipid-rich caveolar domain. Our caveolin labelling experiments at freeze-fracture replica of cryofixed cells resulted in a belt-like caveolin localisation [2]. The glutaraldehyde induced crosslinking of caveolin and other proteins near the caveolar opening might be the reason for the omega-like deformation.

1. G. E. Palade, J. Appl. Physiol 24 (1953) 1424–1436.
2. M. Westermann, F. Steiniger, W. Richter, Histochem. Cell Biol. 123 (2005) 613-620
3. K. Fujimoto, J. Cell. Sci. 108 (1995) 3443–3449

A. Aretz, B. Hermanns-Sachweh, J. Mayer (Eds.): EMC 2008, Vol. 3: Life Science, pp. 357–358, DOI: 10.1007/978-3-540-85228-5_179,

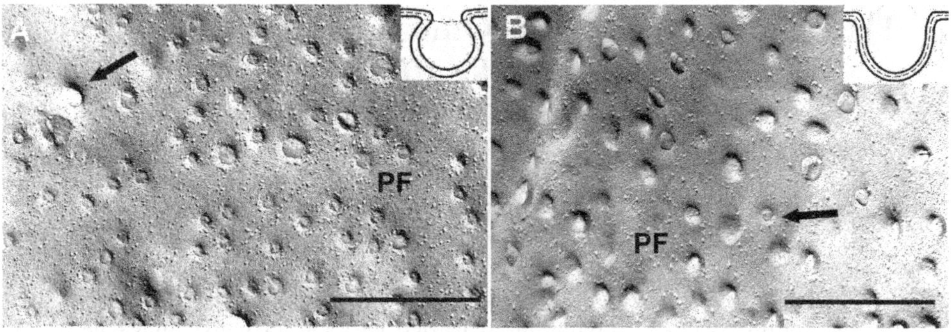

Figure 1. Freeze-fracture micrographs of the plasma membrane P-face (PF) in glutaraldeyde fixed (**A**) and cryofixed (**B**) 3T3 mouse fibroblast cells. In (**A**) almost all caveolae are cross fractured through their pores, completely replicated caveolae are rare (arrow). In (**B**) most caveolae are replicated showing different stages of invagination, cross fractured caveolae are infrequent (arrow). The insets show the assumed caveolar shape, the dashed line demonstrates the run of the fracture plane. Bars: 500 nm

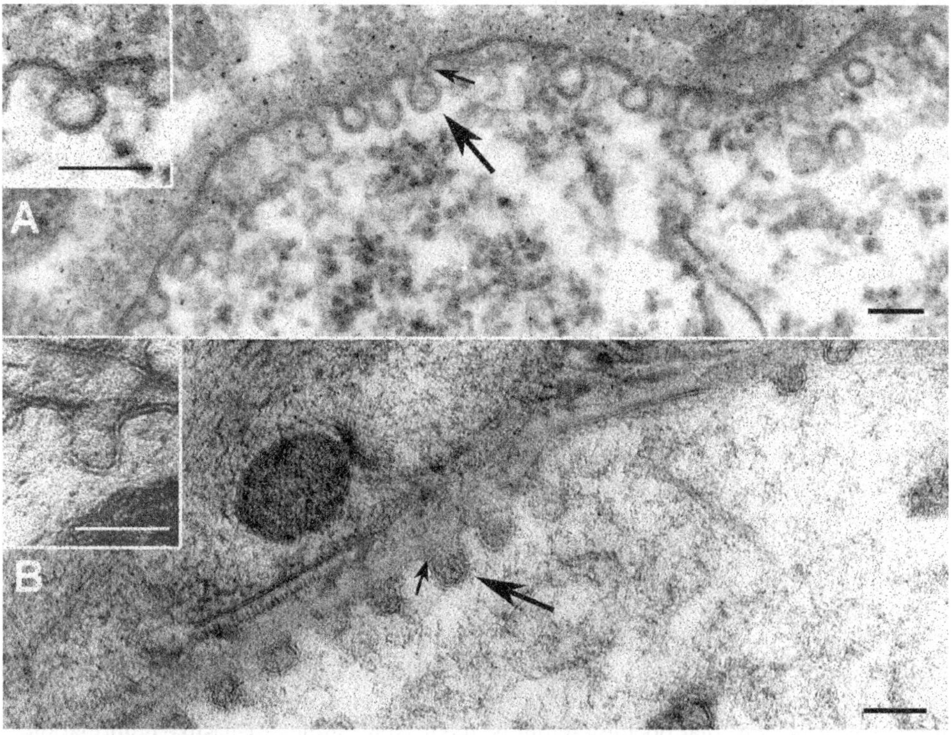

Figure 2. The shape of caveolae in ultrathin sections of endothelial cells of the heart is omega-like after glutaraldehyde fixation and embedding at room temperature (**A**, inset and arrows) and cup-like after cryofixation by high-pressure freezing combined with freeze substitution and low temperature embedding (**B**, inset and arrows). Bars: 100 nm

Integrated laser and transmission electron microscopy

A.V. Agronskaia[1], J.A. Valentijn[2], L.F. van Driel[2], B.M. Humbel[3], A.J. Verkleij[3], A.J. Koster[2], H.C. Gerritsen[1]

1. Molecular Biophysics Group, Utrecht University, Princetonplein 1, 3508 TA Utrecht, the Netherlands
2. Molecular Cell Biology – Electron Microscopy, Leiden University Medical Center, Albinusdreef 2, 2333 ZA Leiden, the Netherlands
3. Cell Biology, Utrecht University, Padualaan 8, 3584 CH Utrecht, the Netherlands

a.v.agronskaia@phys.uu.nl
Keywords: correlative microscopy, scanning fluorescence microscopy, transmission electron microscopy

Fluorescence microscopy (FM) is one of the most commonly used imaging techniques in biological and biomedical studies due to its high sensitivity and possibility of selective imaging of labeled features in the specimen. One of the restrictions of standard FM is the limited spatial resolution that is usually in the order of 0.3 μm. Special optical microscopy techniques can push resolution to 50 – 100nm, however it is still far from resolution needed for study of, for instance, the organization of assemblies of proteins and lipids in biological specimens. Such ultrastructural studies require nm resolution that at present can only be achieved by electron microscopy (EM).

Although the spatial resolution of EM is superior its field of view is limited: a resolution of 1 nm can only be realized when small (2x2 μm^2) areas are imaged. Consequently, the study of rarely occurring events in cells or tissues is extremely tedious and time consuming. This limitation has motivated researchers to embark on the development of imaging methods that combine FM and EM – correlative microscopy.

At present correlative microscopy is carried out in two separate set-ups. FM is employed to identify and locate features of interest and EM is used to map their ultra-structure. To carry out such experiments using two separate instruments is time consuming and prone to errors.

To improve the success rate of correlative microscopy and speed up the acquisition procedure, we developed a laser scanning fluorescence microscope that is integrated in a standard transmission electron microscope (TEM): fluorescence module is mounted on one of the side ports of TEM. Both microscopes share the same single tilt specimen stage. For fluorescence microscopy operation the stage is tilted over 90 degrees to position the specimen perpendicular to the optical axis of the fluorescence microscope; the microscope is moved towards the sample between the magnetic poles of the TEM; and focusing is accomplished by adjusting the Z-axis of the stage. For TEM operation the optical module is retracted from the space between the pole shoes and the stage is rotated back to the 0 degrees position.

The fluorescence microscope is equipped with a 473 nm solid-state laser and an avalanche photo diode (APD) detector operating in photon counting mode. Fluorescence images can be recorded at rates of up to 1 frame per second. In practice fluorescence

A. Aretz, B. Hermanns-Sachweh, J. Mayer (Eds.): EMC 2008, Vol. 3: Life Science, pp. 359–360,
DOI: 10.1007/978-3-540-85228-5_180,

image acquisition times are usually on the order of 10 s. Entire 3 mm diameter EM grids can be imaged with the FM module in about 10 minutes. Switching to TEM mode and moving to a specific region of interest identified by FM is limited by the speed of the specimen stage and takes about 10 s.

Optical characteristics of the fluorescence microscope are verified using a specimen containing 0.2 μm diameter 'Yellow-Green' fluorescent latex beads: The optical resolution of the microscope is 0.55 ±0.03 μm, the aberration free filed of view is at least 300x300 μm^2. Correlation of the positions in the fluorescence images and the TEM images is easy and reproducible (within ±0.5 μm for both X and Y directions).

The potential of the integrated approach is demonstrated on specimens of rat intestine cells labeled with Alexa Fluor 488 conjugated to wheat germ agglutinin and on rat liver peroxisomes immunolabeled with anti-catalase antibodies and secondary Alexa Fluor 488 antibodies and 10 nm protein A-gold.

Correlative 3D microscopy: CLSM and FIB/SEM tomography used to study cellular entry of vaccinia virus

M.S. Lucas[1], P. Gasser[1], M. Günthert[1], J. Mercer[2], A. Helenius[2] and R. Wepf[1]

1. Electron Microscopy ETH Zurich (EMEZ), CH-8093 Zurich, Switzerland
2. Institute of Biochemistry, ETH Zurich, CH-8093 Zurich, Switzerland

miriam.lucas@emez.ethz.ch
Keywords: Virus entry, correlative microscopy, FIB/SEM

The development of new anti-viral strategies requires detailed information on the replication cycle of viruses. In this study we focused on the early steps of viral interaction, the surface dynamics after binding to non-infected cells. We used the more abundant infectious form of the virus, the intracellular mature virus (IMV). IMVs have a dumbbell-shaped core, a single lipid bilayer, and a size of 360 nm in the longest dimension [1]. IMVs have been observed to bind to filopodia, actin-containing, finger-like cell protrusions. Live cell imaging showed that fluorescent virus particles associate with filopodia, and glide along them to the cell body [2]. They also induce the extrusion of large, transient membrane blebs, which upon retraction cause endocytic internalization of the virus [3].

To study membrane interactions and membrane structures at higher resolution requires extending live cell imaging to electron microscopic investigation – ideally on the identical sample. And as the cell surface with protruding filopodia, membrane blebs and bound virus particles constitutes a rather complex three-dimensional object, effective analysis of the ultrastructure also involves three-dimensional imaging (Fig. 1).

To achieve this, HeLa cells were cultivated on sapphire discs [4], infected with vaccinia virus (WR) and fixed by HPF. The samples were then dehydrated by freeze-substitution in the presence of uranyl acetate and fluorescent dyes, followed by embedding in resin. This preparation protocol combines optimal preservation of ultrastructure with the possibility of a true pre-selection of a region of interest (ROI), and 3D reconstruction of the resin-embedded sample by confocal laser scanning microscopy (CLSM) prior to cutting sections of the very same ROI for investigation of the ultrastructure by TEM [5]. 3D investigation of structures of interest is usually realized by serial sectioning for TEM imaging and 3D reconstruction.

A completely new approach has become available by the establishment of FIB/SEM 3D imaging, initially established in material science. This method enables a semi-automated collection of consecutive 20 nm section planes directly from the resin block, without the preparative effort of serial sectioning. With this approach we were able to elucidate the complexity of virus-cell interactions at the early stages of the binding and entry process (Fig. 1). FIB/SEM, although an invasive method, has the additional benefit of pinpoint extraction of ultra-thin TEM lamella for further investigation at higher magnification in a (S)TEM. By combination with CLSM 3D imaging not only the time of searching a ROI is shortened, but also a real 3D correlation on one and the same structure becomes possible in different imaging modes and at different resolution levels.

A. Aretz, B. Hermanns-Sachweh, J. Mayer (Eds.): EMC 2008, Vol. 3: Life Science, pp. 361–362,
DOI: 10.1007/978-3-540-85228-5_181,

1. J. Mercer and A. Helenius, Science (in press).
2. B. Moss in "Fields' Virology", ed. D. M. Kniepe and P. M. Howley, (Lippincott Williams & Wilkins, Philadelphia) (2007), p. 2905.
3. M. J. Lehmann et al., J. Cell Biol. **170** (2005), p. 317.
4. S. Reipert et al., J. Microsc. **213** (2004), p. 81.
5. S. S. Biel et al., J. Microsc. **212** (2003), p. 91.

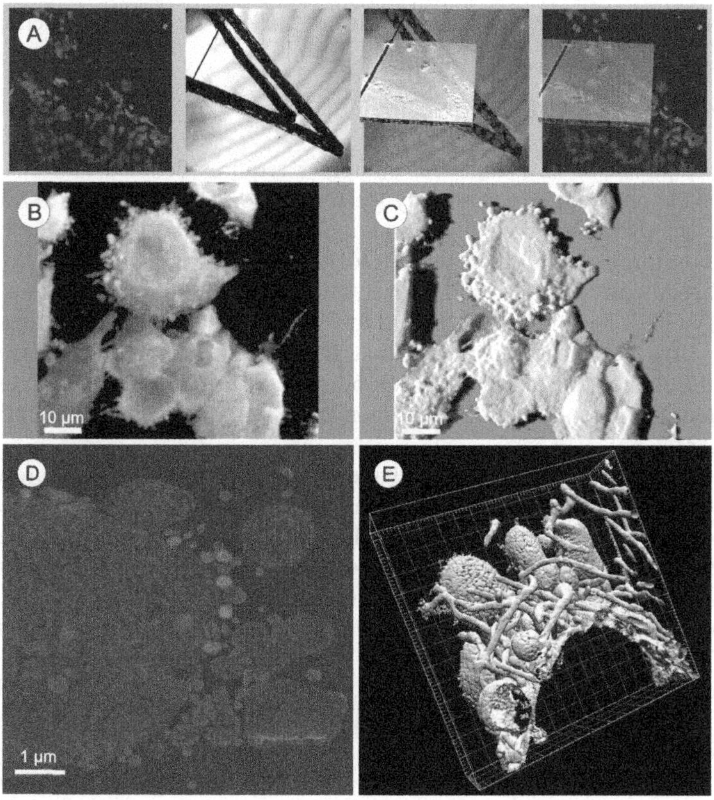

Figure 1. (A) Pre-selection of ROI is accomplished by introducing a fluorescent dye during freeze-substitution, enabling 3D investigation of the resin-embedded samples by CLSM (B+C). Surface features of the resin block can be recognized in CLSM as well as SEM to determine the ROI for FIB/SEM. (D) SEM image of perpendicular cut into the resin block generated by FIB, showing a section of a HeLa cell infected with vaccinia virus. Virus particles appear as bright roundish forms at the cell membrane. The large spherical structures are membrane blebs, the smaller round forms are filopodia. The spatial context of these structures becomes visible only in 3D (E). The 3D reconstruction demonstrates the complexity of the cell surface with filopodia and blebs.

Strategies for the morphological analysis of hydrated and life-like preserved biomedical material in the SEM

H. Hohenberg[1], R. Reimer, M. Warmer and B. Holstermann

1. Department for Electron Microscopy and Micro-Technology, Heinrich-Pette-Institute for Experimental Virology and Immunology, Martinistr. 52, D-20251 Hamburg, Germany

hohenberg@hpi.uni-hamburg.de
Keywords: hydrated surfaces, ESEM, Kryo-SEM, biomedical imaging, correlative LM-EM

Up to now the morphological investigation and analysis of fully hydrated biomedical samples in the SEM or the Environmental SEM (ESEM) can be realized by three different techniques:

(1) The ESEM "Wet-Mode"- technology, which allows the investigation of fully hydrated biological surfaces of unfixed and dynamic specimen.

(2) The structural analysis of hydrated but chemically fixed specimens, stained with heavy metals, realized by separating the sample from the ESEM vacuum system applying a closed chamber for specimen encapsulation. Here a very thin electron-transparent membrane at the top of the chamber likewise serves as interface and specimen support (Quantomix^{TM}-technology).

We introduce an alternative method avoiding the application of a specimen chamber by exactly adjusting the suitable temperature and pressure parameter in the ESEM with the intent to prevent specimen dehydration.

(3) And finally, the investigation of frozen-hydrated material after cryo-fixation (preferably by high-pressure freezing), deep-temperature manipulation (freeze-fracture or cryo-sectioning) and cryotransfer into the Cryo-SEM.

Beside the fact that each of the three specified techniques allows the imaging of fully hydrated material in an SEM, the spectrum and quality of morphological details that can be investigated and analyzed by each specific technique differs significantly.

The ESEM "Wet-Mode"-technique [1] is a fast analysis technology and enables -in a period of 5 minutes- to image the surfaces of unfixed, unstained, dynamic, living and fully hydrated biological and clinical material (see fig.1), but shows resolution/ magnification limits due to the ESEM-specific signal generation, partial dehydration and possible beam damages of the specimen at higher magnification. Moreover, this ESEM technology does not allow the direct investigation of the interior of biomedical-material.

The Quantomix^{TM}-technology [2] in contrast enables one to image cytoplasmatic components of cells cultivated on an electron-transparent membrane but can not precisely depict cell surfaces. Nevertheless, the investigated material is fixed and heavy metal stained in order to generate back-scattered electrons from a defined sample volume positioned proximal to the membrane. The achievable fine-structural resolution of stained material is generally better compared to the resolution that can be reached

A. Aretz, B. Hermanns-Sachweh, J. Mayer (Eds.): EMC 2008, Vol. 3: Life Science, pp. 363–364,
DOI: 10.1007/978-3-540-85228-5_182,

with unstained material in the ESEM under "Wet-Mode" conditions. This applies also to tissue material that is investigated without using a separating chamber.

The investigation of frozen-hydrated cellular material in the Cryo-SEM after cryoimmobilization permits to analyze material showing life-like structural integrity. Moreover, the prepared surfaces can be investigated at high resolution (see fig. 2). However, it should be mentioned that the analysis of frozen-hydrated material in a Cryo-SEM is based on sophisticated cryo-technology like: a) high-pressure freezing, b) deep-temperature specimen preparation in order to investigate the interior of cells and tissues (freeze-fracture for more randomly dissected surfaces and cryo-sectioning for the precise and specific dissection of block surfaces). c) The cryopreparation chain is completed by contamination-free cryo-transfer techniques and Cryo-SEM at deep temperatures [3].

In our contribution we will present the potentials and limits of each SEM imaging method in relation to specific application fields and compare the SEM techniques with correlative light and TEM techniques.

1. N. Franz et al., J. Mat. Sci. **41** (2006) pp. 4561-4567
2. I. Barshak et al., Ultrastructural Pathology **8** (2004) pp. 28-31
3. M. Richter et al., Skin Pharmacol. Physiol. **17** pp. 246-257
4. The ESEM-project was supported by the DFG

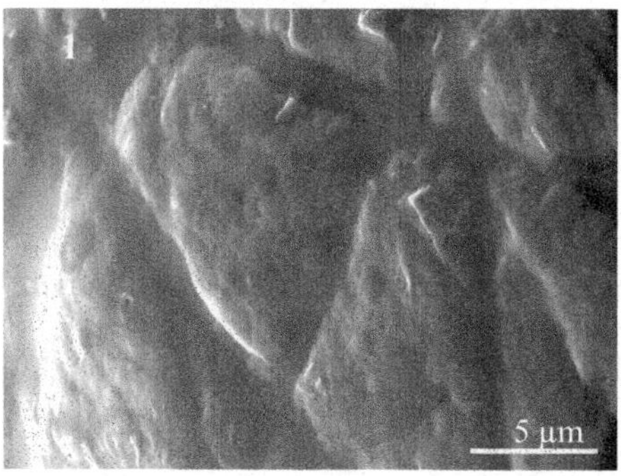

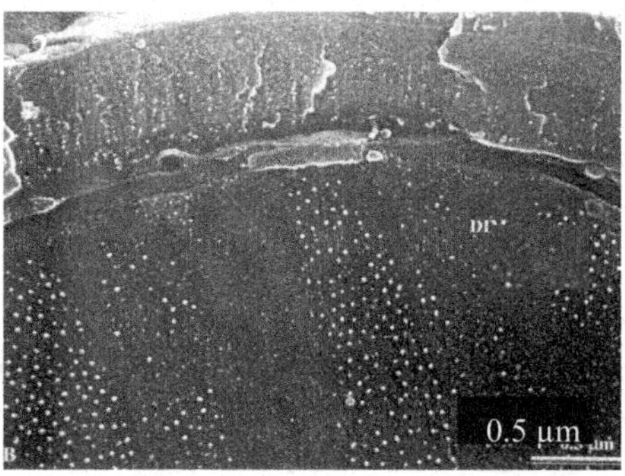

Figure 1. Surface of a native human skin biopsy in the ESEM.
Figure 2. Outer muscle of a high-pressure frozen and freeze-fractured Nematode in the Cryo-SEM.

Denaturation of metaphase chromatin plates observed by transmission electron microscopy and atomic force microscopy in aqueous solution

I. Gállego[1], X. Sisquella[2], X. Fernàndez-Busquets[3] and J.R. Daban[1]

1. Departament de Bioquímica i Biologia Molecular, Facultat de Biociències, Universitat Autònoma de Barcelona, 08193-Bellaterra
2. Plataforma de Nanotecnologia, Parc Científic de Barcelona, 08028-Barcelona
3. Biomolecular Interactions Team, Grup de Nanobioenginyeria, Institut de Bioenginyeria de Catalunya, 08028-Barcelona

Isaac.gallego@campus.uab.cat
Keywords: chromatin structure, metaphase chromosome structure, atomic force microscopy

It is currently considered that metaphase chromosomes are formed by folded chromatin fibers [1]. Nevertheless, chromatin fibers are only observed when metaphase chromosomes are treated with solutions of very low ionic strength; in metaphase ionic conditions, chromosomes are very dense and chromatin is apparently folded forming plate-like structures [2]. Metaphase chromosomes from HeLa cells were purified on sucrose gradients containing 5 mM Mg^{2+}, dialyzed extensively to remove sucrose and finally mechanically disrupted by rapid passage through a syringe needle. A TEM micrograph corresponding to typical chromatin plates from metaphase chromosomes prepared following this procedure is presented in Figure 1A. When the plates spread on carbon-coated grids were treated with high concentrations of NaCl, before crosslinking with glutaraldehyde, we observed denaturation of these planar structures leading to different levels of unfolding; in some micrographs we observed small granules having the diameter corresponding to single nucleosomes or aggregates that may contain few nucleosomes (see Figure 1B). AFM imaging was performed using an Asylum MFP-3D system working in AC mode. Uncrosslinked samples were adsorbed on freshly cleaved mica and observed in the presence of a solution containing 5 mM Mg^{2+}. The best images were obtained under a low resonant frequency (~12 kHz) and increasing the setpoint after engaging, in order to retract the piezo at maximum without losing the interaction between tip and plate. An example of an AFM image of chromatin plates is presented in Figure 2A. After the addition of approximately 1 volume of 4 M NaCl, the same structures can be seen in Figure 2B completely denatured. Our results indicate that the plates obtained from metaphase chromosomes show the typical behaviour expected for structures formed by condensed chromatin. The resolution of TEM images is higher than that obtained with AFM. However, AFM has allowed us to perform direct monitoring of individual plate unfolding in real time in aqueous environment.

1. J.R. Daban, Biochem. Cell Biol. **81** (2003) 91.
2. J.M. Caravaca, S. Caño, I. Gállego and J.R. Daban, Chromosome Res. **13** (2005) 725.
3. Work supported in part by grants BFU2005-03883 and CSD2006-00012.

A. Aretz, B. Hermanns-Sachweh, J. Mayer (Eds.): EMC 2008, Vol. 3: Life Science, pp. 365–366,
DOI: 10.1007/978-3-540-85228-5_183,

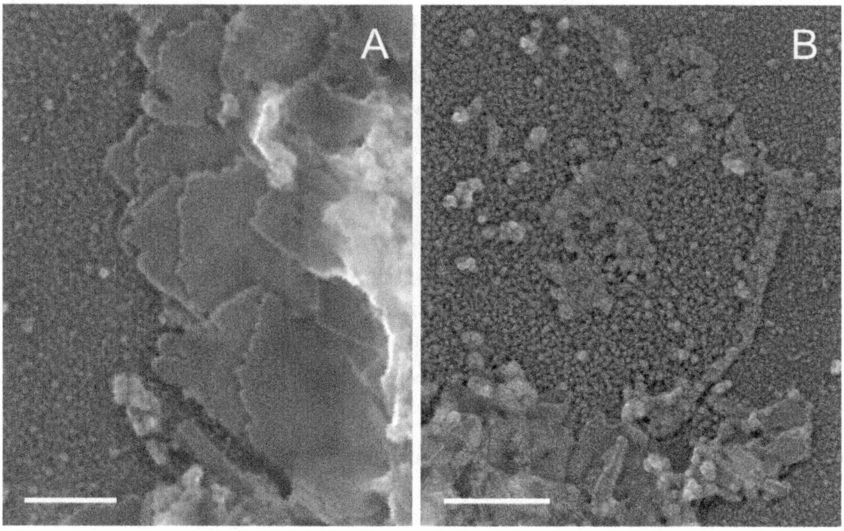

Figure 1. (A) TEM micrograph of metaphase chromatin plates. (B) A similar sample was denatured with 2 M NaCl. Grids were rotary-shadowed. Bars: 200 (A), 500 (B) nm.

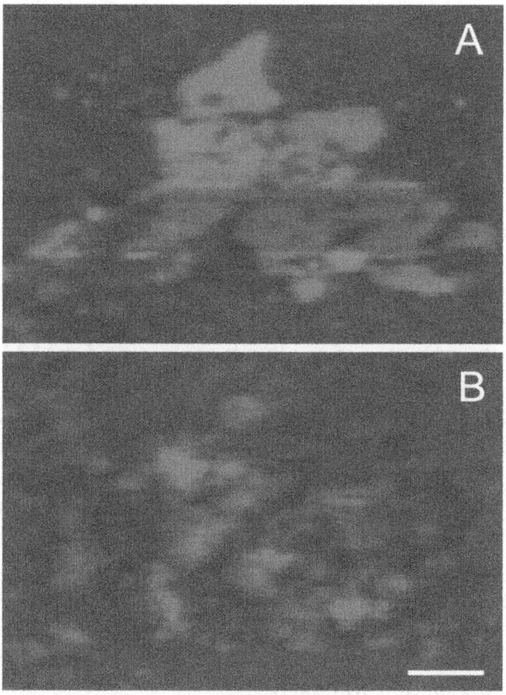

Figure 2. AFM topographic image of an uncrosslinked chromatin plate before (A) and after (B) the addition of NaCl. Bar: 500 nm.

Discovery of a Nuclear Structure in UVC-induced Apoptotic Cells by Integrated Laser Electron Microscopy

B.M. Humbel[1], M.A. Karreman[1], A.V. Agronskaia[2], A.J. Verkleij[1], H.C. Gerritsen[2], A.F.M. Cremers[1]

1. Electron Microscopy and Structure analysis, Cellular Architecture and Dynamics, Utrecht University, Padualaan 8, NL-3584 CH Utrecht, The Netherlands
2. Molecular Biophysics, Utrecht University, Princtonplein 1, NL-3584 CC, Utrecht, The Netherlands

B.M.Humbel@uu.nl
Keywords: Integrated Laser Electron Microscope (ILEM), Apoptosis, HUVEC, Cell Nucleus, UV damage

Electron microscopy has the great advantage to analyse cellular interactions at high resolution. In addition, subcellular structures and organelles can directly be seen, unequivocally identified and correlated with the labelled proteins. Labelling and localisation studies are easy and straightforward for simple, uniform samples e.g., culture cells [1, 2]. In complex tissue, however, finding the labelled area of single events becomes cumbersome [3, 4]. Gold labels can only be seen at rather high magnifications of 15.000 to 20.000 times, where only parts of a cell are in the field of view. To find a few gold particles on an electron microscopy grid is comparable to find a few coins in a soccer field.

To overcome these difficulties ultra-thin sections are cut and mounted alternatively on glass coverslips and EM grids. Both sections are labelled for the protein of interest detected either with a fluorescence marker or with gold. The expressing cells are easily spotted in the fluorescence microscope and with the help of landmarks the same area can be analysed in the electron microscope [3, 4].

To improve the throughput and to directly correlate fluorescence microscopic data with electron microscopic images a setup containing both a light and an electron microscope would be favourable. Recently, such an integrated laser electron microscope has been realised at Utrecht University [5].

Here, we demonstrate the advantages of this approach by studying early apoptotic responses of UVC-irradiated endothelial cells. To identify early apoptotic cells Tokuyasu cryo-sections [6] were picked-up on Formvar / carbon coated elevtron microscopy grids and labelled for cleaved caspase 3, a marker for apoptosis [7] using Alexa 488. Then the sections were embeded in methyl cellulose [6]. Only a few cells labelled for cleaved caspase 3. With the integrated light-electron microscope [5] we searched for the cleaved caspase 3 positive cells in the fluorescence microscopy mode. Thereafter, the same cells were investigated at high resolution in the electron microscope mode.

In all cells expressing cleaved caspase 3, we found an electron dense nuclear body, which so far has not been described in UV stressed cells. Morphologically it is clearly

A. Aretz, B. Hermanns-Sachweh, J. Mayer (Eds.): EMC 2008, Vol. 3: Life Science, pp. 367–368, DOI: 10.1007/978-3-540-85228-5_184,

different to nucleoli, Cajal bodies or interchromatine granules. The body is already present in very early stages of apoptosis, when the DNA distribution and the overall morphology of the nucleus are not yet affected. Preliminary analysis suggests that the newly discovered body contains RNA.

1. G.R. Bullock and P. Petrusz, eds. *Techniques in Immunocytochemistry*. Vol. 1-3. 1985, Academic Press: London.
2. A.J. Verkleij and J.L.M. Leunissen, eds. *Immuno-Gold Labeling in Cell Biology*. 1989, CRC Press: Boca Raton.
3. H. Schwarz, in *Electron Microscopy 1998, ICEM 14*, H.A. Calderón Benavides, M.J. Yacamán, L.F. Jiménez and J.B. Kouri, Editors, (Institute of Physics Publishing, Bristol, Philadelphia) 1998, p. 865-866.
4. H. Schwarz and B.M. Humbel, in *Electron Microscopy: Methods and Protocols*, J. Kuo, Editor, (Humana Press Inc, Totowa, NJ) 2007, p. 229-256.
5. A.V. Agronskaia, J.A. Valentijn, L.F. van Driel, C.T.W.M. Schneijdenberg, B.M. Humbel, P.M.P. van Bergen en Henegouwen, A.J. Verkleij, A.J. Koster and H.C. Gerritsen, (2008), p. in preparation.
6. K.T. Tokuyasu, J. Cell Biol., **57** (1973), p. 551-565.
7. T.R. Dunkern, G. Fritz and B. Kaina, Oncogene, **20** (2001), p. 6026-6038.
8. The project is supported by the Dutch Technology Foundation STW, applied science division of NWO and the Technology Program of the Ministry of Economic Affairs, by the European Network of Excellence 'Three Dimensional Electron Microscopy', FP6, and the Dutch Cyttron consortium.

Analysis of biomineral formation in three-dimensional micro-mass stem cell cultures

L. Lammers[1], U. Meyer[2], J. Handschel[2], C. Naujoks[2], H.P. Wiesmann[1]

1. Department of Cranio-Maxillofacial Surgery, University Hospital Münster[1], Waldeyerstr. 30, 48149 Münster
2. Department of Cranio-Maxillofacial Surgery, Heinrich-Heine-University Düsseldorf, Moorenstr. 5, 40225 Düsseldorf

Lydia.Lammers@ukmuenster.de
Keywords: stem cells, biomineralisation, 3D cell culture

In our working group a new technique for an improved three-dimensional (3D) cell culture has been developed. Cells self-assemble into a spheroidal cell-agglomerated micro-mass culture (sphere) [1, 2]. These cell-spheres consist of cells and their protein-containing extra-cellular matrix (ECM). In the present study the influence of nutrient-additives on the biomineral formation within the 3D stem cell culture system has been investigated.

The additives, given with the medium were dexamethasone, ascorbic acid and beta-glycerophosphate (DAG), which are known to promote mineralisation in osteoblast-like and stem-cell cultures [3, 4].

The features of biomineral formation have been analysed by a combination of several techniques. Light microscopy (LM), histological staining methods, as well as SEM, TEM and Raman-spectroscopy were applied to obtain elemental and structural information for an identification and characterisation of newly formed minerals within the spheres.

SEM-investigations showed huge mineralisation in spheres “Figure 1a)-e)”.
Small mineral matrix vesicles (MV) were seen in SEM and TEM, which grew together to bigger mineral agglomerates “Figure 1b)-e), g), h)”.

There is a change in size and crystallinity of mineral aggregates with time in newly formed minerals. There seems to be a small difference in the Mg composition between cultures fed with and cultures fed without DAG containing medium. The minerals in micro-mass cultures reveal features of early biomineralisation seen in apatite-like minerals of bone-tissues “Figure 1 f), i), j)”.

1. Anderer U, Libera J, J Bone Miner Res **17(8)** (2002) 1420-1429
2. J.Handschel, R. Depprich et al, Head & Face Medicine **3(4)** (2007) 1-4
3. N.R. Jorgensen, Z. Henriksen F. et al, Steroids **69** (2004), 219-226
4. M. Jaeger, M. Sager et al, Orthopäde **33 (12)** (2004) 1361-1372

A. Aretz, B. Hermanns-Sachweh, J. Mayer (Eds.): EMC 2008, Vol. 3: Life Science, pp. 369–370, DOI: 10.1007/978-3-540-85228-5_185,

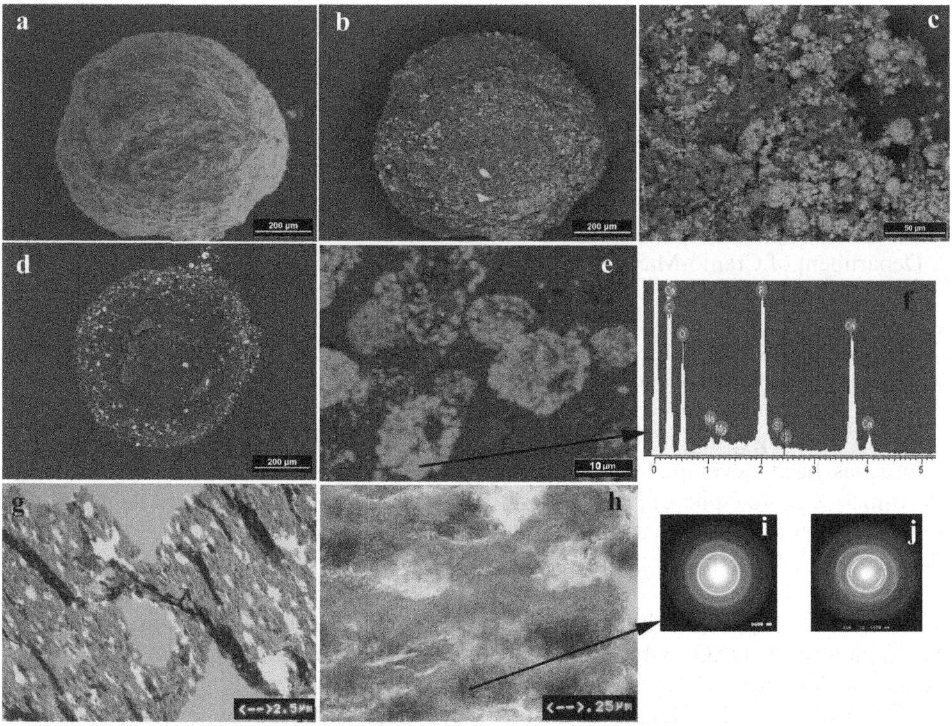

Figure 1. a) and **b)** SEM overviews of 14 day old stem cell sphere, in SE-mode (a) and in BSE-mode (b). Minerals appear lighter grey and can easily be distinguished from the darker cell matrix. **c)** Higher magnification of the sphere in BSE-mode, where minerals can be observed to be close connected to the cellular matrix. **d)** and **e)** embedded in Epofix and polished sphere to obtain microanalysis. Mineralisation starts within matrix vesicles, these form small globular structures, later on growing together to bigger mineral agglomerates (e). **f)** EDX-spectra taken at 10 KV showed an elemental composition of calcium and phosphor similar to hydroxyl-apatite, but with some amounts of magnesium. Unstained TEM images (**g, h**) showing the micro-structure of the polycrystalline mineral vesicles. **i)** Diffraction patterns of the minerals appear to be similar to diffraction pattern from human calvaria bone (**j**).

Comparative structural studies of highly-filled polymethacrylates with light microscopy, SEM, TEM and AFM

C. Lurtz, L. Schmitt, D. Behrend, K. Sternberg, K.-P. Schmitz

Institute for Biomedical Engineering, Medical Faculty, University of Rostock, Ernst-Heydemann-Str. 6, 18057 Rostock, Germany

claudia.lurtz@uni-rostock.de
Keywords: dental filling composite, ultrastructure, analysis

Highly-filled (> 80 w/w %) polymethacrylates are commonly used as restoration materials in dentistry. Required properties for handling and durability of dental filling composites are realized by the addition of different sized fillers and influenced by the distribution of the fillers. Further development of dental filling materials assumes detailed knowledge of the ultrastructure regarding surface and bulk, determined by different microscopic methods. In this context the aim of our study was the investigation of dental filling composite's ultrastructure by light, scanning and transmission electron microscopy as well as by atomic force microscopy. Scanning electron microscopy (SEM) has been frequently used to characterize the structure of dental filling composites [1, 2, 3]. Even if transmission electron microscopy (TEM) is a common microscopic method, it is rarely used for analysis of ultrastructure of highly-filled materials, because of the high preparation effort [4]. Atomic force microscopy (AFM) became more public for the investigation of the surface of dental filling composites in the last few years [3, 5].

We have investigated dental filling composites, which are clinically used, composed of dimethacrylates (Bis-GMA, TEGDMA, UDMA) and inorganic fillers (Ba- or Sr-glass, SiO_2). Specimens were prepared by casting in PTFE moulds and cured by light (λ_{max} 470 nm, 750 mWcm^{-2}, 20 s). Reflection light microscopy was performed on thin sections (30-50 µm) with the Lext OLS 3000 (Olympus, Hamburg). Additionally, SEM (XL 30 ESEM, Phillips Eindhoven) was used to examine the ultrastructure on fracture surfaces after three-point bending test. Therefore, fracture surfaces were sputtered with an approximately 10 nm thick layer of gold (Agar Sputter Coater 109, Plano Wetzlar). Ultrastructure analysis by TEM (LIBRA 120, Zeiss Oberkochen; acceleration voltage 120 kV) carried out on ultra thin cuts of 60-90 nm thickness (Ultracut S, Leica Wetzlar), which were placed on Formvar® coated copper grids. Scanning probe microscopy by AFM (Nanowizard® I, JPK Instruments AG Berlin, CSC37 Mikromasch cantilever) was realized in contact-mode on prepolished test samples (SiC-paper, FEPA 4000).

The results of morphological investigations of the dental filling composites are exemplified on the nanohybrid composite Grandio® (VOCO, Germany) (Fig. 1). Reflection light microscopy (Fig. 1 (a)) is used to get a first overview about composite ultrastructure as a starting point to further structural analysis. Examination of SEM micrographs leads to a more detailed microscopic view and is suitable to define fracture characteristic and the spherical morphology of filler particles. SEM analysis of

A. Aretz, B. Hermanns-Sachweh, J. Mayer (Eds.): EMC 2008, Vol. 3: Life Science, pp. 371–372,
DOI: 10.1007/978-3-540-85228-5_186, © Springer-Verlag Berlin Heidelberg 2008

Grandio® indicates adhesive failure between matrix and fillers and shows a homogeneous distribution of sharp-edged glass macrofillers (Fig. 1 (b)). TEM turned out to be a very useful method to investigate the ultrastructure of dental filling composites. It provides detailed information in the micro- and nanoscale, especially about different sizes and shapes of fillers and the filler size distribution. TEM micrograph of Grandio® demonstrates a high amount of nanofillers ($\leq$ 50 nm) and an evenly and narrowly distribution of different sized fillers (Fig. 1(c)). The AFM micrograph of Grandio (Fig. 1 (d)) presents a three-dimensional view of the different sized fillers and their distribution. AFM is a valuable technique to assess and verify the results of the TEM analysis, especially to differ cutting artefacts and gas inclusions.

1. K.K. Choi, J.L. Ferracane, G.J. Ruy, et al., Operative Dent **29-4** (2004), pp. 462-469.
2. C.P. Turssi, J.L. Ferracane, L.L. Ferrancane, J Biomed Mat Res **78B-1** (2006), pp. 196-203.
3. A. Kakaboura, M. Fragouli, C. Rahiotis, N. Silikas, J. Mater Sci **18** (2007), pp. 155-163.
4. B. Van Meerbeek, A. Dhem, M. Goret-Nicaise, et al., J Dent Res **72-2** (1993), pp. 495-501.
5. N. Silikas, K. Kavvadia, G. Eliades, D. Watts, American J Dent **18-2** (2005), pp. 95-100.
6. The authors would like to thank Prof. L. Jonas, M. Schröder and A. Cordes for assistance.

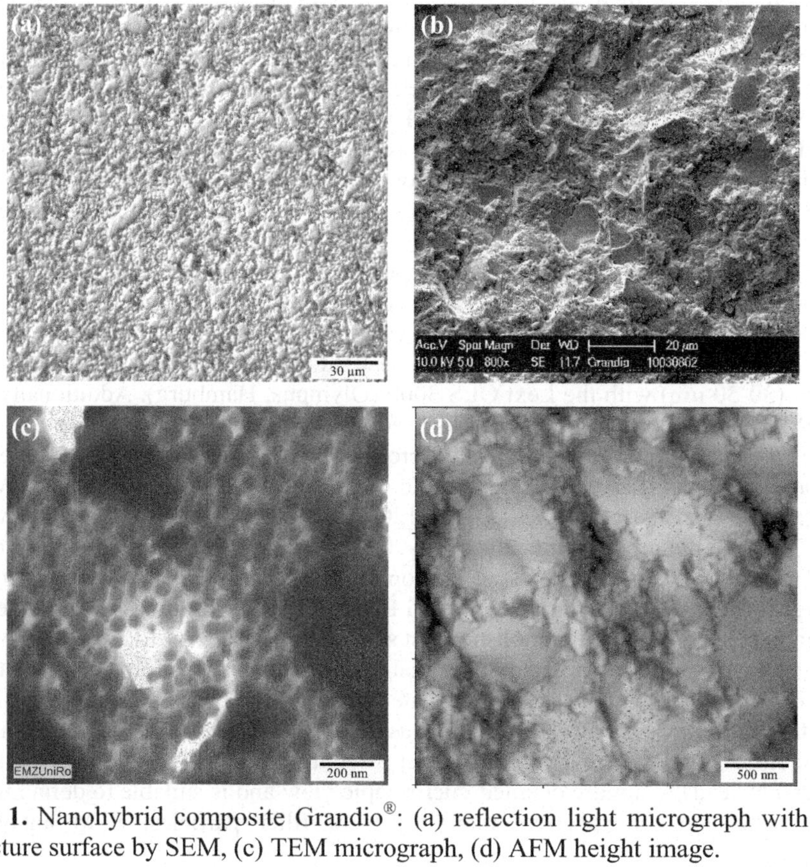

Figure 1. Nanohybrid composite Grandio®: (a) reflection light micrograph with DIC, (b) fracture surface by SEM, (c) TEM micrograph, (d) AFM height image.

Cell surface characteristics as reporters for cellular energy state

H. Pluk[1], H. Croes[1], B. Lich[2], B. Wieringa[1] and J. Fransen[1]

1. Department of Cell Biology, NCMLS, Radboud University Nijmegen Medical Centre, PO Box 9101, 6500 HB, Nijmegen, The Netherlands
2. FEI Company, PO Box 80066, 5600 KA, Eindhoven, The Netherlands

h.pluk@ncmls.ru.nl
Keywords: cellular energetics, cytoarchitectural morphology, ESEM

Adequate homeostasis of cellular energetics is important for proper cell functioning. Phenomena like adaptation to environmental changes, differentiation of cells, stress tolerance and the architectural dynamics of cells are all critically dependent on the localised availability of ATP, the main carrier of chemical energy. Mitochondria as the major supplier for ATP are, therefore, central to the control of these processes. During energy stress, cells may remodel their mitochondrial network in accordance to the metabolic needs. In parallel, changes in the cell energy state also may result in dynamic adaptations of the acto-myosin cytoskeleton. Since cell surface characteristics like filopodia, lamellipodia, ruffles and phagosomes are largely determined by the acto-myosin cytoskeleton, we hypothesize that the cellular energy state can be reflected by the surface morphology of the cell.

To study how cellular energetics and surface characteristics are interconnected, we apply state-of-art scanning electron microscopy techniques of fixed non-coated tissue culture cells both in wet (ESEM) as well as semi-dry (low vacuum) conditions. We try to link mitochondrial morphology directly to cell surface characteristics of the same cell by combining backscatter or (wet)STEM imaging of actin and mitochondria with SE detection of cell surface structures. In addition, we use correlative microscopy combining fluorescence imaging of the mitochondrial network with SE detection (Figure 1).

Growing of cells on conductive glass slides enables us to easily visualise by wetSEM the surface of a fixed, hydrated monolayer of cells, like mouse embryonic fibroblasts or mouse macrophages. Surprisingly, surface features of hydrated cells appear not to be influenced very much by decreasing the vapour pressure, allowing cells to dry in the SEM chamber. Low vacuum backscatter electron detection reveals internal structures of these cells like mitochondria and (actin-) cytoskeletal structures. Interestingly, cellular information obtained from backscatter electron detection is not only dependent on SEM voltage settings but also on glass slide coating properties.

To validate our ideas on the complex relationship between cellular energetics and the regulation of cytoarchitectural morphology, we also started to work on specific enzyme systems. CK-B (Creatine Kinase-Brain type) is one of the enzymes that play a key role in the buffering and transfer of cellular energy, providing microcompartments of the cell with preferential ATP-access. Next to the above approaches, we therefore used 'classical' SEM imaging of dried, gold coated samples to determine if cells that overexpress CK-B display altered cell surface characteristics. Indeed, we observed that

A. Aretz, B. Hermanns-Sachweh, J. Mayer (Eds.): EMC 2008, Vol. 3: Life Science, pp. 373–374,
DOI: 10.1007/978-3-540-85228-5_187,

the elaborate surface morphology of macrophages appear to change when the level of CK-B expression is modulated. Our next step will be to develop and introduce new image analysis methods to obtain quantitative data on these changes.

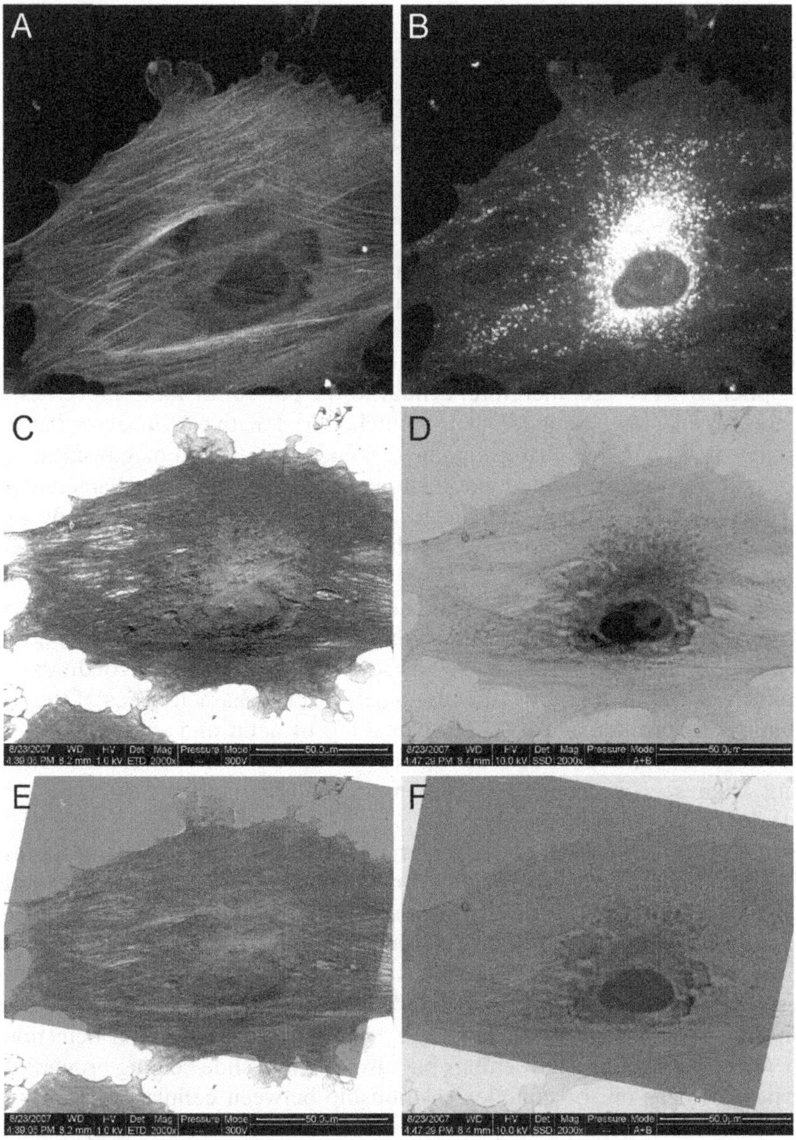

Figure 1. Example of correlative imaging of YFP-actin expressing mouse embryonic fibroblasts. **A.** YFP-actin fluorescence showing the actin cytoskeleton, **B.** Mitochondria visualised by Mitrotracker red fluorescence, **C.** Surface features detected by SE imaging, **D.** Backscatter imaging revealing internal ultrastructural features like mitochondria and (actin-) cytoskeleton, **E.** Overlay of A and C, **F.** Overlay of B and D.

Analysis of multimodal 3D microscopy measurements

M. Ritter[1], Axel Kranzmann[1], M. Hemmleb[2]

1. Bundesanstalt für Materialforschung und –prüfung (BAM), FG V.1, Unter den Eichen 87, D-12205 Berlin, Germany
2. m_2c microscopy measurement & calibration, Alt Nowawes 83a, D-14482 Potsdam, Germany

Martin.Ritter@bam.de
Keywords: Multimodal data fusion, coordinate registration, 3D microscopy

The fusion of spatial data generated by different microscopes allows an integrative and comprehensive characterization of 3D geometries and their characteristics.

For the comparison of multimodal measurement data a 3D linear transformation based on reference markers has been applied [1]. The test structures used were a micro-lens array, and, a focused ion-beam (FIB) metal deposited step pyramid that has been applied on the same sample. In order to be able to obtain defined measurement results, the test structures were tagged with ring-like reference markers created by FIB milling: 121 markers with a radius of 8 µm on the lenses, and 56 markers with a radius of 2 µm on the step pyramid. The test structures were then measured with a scanning probe microscope (SPM), two white-light interferometers (WLI) and two confocal microscopes (cM). The pyramid measurements have been labeled P1 to P5, the lens measurements L1 to L2 (table I).

Table I. Used measurement data of the step pyramid structure and micro-lenses.

Device	Provider	Calibration Measurements (µm)				Lens Measurements (µm)			
		Name	x-range	y-range	z-range	Name	x-range	y-range	z-range
WLI	QFM	P1	179,856	165,531	9,062	L1	366,511	366,493	20,645
AFM	IFF	P2	103,703	96,103	7,746	L2	101,896	95,950	5,106
cM	IFF	P3	254,567	190,823	7,714	L3	626,484	614,864	41,641
cM	ITO	P4	225,792	165,413	8,080	L4	437,168	886,116	20,088
WLI	ITO	P5	146,076	346,449	10,512	L5	628,000	628,000	46,400

The analysis of the microscope data is based on coordinate measurements of the reference markers. A calibration software automatically locates the centre of the markers with sub-pixel accuracy [2]. Comparison of homologous marker coordinates not only provides the relative spatial orientation, but also aberrations in scale and orthogonality of the various measurements. The underlying transformation model of the coordinate registration by parameter estimation uses 12 degrees of freedom: 6 parameters for the spatial orientation, 3 scale parameters (c_x,c_y,c_z), and, for the first time, 3 coupling factors (c_{xy},c_{xz},c_{yz}) that describe shear between the three coordinate axes. Table 2 shows the scale and coupling factors between the step pyramid measurements. Difference of the lateral scale factors (c_x, c_y) and lateral coupling (c_{xy}) is 4 % at highest. However, the vertical scale factor (c_z) differs up to 9 %. Even larger deviations up to 20 % can be seen

A. Aretz, B. Hermanns-Sachweh, J. Mayer (Eds.): EMC 2008, Vol. 3: Life Science, pp. 375–376,
DOI: 10.1007/978-3-540-85228-5_188,

for the coupling factors between the lateral and vertical coordinate directions (c_{xz}, c_{yz}). Table 2 also shows the scale and coupling factors when comparing the lens measurements. The deviations of the lateral parameters and the coupling factors are of the same magnitude as in the pyramid measurements, with the exception of one outlier (L2). While the lateral parameters fit quite well, the vertical coupling factors vary up to 23%, the scale factors in z-direction even up to 29 %.

Table II. Scale and coupling factors of pyramid and lens measurements

	cx	Cy	cz	cxy	cxz	cyz		cx	cy	cz	cxy	cxz	cyz
P1/P2	1,02	1,02	1,09	0,05	0,12	-0,17	L1/L2	1,08	0,96	1,07	0,03	-0,39	0,13
P1/P3	1,00	0,99	1,00	0,00	0,05	-0,06	L1/L3	1,00	1,00	1,13	0,00	0,06	0,04
P1/P4	0,97	1,03	1,03	-0,01	0,20	-0,12	L1/L4	1,00	1,00	0,98	0,00	-0,01	0,03
P1/P5	1,01	1,00	1,01	0,01	-0,09	-0,1	L1/L5	1,00	1,00	0,96	0,00	0,05	0,01
P2/P3	0,98	0,98	0,91	-0,04	-0,07	0,13	L2/L3	0,94	1,04	1,09	-0,04	0,68	-0,13
P2/P4	0,96	1,01	0,94	-0,05	0,02	0,15	L2/L4	-	-	-	-	-	-
P2/P5	0,99	0,99	0,93	-0,04	-0,20	0,11	L2/L5	0,93	1,04	0,95	-0,05	0,55	-0,18

The quality of the registration of the homologous coordinates can be defined by the mean point error σ_p that is calculated by the root of the weighted square-sum of the remaining coordinate differences after parameter estimation. In order to evaluate the influence or the coupling factors, the registration using 12 parameters was compared with a 9-parameter estimation, considering only 3 scale factors. The average σ_p in the step pyramid measurements (P1 to P5) using coupling factors was 2.39 times smaller as when only considering the scale factors. Interestingly, the standard deviation of the average σ_p is lowered by a factor of 4.28, indicating that the 12-parameter approach much better describes the geometrical differences between the various microscope data. Such is not the case for the analogous analysis of the lens data (L1, L3-L5). Although, the average σ_p as well is lowered when considering coupling in the parameter estimation, the standard deviation of the average σ_p remains high, indicating flaws in the lens measurements that are not corrected with the linear 12-parameter approach.

Classical calibration procedures normally only consider lateral correction parameters and the scale factor in vertical direction (9 parameters). Results here show: in order to analyze multimodal data, the coupling of all measurement coordinate axes (12 parameter) has to be taken into account for data correction.

1. Ritter M; Dziomba T; Kranzmann A; Koenders L: A landmark based 3D calibration strategy for SPM. Meas. Sci. Tech. 18 (2007), p. 404 – 414
2. microCal Software, m2c microscopy measurement & calibration GbR, Potsdam, Germany, http://www.m2c-calibration.com
3. We want to kindly thank R. Wiesendanger (Institut für technische Optik, ITO, Stuttgart), J. Regin (Institut für Industrielle Fertigung und Fabrikbetrieb, IFF; Stuttgart) and T. Wiedenhöfer (Qualitätsmanagement und Fertigungsmesstechnik, QFM, Erlangen) for providing measurement data. This work has been partly supported by the DFG within the SPP 1159.

Permanent plastid – nuclear complexes (PNCs) in plant cells

T. Selga, M. Selga

1. Laboratory of Plant Cell Biology, Faculty of Biology, University of Latvia, Kronvalda blvd. 4, LV1586 Riga, Latvia

turs.selga@lu.lv
Keywords: plant cell, nucleus, chloroplast

Conventional opinion assumes random distribution of plastids in the plant cell and light regulated movement realised with a help of stromules and actin microfilaments. In several organisms from protists to plants the joining of chloroplasts to the nucleus has been mentioned as a phenomenon [1-3]. However, little is known what plants and tissues and how frequently contain these structures? Whether appearance of PNC in cell depends on the state of differentiation? What is physiologic role of the PNC.

Bright field microscopy was used to analyse appearance of PNC in different vascular plants: a) monocotyledons: *Allium cepa*, *Convallaria majalis, Secale cereale;* b) dicotyledons: *Cucumis sativus, Pisum sativum, Nicotiana tabaccum;* c) gymnosperms: *Juniperus communis*, *Pinus sylvestris;* d) pteridophytes: *Athyrium filix-femina, Equisetum arvense.*

Transmission and scanning electron microscopy we used to characterise structural interaction among chloroplasts and nucleus in *Cucumis sativus, Pisum sativum* and *Nicotiana tabaccum* leaf parenchyma cells.

3D structure of the cell and movement of chloroplasts and PNC was analysed with laser scanning confocal microscope Leica DM RA-2 equipped with a TCS-SL confocal scanning head.

To visualise nuclear envelope we used tobacco plants transformed in Oxford Brookes University where green fluorescent protein GFP what has been fused to rat sialyltransferase located in Golgi bodies, ER and nuclear envelope [4].

Chloroplast-nucleus complex in bright field microscopy resembles a "daisy" like structure (Figure 1a). PNCs were typical for leaf epidermis and parenchyma from the stage of elongation till the disappearance of nuclei during the cell death of all analysed plant species.

SEM and TEM studies proved that three ways of chloroplast attachment to nucleus were possible: 1) direct joining of outer envelope membranes (Figure 1b and 1 c); 2) joining of the chloroplasts reciprocally and to the nucleus by ER; 3) penetration of chloroplast among the surface covings of ragged nuclei, i.e., among wide folds of the nuclear envelope – a specialised region of the rough ER.

Analysis of PNC along "z" axis demonstrated that all nuclei of tobacco cells formed PNC and fluorescence of GFP and chlorophyll at some spots were co-localised (Figure 1d and 1e). The time-lapse study shows, that group of chloroplasts were fixed to the nuclear envelope and oscillates together with nucleus. Second group was attached to

A. Aretz, B. Hermanns-Sachweh, J. Mayer (Eds.): EMC 2008, Vol. 3: Life Science, pp. 377–378,
DOI: 10.1007/978-3-540-85228-5_189, © Springer-Verlag Berlin Heidelberg 2008

nucleus by short (approximately 0.5μm long) joints and changed distance to nucleus during oscillation. Third group were far from the nucleus, at least few micrometers. These chloroplasts actively move towards or apart from the nucleus with velocity of 1 μm per second. Chloroplasts involved in PNC were bigger with brighter chlorophyll fluorescence in comparison with free randomly distributed chloroplasts in cells of yellowing leaves. Mechanically separated PNCs were cultivated in standard cell culture media and were viable for several weeks.

1. Ehara T., Sumida S., Osafune T., Hase E. Interactions between the nucleus and cytoplasmic organelles during the cell cycle of Euglena gracilis in Synhronized Cultures. Part I, Plant and Cell Physiol. 25, 1984, 1133-1146.
2. Gibbs S.P. Nuclear envelope – chloroplast relationships in algae. J. Cell Biol., 14, 1962, 433-444.
3. T. Selga and M. Selga "Chloroplast – nucleus cooperative interaction and change of morphogenesis caused by ethylene" ed. P.Mathis, (Photosynthesis), Kluver Acad. Pub. Dordrecht, Boston, London, vol. 5, 1995, 463-466.
4. Andreeva A.V., Zheng H., Saint-Jore C.M., Kutuzov M.A., Evans D.E., Hawes C.R. Organization of transport from endoplasmic reticulum to Golgi in higher plants. Biochem.Soc.Trans. 28, 2000, 505-512.

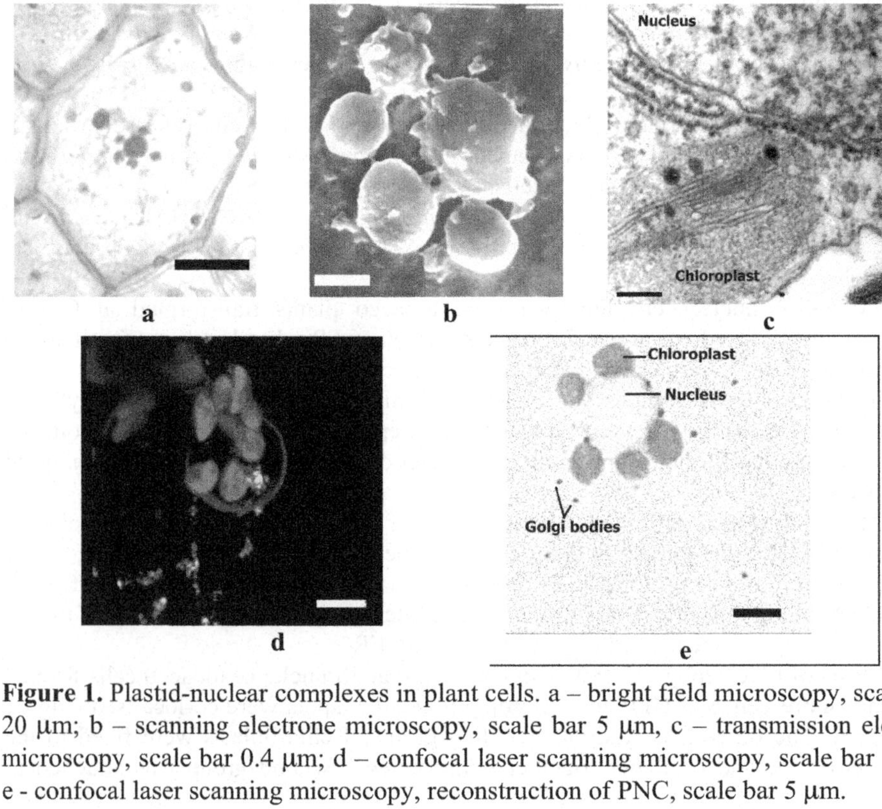

Figure 1. Plastid-nuclear complexes in plant cells. a – bright field microscopy, scale bar 20 μm; b – scanning electrone microscopy, scale bar 5 μm, c – transmission electron microscopy, scale bar 0.4 μm; d – confocal laser scanning microscopy, scale bar 5 μm; e - confocal laser scanning microscopy, reconstruction of PNC, scale bar 5 μm.

Cell imaging by dynamic secondary ion mass spectrometry (SIMS): basic principles and biological applications

Alain Croisy, Jean-Luc Guerquin-Kern & Ting-Di Wu

INSERM U759, Institut Curie Recherche, Laboratoire Raymond Latarjet
Bâtiment 112, Centre Universitaire, 91405 Orsay cedex (France)

alain.croisy@curie.u-psud.fr
Keywords: secondary ion mass spectrometry, cell imaging, analytical microscopy

The development of analytical microscopes based on various physical principles has allowed increasingly precise elementary analysis. Among these various methods, mass spectrometry imaging (ion microscopy) offers remarkable possibilities insofar as it can apply with extreme sensitivity to the quasi totality of the elements and their isotopes. Originally introduced in the early sixties by Castaing & Slodzian [1], this technique is based on the mass spectrometry of secondary ions (SIMS) extracted from the surface of a solid sample under the impact of an energetic beam of primary ions.

In an ion microscope the sample is brought to a potential of several kV. Thus, secondary ions resulting from the impact of the primary beam, and bearing a charge of the same sign as the sample, can be readily extracted through a first electrostatic lens. This secondary beam is then focused and guided to the entrance of a double focusing magnetic sector (mass spectrometer) using several transfer lenses. Secondary ions are then sorted in energy in the electrostatic sector before undergoing specific deviation by the magnetic field according to their m/z ratio (mass/charge) while keeping the topological information of the origin of the emission. Ions with the same m/z ratio can be selected and directed on the exit slit of the mass spectrometer in a beam which could be guided and widened by a set of projection lenses either to a Faraday cylinder for total ion current measurement or to a visualization screen for imaging purpose.

Numerous applications have emerged in such diverse fields as surface analysis, materials science, geochemistry and microelectronics, in which SIMS is now a central analytical tool. However, in spite of sporadic application of this technique in biology, over more than 30 years ago [2], ion microscopy has been for a long time considered only as a marginal method for solving problems in the life sciences, due mainly to poor lateral resolution (1-0.5 μm) and insufficient mass separation power. Many technological and conceptual improvements led to significant progress in both lateral resolving power, *i.e.* at the level of the image itself, and mass resolution, the parameter on which the precision of the analysis depends.

The first breakthrough came with the use of a finely focused primary ion beam rastering across the sample surface [3], thus improving the "theoretical" side resolution to the size of the primary beam. Enhancement of the mass resolving power was achieved by using various mass analyzers (double focusing magnetic sector, quadrupole mass analyzer). However, with such tools, only one specific ion image can be acquired at a time. Because SIMS analysis is a destructive method, this major restriction was

A. Aretz, B. Hermanns-Sachweh, J. Mayer (Eds.): EMC 2008, Vol. 3: Life Science, pp. 379–380,
DOI: 10.1007/978-3-540-85228-5_190,

very limiting in terms of in-depth resolution and sequential images of various species cannot be properly correlated. To overcome this difficulty, classical mass analyzers were tentatively replaced by a time-of-flight (TOF) spectrometer, which can record simultaneously all the secondary ions formed over a wide mass range [4]. However, TOF analysis required a pulsed primary ion beam usually delivering less energy than continuous sources, thus strongly decreasing the secondary ion yield. Actually, this evolution was turned to account for developing "static" SIMS previously introduced by Benninghoven [5], in which measurements are performed with a number of incident ions about one order of magnitude less than the number of atoms at the surface of the sample. Static SIMS is now used to acquire molecular information, (*e.g.* in proteomics studies) although count rates are low and information is restricted to relatively abundant species within the very superficial layers of the target. By contrast, in dynamic SIMS, which is the main purpose of this presentation, the number of incident ions exceeds the number of surface atoms on the sample leading mainly to elemental information with high-count rates, thus permitting trace analysis.

At the end of the eighties, a concerted collaborative effort between the University of Paris-Sud, Orsay (G. Slodzian), the French Space Agency (ONERA, B. Daigne), and CAMECA, a high-tech company developing EPMA and SIMS instruments, was established to design a new ion microprobe with high technical standards in both mass and lateral resolution, as well as in sensitivity. A machine possessing these characteristics was expected to open the way to spectacular new applications in cell biology, pharmacology and medicine. This effort has finally led to the last generation of dynamic SIMS instrument, the CAMECA NanoSims 50™.

This new ionic nanoprobe has a lateral resolution of less than 50nm with primary Cs^+ ions, the ability to detect simultaneously 5 different ions from the same micro volume and a very good transmission even at high mass resolution (60% at $M/\Delta M = 5000$). The decisive capability of this new instrument, and more generally of high resolution dynamic SIMS imaging in biology has been reviewed recently [6], however new observations in the field of isotopic ratio measurements applied to cell biology and microbiology as well as in trace analysis deserves further communication to illustrate the high potential of the technique in Life Sciences.

1. R. Castaing, G. Slodzian, J Microsc **1** (1962), 395-410.
2. P. Galle, Ann Phys Biol Med **42** (1970), 83-94.
3. R. Levi-Setti, G. Crow, Y.L. Wang, Scan Electron Microsc **Pt 2** (1985), 535-52 - G Slodzian, B Daigne, F Girard, Microsc Microanal Microstruct **3** (1992), 99–118.
4. B. T. Chait, K. G. Standing, Int J Mass Spectrom Ion Phys **40** (1981), 185-193.
5. A.Benninghoven, Surf Sci **35** (1973), 427-437.
6. JL Guerquin-Kern, TD Wu, C Quintana, A Croisy, Biochim Biophys Acta (Gen Sub) **1724** (2005), 228-38

Water imaging in cells cryosections by electron energy loss spectroscopy

J. Michel, C. Terryn and G. Balossier

Laboratoire de Microscopie Electronique Analytique INSERM UMR-S 926
Université de Reims Champagne Ardenne. 21, rue Clément Ader 51685 Reims, France.

Jean.michel@univ-reims.fr
Keywords: water, cells cryosections, EELS.

Water is concerned in ionic flows regulation and in cellular volume regulation. Measurement of water content variations in the cell allows to study cellular life phenomena and especially pathologies related to ionic or water channels failing. Moreover the knowledge of water content can be coupled with elemental microanalysis by EDXS to express ionic concentrations in mmol/l. Hence it is an important task to measure water content at a subcellular level. In this way, different methods based on X-ray microanalysis or using the STEM quantitative darkfield intensity on freeze-dried cryosections have been developed. Unfortunately methods based on X-ray microanalysis require an important electron dose which can induce a considerable mass loss in the electron beam sensitive hydrated specimen. Moreover all these methods estimate water content by using freeze-dried cryosections. Thereby they are indirect methods based on the assumption that no differential shrinkage occurs during freeze-drying. In this lecture, we will present a direct method of water mass measurement on hydrated cryosections by EELS.

This method uses the fact that amorphous ice and macromolecular compounds exhibit specific spectral features in the low loss part of electron energy loss spectra. Quantification is achieved by a weighted multiple least-squares fit of reference spectra from pure compounds. The experimental normalised single-scattering distribution for the hydrated specimen S(E) is expressed as a linear sum of the normalised single scattering distributions S_i (E) from its separate compounds. $S(E)=\sum_{i=1}^{M} a_i S_i(E)$.

Single scattering distributions are obtained thanks to the Fourier-log deconvolution process and consequently the method is independent of the thickness of the cryosections. The experimental spectrum can be reconstructed from the ice reference spectrum and some macromolecular reference spectra. Then, in the frozen hydrated cryosection, the local water content measurement can be obtained after some derivation using the different a_i coefficients and the corresponding molecular mass cross sections.

Moreover the method can be implemented in the spectrum-imaging mode . An EELS spectrum is collected from each pixel in a defined STEM image area. Acquisition result is a data cube in which an entire EELS raw spectrum is stored for each (x,y) position. Water contents can then be presented as maps to be correlated with the freeze-dried STEM image (Figure 1) . Moreover, adding up the different measurements in a

A. Aretz, B. Hermanns-Sachweh, J. Mayer (Eds.): EMC 2008, Vol. 3: Life Science, pp. 381–382,
DOI: 10.1007/978-3-540-85228-5_191,

particular cell compartment allows to obtain averaged water mass content. We will present different applications concerning water mass content measurements from test specimens cryosections (human erythrocytes, Saccharomyses) and from rat hepatocytes cryosections. We will discuss the use of this method to perform the mapping of water and ions within the different nuclear compartments in order to understand how water and ions repartition and quantity are modified as a function of cell activity changes due to modification of cell environment.

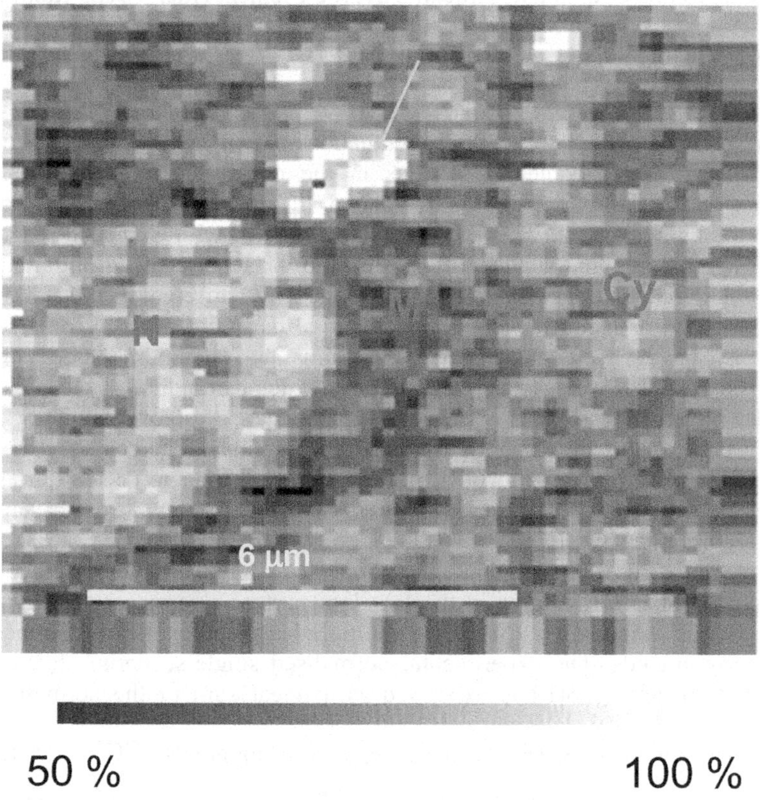

Figure 1. Parametric image of water mass content in rat hepatocyte cryosection.
N : nucleus area; M : mitochondria area; Cy: cytoplasm area. The green arrow shows a spurious ice crystal on the cryosection

Study of the cellular uptake of Pt nanoparticles in human colon carcinoma cells

H. Blank[1], R. Schneider[1], P. Brenner[1], D. Gerthsen[1], M. Esselen[2], D. Marko[2], M. Crone[3], M. Türk[3], T. Muller[4], and S. Bräse[4]

1. Laboratorium für Elektronenmikroskopie
2. Institut für Angewandte Biowissenschaften, Abteilung Lebensmitteltoxikologie
3. Institut für Technische Thermodynamik und Kältetechnik
4. Institut für Organische Chemie
1.-4. Universität Karlsruhe (TH), D-76128 Karlsruhe

blank@lem.uni-karlsruhe.de
Keywords: Pt nanoparticles, cellular uptake, FIB, SEM

Nanoparticles might enter the food chain, e.g. as a contaminant on the surface of food plants. However, there is only little knowledge about their uptake into cells of the gastrointestinal tract. Therefore, it is of general interest to study the correlation between particle properties like size, shape, and crystal structure on the one hand and biological responses on the other (see, e.g., [1]).

Since a substantial DNA damage can be observed after incubation of the human colon carcinoma cell line (HT29) with Pt nanoparticles, this cell line has been chosen as a model system. In order to control the particle properties, Pt nanoparticles were produced on β-cyclodextrin (β-CD) by supercritical fluid reactive deposition (SFRD) [2]. Depending on the organometallic complexes and process conditions used, SFRD enables the deposition of pure and highly dispersed Pt nanoparticles with distinctly different sizes and shapes. This is illustrated in Figs. 1a-c, showing as examples a single decahedral nanoparticle with about 40 nm in extension, an agglomerate of very small (~5 nm) irregularly-shaped particles, and approximately 100 nm sized Pt particles with predominantly cuboctahedral shape.

The cellular uptake was studied in dependence on the mean particle size, particle concentration, and time of incubation. For scanning electron microscopy (SEM) and focused ion beam (FIB) milling HT29 cells grown on Transwell membranes were prepared by fixation, dehydration and critical-point drying (cf. [3]). Fig. 2a shows a secondary-electron (SE) image of a single HT29 cell after 24 h Pt (1 $\mu g/cm^2$) incubation. The surface of the HT29 cell is characterized by microvillis and some phagocytotic protrusions. In the backscattered-electron (BSE) image of the same cell (Fig. 2b) agglomerates of Pt particles can be revealed on the cell surface (cf. arrows). This HT29 cell was subsequently cut section by section with the focused Ga^+ ion beam in the FIB instrument and imaged by BSE. In this way, Pt particles were found in different depths of the cell interior (cf. Fig. 2c), proving their cellular uptake. Moreover, as visible in Fig. 2d the presence of platinum was detected by energy-dispersive X-ray spectroscopy (EDXS). Transmission electron microscopy (TEM) investigations are in progress to localize Pt nanoparticles in thin sections of HT29 cells.

A. Aretz, B. Hermanns-Sachweh, J. Mayer (Eds.): EMC 2008, Vol. 3: Life Science, pp. 383–384,
DOI: 10.1007/978-3-540-85228-5_192,

1. Nanotoxicology - Interactions of Nanomaterials with Biological Systems. eds. Y. Zhao, H.S. Nalwa (American Scientific Publishers, Stevenson Ranch, CA), 2006.
2. F. Cansell, C. Aymonier, A Loppinet-Serani, Current Opinion in Solid State & Materials Science **7** (2003), p. 331.
3. E. Cohen, I. Ophir, and Y.B. Shaul, J. Cell Science **112** (1999), p. 2657.

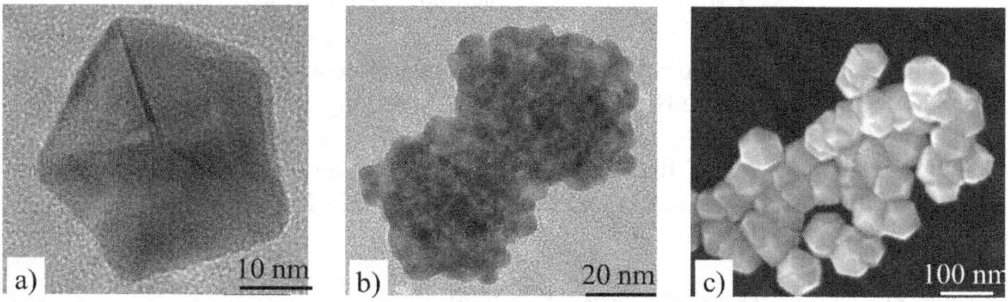

Figure 1. TEM images of a) a decahedral Pt particle, b) an agglomerate of small irregularly-shaped particles, c) agglomerated large particles. They were produced on β-CD under different SFRD conditions.

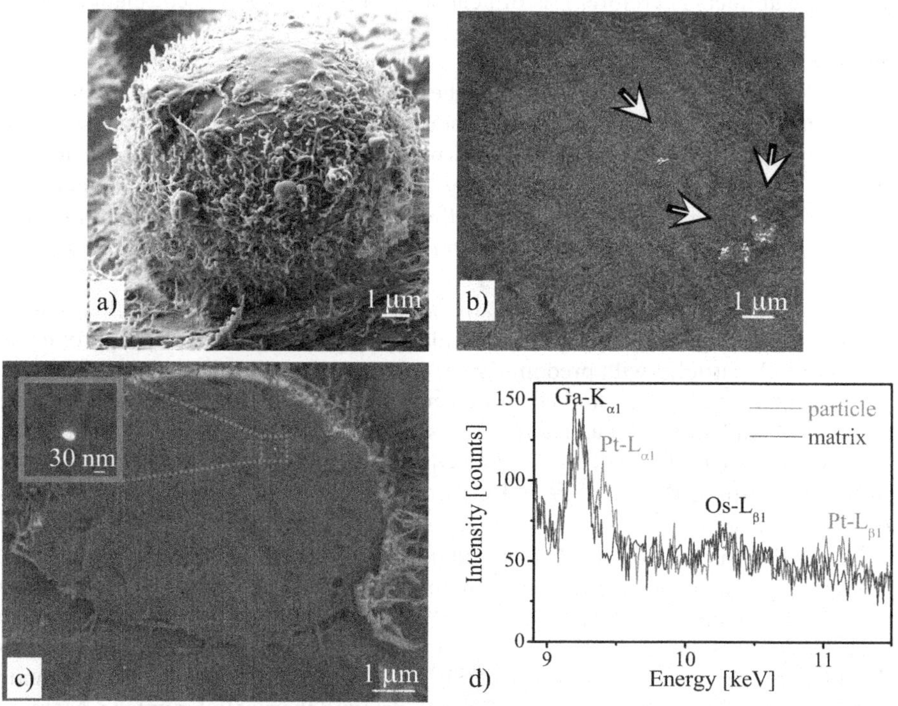

Figure 2. a) SE image of an individual HT29 cell, b) same cell imaged with BSE, c) detection of Pt particles in the cell interior of HT29 by BSE imaging, d) corresponding EDX spectra of the particle and the surrounding matrix.

Ion spectrometry and electron microscopy correlative imaging for accurate comprehensive trace element localization at subcellular level, in normal and pathological keratinocytes

J. Degrouard[1,2], E. Larras-Regard[1] and D. Jaillard[1]

1. UMR 8080 CNRS, Université Paris-Sud, 91405 Orsay cedex France
2. LPS UMR 8502 CNRS, Université Paris-Sud, 91405 Orsay cedex, France

degrouard@lps.u-psud.fr
Keywords: SIMS, TEM, keratinocyte

Due to a sensitivity close to 1 ppm, Secondary Ion Mass Spectrometer (SIMS) imaging has made currently possible localized analysis of trace elements at tissular level. Improvements in spatial resolution permitted a better recognition of trace element cellular targets. In spite of a relative high resolution (50 nm), dynamic SIMS imaging remains far behind Transmission Electron Microscopy (TEM), and it is still difficult to identify with certainty the various cellular compartments observed with SIMS imaging.

The use of these two complementary methods, in terms of correlative imaging, was applied to the follow-up of keratinization process in normal and pathological cultured keratinocytes before trace element localized analysis. The culture medium contained lithium gluconate (1.6, 3.0 and 5.0 mM/l) during an exposition of 24 or 48 h. The optimization of results for diffusible ions required cryopreparation of samples (Rapid freeze fixation, freeze substitution) and resin embedding. First attempts were performed on serial sections: ultrathin sample sections (60 nm) for TEM and thin sections (1μm) for SIMS. More accurate correlations required the study of a same sample section, both with TEM and SIMS instruments. So, a slight modification of SIMS sample holder was performed and allowed a direct analysis of sample sections (200 nm) when deposited on the grid previously observed by TEM.

A specific treatment of the sample avoided undesired charge accumulation. Positive and negative secondary ion images were performed. Sulfur (S-) images were helpful for mapping the sulfur-rich precursor of keratin and interpreting keratohyalin compartment observed by TEM, in normal and pathological cells "Figure 1". The correlation with Li+ distribution was established "Figure 2".

Such SIMS and TEM new approaches are promising for comprehensive biodistribution of trace elements at subcellular level, through physiological and pathological process.

1. We acknowledge Laboratoire « Assemblages Moléculaires: Modélisation et Imagerie SIMS » Université Rouen, France ; Centre de Recherche Public Gabriel Lippmann, Sciences et Analyse des Matériaux, Luxembourg ; Laboratoire de Dermatologie, CHU Nantes, France and Laboratoire Pharmaceutique Labcatal, France.

A. Aretz, B. Hermanns-Sachweh, J. Mayer (Eds.): EMC 2008, Vol. 3: Life Science, pp. 385–386,
DOI: 10.1007/978-3-540-85228-5_193,

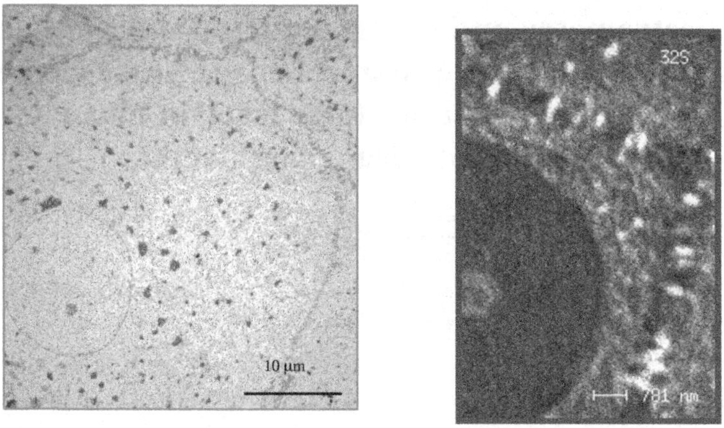

Figure 1. TEM and SIMS imaging in keratinocyte (*stratum granulosum*)

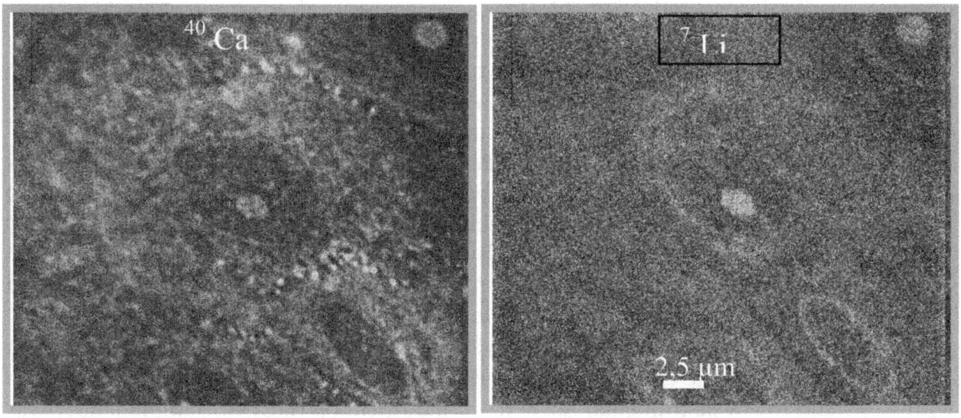

Figure 2. Ca^{++} and Li^{+} imaging in treated psoriasic epidermis (3 mM Li/l- 48h)

Calcium Spark Detection and Analysis in Time Series of Two-Dimensional Confocal Images

K. Hammer[1], A. Flockerzi[1], A. Zeug[2], P. Lipp[1] and L. Kaestner[1]

1. Institute for Molecular Cell Biology, Saarland University, 66421 Homburg/Saar
2. DFG Research Center Molecular Physiology of the Brain, Department of Neurophysiology and Cellular Biophysics, University Göttingen, 37073 Göttingen

k.hammer@uks.eu
Keywords: calcium sparks, confocal microscopy, 3D-analysis

Calcium sparks are highly localized, brief calcium transients reflecting calcium release from the sarcoplasmic reticulum. They represent elementary release events during the excitation-contraction coupling in cardiac myocytes. Under pathophysiological circumstances, such sparks can modify membrane currents and alter the proper function of myocytes - potentially leading to disturbances of the heartbeat. Thus, the analysis of calcium sparks can help to understand the molecular mechanisms underlying cardiac pathologies. In the past, calcium sparks were solely analyzed in line scan images consequently neglecting their 2D spatial properties and distribution inside the cell.

During recent years the increased performance of confocal microscopes enabled high spatial and temporal resolution imaging. This allowed recording of calcium sparks in adult mouse ventricular myocytes as time series in two-dimensional confocal images at acquisition rates exceeding 200Hz.

We developed a three-dimensional approach for such analysis using locally "derivative-like" functions for spark-detection allowing the algorithm to be suitable for temporal and spatial analysis of calcium sparks. We deliberately avoid user-operations by calculating all critical parameter values. All detection criteria are automatically adapted to inherent experimental parameters such as frame rate and pixel size (i.e. camera binning, objective magnification). This method ensures largely hardware-independent operation of our algorithm.

As a first step, we developed a proper method for cell recognition (Figure1) before performing detection of calcium sparks. For user-friendly evaluation of the final results each spark can be visualised as original data and as a 3D-gaussian function (Figure 2) thereby data can be doubled checked. Additionally the experimental and the Gaussian fit are used to calculate the spark parameters (e.g. amplitudes, decay time, frequency, spatial spread…) for further analysis.

This novel approach enables us to automatically analyze local calcium signals and thus fosters the admission of such algorithms for high-content screening.

A. Aretz, B. Hermanns-Sachweh, J. Mayer (Eds.): EMC 2008, Vol. 3: Life Science, pp. 387–388,
DOI: 10.1007/978-3-540-85228-5_194,

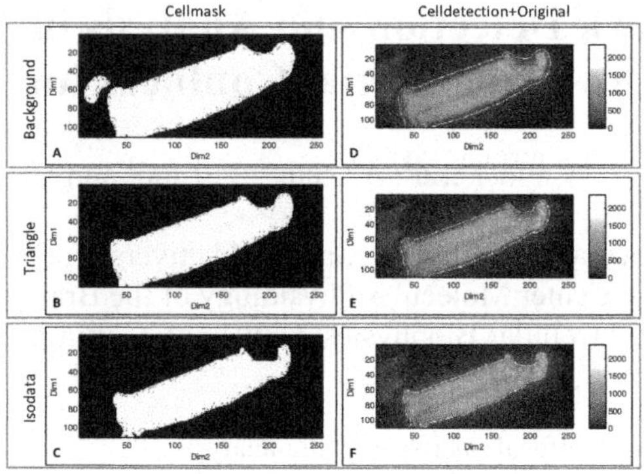

Figure 1. Different thresholding methods for cell detection. (A,D) background method sub-optimal results – here: detection of the cell fragment on the left side. (B,D) triangular method – improved recognition. (C,F) isodata method - optimized cell shape recognition.

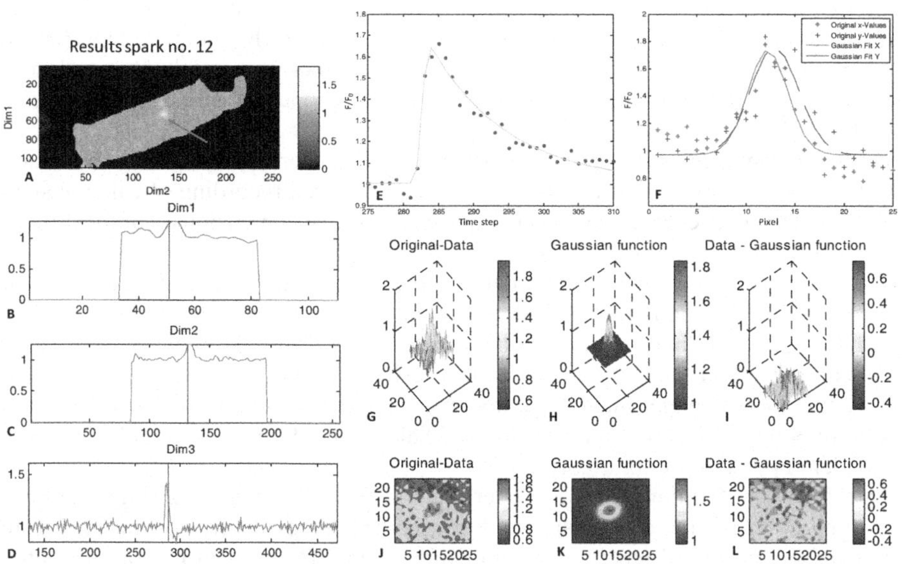

Figure 2. Visualization of the results. **A-D:** "Derivative-like" filter data and corresponding plots at the spark center (red arrow). **E:** Convolution fitting curve (green line) together with the original data (blue dots). **F:** Gaussian fitting curve and the corresponding original data points. **G,J:** 3D plot of the original data. **H,K:** 3D plot of the Gaussian fit. **I,L:** Original values subtracted by the fitted data.

Visualisation of the attachment, possible uptake and distribution of technical nanoparticles in cells with electron microscopy methods

Armin Springer[1], Wolfgang Pompe[1], Susanne Bastian[2], Maria Iwe[2], Hrissanthi Ikonomidou[2], Wibke Busch[3], Dana Kühnel[3], Annegret Pothoff[4], Volkmar Richter[4], Michael Gelinsky[1]

1. Max Bergmann Center of Biomaterials, Institute of Materials Science, Technical University Dresden, Budapester Str. 27, 01069 Dresden, Germany
2. Department for Neuropediatrics, University Children´s Hospital Carl Gustav Carus, Technical University Dresden, Fetscherstr. 74, 01307 Dresden, Germany
3. Helmholtz Centre for Environmental Research, Permoserstr. 15, 04318 Leipzig, Germany
4. Fraunhofer Institute of Ceramic Technologies and Systems, Winterbergstr. 28, 01277 Dresden, Germany

armin.springer@nano.tu-dresden.de
Keywords: technical nanoparticles, INOS, toxicity, EDX, skin cells

Technical nanoparticles, which are much smaller than human cells, are beginning to be used in a wide variety of products. But there is little known yet about the possible impacts of released nanoparticles on the human organism and the environment. With *in vitro* cell systems and electron microscopical methods the possible uptake and cell toxicity should be investigated.

Therefore, the research project "Identification and Assessment of the Effects of Engineered Nanoparticles on Human and Environmental Health" (INOS), funded by the German Federal Ministry of Education and Research (BMBF) should investigate the behaviour and possible toxicity of technical nanoparticles in *in vitro* cell systems. This research also should help to understand how the particles interact with cells of the skin, the lungs, intestine or nervous system and whether they cause damage or have negative effects on general cell functions. Besides others one main question is how nanoparticles attach, enter and distribute into the cells. In order to identify nano-sized particles on and in cells, a simple morphological analysis with conventional electron microscopy methods is not always sufficient [1] due to the possible risks of misinterpretation. Therefore, an additional reliable identification method is necessary. One technique which allows a secure identification of nanoparticles with metallic or ceramic properties is elemental analysis with energy dispersive X-ray spectroscopy (EDX) performed in the scanning electron microscope (SEM). We demonstrate a proper identification of tungsten carbide (WC) nanoparticles both attached at the surface (Figure 1, A-D) and distributed within the cells (Figure 1, E, F) of an *in vitro* skin cell culture. Further SEM investigations also give evidence of a discontinuous distribution of WC nanoparticles within the cells. These nanoparticles are partly encapsulated with a membrane or in a membranous organelle which indicates a possible uptake by phagocytosis as Stearns,

A. Aretz, B. Hermanns-Sachweh, J. Mayer (Eds.): EMC 2008, Vol. 3: Life Science, pp. 389–390,
DOI: 10.1007/978-3-540-85228-5_195,

Paulauskis and Godleski [2] documented for ultrafine TiO_2 particles. So far, no WC nanoparticles have been found within the cell nucleus (Figure 1, E).

1. Gatti, A.M., Montanari, S., Monari, E., Gambarelli, A., Capitani, F. and Parisini, B., Journal of Materials Science: Materials in Medicine, **15(4),** (2004).
2. Stearns, R.C., Paulauskis, J.D. and Godleski, J.J., American Journal of Respiratory Cell and Molecular Biology, **24(2)**, (2001).

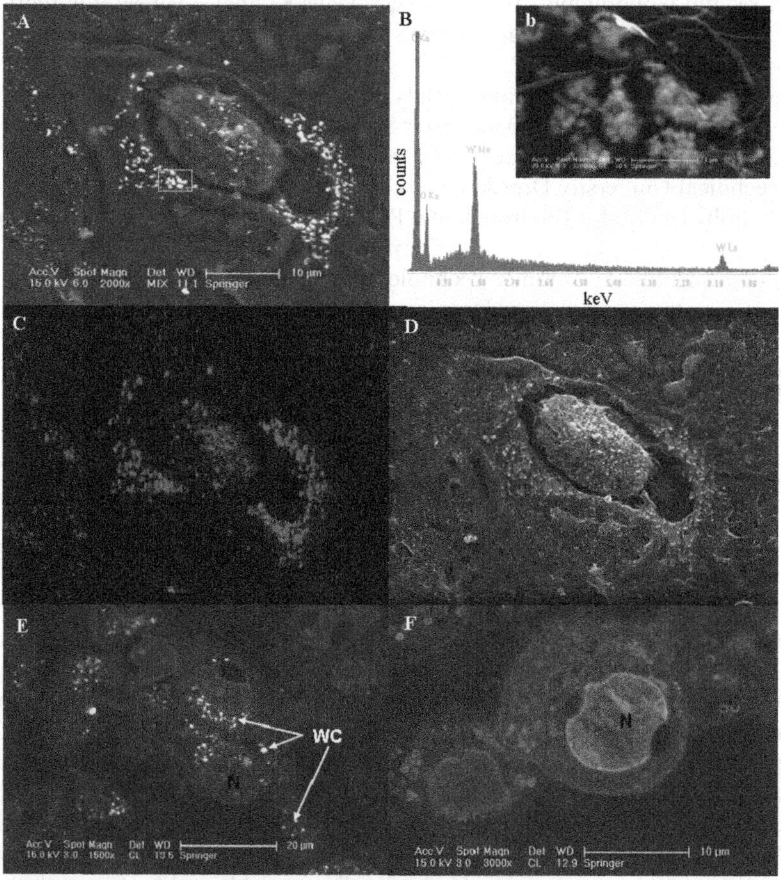

Figure 1. A: Overview with mixed BSE/SE (approx. 70%:30%) signal: The probable WC particles on the cell surface are detectable with this type of mixed signal. **B:** EDX spectra of the area shown in **A** (enlargement in **b**): besides C and O the W (Mα-line) representing peak is very prominent. In the analysed area W (and WC) is therefore definitely present. **C:** Elemental mapping of W distribution of **A**: WC particles and/or agglomerates are proper identified. **D:** Merged pictures **A** and **C**: The W distribution is congruent with the bright BSE/SE signals of WC nanoparticles/agglomerates on the surface of the skin cell. **E:** Block face image (BSE) of embedded skin cells with prominent WC particles inside the cell (not in the nucleus).
F: Control without WC. N = nucleus; WC = tungsten carbide

Author Index

A

Aach, T. 187
Adams, S.R. 337
Adler, G. 91, 97
Adriaensen, D. 197
Aebersold, R. 7
Agronskaia, A.V. 359, 367
Akkuş, S 257
Al-Amoudi, A. 301
Aleo, M.F. 289
Altmann, Th. 177
Alvang, R. 55
Alvarez-Otero, R. 127
Amler, E. 199
Amoroso, A. 233
Apaydin, N 255
Apel, C. 253
Arana-Chavez, V.E. 159
Ardelean, A. 109, 115, 149, 181
Arias, R. 223
Arias-Gonzalez, J.R. 219
Arndt-Jovin, D.J. 335
Arnold, P. 13, 37
Arrebola, F. 235, 259
Arroyo, R. 131
Asar, M. 251
Ayadi, A. 303
Aytaç, R 257

B

Baes, M. 291
Bálint, Zs. 101
Balossier, G. 381
Bannert, N. 237, 271
Barbieri, E. 309
Barboza, L. 207
Barbu-Tudoran, L. 105, 109, 111, 149
Bárcena, M. 63
Barlas, A. 353
Basciano, L. 171
Bastian, S. 389
Baterlink, W. 63
Battistelli, M. 239
Baumeister, W. 1, 31, 69, 325
Bayar, M.A. 245
Beck, F. 31
Beck, M. 7
Becker-Pauly, C. 13
Behrend, D. 371
Beil, M. 91
Beimforde, N. 237
Beles, D. 111
Benabdelhak, H. 61
Benarous, R. 45
Berghe, T.V. 189
Berrueta, L. 207
Bettoni, F. 289
Biagiotti, L. 309
Bianchini, P. 199
Bielniková, H. 343
Biggiogera, M. 305, 327
Biró, L.P. 101
Blacha, A. 177
Blank, H. 383
Bleck, C. 227
Boczkowski, J. 243
Boekema, E.J. 49
Boekema, E.J. 59
Bohn, S. 85
Boisset, N. 17, 27, 33
Boller, K. 107
Bornemann, J. 241
Borzì, R. 239
Bos, E. 89
Bos, E. 339
Bosch, B.J. 63
Bosshart, P. 3
Botelho, M.G. 335
Böttcher, B. 5
Braeckmans, K. 155

Braet, F. 67
Bräse, S. 383
Brenner, P. 383
Briones, M.J.I. 127
Brouns, I. 197
Brun, N. 243
Bruno, C.M. 233
Büchler, K. 21
Buers, I. 165
Bui, K.H. 25
Burattini, S. 309
Burdíková, Z. 199
Busch, W. 389
Bussy, C. 243
Buzadzic, B. 123

C

Caarls, W. 335
Cachita, D. 109
Cachiţă, D. 149
Cachiţă-Cosma, D. 111
Caeiro, M.F. 103
Caiolfa, V.R. 157
Čakić-Milošević, M. 113, 123
Callebaut, I. 17, 27, 33
Cambedouzou, J. 243
Cambi, A. 335
Can, B. 285
Cañizares, F.J. 235, 259
Čapek, M. 199
Carasco, C. 221
Carazo, J.M. 15
Cardoso, M. C. 167
Carrasco, C. 217
Carrascosa, J.L. 11, 89, 215, 219, 223
Carrière, C. 17, 27
Castellanos, M. 221
Castellini, H. 171
Çelik-Özenci, Ç. 251
Chaisri, U. 143, 145
Chaithong, U. 145
Chaiwong, T. 143, 145
Chami, M. 341
Chernyavskiy, O. 201
Chuang, C.K. 41
Cisterna, B. 305
Consolo, M. 233
Costa, G. 103
Costagliola, A. 203
Cottevieille, M. 33
Couture-Tosi, E. 139
Cox-Brinkman, J. 297
Craciun, C. 105, 109, 111, 115, 149
Cremers, A.F.M. 367
Croes, H. 373
Croisy, A. 379
Crone, M. 383
Cubero, M.A. 235, 259
Cuesta, I. 15
Curry, A. 229
Cyrklaff, M. 69, 85
Czerny, K. 263, 273

D

Daban, J.R. 365
Daffé, M. 341
D'Amico, F. 233
Danino, D. 51
Dass, M. 19
Davilla, S. 53
de Bruijn, H.S. 213
de Haas, F. 73, 95
de Pablo, P.J. 217, 221
de Pedro, M.A. 307
De Proost, I. 197
De Smedt, S.C. 155
Deckers, W. 189
Deerinck, T.J. 337
Degrouard, J. 385
Deibler, M. 225
Deike, S. 247
Demeester, J. 155
D'Emilio, A. 239, 309
Depmeier, W. 137
Depoix, F. 21, 39, 57
Depoix, F. 13, 37

Deschout, H. 155
Deutsch, E. 7
Devos, M. 189
Devys, D. 317
Diaspro, A. 155
Dieker, J. 305
Dieriks, B. 249
Dijk, F. 265
Dobrucki, J.W. 153
Donato, T.A.G. 159
Dubochet, J. 69, 341
Dubochet, J. 75, 227
Ducruix, A. 61
Dumas, D. 171
Dyba, M. 331

E

Eduardo, C. P. 253
Ehara, T. 133
Eimer, S. 71, 331
El Hili, A. 303
Ellinger, A. 87
Ellisman, M.H. 337
Engberts, J.B.F.N 49
Engel, A. 3
Er, H. 251
Espeel, M. 291
Esselen, M. 383
Esteves-Oliveira, M. 253
Evirgen, O 255, 257

F

Facchini, A. 239
Falcieri, E. 239, 309
Fernández, J.J. 89
Fernández-Busnadiego, R. 69
Fernàndez-Busquets, X. 365
Fernández-Segura, E. 235, 259
Ferraguti, F. 315
Ferrando-May, E. 225
Ferri, P. 309
Festjens, N. 189
Filimonenko, V. 119
Filová, E. 199
Flockerzi, A. 387
Florquin, S. 297
Foglia, A. 117
Fontana, J. 349
Fouquet, W. 331
Fourmentin, E. 175
Frangakis, A. 301
Frangakis, A.S. 61
Fransen, J. 373
Frederik, P. 67
Frederix, P. 3
Friedl, P. 99
Frischknecht, F. 85

G

Gago-Duport, L. 127
Gaietta, G.M. 337
Gállego, I. 365
Gasser, P. 361
Gaßler, N. 241
Gatsogiannis, C. 21, 39
Gaztelumendi, N. 131
Gebauer, W. 57
Geerts, W. 67
Geerts, W.J.C. 65
Geerts, WJ 83
Gelderblom, H. R. 261
Gelinsky, M. 389
Gerritsen, H.C. 213, 359, 367
Gerthsen, D. 383
Gibson, B. 75
Giepmans, B.N.G. 265, 337
Gnäegi, H. 89
Godsave, S. 339
Gonzalez, S. 353
Gottikh, M. 45
Granja, T. 103
Griffiths, G. 341
Griffiths, G. 227
Grigolato, P. 289
Grodet, A. 243

Grubic, M. 123
Grundberg, B. 55
Gruska, M. 325
Guardeño, L.M. 23
Guerquin-Kern, J.-L. 379
Guescini, M. 309
Günthert, M. 361
Gürster, S. 321

H

Hagen, G. 335
Hammer, K. 205, 387
Han, H.-M. 75
Handschel, J. 369
Harris, M. 47
Hause, G. 357
Hausser, I. 287
Hax, W. 47
Hebert, H. 55
Heck, A.J.R. 215
Hegermann, J. 71
Heijnen, H. 73
Helenius, A. 361
Hellström, J. 209
Hemmleb, M. 375
Heper Okçu, A 257
Heresanu, V. 243
Hermanns-Sachweh, B. 241
Hernando, I. 23, 275, 277
Hertzberger, LO 83
Hild, S. 185, 333
Hobot, J. A. 311
Hoebe, R.A. 169
Hofmannová, J. 119
Hofnagel, O. 165
Hohenberg, H. 363
Hollak, C.E.M. 297
Holland, G. 237
Holstermann, B. 363
Hormeño, S. 219, 223
Houben, D. 339
Houssin, C. 341
Hovnanyan, K. 121
Hovnanyan, M. 121
Hozák, P. 119, 345
Huber, H. 137, 321
Hucek, S. 323
Hudoklin, S. 77
Humbel, B.M. 359, 367
Humbel, BM 83
Humbel, B.M. 65

I

Ibarra, B. 219, 223
Ikonomidou, H. 389
İlkay, H 255, 257
Illy, E. 209
Ingram, P. 53, 231
Inoue, Y. 93
İşcan, A. 269
Ishikawa, T. 25
Ishino, Y. 29
Isla, N. de 171
Iwe, M. 389

J

Jaillard, D. 385
Jakobs, S. 191
Janáček, J. 345
Janda, P. 345
Jankovic, A. 123
Jansen, H. 339
Jap, B.K. 41
Jędrych, M. 263
Jetten, M.S.M. 65
Jezernik, K. 77
Jezernik, K. 161
Jodłowska-Jędrych, B. 263, 273
Johnson, L. 27
Jokitalo, E. 347
Jonas, L. 313
Jonic, S. 27, 33
Joshi, N.V. 207
Joss, G. 169

K

Kaestner, L. 205, 209, 387
Kajimura, N. 29
Kalicharan, D. 265
Kantelhardt, S. 335
Kapitein, L.C. 217
Kapoor, T.M. 217
Karlsson, H. 209
Karreman, M.A. 367
Kaufmann, W.A. 315
Kertész, K. 101
Kessler, R.W. 281
Kifer-Wysocka, E. 273
Kiyohara, N. 133
Kizilyaprak, C. 317
Klijn, J. E. 49
Klingl, A. 137
Klompmaker, S.H. 59
Knaus, H.-G. 315
Kodama, J. 93
Koeck, P.J.B. 55
Kölsch, A. 187
Koltovoy, N. 173, 267
Kong, G.W.K. 69
König, M. 247
Korać, A. 113, 123
Korac, B. 123
Korgun, E.T. 251
Korinek, A. 31
Köse, S.A. 245
Koster, A.J. 63, 359
Koster, AJ 83
Kothe, E. 129
Krakhmalev, V. 179
Krämer, B.K. 287
Kranzmann, A. 375
Kreft, M.E. 161
Krisp, H. 79, 97
Krysko, O. 291
Kubínová, L. 199, 201
Kudryashev, M. 85
Kühnel, D. 389
Kujala, P. 339
Kurahashi, H. 145
Kurth, R. 237
Kusel-Fetzmann, E. 151
Kutlu, M. 269
Kwok, B.H. 217

L

Ladhoff, A.M. 293
Lakämper, S. 217
Lambert, O. 61
Lammers, L. 369
Lampert, F. 253
Landfester, K. 19
Landwehr, S. 81, 97, 141
Lange, V. 7
Lanone, S. 243
Larivière, D. 175
Larquet, E. 17, 27, 33
Larras-Regard, E. 385
Laue, M. 237, 271
Lavazza, A. 289
Lazar, Z. 353
Lebbink, MN 83
LeFurgey, A. 53, 231
Leis, A. 325
Leisinger, S. 355
Leitenstorfer, A. 225
Lembrechts, R. 197
Lemke, C. 357
Lepper, S. 85
Leube, R.E. 187
Levron, C. 125
Li, L.-T. 163
Lich, B. 373
Liebl, D. 227
Lin, C.-Y. 163
Lindman, M. 347
Lipp, P. 205, 209, 387
Llorca, E. 23
Llorca, O. 35
López-Iglesias, C. 349
Lucas, F. 351
Lucas, M.S. 351, 361

Lučić, V. 69
Lück, S. 91
Luecken, U. 47
Lurtz, C. 371
Lütz-Meindl, U. 319

M

Maghraoui, S. 303
Malaponte, G. 233
Malatesta, M. 305, 327
Malmström, J. 7
Manders, E. 169
Maninová, M. 119
Manova, K. 353
Marco-Molés, R. 275
Márk, G. 101
Markl, J. 21
Marko, D. 383
Marschang, R. 103
Martinez, E.F. 159
Masudi, S.M. 211
Mateu, M. G. 221
Matysiak, W. 263, 273
Mauceri, B. 233
Maurer, U. 69
Maximova, E. 177
Mayanagi, K. 29
Mazza, D. 155
Mazzarino, M.C. 233
Megens, R.T.A. 195
Meißlitzer-Ruppitsch, C. 87
Meissner, U. 13, 37
Méndez, J. 127
Mercer, J. 361
Mertel, S. 331
Mertens, T. 81, 97
Mertens, Th. 141
Merz, A. 227
Merz, T. 281
Meyer, C. 321
Meyer, U. 369
Michel, F. 45
Michel, J. 381
Micunovic, K. 123
Mihali, C. 149
Mihali, C.V. 105
Milla, M. 319
Mironov, A.A. 161
Mock, M. 139
Moeller, A. 13, 37, 39, 57
Möller, L. 237
Mor, A. 51
Morandini, F. 289
Moras, D. 45
Morikawa, K. 29
Morin, J. A. 223
Mornon, J.-P. 27
Moscicka, K. 59
Movassagh, T. 25
Müller, D. 321
Müller, M. 355
Muller, S. 305
Müller, S.A. 3
Muller, T. 383
Musyanovych, A. 19

N

Naujoks, C. 369
Navarrete, P. 259
Nebesarova, J. 323
Nebesářová, J. 343
Neri, S. 233
Neumüller, J. 77, 87
Neuner, C. 137
Nickell, S. 31
Nietzsche, S. 129, 357
Nogués, C. 131
Nuñez-Ramirez, R. 15

O

Oberhofer, M. 205
Oğuz, E.O. 285
Oikawa, T. 93
Oiwa, K. 25
Olivotto, E. 239

Olson, J.K. 145
Oostergetel, G.T. 63
Osafune, T. 133
oudeEgbrink, M.G.A. 195
Owald, D. 331

P

Pagani, S. 239
Paiziev, A. 179
Palero, J.A. 213
Papp, T. 103
Parthoens, E. 189
Pavelka, M. 77, 87
Pearl, L.H. 35
Pehau-Arnaudet, G. 139
Pérez-Munuera, I. 23, 275, 277
Peterman, E.J.G. 217
Peters, J. 41
Peters, P.J. 89
Peters, P.J. 339
Petrovic, V. 123
Philimonenko, A.A. 345
Pierson, J. 89
Pierson, J. 339
Pintelon, I. 197
Plekhova, N.G. 135
Plitzko, J.M. 31, 325
Pluk, H. 373
Pogorelov, A.G. 279
Pogorelova, M.A. 279
Pompe, W. 389
Post, JA 83
Posthuma, G. 299
Pothoff, A. 389
Pribac, C. 181
Prosperi, E. 305
Puig, A. 275, 277
Purhonen, P. 55
Pustovalov, E.V. 135

Q

Quiles, A. 23, 275, 277

R

Rachel, R. 137, 321
Rampichová, M. 199
Ranck, J.-L. 139
Rebner, K. 281
Recuero-Checa, M.A. 35
Reimer, R. 363
Reyes 207
Riches, J. 67
Richter, S. 283
Richter, V. 389
Richter, W. 129, 357
Rigort, A. 325
Riquelme, B. 171
Risco, C. 349
Ritter, M. 375
Rivera-Calzada, A. 35
Robenek, H. 165
Robenek, M. J. 165
Robles, M.A. 235, 259
Rockel, B. 41
Rodella, L.F. 289
Rodríguez-González, J. B. 127
Romanowska-Sarlej, J. 273
Romih, R. 77
Romin, Y. 353
Roos, W.H. 215
Rosa, A. 121
Rottier, P.J.M. 63
Ruff, M. 45
Ruíz, S. 349
Ruppenthal, S. 205

S

Saçan, M. 251
Sachse, M. 139
Sagalowicz, L. 43
Sailer, M. 91
Sakakibara, H. 25
Salas, M. 223
Salleh, N.I.M. 211
Saltan, N. 269

Sani, M. 339
Saran, Y. 285
Sargın, A.K. 285
Sartori, A. 85, 325
Sasov, A. 183
Satchell, S. 67
Sato, A. 93
Sawa, Y. 93
Scarzello, M. 49
Schaap, I.A.T. 217
Schäd, S. 313
Schauflinger, M. 141
Scheres, S.H. 15
Schmalz, G. 137
Schmidt, A. 7, 129
Schmidt, C.F. 215, 217
Schmidt, V. 91
Schmitt, L. 371
Schmitz, K.-P. 371
Schneider, R. 383
Schoenmakers, R. 95
Schönfeld, K. 107
Schraner, E.M. 355
Schroeder, J.A. 287
Schultz, P. 45, 317
Schwarz, H. 307, 329
Selga, M. 377
Selga, T. 377
Semiz, O 255, 257
Serrano, M.M. 235
Seufferlein, T. 79, 97
Severs, N. J. 165
Seyit, G. 41
Shigemoto, R. 315
Sigrist, S.J. 331
Siriwattanarungsee, S. 143
Sisquella, X. 365
Sjollema, K.A. 265
Skamnaki, V. 27
Skarmoutsou, E. 233
Slaaf, D.W. 195
So, Peter T. C. 193
Somova, L.M. 135
Spagnolo, L. 35
Spanu, M. 105, 115
Spedito, A. 327
Spehner, D. 317
Springer, A. 389
Stacchiotti, A. 289
Steiniger, F. 357
Sterenborg, H.J.C.M 213
Sternberg, K. 371
Steven, A.C. 215
Stevens, M. 291
Stierhof, Y.-D. 329
Stivala, F. 233
Stocchi, V. 309
Stöcker, W. 37
Stöger, B. 319
Stoltz, JF 171
Storms, M. 47
Strous, M. 65
Štrus, J. 185
Stuart, M.C.A. 49
Suh, S. 353
Sukontason, K.L. 143, 145
Sukontason, K. 143, 145
Suzuki, H. 29
Svergun, D. 33
Sylvestre, P. 139

T

Takahashi, I. 293
Takama, M. 293
Tama, F. 9
Tapasztó, L. 101
Taveau, JC. 61
Tekaya, L. 303
Tekiner, A 245
Terryn, C. 381
Thiel, H.-J. 247
Thomm, M. 137
Timmermans, J.-P. 197, 203
Tong, T. 353
Tönjes, R.R. 107
Tora, L. 317
Tosti, M. 295

Träutlein, D. 225
Trépout, S. 61
Troyer, D. 165
Tsekos, I. 147
Tsien, R.Y. 337
Turcus, V. 109
Turcuş, V. 149
Türk, M. 383
Turkekul, M. 353

U

Udvardi, M. 177
Uetrecht, C. 215
Ukropina, M. 113, 123

V

Valentijn, J.A. 359
Valpuesta, J.M. 219
van den Bergh Weerman, M.A. 297
van der Ploeg van den Heuvel, A. 213
van der Want, J.J. L. 265
Van der Wel, N. 339
van Donselaar, E.G. 65, 83
van Driel, L.F. 359
Van Nassauw, L. 203
van Niftrik, L. 65
van Nispen tot Pannerden, H. 73
Van Oostveldt, P. 169, 249
van Zandvoort, M.A.M.J. 195
van Zon, M. 339
Vancova, M. 323
Vancová, M. 343
Vandenabeele, P. 189
Vanecek, J. 323
Vanlangenakker, N. 189
Vanoni, M.A. 33
Vasilijevic, A. 123
Vecerdea, A. 205
Vedder, A.C. 297
Venien-Bryan, C. 17, 27
Vercauteren, D. 155
Verkleij, A.J. 65, 83, 359, 367
Verklijk, A. 63
Vértesy, Z. 101
Vetterlein, M. 87
Vigneron, J.P. 101
Vihinen, H. 347
Vogt, T. 287
Voigt, A. 95
von Einem, J. 81, 141
von Gegerfelt, D. 209
von Ruhland, C.J. 311
von Wichert, G. 79
Voorhout, W. 95
Vos, M. 339
Vos, W.H. De 169, 249

W

Wagner, R. 47
Walian, P.J. 41
Walther, P. 19, 79, 81, 91, 97, 141
Walther, P. 227
Wang, D. 59
Warmer, M. 363
Watanabe, I. 133
Watts, N. 215
Weidinger, M. 151
Weingart, C. 287
Weinger, J.S. 217
Weiss, C. 87
Wepf, R. 351, 361
Wernsdorf, P. 283
Westermann, M. 129, 357
Wichmann, C. 331
Wieringa, B. 373
Wiesmann, H.P. 369
Wijburg, F.A. 297
Wild, P. 355
Windoffer, R. 187
Wingfield, P. 215
Wirth, R. 321
Wisse, E. 67
Wojcik, K. 153

Wölfel, M. 351
Wöll, S. 187
Wu, T.-D. 379
Wuite, G.J.L. 215
Würflinger, T. 187

Y

Yakushevska, A. 95
Yücel, O 245

Z

Zamai, M. 157
Zeug, A. 387
Zeyrek, T 255
Ziegler, A. 185, 333
Ziserman, L. 51
Žnidaršič, N. 185
Zuber, B. 69, 341
Zuber, B. 227

Subject Index

1

3D analysis 387
3D cell culture 369
3D EM 13, 37
3D Imaging 47
3D microscopy 375
3D visualization 175
80s Ribosome 89

A

AAA+ ATPase 29
absorption spectrum 267
ACC 127
actin-binding proteins 119
actinin 119
active zone architecture 331
A-DNA 219
AFM (atomic force microscopy) 3, 217, 221, 365
ageing 255
algae 319
alizarin red 211
aluminium 303
amniochorionic membrane 257
amyloid 241
analysis 371
analytical 379
anammox bacteria 65
anatomy 111
anthocyans 109
anti-DNA mitochondrial antibody 133
apical vesicles 93
apoptosis 189, 239, 367
arabidopsis 177
atherosclerosis 195
ATP depletion 305
autofluorescence 199
autofluorescence imaging 213
automated data analysis 69
automation 47
autophagy 113
Autrata BSE YAG detector 323

B

bacillus anthracis 139
basement membrane 257
beam damage 75
beta cells 265
betaretrovirus 237
biology 231
biomaterials 215
biomedical imaging 363
biomineralisation 369
biomineralization 127, 333
birefringence 253
blood cells 105
brown adipocytes 113, 123
bystander effect 249

C

cadherins 301
caenorhabdits elegans 71
calcium 17, 185
calcium imaging 197
calcium sparks 387
calibration 201
cancer 99, 237, 269
carbon films 343
carbon nanotubes 243
cardiac tissue 241
cardiomyocyte 279
catalase crystals 75
caveolae 357
cell 173, 267
cell biology 183
cell cycle 169
cell death 259
cell differentiation 171
cell dynamics 87
cell identification 173

cell imaging 379
cell nucleus 367
cell preservation 235
cell transport 67
cells cryosections 381
cell-surface interaction 137
cellular energetics 373
cellular uptake 19, 383
cellulose microfibrils 147
CEMOVIS 75, 89, 139
Cestoda 125
chemical imaging 281
chicken organs 115
chiral self-assembly 51
chloroplast 377
chondrocyte 239
chromatin 153
chromatin structure 365
chrysomya megacephala 143, 145
CLEM 169
CLSM 131, 211
CO_2 laser 253
coating method 343
collagen 73, 195
colloidal gold nanopartikel 323
compartmentalization 65
conduit arteries 255
confocal microscopy 153, 163, 197, 201, 205, 387
contrast transfer function (CTF) 33
coordinate registration 375
coronavirus 63
correlated microscopy 265, 337
correlation 85
correlative LM-EM 363
correlative microscopy 67, 325, 351, 359, 361
corynebacterineae 341
CR-real time 309
cryo electron microscopy (cryo EM) 11, 15, 25, 27, 33, 39, 41, 45, 49, 73, 89, 325, 341
cryo electron tomography (cryo ET) 7, 61, 63, 69, 85, 339
cryo electron tomography of vitreous sections 301
cryo field emission SEM 323
cryo fluorescence microscopy 325
cryo microscopy 47
cryomethods 345
cryo preparation 137, 321
cryo SEM 23, 363
cryo substitution 317
cryo TEM 13, 37, 43, 51
cryotomogram segmentation 175
crystal distortion 55
cuticle 185
cytoarchitectural morphology 373
cytogenetic 281
cytometry 193
cytoplasmic potassium 279
cytoskeleton 67, 91, 187

D

dendritic cells 335
dental enamel 253
dental filling composite 371
dental fissure 211
desmosomes 301
diagnostic EM 261
differentiation 161
DNA 153
DNA damage 225
DNA integration 45
DNA methylation 167
DNA repair 35, 225
DNA replication 29, 167, 223
DNA-PK 35
Drosera 149
Drosophila 331

E

earthworms 127
EDX 389
EDX/EELS 313
EELS 287, 319, 381
EELS analysis 243

EFTEM-TEM tomography 93
EGF receptor 335
elastic lamella 255
electron crystallography 1, 55
electron microscopy 71, 73, 229, 297, 299, 305, 317
electron probe microanalysis 279
electron probe X-ray microanalysis 259
electron tomography 1, 65, 83, 87, 349
elemental analysis 313
EM thin sections 227
EM tomography 77
embryoid bodies 131
emerging pathogens 229
endoplasmic reticulum 347
entamoeba 121
enteric glia 203
envelopment 81
epon 355
erythrocytes binding 117
ESEM 363, 373
euglena 133
euglenophyceae 151
event-streaming 53

F

fenestra 67
FIB 383
FIB/SEM 361
filamin 119
filopodia 117
fitting 27
fixation 357
flagella 25, 321
flexible fitting 9
FLIM 157
fluorescence correlation spectroscopy 171
fluorescence microscopy 79, 155, 201
fluorescence photobleaching 201
force spectroscopy 215
FOXP3 233
FRAP 155
freeze fracture 147
freeze substitution 97, 347, 349
freeze-fracture immunocytochemistry 165
freeze-substitution 355
FRET 157
function 17, 27, 33
fusiform vesicles 77

G

gadolinium 287
gastropoda 57
gemini surfactant 49
genital chamber 143
GFP 79
γH2AX 249
gingival fibroblasts 159
glucose oxidase 277
glutamate synthase 33
Golgi 161
Golgi apparatus 77, 87, 337
granules 73

H

hairs 149
HCMV 81, 141
HCV 233
heart biopsy 241
heavy metal resistance 129
heavy metals 319
hemocyanin 21, 39
hemoglobin 57
herpesvirus 355
HERV 237
heterosis 177
high content sc 249
high pressure freezing 19, 71, 79, 81, 97, 141, 317, 329, 331, 349, 351
high throughput cryo electron microscopy 31
histone 153
HIV 45

HIV matured model 293
HIV-1 117
homology modeling 27
homology modelling 17, 33
HpPex20p 59
HpPex5p 59
human caries 211
HUVEC 367
hybrid 177
hybrid structure 41
hydrated surfaces 363
hydrophobic cluster analysis HCA 17, 27
hypometabolism 327
hypothermia 235
hypothermic preservation 259

I

image classification 15
image processing 137, 173, 187
imaging 53
immun localization 321
immunocytochemistry 77, 305, 311
immunoelectron microscopy 107
immunogold 315
immunogold labeling 345
immunohistochemistry 233
immunolabeling 299, 307, 329, 355
in situ molecular detection 353
in vitro 197
in vitro cultivation 149
in vitro culture 181
in vivo 195
increased sample size 351
infectious diseases 261
inmunofluorescence 207
INOS 389
integrase 45
Integrated Laser Electron Microscope (ILEM) 367
intercalator 153
interference 179
intermediate filaments 91
iridoviruses 103
islets 265
IUGR 251

K

karyotyping 281
keratinocyte 385
kidney 273, 303

L

laser technology 209
lead 289
LEDGF 45
leptospirosis 7
life cell imaging 187, 209
light microscopy 179
light polarized microscopy 253
lipid droplets 165
liposomes 109
liquid crystals 43
live cell image 163
live cell imaging 227
live cell microscopy 167
live tissue imaging 213
liver 233
liver sieve 67
lizards 103
long term culture 205
low voltage transmission electron microscope 343
lymphocyte 207
lysine 51
lysosomal storage disease 297

M

macromolecular complexes 15
macrophages 135, 227, 243
magnesium 185
magnetic field 207
magnetic nanoparticle 269
malaria 85
mass spectrometry 7
medical diagnostics 313

membrane domains 83
membrane proteins 55
membrane reconstitution 61
membrane staining 347
meprin 13, 37
mercury 289
mesenchymal stem cells 171
metabolic disorders 291
metalloprotease 13, 37
metaphase chromosome structure 365
methotrexate 285
microanalysis 231
microdomain 315
micromass 239
microscopy 193, 303
microspectrophotometer 173
microsporidia 229
microtriches 125
microtubules 73
mineralizing factors 159
minor groove binder 153
missing cone 55
mitochondria 113, 133, 163, 191, 209
mitosis 337
molecular mechanics 9
molecular model 21
monoglycerides 43
morphogenesis 283, 307
morphology 107
mosaic microscopy 265
motor proteins 217, 305
Moult 333
mouse 203
mouse kidney 285
movement cryoultramicrotomy 121
MRF localization 309
multidrug resistance 61
multimodal data fusion 375
multiphoton 225
multiphoton microscopy 213
multispectral imaging 173
murine hepatitis virus 63
mycobacteria 227, 341
myocyte subcellular structure 205

N

nano-indentation 221
native membrane 3
natural photonic crystals 101
nautilus pompilius 39
necrosis 189
negative stain technique 293
nephrogenic fibrosis 287
NHEJ 35
nickel 129
Nicolas Boisset 11
nitrogen mustard 245
non-covalent interactions 157
nonlinear microscopy 213
nonlinear spectral imaging 213
nucleolus 305
nucleus 377
nucleus cytoplasm export 305

O

optical biopsy 213
optical force microscopy 223
optical microscopy 353
optical nanoscopy 191
optical traps 217
osteoarthritis 239

P

pancreas 79, 265
particle enrichment 271
PAT proteins 165
peptide nanotube 51
peroxisomes 291
pestiviruses 247
phagocytosis 227
phase behaviour 49
phosphorylase kinase 17, 27
phylogeny 125
pistia 111
placenta 251, 257
plant cell 377
plant extracts 115

poly-L-lactide nanoparticles 19
polysaccharides 23
Portugal 103
poxviruses 283
proanthocyanidin 245
procaine 109
profile comparisons 17
protein complex 339
protein translocation 209
proteolysis 99
Pt nanoparticles 383
pulp cells 159
pulsed electric fields 275

Q

quantitative microscopy 201
quantum dot markers 329

R

raman spectroscopy 333
rapid diagnosis 261
rapid embedding 271
rat 251
rat brain 245
red algae 147
red blood cells 179
rehydration method 299
remote relationship 17
reserve mobilization 177
resins 311
resolution extension 55
retrovirus 107
reverse transcriptase 237
ribbon 51
ribosomal subunits 305
RNA precursors 327
RNase P 5
rofecoxib 273

S

sacculi 307
sauces 23
SCA7 317
scanning electron microscopy 143, 145, 277
scanning fluorescence microscopy 359
second harmonic generation 213
secondary ion mass spectrometry 379
SEM 111, 131, 183, 255, 295, 383
SEM tomography 91
seminiferous tubule 353
SHG imaging 199
shrink-wrap algorithm 55
SIMS 385
single cell absorption spectroscopy 173
single molecule 219, 223
single molecule imaging 335
single particle analysis 1, 5, 29, 39
single particle electron microscopy 59
single particle reconstruction 55
SIPcharts 201
skeletal muscle cells 309
skin cells 389
slices 197
small-angle X-ray scattering (SAXS) 33
soft X-ray imaging 183
spatial statistics 345
species characteristics 151
spectral imaging 213
spectrin 119
spore 139
SPT 155
stable transfectants 169
STED 191, 331
STEM 3, 79, 81
stem cells 369
STEM tomography 97
stimulation of ovulation 263
streptomyces 129
stress proteins 289
structural color 101
structural transition 219
structure 17, 27, 33, 357

subcellular 315
subcellular localization of viral proteins 247
supramolecular structures 175
surface layers 137
synapse 69

T

T-cell syncytia 117
technical nanoparticles 389
tegument 125
tegument protein pUL71 141
telomeres 169
TEM (Transmission Electron Microscope) 79, 95, 251, 269, 295, 347, 385
TEM and STEM 243
template matching 83
terminal complexes (TCs) 147
testes 145
TFTC/STAGA 317
therapy 297
three-dimensional (3D) reconstruction 11, 27, 33
three-dimensional cryo electron microscopy 21, 57
time laps microscopy 189
tissue 193
Titan Krios 31
TOM toolbox 31
tomography 25, 47, 67, 73, 89, 95, 311
TOVIS 89
toxicity 389
transcription 327
transglutaminase 277
transmission electron microscopy 143, 145, 275, 321, 359
TRF2 169
trichinella 295
trigonella 181
triiodothyronine 123
tripeptidyl peptidase II 41
tuft cells 93
two-photon microscopy 195, 201
type I collagen 199
type II collagen 199

U

UCP1 123
ultracentrifugation 271
ultrastructural 121
ultrastructural changes 235
ultrastructure 105, 115, 135, 151, 273, 285, 293, 345, 371
urothelial cell 161
urothelium 77
UV damage 367

V

vinculin 119
virion morphogenesis 247
virus 135, 215
virus entry 361
virus mechanics 221
vitrobot 47
voltage-gated delayed-rectifier K+ channel 203
volumetric averaging 89
vWF 73

W

water 381
Western Blot 309
white adipocyte 123
whole liquid egg 275
wing scales 101

X

Xplore3D 95
X-ray microanalysis 235
X-ray microscopy 183
X-rays 53, 231
zebrafish 291
zona pellucida 263
zoonosis 283

The manufacturer's authorised representative in the EU is Springer Nature Customer Service Centre GmbH, Europaplatz 3, 69115 Heidelberg, Germany. If you have any concerns regarding our products, please contact ProductSafety@springernature.com

Printed and bound by CPI Group (UK) Ltd, Croydon, CR0 4YY
07/01/2026
02030921-0013